D0991733

HOWE·LIBRARY
HANOVER
NEW HAMPSHIRE

THE MISTAKEN EXTINCTION

THE MISTAKEN EXTINCTION

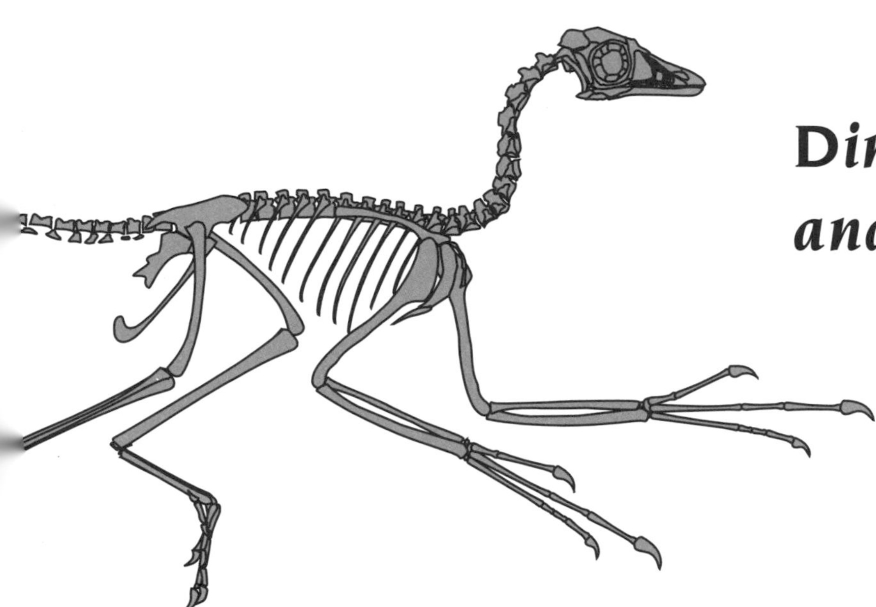

Dinosaur Evolution and the Origin of Birds

Lowell Dingus
American Museum of Natural History

Timothy Rowe
The University of Texas at Austin

W. H. FREEMAN AND COMPANY
NEW YORK

Cover and text designer: Victoria Tomaselli

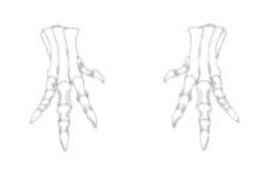

LIBRARY OF CONGRESS CATALOGING-IN-PUBLICATION DATA
Dingus, Lowell.
The mistaken extinction: dinosaur evolution and the origin of birds/
Lowell Dingus and Timothy Rowe.
p. cm.
Includes bibliographical references (p. –) and index.
ISBN 0-7167-2944-X.—ISBN 0-7167-3227-0 (academic version)
1. Dinosaurs. 2. Evolution (Biology) 3. Birds—Origin.
I. Rowe, Timothy, 1953– . II. Title.
QE862.D5D4928 1998
567.9—dc21
97-35749
CIP

Printed in the United States of America.

First printing, 1997

CONTENTS

Foreword vii

Acknowledgments xi

Prologue xiii

PART ONE
The Search for the Smoking Gun 1

CHAPTER ONE
The Seductive Allure of Dinosaurs 3

CHAPTER TWO
Earlier Extinction Hypotheses 11

CHAPTER THREE
Contrasting Volcanic and Impact Hypotheses 21

CHAPTER FOUR
Enormous Eruptions and Disappearing Seaways 31

CHAPTER FIVE
The Fatal Impact 43

CHAPTER SIX
Direct Evidence of Catastrophe 67

CHAPTER SEVEN
Patterns of Extinction and Survival 75

CHAPTER EIGHT
Our Hazy View of Time at the K-T Boundary 91

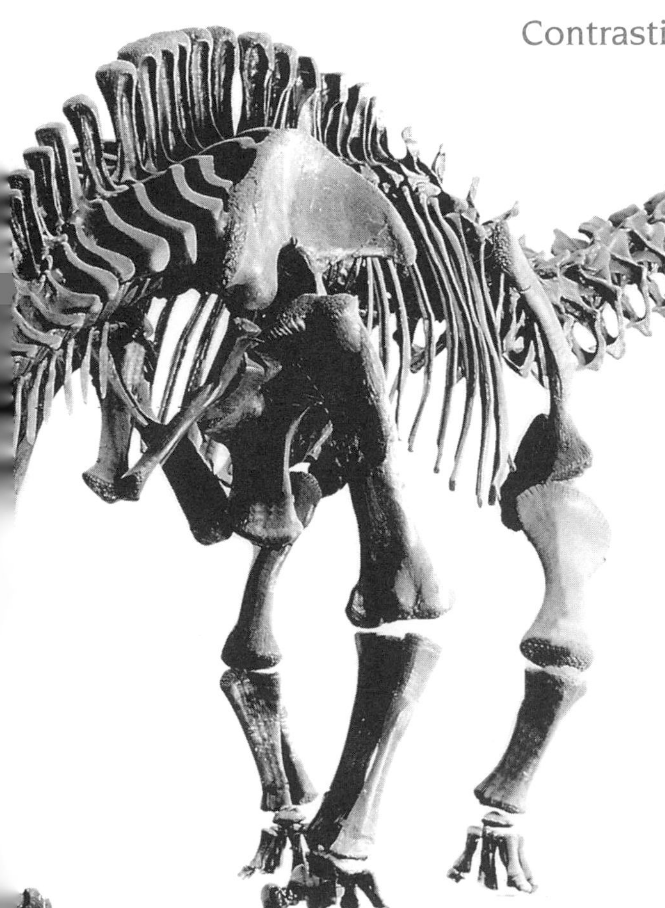

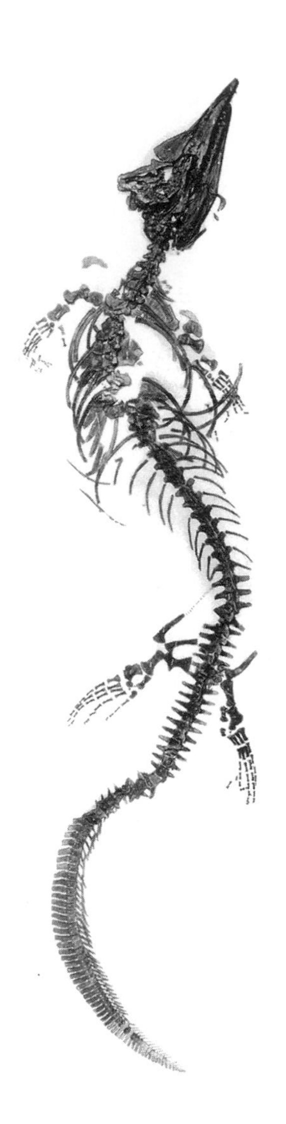

PART TWO
Dead or Alive? 105

CHAPTER NINE
Living Dinosaurs? 107

CHAPTER TEN
Dinosaurs Challenge Evolution 125

CHAPTER ELEVEN
Dinosaurs and the Hierarchy of Life 141

CHAPTER TWELVE
The Evolutionary Map for Dinosaurs 169

CHAPTER THIRTEEN
Death by Decree 195

CHAPTER FOURTEEN
The Road to Jurassic Park 207

CHAPTER FIFTEEN
Crossing the Boundary 229

CHAPTER SIXTEEN
Diversification and Decline 245

CHAPTER SEVENTEEN
The Real Great Dinosaur Extinction 265

CHAPTER EIGHTEEN
The Third Wave 285

Epilogue 297

Notes 303

Credits 319

Index 323

FOREWORD

Tim Rowe and Lowell Dingus ask, "What really happened to the dinosaurs?" If you think you know the answer, you're in for more than one surprise. That question, like this book, has two parts. First, what is a dinosaur? And second, why did all the big ones disappear some 65 million years ago? The authors explain how these deceptively simple questions have spawned more than a few twists, turns, kinks, and knots in our conceptual landscape during the 156 years since the name "Dinosauria" was first coined. But this book isn't limited to the voracious behemoths of childhood fantasy and Hollywood film. Lowell and Tim's real achievement is to place dinosaurs in a sweeping panoply of ideas that have shaped the history, philosophy, and science underlying our sense of our own place in the biological world.

However much they may change through time, organisms owe their very existence to the process of descent. So it's refreshing to finally see birds treated as simply one among the disparate and diverse lineages known as dinosaurs. Few will find this thesis particularly surprising in a modern perspective on dinosaur evolution. The surprising part is that for more than a century there's never been a better explanation for all the data than that birds *are* dinosaurs. So what have we been arguing about for all that time? Tim and Lowell explain in straightforward terms how so little of this controversy is data driven. Its persistence depends instead on the assumptions underlying the competing explanations of these data. The authors remind us once again that although science may be objective, scientists are people, and they aren't always so inclined.

Tim and Lowell argue that a preference for complex answers, in cases where a simple one would do, may have hindered some pre-evolutionary scientists from seeing the connection between birds and other dinosaurs. What is more striking is that one of the first extinct dinosaurs ever found, *Archaeopteryx lithographica*, had feathers and could fly. The Darwinian Revolution should have rectified that error more than a century ago, but we are only now beginning to appreciate its full consequences. They range from the mundane—members of the Audubon Society and Dinosaur Society actually belong to the same club, to the profound—only *some* dinosaurs became extinct 65 million years ago. And this is why Lowell and Tim can say that more dinosaur species may be threatened by humans today than by anything they ever faced in the past.

Tim and Lowell are in a singular position to recount the grand tale of the dinosaurs. They've made original contributions to the field, with Lowell working on the Cretaceous extinction and Tim on early dinosaur phylogeny. They also had the good fortune to have participated in some key events in the recent history of dinosaur paleontology while graduate students at Berkeley. Lowell and Tim entwine their graduate-school years at Berkeley throughout the story and that personal touch makes this book all the more appealing.

The three of us shared those years at Berkeley, so I can say with some force that you'll learn a lot from Lowell and Tim and enjoy the process; I know I did. Lowell was the soft-rock geologist, and his insights—amply demonstrated in this book—into the interplay between Earth and Life history influence my thinking to this day. For someone who has struggled simply to find his position on a topographic map, I will always be in awe of Lowell's unerring ability to correct errors in that same map by simply observing the lay of the land upon which we were standing. He also had an uncanny ability to look at small, exposed sections of sedimentary rock layers and thereby visualize the entire three-dimensional subsurface structure of those layers.

The first part of this book stems from Lowell's Ph.D. dissertation on the mass extinction at the end of the Cretaceous. It's presented in the form of a "whodunit" in which Lowell presents both the prosecution and the defense: A difficult line to walk, to be sure, but one for which Lowell is well-suited. He has the scholar's grasp of the primary data and a deep understanding of the commensurate analytical tools. But it's Lowell's flair for storytelling that enables nonspecialists to weigh the merits of proposed answers to one of the most enduring enigmas in paleontology. He begins by taking us on a review of past ideas about dinosaur extinction. Then he carefully feeds us the geological concepts necessary to appreciate what's at stake. That also helps us to see current conflicts in light of broader questions, such as the relative importance of gradual vs. catastrophic events in shaping the history of Life. He then makes the cases for the modern rivals in this debate, from familiar and ordinary Earth-shaping forces to rare asteroid impacts. Finally, he provides a thought-provoking discussion of the circumstances that will limit our ability to discriminate among alternative answers to events in the deep-time of Earth history.

The second part of this book is a condensed version of Tim Rowe's course on dinosaur evolution from the University of Texas at Austin. He introduces us to the major lineages of dinosaurs, setting them into the larger themes of Earth's history and Life's genealogy. And he throws in sections on diverse topics including human evolution, "Jurassic Park," and Pleistocene extinctions for the sake of completeness. Tim provides an especially lucid, wide ranging, and interesting account of a broad array of conceptual issues in evolutionary biology as they relate to dinosaurs past and present. Against that backdrop we can fully appreciate the irony that extinct dinosaurs were first used in an attempt to discredit the Theory of Descent, and later used to support it.

Tim and I share an intense interest in vertebrate anatomy. Anatomists must endure long apprenticeships to acquire the knowledge and skills necessary to produce original contributions in that venerable field. Tim and I spent endless late-night hours pouring over the fabulous collections housed in the University of California Museum of Paleontology. Those forays led us ultimately to study and revise the evolutionary histories of the two great branches of land egg-laying animals for our doctorate dissertations. Tim worked on synapsids, living mammals and their extinct relatives, and I worked on sauropsids, living reptiles and their extinct relatives.

Tim is a perceptive morphologist committed to the quest for the anatomical clues that dino detectives use to solve their phylogenetic puzzles. His broad interests and acute powers of observation, coupled with our shared penchant for barbecuing "birds 'n dogs" (chicken and Italian sausage), led him to recognize a new hip muscle, the cuppedicus, in living birds. This "tasty little morsel" arises from a concavity on the hip bone lying just in front of the hip socket. It is yet one more piece of evidence linking birds to the other tetanurine theropod dinosaurs that possess this concavity, which Tim named the cuppedicus fossa.

I share with Lowell and Tim an abiding love of field work—if for no other reason than to escape the phones, e-mail and faxes at the office—and we've shared a few adventures in the field together. My last foray with Lowell was field work at its best. He'd been commissioned to secure a *Triceratops* for an exhibit at the California Academy of Sciences and had invited me to join his field crew for a few weeks. There's little in the way of science involved in getting a fossil *Triceratops* out of the ground; it's simple pick and shovel work under a hot sun. We were infinitely grateful that Lowell had the presence of mind to select a specimen overlooking Ft. Peck reservoir. At the end of each day it was off to the lake to strip and dip and scrub the crust away. And each night found Lowell and me at water's edge, determined to bring one of the lake's mighty Channel catfish to our grill. No such luck; all we could manage was some poor old sucker-fish, which provided good sport if poor eating.

Tim and I have spent more time in the field together, but little of that has involved hunting fossils. Our quarry, the diverse lizards of the American Southwest, were pursued instead with crow bars, giant rubber bands, and fishing poles tipped with nooses fashioned from waxed dental floss. But that was during the day, and nightfall found us cruising desert roads at dead slow in search of the nocturnal lizards (especially the long, limbless, carnivorous variety, viz., snakes) who are prone to linger on the warm blacktop after dark.

Tim is an accomplished cellist—which is to say that his musical tastes seldom pass into the twentieth century. So each of our trips is linked in my mind to a particular musical masterwork from the distant past. I'll never forget field dressing specimens on the rim of the Grand Canyon at dusk to the strains of an exquisite Chinese opera. Our expeditions have had us wading in frigid waters in search of larval Pacific Giant salamanders, weathering springtime sandstorms for a chance at dune-dwelling sand lizards, and levering up granite boulders in intense heat up along the Mogollon Rim in search of the elusive Arizona Night lizard.

So saddle up, and let Lowell and Tim take you on an adventure of the mind that will forever change the way you look at dinosaurs. You needn't be a dinosaur aficionado to delight in this marvelous book. Anyone remotely interested in the world around them will find something in it to relish, ponder, and wonder at.

Jacques Gauthier
Curator of Vertebrate Paleontology
Yale Peabody Museum
Professor of Geology and Geophysics
Yale University

September, 1997

ACKNOWLEDGMENTS

Many people have contributed to this book and we gratefully acknowledge their assistance throughout the endeavor. The community of faculty, staff, and fellow students that we shared at the University of California fostered our basic perspective, and much of the information we present was first brought to our attention by members of this community. Chief among these are Walter Alvarez, David Archibald, William Clemens, Kevin de Queiroz, Jacques Gauthier, Howard Hutchinson, Kevin Padian, Carl Swisher, and David Wake. All these people have made contributions that extend from the germination of this book to its completion. Chris Brochu, Luis Chiappe, Jim Clark, Matt Colbert, Ernest Lundelius, John Merck, Jr., and Mark Norell, provided ideas and highly constructive criticisms as we prepared the manuscript. In addition, many colleagues offered encouragement and helpful comments. Without implying that they or the aforementioned necessarily agree with our conclusions, we gratefully acknowledge the reviews provided by Philip Bjork, Philip Currie, Steven D'Hondt, Peter Dodson, David Fastovsky, Nicholas Geist, Louis Jacobs, Kirk Johnson, David Martill, Michael Novacek, and David Weishampel. Essential to the book's development has been the unwavering interest and sound advice of our representative Sam Fleishman. We also thank Sarah Wilson for her help assembling many of the illustrations. Coco Kishi and John Merck, Jr., produced the digital skeletal illustrations used throughout the book. Sharon Bruyere and Egan Jones assisted in preparing these figures.

Vertebrate paleontology would be impossible without museum research collections. Many different museums house the materials we describe, but we owe special thanks to the curatorial staffs who maintain the fossil and modern osteological collections of the American Museum of Natural History, the University of California Museum of Paleontology, the UC Museum of Vertebrate Zoology, and the Texas Memorial Museum. Historical research was supported by the libraries of the American Museum of Natural History, the Langston Library of Vertebrate Paleontololgy and the Walther Geology Library of the University of Texas, the Museum of Comparative Zoology Library, Harvard University, and the John Crerar Library of the University of Chicago. This work was supported in part by a grant from the National Science Foundation (USE-9156073), by the University of Texas Geology Foundation, and the UT Center for Instructional Technologies.

Finally, we would like to thank our editors, Holly Hodder and Diane Cimino Maass, and all the other staff members at W. H. Freeman and Company who worked so tirelessly to bring our vision of this book into reality. Most of all, we want to express our gratitude for the understanding and support provided by Elizabeth Chapman and Elizabeth Gordon, who patiently tolerated our antics while we struggled for a decade to write the book, as well as Larry Dingus, who has always steadfastly refused to accept that tyrannosaurs could fly.

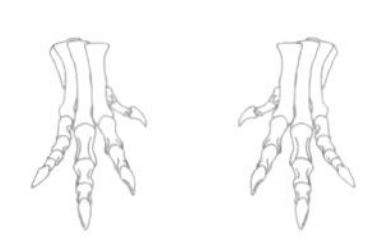

PROLOGUE

Most vertebrate paleontologists have been hooked on paleontology since they were kids, and we are no exception. The lure of spending our lives studying fossils of dinosaurs, or other ancient animals with backbones, was something that neither of us could shake. In a sense, we've never had to grow up and get "real jobs" like so many of our other childhood friends. During the field-work season, we can travel to exotic locales, where the chance of finding an exquisite fossil of some previously unknown animal always lies just around the next rocky outcropping. We feel very fortunate to have jobs where the treasure hunts of our childhood still continue.

Of course, it's not all fun. Like our paleontolgical colleagues before us and the students behind us, we have all endured countless courses and endless examinations, knowing full well that there might not be a job for us at the end of our educational travails. But, it's worth the risk. We all share the goal of spending our lives discovering and studying these ultimate antiques of evolutionary history. So friendships are formed that often last a lifetime, especially during graduate school when all one's skills and determination are put to the test.

It was during this phase of our lives, in the late 1970s and early 1980s, that we became close friends as classmates in the Department of Paleontology at the University of California, Berkeley. Both of us were drawn to Berkeley by the romance of studying dinosaurs at an institution with a long and distinguished paleontological heritage.

At the outset, it didn't seem to us that our Ph.D. projects had much in common. Tim focused his research on clarifying evolutionary relationships among mammals and reptiles, including dinosaurs. Lowell focused his studies on evaluating the geological and paleontological aspects involving the disappearance of dinosaurs 65 million years ago. Yet, as we completed our dissertations and moved on to other jobs, we realized that our efforts had dealt with the same issue, albeit from different evolutionary angles—dinosaur extinction. *The Mistaken Extinction* presents our personal perspective on this intriguing problem in Earth and evolutionary history.

We present the issues surrounding dinosaur extinction as though they are elements in a scientific detective story; following a trail of geologic and paleontologic clues toward a solution. After reviewing some earlier ideas put forward to explain the demise of the dinosaurs, Part I presents two scenarios to contrast the most prevalent current explanations for the extinctions at the end of the Cretaceous Period. Part II then examines whether all the dinosaurs really became extinct 65 million years ago. Over the course of this book, it will become clear that the questions being raised today actually have their roots in the debates that raged within the scientific community in the nineteenth century, when Darwin's theory of evolution first burst upon the scene.

As graduate students and later as professionals, we have each had the privilege of participating in these fascinating scientific investigations. The experience has taught us invaluable lessons, not only about the history of life but also about how science is conducted. Our studies, tempered by the insightful contributions of hundreds of other dedicated scientists, have led us to form our own conclusions about the mysteries of dinosaur extinction. But in the end, our goal is to lay out the evidence involved in these investigations so you can judge for yourself what really happened to the dinosaurs.

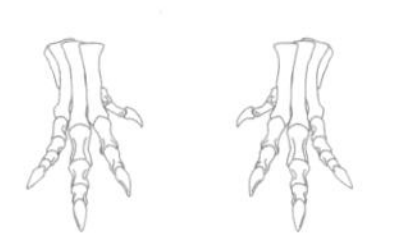

PART ONE

The Search for the Smoking Gun

CHAPTER ONE

The Seductive Allure of Dinosaurs

Paleontology—the study of fossils from prehistoric animals and plants—has experienced a surge in popular interest over the last few decades. Dinosaurs, however, have never fallen out of favor. Since Richard Owen, the famous British anatomist, first coined the name "Dinosauria" in 1841, dinosaurs have captured the curiosity of both paleontologists and the general public. Such attention is well deserved because dinosaurs are widely recognized to have ruled the continents for over 150 million years.

Public relations events were thrown into high gear soon after Richard Owen first recognized the group Dinosauria. In the early 1850s, as part of the opening for an exhibition of dinosaur models at Sydenham Park in London, Owen held the special-invitation dinner party depicted here. His twenty guests dined inside the partially built model of an *Iguanodon* which, along with others still on view today, was sculpted by Waterhouse Hawkins.

As paleontologists, one of the questions that we are most frequently asked is "Why are dinosaurs so popular?" There is no single, all-encompassing answer, but we can offer a few explanations. Dinosaurs inspire the imaginations of young and old alike. For many youngsters, like we once were, these magnificent animals represent our first introduction to science. The immense size of many dinosaurs, their strange appearance, and incredible age all contribute to their mystique. They seem to be both stranger and larger than life, like monsters out of science-fiction stories, but their fossilized bones prove that they really lived. To us, that's the key: dinosaurs are not just imaginary. The events of their lives were every bit as real and vital as our own.

Our curiosity about dinosaurs drives us to want to know everything about them. In this pursuit, however, our love of science fiction runs headlong into the methods of scientific research and the limits of scientific knowledge. Fossils can help us solve many mysteries about dinosaur history. The size of their skeletons gives us a good idea about how large these animals were. By studying the shape of their bones and the structure of their skeletons, we can establish which dinosaurs were close and which were distant evolutionary relatives.

For example, fossils of the earliest known dinosaurs provide evidence of how dinosaurs differ from their reptilian cousins. By studying the structure of the fossilized bones in their hind legs and hips, we have found features which indicate that dinosaurs walked upright, with their legs extending straight down from the hips to the ground. This erect posture is one of the evolutionary innovations that makes a dinosaur a dinosaur, and it is by identifying such features that we can trace the sequence of dinosaur evolution.

All later dinosaurs inherited this new posture from the first dinosaur. Two major lineages sprang from this common ancestor: Saurischians contain all the commonly recognized carnivorous dinosaurs, such as *Tyrannosaurus* and *Allosaurus*, as well as the largest dinosaurs, such as *Apatosaurus* and *Diplodocus*. Ornithischians include the armored dinosaurs, such as *Stegosaurus* and *Ankylosaurus*; the duckbills; the horned dinosaurs, such as *Triceratops*; and the thick-skulled pachycephalosaurs.

In addition to establishing evolutionary relationships, fossils can provide some information about the behavior of dinosaurs. Sequences of fossil footprints, called trackways, confirm that dinosaurs moved around in an upright or erect posture.

Besides evolutionary relationships and some behaviors, the rocks in which dinosaur fossils are preserved provide clues about what their environment might have been like and how long ago they lived.

The evidence for measuring the vast spans of geologic time involved in dinosaur evolution comes from the layers of volcanic rock preserved among the beds that contain the dinosaur fossils. Some of these volcanic layers are composed of minerals formed, in part, by radioactive atoms. Once the lava erupted and cooled, radioactive "parent" atoms began breaking down, or decaying, into "daughter" atoms of different composition at a constant rate. This rate can be measured experimentally with sophisticated equipment. In essence, to determine when the volcanic rock was formed, one must measure the proportion between the parent atoms and daughter atoms contained in the volcanic rock. Then, because we know the rate at which parent atoms decay, an age can be calculated. This process is called radioisotopic dating. Such calculations have led us to the realization that the evolutionary roots of dinosaurs stretch back hundreds of millions of years.

The age of the era during which dinosaurs ruled the Earth is truly mind-boggling.[1] Most of us are used to dealing in time scales ranging from days to decades, but the dinosaurs, as we commonly recognize them, lived more than 60 million years before any people were around. The roots of our own human ancestry stretch back only 4 or 5 million years. However, the Mesozoic Era, often termed the Age of Dinosaurs,

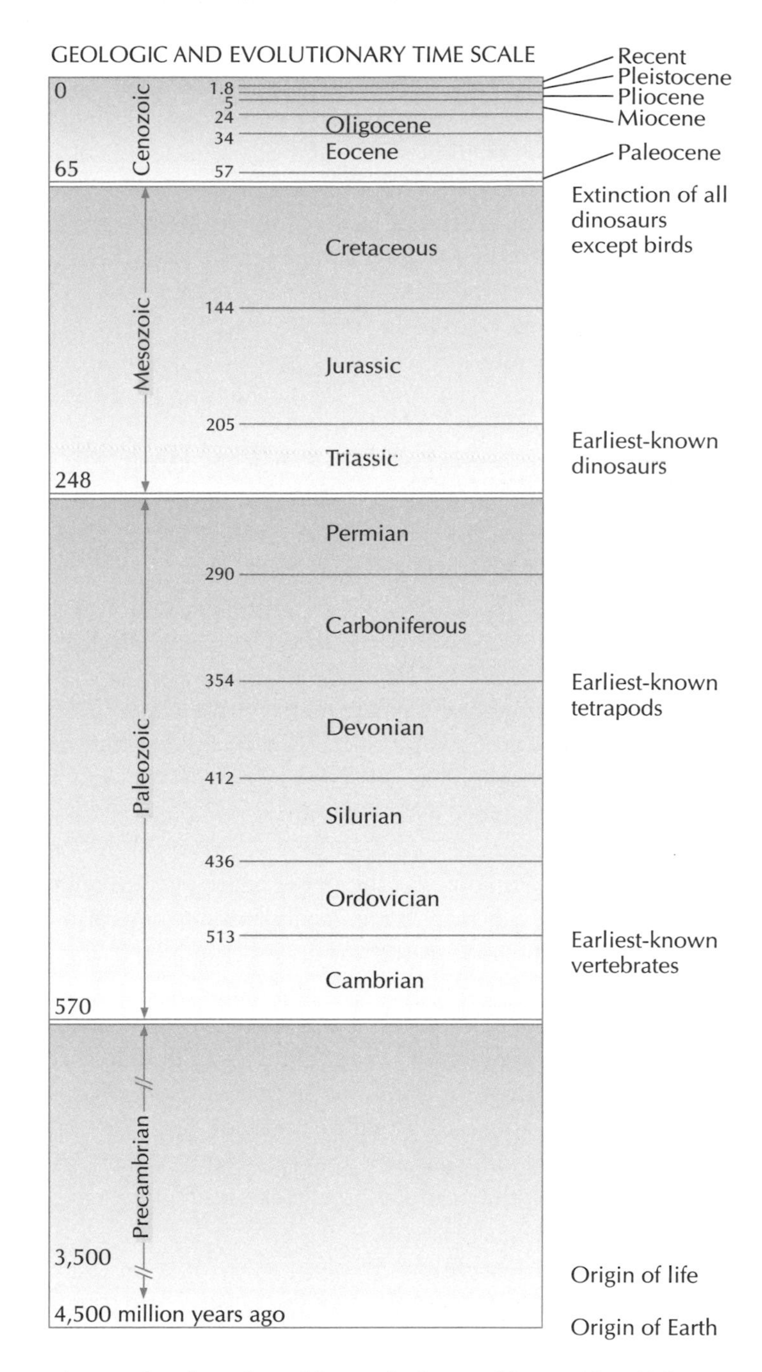

Time scale of Earth and its evolutionary history identifying major eras and periods. Major events in the history of life are listed along the right side. The earliest known humans originated near the start of the Pliocene, less than 5 million years ago. The boundary between the Mesozoic and Cenozoic Eras corresponds to the boundary between the Cretaceous and Tertiary Periods. This is often abbreviated as the K-T boundary. The "K" stands for Cretaceous, which in German is called "Kreide." The Tertiary Period contains all the time intervals from the Paleocene through the Pliocene.

stretched from about 250 million years ago up to about 65 million years ago. This era lasted more than 35 times longer than the entire evolutionary history of humans to this point. If a human generation is assumed to average 20 years, then 9,250,000 generations of parents and children could be fit into the Age of Dinosaurs.

Taking a closer look at the geologic time scale, the earliest known dinosaurs lived near the middle of what geologists call the Triassic Period—about 228 million years ago. The Triassic was the earliest period of the Mesozoic Era. The next period, called the Jurassic, extends from about 205 million years ago to about 144 million years ago. It was during this period that dinosaurs achieved their maximum size in the form of the titanic sauropods, such as *Apatosaurus*. Dinosaurs continued to dominate the continents during the final period of the Mesozoic, called the Cretaceous. This period saw the evolution of the most fearsome terrestrial predators the world has ever seen, like *Tyrannosaurus*, as well as the evolutionary bursts that produced a dazzling diversity of duckbills and horned dinosaurs. Then, 65 million years ago, at the end of the Cretaceous Period, all the commonly recognized kinds of dinosaurs went extinct. These spans of time are essentially incomprehensible to animals like us, living lives that last only about 70 years.

Yet, dinosaurs are actually newcomers on the stage of Earth and evolutionary history. The Earth was formed about 4.5 billion years ago, and the earliest fossils of single-celled life date back to between 3.5 and 4.0 billion years ago. Vertebrates—animals with backbones—originated about 500 million years ago. So, dinosaurs arose about half way through the evolutionary history of vertebrates. Humans, on the other hand, evolved within the last one percent of vertebrate history. Paleontologists—scientists who study fossil remains of ancient life—have accomplished an amazing feat by tracing evolutionary lineages so far back into the ancient past.

Fossils and the rocks containing those fossils reveal incredible perspectives on the history of dinosaurs. Yet, there is a multitude of intriguing questions that the fossils and rocks cannot help us answer with any degree of certainty. For example, the fossil skeletons provide no evidence for us to judge what color the living animals were or what kind of sounds they made. Also, there are many questions about how these animals behaved that are not easily interpreted from the fossils.

Living with these limitations is often frustrating for paleontologists. As scientists, we can only ask questions of the fossil record that we can test with the available evidence. Because all the dinosaurs like *Tyrannosaurus* and *Triceratops* are extinct, we can't go out to the field and observe how they behaved. Consequently, we can't be certain how *Stegosaurus* used the ornate set of plates on its back. Nor do we know for sure whether the duckbills used their crests for amplifying their vocalizations: it's certainly possible, but there is no unequivocal evidence to support this hunch.

Nonetheless, it is easy to be seduced by the spectacular skeletons. The limitations imposed by the evidence in the rock and fossil records haven't stopped some paleontologists from imaginatively speculating about the use of these unusual structures and many other aspects of dinosaur behavior. The public desperately wants answers to all their questions about dinosaurs, and some paleontologists have been unable to resist the urge to oblige and speculate. The literature of both the scientific and popular press is replete with intriguing and outrageous claims about how dinosaurs lived and died.

As a result, conflicting claims about dinosaurs, fueled by the public frenzy over blockbusters like *Jurassic Park*, have led to numerous public debates among paleontologists. Were dinosaurs "warm-blooded"? Did they take care of their young? And within the present context: What caused so many dinosaurs to become extinct 65 million years ago?

Our charge in this book is to investigate all the major issues surrounding dinosaur extinction. What do we really know about the extinction of dinosaurs at the end of the Cretaceous Period? To try and solve this case, which is somewhat like an ancient murder mystery, we must closely examine the evidence available in the rock and fossil records. That's the subject in Part I of this book.

Our review of extinction hypotheses will basically follow the order in which they were proposed. To begin, we will review some earlier ideas or hypotheses that sought to explain the cause of dinosaur extinction. In retrospect, some seem comical, but others are more sophisticated. Our goal will be twofold: First, we will try to distinguish between hypotheses that are scientific and those that are not. In pragmatic terms, this distinction depends on whether there is evidence available to test the hypothesis. If the rock and fossil records contain evidence that can be used to test the proposed cause, then the hypothesis is considered to be scientific. If the hypothesis cannot be tested with evidence present in the rock and fossil record, then, for the time being, it falls outside the realm of science. Second, for each scientific extinction hypothesis, we will try to evaluate whether the evidence preserved in the rock and fossil records is consistent or inconsistent with the proposed cause of extinction. This exercise will be used to illustrate how scientific methods work by testing new ideas against the available evidence. In the end, the hypothesis that is consistent with, or explains, more evidence than any other is deemed to be the most likely. This rule for making decisions, often called Occam's Razor or the Principle of Parsimony, forms the foundation for choosing between competing scientific hypotheses.

Next, we will investigate the current debate about dinosaur extinction by exploring clues in the rock record relating to two of the most momentous geological and astronomical events in Earth history. The first event began to be seriously discussed in 1972 as a cause for the Cretaceous dinosaur extinctions. It involves the second largest known episode of volcanic activity ever inflicted on the continents of our planet, along with potentially associated changes in the location and configuration of the seas. Massive piles of lava flows that now cover vast areas of India represent the evidence for these eruptions at the end of the Cretaceous. Because of the enormous amounts of gases and other pollutants erupted during this event, the Earth's environment could have been severely damaged. Could this have caused the extinction of dinosaurs?

Or was that extinction caused by the Earth-shattering impact of a large comet or asteroid near present-day Yucatan? This stunning hypothesis was put forward in 1980 by some of our colleagues in the Geology and Geophysics Department when we were graduate students at Berkeley. Their scenario suggested that the impact blasted a cloud of debris into orbit, which completely enveloped the Earth. This cloud cut off all light from the Sun for several months, preventing plants from photosynthesizing. With the death of most plants, the herbivorous dinosaurs died out, and with the extinction of their plant-eating prey, so did the carnivorous dinosaurs.

These are both sensational scenarios. Almost immediately after the asteroid hypothesis was published in 1980, both the press and the public became as engrossed in the issues as our scientific community at Berkeley did. As students of professors involved in the debates, we were quickly drawn into the middle of the ensuing exchanges between the proponents of these two competing hypotheses. They were rather daunting but extremely exciting times because we realized that an important phase of evolutionary history might be rewritten right before our very eyes. At stake was the stately and gradualistic view of change that had dominated geological and evolutionary thinking for the last two centuries. Did catastrophic change, like that predicted by the impact hypothesis, really drive the history of life as some scientists in the 1800s had argued?

Over 15 years have now passed since the impact hypothesis was proposed, but still the debate over the cause of the Cretaceous extinctions continues. What new evidence to refute or support the competing ideas has been discovered? By searching for clues in the rock record, we will first try to establish whether these extraordinary events really happened. Then, we will try to evaluate the effects that these two events had on dinosaurs and other contemporary life forms, by looking at the fossil record to determine which groups of organisms went extinct and which groups survived.

One key to deciding whether the eruptions or the impact was responsible lies in the proposed durations for these events. Most of the eruptions are thought to have occurred over a period of 500,000 years leading up to the end of the Cretaceous. The effects of the impact, however, are thought to have lasted only between a few months and a few thousand years, at most. Can we tell time precisely enough to distinguish between the effects of these events in rock and fossil records that are 65 million years old? Our ability to do so may well determine whether we can assign responsibility for the extinctions to either of the two proposed causes.

However, the debate over the cause of the Cretaceous dinosaur extinctions was, by no means, the only argument about dinosaur extinction going on at Berkeley when we were students. In Part II of the book, we will turn to investigate a rather startling question that some of our paleontological classmates were beginning to pose: Did all the dinosaurs really become extinct at the end of the Cretaceous? When we first entered college, everyone agreed that they did. Yet, by the time we got to graduate school at Berkeley, we were shocked to find out that this long-accepted evolutionary fact might not be correct.

This debate was catalyzed by a new method of determining evolutionary relationships, called cladistics. Its goal is to sort out smaller lineages within larger families of organisms by rigorously identifying unique anatomical features that the different lines shared. In essence, members of a lineage that share a unique feature, like the upright posture of dinosaurs, are descendants of the first animal in which the unique feature evolved. By following the evolutionary trail of anatomical clues from larger lineages into smaller lineages, such as from Dinosauria into saurischians and ornithischians, one can reconstruct the sequence in which different features and lineages evolved.

Applying the cladistic approach to the evolution of vertebrates generated some results that contradicted several long-held evolutionary notions. Consequently, it was very controversial. It ruffled the feathers of many prominent evolutionary biologists. One of our landlords when we studied at Berkeley was intimately involved in these debates. Gareth Nelson, a Curator of Ichthyology at the American Museum of Natural History in New York, was the most prominent and influential proponent of cladistics in the United States. In order to get a break on our rent, we helped him do renovations around the house, and while working, we'd pass the hours discussing cladistics.

One of the most radical results of cladistic analysis suggested that living birds descended from small carnivorous dinosaurs. This analysis formed the basis for the dissertation by Jacques Gauthier, one of our classmates and roommates. Actually, this result was not a totally new idea. Back in the mid-1800s, shortly after Charles Darwin had published his seminal work on evolution, the scientific community was rocked by the discovery of *Archaeopteryx*, a Late Jurassic animal that exhibited many reptilian features, but also had feathers like a bird. Eventually, most scientists accepted that *Archaeopteryx* represented the earliest-known bird, but debate continued over whether birds evolved from dinosaurs or other reptiles. More than 100 years later, our roommate had run headlong into the same evolutionary quagmire. As with the debate over Cretaceous extinctions, the intellectual exchanges were electrifying to incoming Ph.D. students like us. We might actually witness the solution of a long-running evolutionary mystery: From what ancestor did birds really evolve?

Did we witness that solution? To judge, we'll start back at the beginning of Life on Earth. By following the trail of anatomical clues leading from the origin of early life forms through the first vertebrates and on into dinosaurs, we'll seek to discover whether any of the organisms that survived the extinctions at the end of the Cretaceous descended from dinosaurs. Are birds such survivors? Or did they descend from early relatives of crocodiles, as many paleontologists have argued over the last century? If birds did evolve from dinosaurs, they would be members of the dinosaur's evolutionary lineage. Therefore, dinosaurs would not really be extinct . . . at least not yet.

Although we will be frank about our own conclusions concerning dinosaur extinction, the point of the book is to lay out the evidence in these debates so that you can make your own judgments. In setting out the evidence, we have tried to be fair to the competing ideas, although we realize that some of our colleagues have come to conclusions different from ours. More importantly, by reading on, you will have the privilege to serve on the same scientific jury that we and our colleagues have—deliberating over some of the most fascinating mysteries in the history of Life.

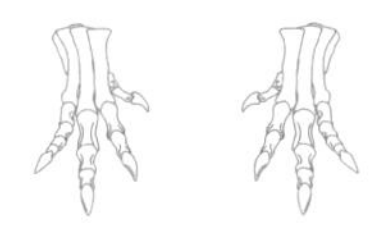

CHAPTER TWO

Earlier Extinction Hypotheses

When it comes to the extinction of the dinosaurs, we really shouldn't feel too badly about it because it's well known that dinosaurs majestically ruled the land for over 150 million years. Given that we humans are on the verge of annihilating our species after only about 100,000 years, it's not as though the dinosaurs got cheated. Conventional paleontological wisdom teaches that extinction is as inevitable as death and taxes. Looking back at the fossil record, paleontologists point out that less than one percent of all the species that have ever lived are probably still alive today. So, with a stiff upper lip, paleontologists not only have come to accept the extinction of dinosaurs but to revel in their mysterious disappearance.

This revelry has long constituted one of the most entertaining, intriguing, and controversial pursuits in science. In trying to solve the mystery, hundreds of scientists from paleontologists to astrophysicists have paid their money and taken their chances by putting their hypotheses in print. Some of the earlier explanations seem rather fanciful, whereas others appear more reasonable. Most earlier hypotheses were predicated on relatively gradual, Earth-based causes. However, some have been related to relatively sudden, extraterrestrial events. Investigations into dinosaur extinction intensified in 1980, when scientists at Berkeley proposed that the impact of an extraterrestrial body was responsible. But before tackling the more demanding technical evidence involving impacts and volcanic events, let's ease into the topic by informally evaluating a small sample of earlier explanations. Then, we'll focus on the hypotheses involved in the more recent debate.

It is not our intention to deal with all the hypotheses that have been proposed to explain dinosaur extinction or to criticize the individuals that proposed these earlier ideas. Consequently, personal references are not provided. A more comprehensive listing of these and other early extinction hypotheses can be found in the references listed in the back of the book.[1] The goal here is to illustrate how the scientific method of proposing and testing ideas actually works.

Throughout the review of extinction hypotheses, our emphasis as scientific detectives of ancient history will be on two questions: First, *how can we scientifically test these hypotheses for the extinction of dinosaurs with the evidence available in the rock and fossil records*? Second, *which hypotheses are refuted and which are supported by most of that evidence*? The first question can be used to separate scientific explanations from unscientific ones. Because, if we're going to assume the role of scientific detectives in solving this mystery, we must propose ideas or hypotheses that we can test with the evidence available in the geologic and paleontologic records. The second question must be applied to decide which alleged cause of extinction is the most likely culprit. That will be the one that explains the most evidence found in the rock and fossil records. In the end, it is impossible to *prove* which, if any, of these hypotheses is correct. However, it is possible to show which hypotheses are inconsistent, or fail to explain, the available geologic and paleontologic evidence. Now, let's get on to the allegations and the evidence in the case.

The following scenarios exemplify extinction hypotheses that are almost impossible to test directly with the clues available in the rock and fossil records. Thus, whether they actually happened or not, they remain unsatisfying explanations to the scientific community.

ALL FILLED UP, BUT NO WAY TO GO?

The Cretaceous, or last period of the Age of Dinosaurs, saw the rise and proliferation of angiosperms—more commonly known as flowering plants. Along with this evolutionary burst came a decrease in the predominance of ferns. One dietary ramification of this floral change might have been a decrease of fern oil in the diet of plant-eating or herbivorous dinosaurs. This could have led to acute constipation which exterminated all the herbivorous dinosaurs. In turn, the meat-eating or carnivorous dinosaurs may have died out after the extinction of their herbivorous prey, which presumably had served as their primary food source.

Can we test this idea with evidence from the fossil record? The stomach contents of dinosaurs are not usually preserved inside fossil skeletons. Thus, we cannot directly track dietary changes through time with any degree of continuity. Another way to get at the question would be to study fossilized dinosaur dung, or coprolites. This is an active area of contemporary dinosaur research that has the potential to shed new light on the diet of dinosaurs. To date, however, research on coprolites has been limited, and they are often difficult to attribute to a particular kind of dinosaur. Consequently, this hypothesis is almost impossible to test with the currently available fossil evidence—whether or not it's true.

TERMINAL SNEEZES AND SNIFFLES?

Another hypothesis has been associated with the proliferation of flowering plants, but it was offered more in jest. The idea is that more flowers meant more pollen, and more pollen meant more hay fever—so much so that all the dinosaurs died from it. Allergic drowning would indeed have been a terrible fate for such Herculean life-forms.

Like stomach contents, however, the nasal tissues of dinosaurs are not preserved as fossils. So no direct evidence for judging the hypothesis exists in the rock and fossil records.

Even if one assumes that hay fever was the cause, however, those of us with allergies might wonder why our mammalian ancestors and other late Cretaceous reptiles were not similarly afflicted. Besides, many groups of herbivorous dinosaurs, such as duckbills and horned dinosaurs, were proliferating toward the end of the Age of Dinosaurs, apparently undaunted by the increased levels of pollen. So again, the idea is not directly testable with any fossil evidence related to sinus problems, and what little indirect evidence there is refutes the hypothesis.

MORE MALADIES

Plagues have long haunted humanity and, presumably, life in general. Even today, we are again reminded of our vulnerability through the rapid spread of AIDS. So it comes as no surprise that epidemics have often been implicated in the demise of the dinosaurs.

But problems exist with testing this scenario scientifically. The most significant problem is finding evidence of disease in fossil organisms. Most diseases affect the soft anatomy: the organs, muscles, nerves, and circulatory system. However, the symptoms of diseases that might have affected the soft anatomy of dinosaurs are not evident in the fossil record because soft tissues are not preserved. Only the harder parts of the body, the bones and teeth, are commonly fossilized. Some diseases, such as arthritis and bone cancer, do leave

Epidemics have been cited as a possible cause of dinosaur extinction. However, no evidence for such an epidemic is preserved in the fossil skeletons.

their mark on the fossilized bones. But there is no evidence in the late Cretaceous bones or teeth of dinosaurs to indicate such a pandemic cause of extinction. So, diseases may well have played a role in the death of some individual dinosaurs, but we cannot directly test whether some disease of the soft anatomy caused the extinction of dinosaurs.

Nonetheless, diseases within the soft tissues of vertebrates tend to be pretty specific. They usually attack only one species or a few closely related species. Just in the midcontinent of North America, fossils representing nineteen species and at least eight major lineages of dinosaurs have been recovered from rocks deposited just before the end of the Cretaceous.[2] Indirectly, therefore, it seems unlikely that one massive epidemic could eradicate a group as diverse as the late Cretaceous dinosaurs.

AN EYE FOR AN EYE?

This hypothesis suggests that dinosaurs became plagued with cataracts, which rendered them blind and led to their extinction. This assertion is both hard to believe and impossible to test directly, because no fossilized cataracts from dinosaurs are known.

A MATTER OF DEGREE?

The sex of baby alligators, which are close living relatives of dinosaurs, is determined by the temperature at which the egg incubates in the nest. A temperature of less than 86 degrees Fahrenheit (30 degrees Centigrade) results in a female. A temperature of greater than 93 degrees Fahrenheit (34 degrees Centigrade) results in a male—which leaves a curious "no man's land" in the middle. Consequently, a climatic change at the end of the Age of Dinosaurs might have resulted in the birth of all male or all female dinosaurs. This, obviously, would have posed some serious problems for reproduction.

Unfortunately, the temperature of the climate in which the dinosaurs lived is difficult to determine that precisely, especially the temperature inside a nest. Evidence from fossil plants suggests a warming trend at the end of the Cretaceous, at least in the middle of North America.[3] Yet precise temperatures cannot be determined, making this scenario impossible to evaluate directly with the available rock and fossil evidence.

If this hypothesis were true, however, one has to wonder why so many dinosaurs died out, while some Late Cretaceous crocodiles survived. Indirectly, this hypothesis fails to account for the survival of some similarly adapted animals that were alive at the time.

THOSE LITTLE THINGS?

Paleontologists who study fossil mammals are often proud to implicate their "pets" within the context of dinosaur extinction. This hypothesis suggests that dinosaurs were eradicated by early mammals, either through direct competition for resources or because the mammals ate all their eggs. This idea is impossible to test directly with available geologic and paleontologic evidence. Such complex ecological scenarios are often difficult for contemporary ecologists to get a handle on. What's more, modern ecologists usually work in local settings with living species that they can observe—not on a global scale requiring analyses of a very incomplete fossil record that is 65 million years old. We have no direct evidence about what Late Cretaceous mammals ate or how they affected the balance of resources.

Indirectly, fossils show that all the known forms of Late Cretaceous mammals were no larger than modern beavers. So it's a bit difficult to imagine them being a serious threat to dinosaurs such as *Tyrannosaurus* and *Triceratops*. Even if they were, mammals had coexisted with dinosaurs for at least 100 million years before the end of the Cretaceous. So the dinosaurs had more than demonstrated their ability to cope with small mammals over the long run. Finally, one is left wondering, if those little, rat-sized mammals were so rough on the dinosaurs and their eggs, how did crocodiles, lizards, turtles, and other reptiles survive? Thus, the available indirect evidence appears to exhonerate our early relatives.

Proposals that dinosaurs became sexually frustrated, became suicidal, or succumbed competitively to leaf-eating caterpillars represent other hypotheses that are difficult, if not impossible, to test with the available geologic and paleontologic evidence. In contrast, many hypotheses seeking to explain dinosaur extinctions have been directly tested and apparently refuted. A small sample follows.

Minute, primitive mammals coexisted with dinosaurs for over 100 million years before the end of the Cretaceous Period. Did they represent a competitive threat to dinosaurs?

BIG, BIGGER, BIGGEST . . . BOOM?

One notion that has been tested and refuted is the idea that dinosaurs just got too big for their own good and became extinct as a result of their enormous size. This doesn't jibe with the known fossil record for a couple of reasons. First, most of the largest dinosaurs known lived and died long before the end of the Cretaceous. These were the Jurassic sauropods, including *Apatosaurus*, *Brachiosaurus*, *Ultrasaurus*, *Supersaurus*, and *Seismosaurus*. They lived about 150 million years ago. The largest dinosaurs that lived at the end of the Cretaceous, such as *Tyrannosaurus*, *Anatotitan*, *Triceratops*, and *Pachycephalosaurus*, were not nearly as large as the Jurassic sauropods. The only known exception was

Argentinosaurus, an enormous sauropod from the Late Cretaceous of South America. Nonetheless, there was not a generally progressive increase in size up to the end of the Cretaceous.

Another piece of fossil evidence refuting this hypothesis is that many relatively small kinds of dinosaurs continued to thrive in the latest Cretaceous, especially small carnivorous dinosaurs such as *Dromaeosaurus*, *Troodon*, and *Saurornithoides*.

THIN, THINNER, THINNEST . . . CRACK?

Eggs of the Late Cretaceous sauropod, *Hypselosaurus*, had unusually thin shells. Similar suggestions have been made for the rich fossil record of eggs at the end of the Cretaceous in southeastern China and possibly

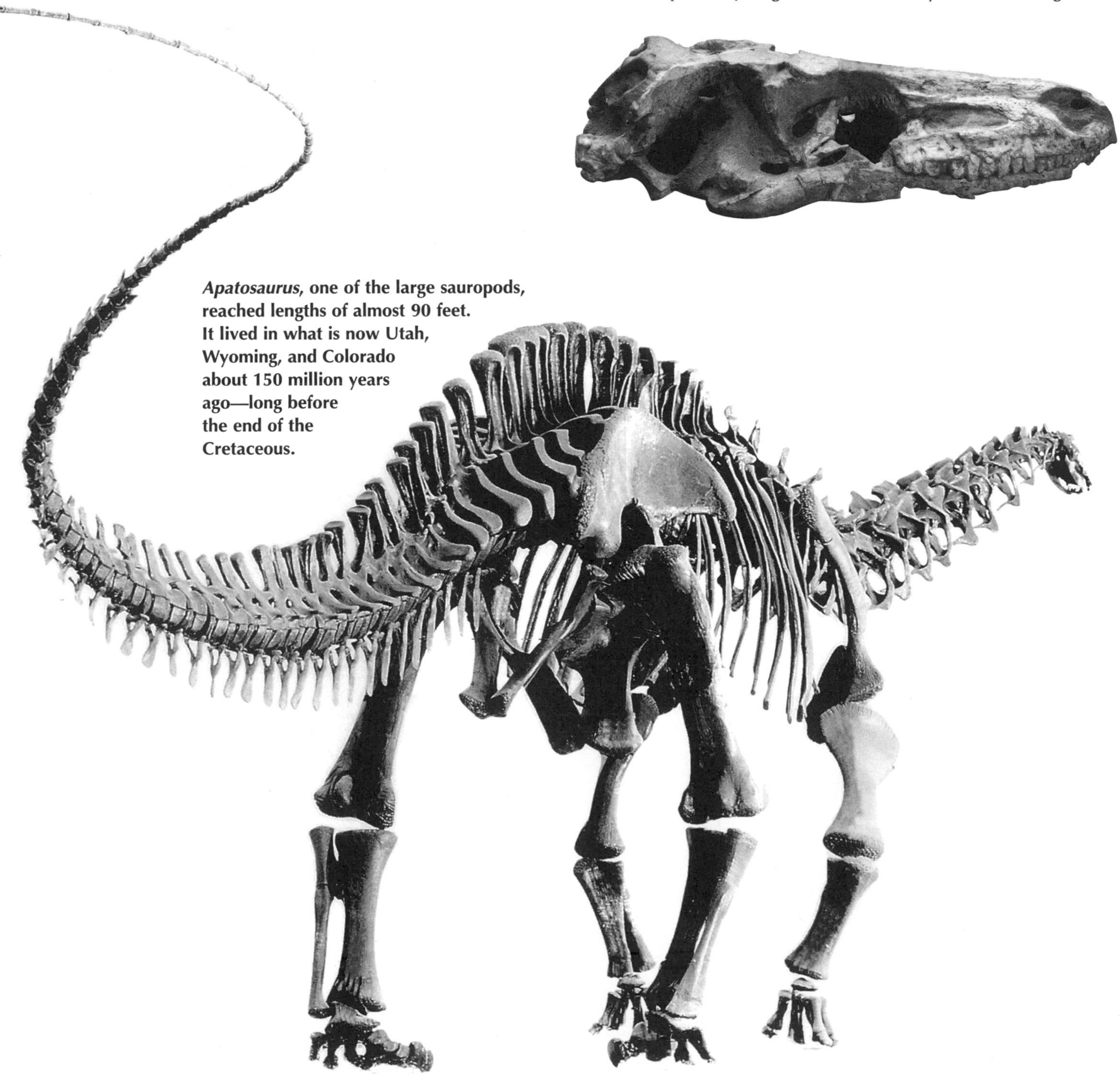

***Saurornithoides*, a small carnivorous dinosaur from Asia, is a fairly close relative of *Troodon*, *Velociraptor*, and *Dromaeosaurus*. It lived near the end of the Cretaceous, about 72 million years ago. Its skull was only 7.5 inches long.**

***Apatosaurus*, one of the large sauropods, reached lengths of almost 90 feet. It lived in what is now Utah, Wyoming, and Colorado about 150 million years ago—long before the end of the Cretaceous.**

India. Did the egg shells of all dinosaurs get so thin toward the end of the Cretaceous that they became inviable, causing the extinction of the whole group?

Within the last fifteen years, many spectacular discoveries of Late Cretaceous eggs and nests of several types of dinosaurs have been made in Montana, Canada, Argentina, China, India, and Mongolia. Nests, eggs, and embryonic remains have been found for duckbills, hypsilophodonts, troodontids, therizinosaurids, and one species related to *Oviraptor*. The newly discovered eggs were found in nests that also preserved skeletons representing embryos and hatchlings. Adolescents were also found near the nests. Thus, the eggs for these species appear to have been quite viable.

The eggs of the sauropod, *Hypselosaurus*, have thin shells for such large eggs, which are 9 inches in length.

DEATH BY DIPOLES?

Throughout the known history of the Earth, our planet's magnetic poles commonly reversed direction such that the magnetic end of the compass pointing North today would have pointed South, as it did at the time of the dinosaur extinction. The reason for these reversals is not well understood, although some ideas will be mentioned in later chapters. Nonetheless, the history or sequence of these reversals has been recorded in, and can be read from, many of the rocks that make up the Earth's crust. Some recent data suggest that when the poles reverse direction, it can happen in a period of as little as a few hundred years.[4] There is some evidence that the Earth's magnetic field plays a role in the migration of some animals, such as monarch butterflies and various birds. Was one of these reversals to blame for the extinction of the dinosaurs?

The available evidence refutes this idea. First, the last dinosaur fossils in many different geographic areas consistently occur near the middle of a stable interval when the Earth's magnetic poles were reversed—not near a

The embryonic bones of an Oviraptor-like dinosaur are preserved within this 4-inch-long egg. The bones represent the first embryo of a carnivorous or theropod dinosaur ever found. The egg was laid about 72 million years ago, near the end of the Cretaceous Period.

boundary when the poles were in the process of reversing position. In addition, the Earth's magnetic poles are known to have reversed themselves more than one hundred times during the Age of Dinosaurs,[5] so why didn't one of those other reversals knock the dinosaurs out? Finally, why would this reversal extinguish dinosaurs and leave other reptiles, such as crocodiles, turtles, and lizards, unscathed? This evidence suggests that a reversal of the magnetic poles was not the culprit.

The preceding examples do not constitute an exhaustive list of extinction hypotheses that can be tested and refuted. Others include the idea that too many predatory types evolved. This can be refuted by examining the ratio between herbivorous and carnivorous dinosaurs through time. This brings us up to the event that was most commonly thought to be the cause of dinosaur extinction before the debate over extraterrestrial impacts and volcanic eruptions began in 1980.

TRIED AND TRUE?

Before the debate about dinosaur extinction focused on impacts and volcanism, conventional wisdom long held that dinosaur eradication was due to climatic changes associated with changes in the configuration of the Earth's seas. Throughout the last part of the Age of Dinosaurs, shallow seaways covered extensive areas of the continents, especially across Europe and North America. North America was essentially cut in half by a sea that extended from the Gulf of Mexico north to the Arctic Ocean. It is now called the Western Interior Seaway. Most of our evidence of Cretaceous dinosaurs and other animals that lived in North America comes from fossils preserved in sediments that were laid down by rivers along the floodplains adjacent to this seaway. The floodplain stretched from the eastern flank of the ancestral Rocky Mountains on the west to the shallow sea on the east.[6] The subsequent uplift of the Rocky Mountains, coupled with erosion by rivers running out of the Rockies, exposed these ancient sediments across a wide swath of the western Great Plains.

Evidence from fossil plants is critical to interpreting the ancient climate of the area. As inferred from living relatives of the fossil plants that lived on that floodplain, the climate was subtropical to tropical and quite equable.[7] The

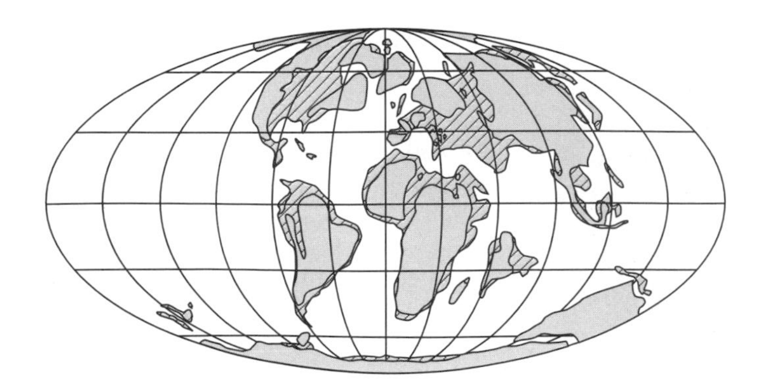

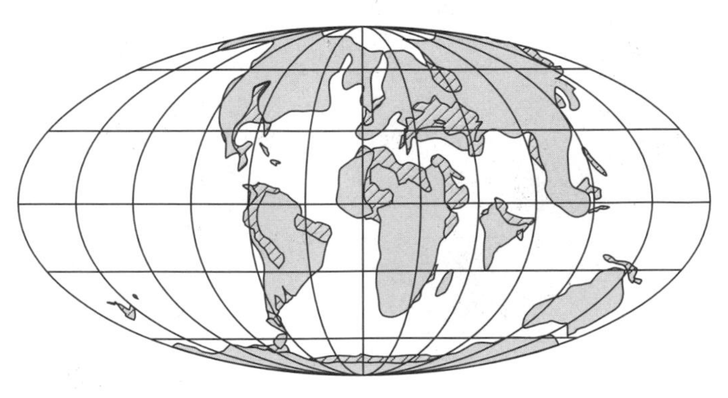

Maps of the Earth showing the distribution of continents (gray), oceans (white), and shallow continental seaways (hatched light green areas) just before the end of the Cretaceous Period *(left)* and at the beginning of the Tertiary Period *(right)*.

days weren't too hot, nor the nights too cold. The summers weren't too warm, nor the winters too frigid. In general, the climate of the Earth was more uniform than it is today, as evidenced by the fact that there were no polar icecaps. Locally in the middle of North America, the seaway also acted like a heat sink to buffer against climatic extremes. Abundant fossil evidence of "cold-blooded" vertebrates, such as crocodiles and amphibians, provide evidence that the temperature didn't drop below freezing, at least for very long. In fact, the climate gradually warmed at the very end of the Cretaceous.[8] Relatives of palms and Norfolk pines (more typical of Late Cretaceous floras found in southern Colorado and New Mexico) migrated north into areas like Montana and North Dakota.

Maps of the western United States illustrating how the Western Interior Seaway retreated to the south and east at the end of the Cretaceous Period. Areas covered by the seaway are shaded in green. The maps are based on geologic evidence showing the geographic extent of marine and floodplain sediments at different times during the late Cretaceous. Moving from (a) to (e), we travel forward in time toward the end of the Cretaceous. All together, the interval represents approximately the last 4 million years of the Cretaceous. The time difference between (d) and (e) is about 1 million years.

However, at the end of the Cretaceous, the seaways pulled back off the continental margins and lowlands into the major ocean basins. No one is exactly sure why, although some ideas will be addressed in later chapters. What is clear is that over a period of 3 or 4 million years, when the seas retreated,[9] climates would have became more extreme: warmer days, cooler nights; hotter summers, colder winters.

Conventional wisdom had long held that the cold-blooded dinosaurs could not tolerate these extreme climatic changes and, as a result, suffered total extinction. On the surface, this scenario seems reasonable, but critics have raised a couple of simple counterpoints.

First, look at some of the cold-blooded animals that survived: snakes, lizards, turtles, crocodiles. Why didn't the freezing winters and torrid summers knock these animals out? Didn't they also rely on the external climate to maintain a livable body temperature? It's hard to see why the physiology of these vertebrates would not be adversely affected, whereas that of the dinosaurs was rendered too crippled to cope. Besides, many paleontologists now argue that dinosaurs were warm-blooded, not cold-blooded, meaning that they might have been able to maintain a fairly constant body temperature regardless of the temperature outside. This is still a controversial issue, and we are still searching for uneqivocal evidence in the cellular structure of fossil bones or the presence of specialized nasal bones called turbinates that may allow us to decide for sure. Furthermore, critics point out that the shallow continental seaways had retreated and advanced numerous times during the Age of Dinosaurs. So why did dinosaurs survive the climatic changes that were associated with those earlier fluctuations but not this one? Was it simply more extreme? Some evidence suggests that it may have been, as we will see in Chapter 4.

We have now reviewed a significant number of earlier extinction hypotheses and found them wanting. They either proved to be untestable or unable to explain all the available geologic and paleontologic evidence. To try and solve the mystery, we must now turn to investigate the two competing hypotheses involved in the more recent extinction debate. But before evaluating all the detailed rock and fossil evidence, let us first provide a better image of what these volcanic and impact events might have been like, assuming that they both occurred. To do so, we will briefly set the Late Cretaceous scene just prior to and during these proposed extinction events.

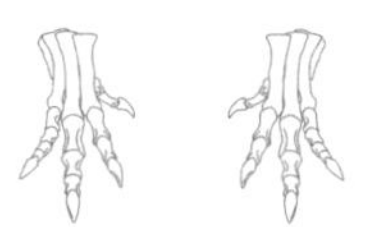

CHAPTER THREE

Contrasting Volcanic and Impact Hypotheses

Our investigation into the extinction of dinosaurs in the Late Cretaceous Period brings us now to two currently competing hypotheses. One involves terrestrially-based causes, which are related to massive volcanic eruptions and the retreat of shallow continental seas that occurred over hundreds of thousands of years. The other hypothesis invokes catastrophic, extraterrestrial causes operating in less than a few thousand years that were

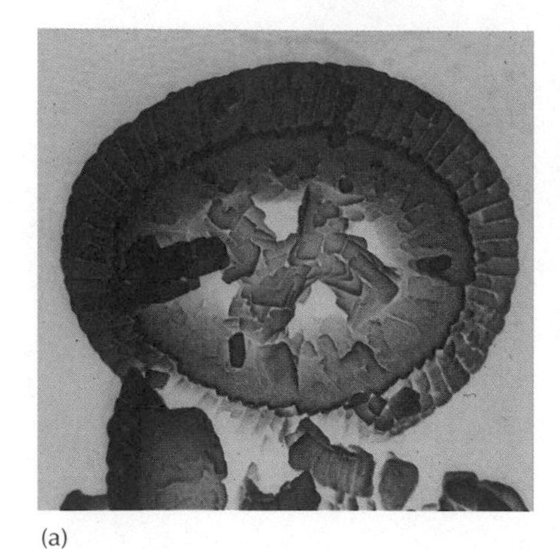

(a)

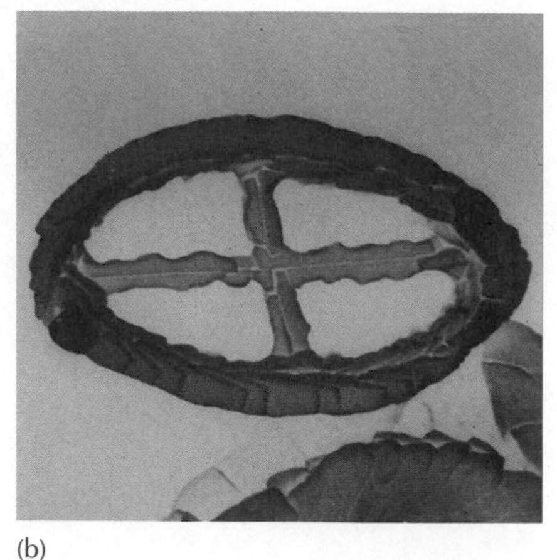

(b)

Microscopic, single-celled, marine organisms like these coccoliths have distinctive fossil shells made of calcium carbonate. Many species, such as (c) and (d), became extinct during the Cretaceous-Tertiary (K-T) transition and were replaced by new forms in the early Tertiary, such as (a) and (b). The evidence for these changes can found in sequences of rocks that span the K-T boundary, like the one at Stevns Klint, Denmark *(shown at left).* The K-T boundary occurs about half way up the cliff. Fossils (a) and (b) are from Tertiary layers of limy chalk above the boundary, whereas (c) and (d) are from layers below the boundary.

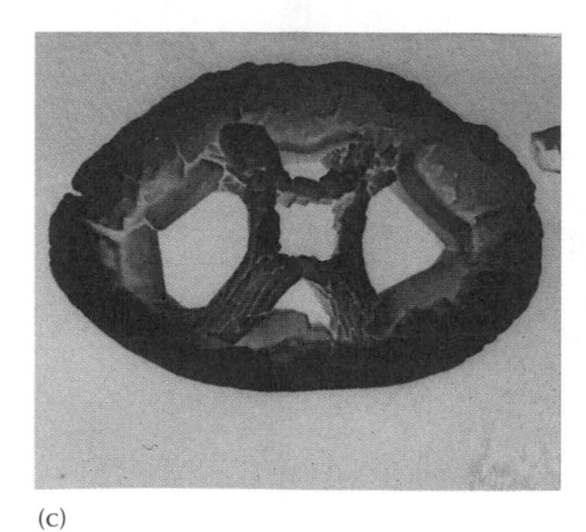

(c)

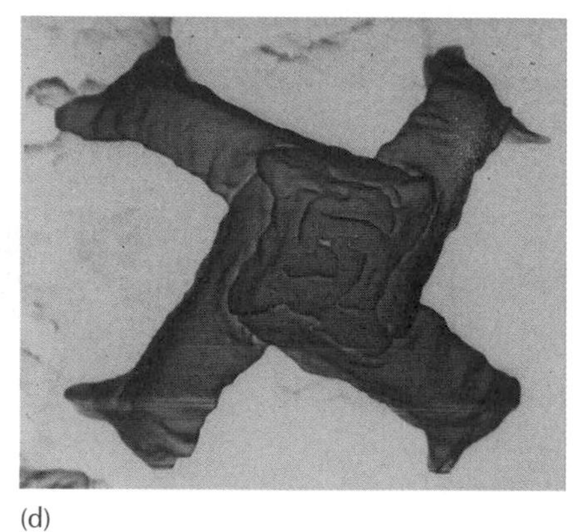

(d)

INOCERAMID

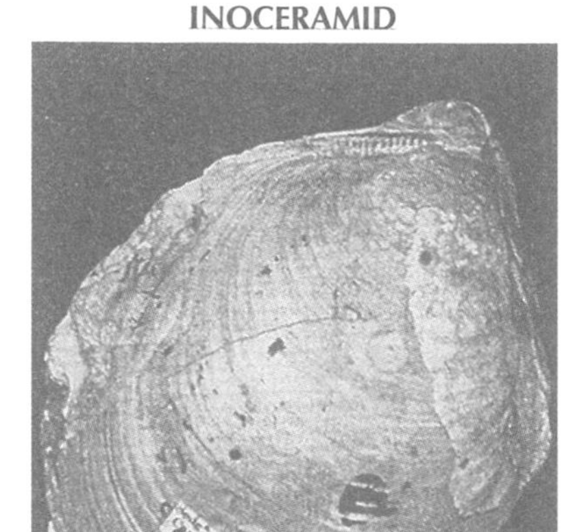

RUDIST

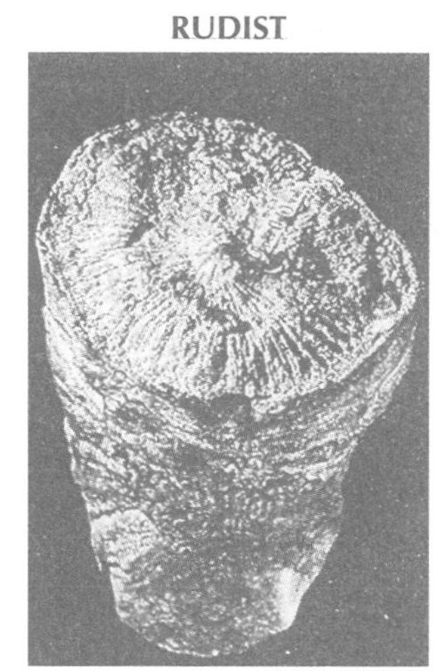

AMMONITES

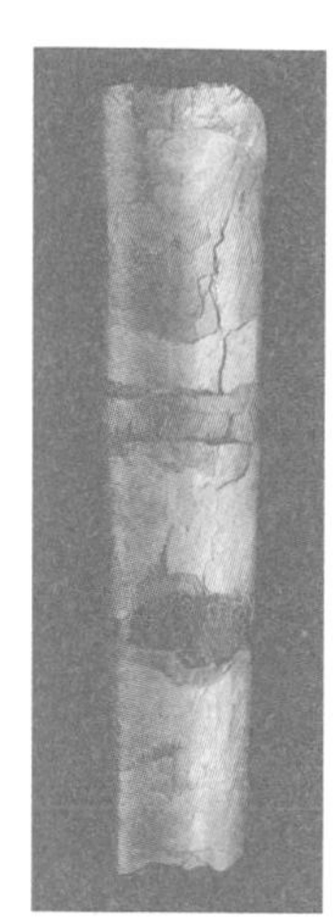

Several kinds of marine animals without backbones were also decimated near the end of the Cretaceous Period, including two kinds of clams—inoceramids and rudists. Also, a whole group of animals called ammonites became extinct. Ammonites were fairly close evolutionary relatives of the living nautilus.

triggered by the impact of a comet or asteroid. Both hypotheses suggest that the extinctions at the end of the Age of Dinosaurs resulted from severe climatic effects and changes. They differ, however, in their identification of the triggering mechanism for those changes.

Are these hypotheses testable? To some extent, elements of both hypotheses can be tested with the available geologic and paleontologic evidence. The real question is, "Are they testable to an extent that will allow us to discriminate between the effects of the two events?" In other words, "Can we distinguish the alleged environmental effects of a catastrophic, extraterrestrially derived mechanism from those of a more gradual, terrestrially triggered one?"

In evaluating the weight of the evidence in this case, there are three important criteria that these extinction hypotheses must pass when tested. First, there must be geologic evidence for the mechanism, whether it is a catastrophic extraterrestrial impact or a series of massive volcanic eruptions that may have been associated with a gradual withdrawal of seaways off the continents.

Second, the hypothesis must explain why the effects of that particular mechanism would have fatally affected the physiology of the organisms that went extinct but not the organisms that survived. This is a very complex problem to address, because we are not just dealing with the extinction of many kinds of dinosaurs. In the oceans, Late Cretaceous species of single-celled plankton were especially hard hit. Also, many multicellular animals such as the nautiluslike ammonites, clams called inoceramids and rudists, plesiosaurs, and ichthyosaurs disappeared near the end of the Cretaceous. On land, the end of the Age of Dinosaurs records the extinction of numerous species of marsupials and some species of rodentlike mammals called multituberculates. Pterosaurs and many other species of reptiles, fish, and lizards also perished. And in some areas, as many as 50 to 80 percent of the plant species apparently became extinct.

Third, each hypothesis must clearly state how long it took for the mechanism to cause the extinctions. Then, we must find evidence in the rock and fossil records to document that the extinctions actually happened as quickly as predicted by the hypothesis. In other words, we must tell time precisely enough to distinguish between the effects of more catastrophic and more gradual mechanisms.

These are tall orders. Yet, only when all three criteria are met can we be assured of a scientifically satisfying solution to the mystery of these extinctions.

PTEROSAUR

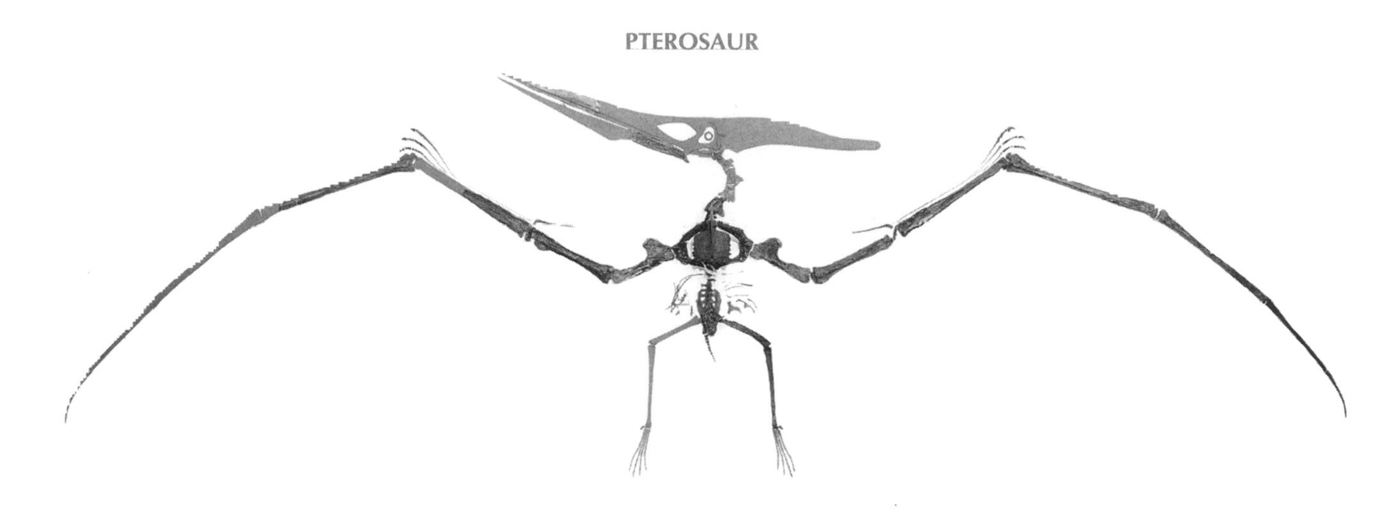

PLESIOSAUR

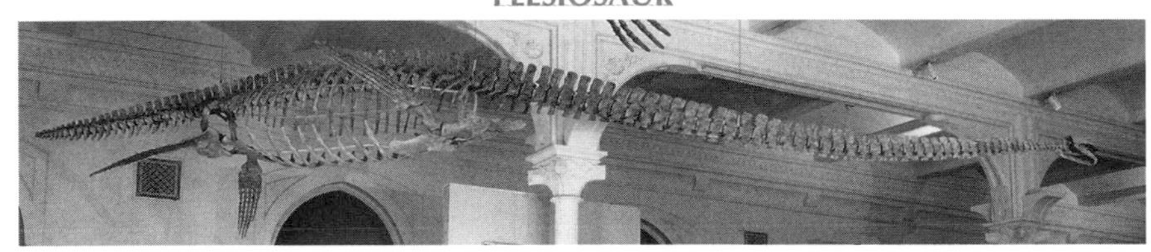

ICHTHYOSAUR

MOSASAUR

Several groups of vertebrates other than dinosaurs suffered extinction during the Late Cretaceous. They include the flying pterosaurs, the marine plesiosaurs, ichthyosaurs, and mosasaurs.

CRETACEOUS PLANTS

TERTIARY PLANTS

Although the extent of plant extinctions varied in different geographic areas at the end of the Cretaceous, numerous species suffered extinction in the middle of North America. Here, fossil leaves from three Cretaceous plants are contrasted with fossil leaves from three Tertiary plants.

To contrast these competing extinction scenarios involving volcanism, seaway retreat, and impact, we might travel back about 65 million years to see what might have happened if each of these events had actually transpired. The setting for each scene will be the long-lost coastal plain along the Western Interior Seaway that extended north from the Gulf of Mexico to connect with the Arctic Ocean. (A detailed listing of the references for the information included in these scenarios will be presented in Chapters 4 to 7.)

TERTIARY POLLEN GRAINS

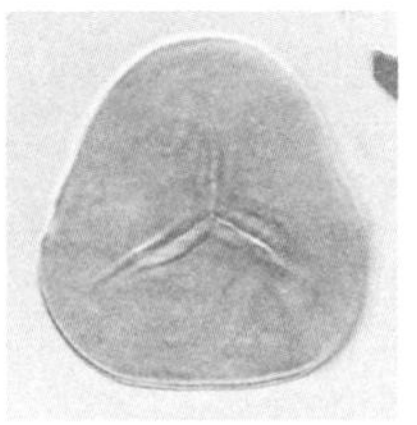

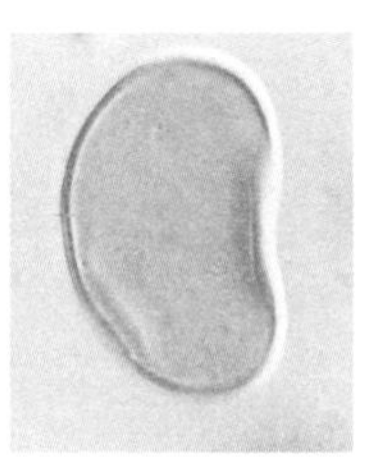

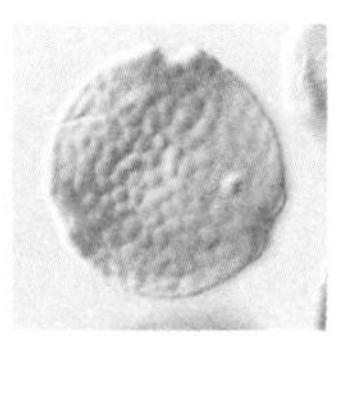

CRETACEOUS POLLEN GRAINS

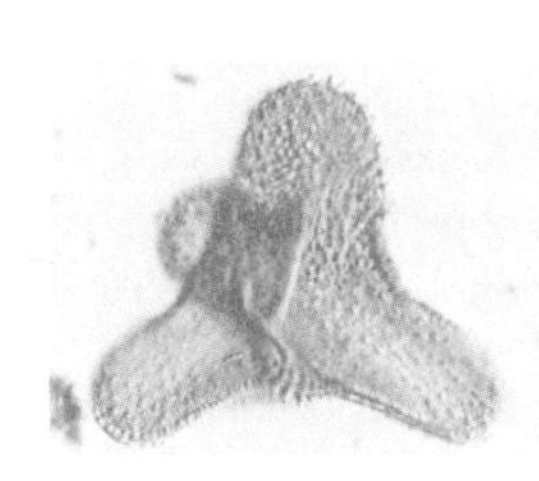

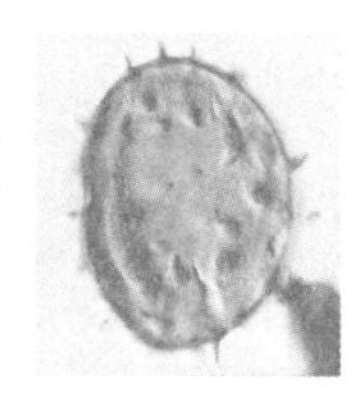

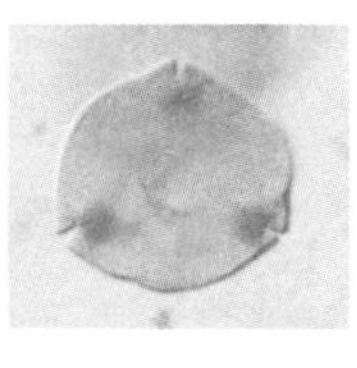

Fossil pollen from plants also can be used as evidence to document extinctions at the K-T boundary in central North America. Here, microscopic views of pollen grains preserved in Cretaceous rocks are contrasted with similar views of Tertiary pollen grains. All specimens are magnified 1,000 times.

(Above left) **During the Late Cretaceous, volcanic ash and toxic pollutants erupted out of volcanoes involved in the uplifting of the ancestral Rocky Mountains.**

(Above right) **Between 66 and 65 million years ago, the landscape of southwest India resembled a toxic calderon of lava erupting from huge fissures in the crust of the Earth. Today, these flows form a huge geologic province called the Deccan Traps.**

A SCENARIO FOR GRADUAL EXTINCTION RESULTING FROM VOLCANISM AND THE RETREAT OF SHALLOW CONTINENTAL SEAWAYS

The last summer day of the Cretaceous dawns like many others over the past half million years. The atmosphere is thick from ash and toxic gases spewed out of volcanoes in the rising ancestral Rocky Mountains to the west, as well as from widespread volcanic eruptions in India, the south Atlantic Ocean, and other areas around the globe. In India alone, these massive eruptions produced about 480,000 cubic miles (2 million km^3)of basaltic lava, making this volcanic event one of the largest ever recorded in the 4.5-billion-year history of the Earth. These flows of lava, piled one on top of the other, are over 1.5 miles (2.4 km) thick in some places. In total, they cover an area larger than

Horned dinosaurs, such as *Triceratops*, were common elements of the fauna along the shore of the Western Interior Seaway. Hundreds of their skeletal remains are preserved in the Hell Creek and Lance Formations of Montana, Wyoming, and the Dakotas, as well as in the Scollard and Frenchman Formations of Alberta and Saskatchewan.

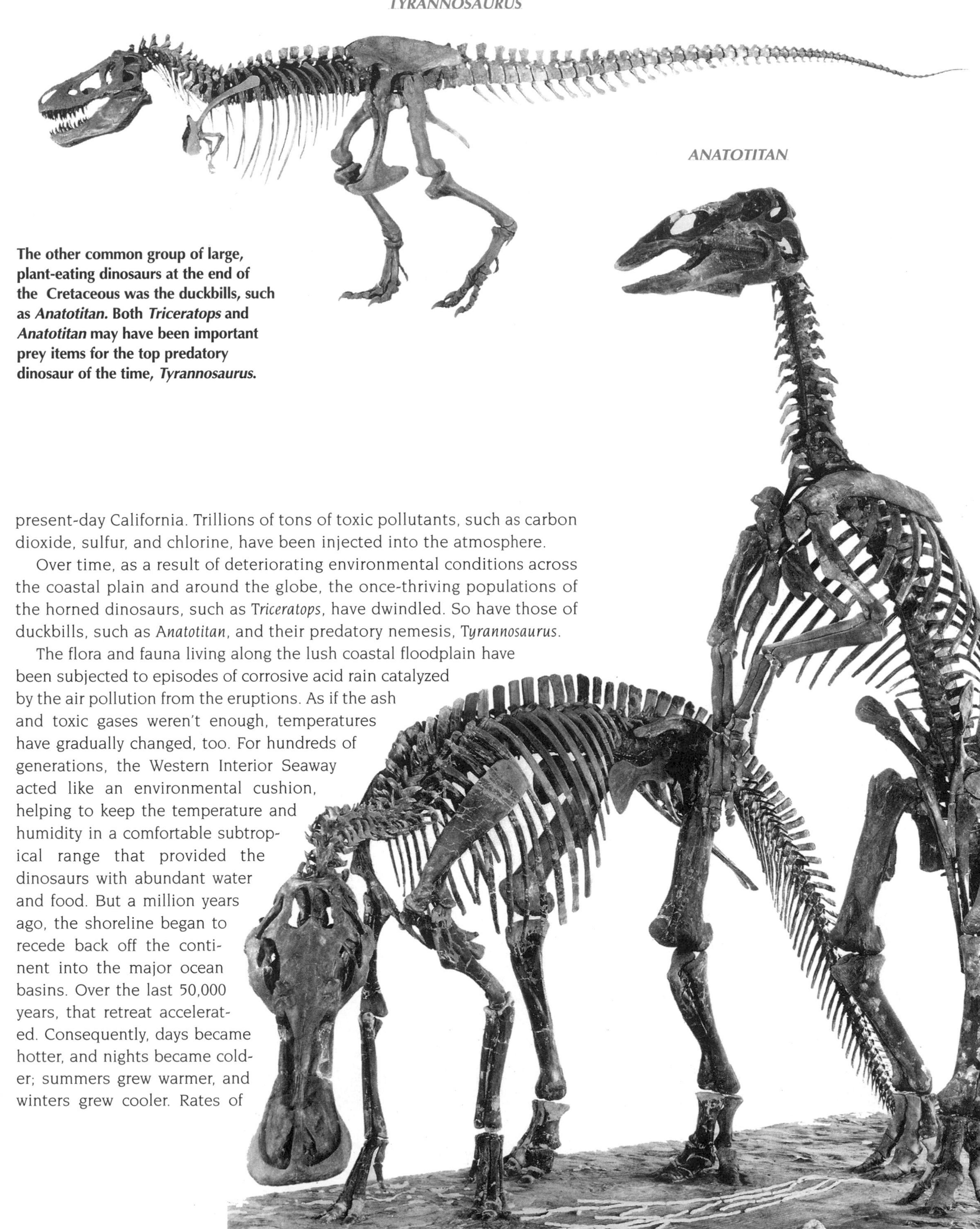

The other common group of large, plant-eating dinosaurs at the end of the Cretaceous was the duckbills, such as *Anatotitan.* Both *Triceratops* and *Anatotitan* may have been important prey items for the top predatory dinosaur of the time, *Tyrannosaurus.*

present-day California. Trillions of tons of toxic pollutants, such as carbon dioxide, sulfur, and chlorine, have been injected into the atmosphere.

Over time, as a result of deteriorating environmental conditions across the coastal plain and around the globe, the once-thriving populations of the horned dinosaurs, such as *Triceratops*, have dwindled. So have those of duckbills, such as *Anatotitan*, and their predatory nemesis, *Tyrannosaurus*.

The flora and fauna living along the lush coastal floodplain have been subjected to episodes of corrosive acid rain catalyzed by the air pollution from the eruptions. As if the ash and toxic gases weren't enough, temperatures have gradually changed, too. For hundreds of generations, the Western Interior Seaway acted like an environmental cushion, helping to keep the temperature and humidity in a comfortable subtropical range that provided the dinosaurs with abundant water and food. But a million years ago, the shoreline began to recede back off the continent into the major ocean basins. Over the last 50,000 years, that retreat accelerated. Consequently, days became hotter, and nights became colder; summers grew warmer, and winters grew cooler. Rates of precipitation also changed. During the warmer times of the year, the temperature rose to uncomfortable levels as the result of global warming generated by the volcanic particulates and aerosols.

These environmental changes decimated the herds of duckbills, and in turn extinguished the fearsome carnivore that preyed on the duckbills, *Tyrannosaurus*. But on this day 65 million years ago, the last *Triceratops*, weakened by the extreme heat of the noonday sun and the choking dust generated by an eruption the night before, rolls over on its side and breathes its last, finally terminating a lineage that had ruled the continents for over 150 million years.

This hypothetical scene from 65 million years ago shows a herd of duckbills watching a comet or asteriod streak through the atmosphere shortly before it impacts near present-day Yucatan.

A pterosaur flies peacefully over the tropical landscape at the moment of impact, as the asteroid or comet smashes into the shoreline of the Yucatan Penninsula at the end of the Cretaceous. A gigantic crater, buried under the surface in that region today, represents evidence for this event.

A SCENARIO FOR SUDDEN CATASTROPHIC EXTINCTION RESULTING FROM IMPACT

Now, try to envision that very same coastal floodplain as if the catastrophic impact hypothesis were true. The summer night is calm and pleasant, punctuated only by the occasional grunts and snorts from a sleeping herd of horned dinosaurs. Through the stand of conifers adjacent to the herd, a bull tyrannosaur slowly and intently stalks toward a plump and peacefully sleeping juvenile near the herd's edge.

Just as the bull edges within striking distance, the southern sky is ignited by a meteor many times brighter than the Sun. It streaks across the Caribbean sky at somewhere between 50,000 and 150,000 miles per hour (80,000 and 250,000 km/h), ripping completely through the Earth's atmosphere in a matter of a few seconds. The whole herd awakens instantly and, in startled panic, stampedes in the opposite direction, ignoring the tyrannosaur fleeing in their midst.

To the south, the Earth is rocked by the shockwave from the tremendous impact near present-day Yucatan. The 6-mile-wide (10 km) meteorite hits near the shoreline of the Gulf of Mexico generating explosion temperatures of several thousand degrees and pressures equal to 1 million times those normally created by the weight of the Earth's atmosphere. It initially excavates a crater between 50 and 60 miles (~80 and 100 km) across and between 13 and 25 miles (~21 and 40 km) deep. An earthquake of magnitude 13 is generated—1 million times greater than the strongest earthquake ever recorded in human history. The explosion is equivalent to exploding 10,000 times the number of nuclear weapons contained in all the world's arsenals at the peak of the Cold War. It blasts 5,000 cubic miles (~21,000 km^3) of material out of the crater, 400 cubic miles (~1,700 km^3) of which is launched into orbit around the Earth at a velocity equal to 50 times the speed of sound. This is 1,200 times the amount of volcanic ash that was erupted out of Mount St. Helens. Part of the Earth's atmosphere is literally blasted away by the ejected debris.

The destructive force of the impact along with the 250- to 300-foot-high (75- to 90-m-high) tsunamis that it triggers instantly decimate organisms living nearby. Within a few hours, the tsunamis (commonly called tidal waves) strike other shorelines around the Gulf of Mexico, eradicating near-shore communities living hundreds or even thousands of miles from the impact point.

As orbiting particles reenter the remaining atmosphere, they heat it to levels comparable to "broiling" temperatures in your oven. Wild fires are ignited across several regions of the globe. Even at a distance of several thousand miles, the environment along the Western Interior Seaway is severely stressed.

Then, over the next several months after this initial incineration, the atmosphere becomes so choked with debris that no sunlight can penetrate to the ground. Without the warmth and rays of the Sun, temperatures plummet below freezing for one to six months. The surviving plants on which *Triceratops* fed, soon die. With the death of its food source, the *Triceratops* herd perishes, and with the death of the herd, so does the last *Tyrannosaurus* in these gloomy last moments of the Cretaceous. Though dinosaurs had dominated the Earth for 165 million years, they disappeared in a geological instant of global cataclysm.

These are both riveting, sensational scenarios. But in our role as scientific detectives, can we find evidence in the geological and paleontological records to confirm that they really happened? That quest will be the focus of the next three chapters.

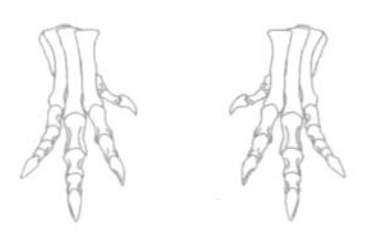

CHAPTER FOUR

Enormous Eruptions and Disappearing Seaways

An extinction scenario based on a massive episode of volcanic activity at the end of the Cretaceous Period has been developed from many well-documented pieces of geological evidence. Recent revisions to this scenario have attempted to link this volcanism with the dramatic environmental changes that may have accompanied the retreating continental seaways. Over the last half of the twentieth century, a number of Earth processes (such as volcanism, earthquakes, and the configuration of the oceans and continents) have been tied to the mechanisms of plate tectonics. Plate tectonics emerged as the predominent model for explaining changes on the Earth and has revolutionized the way that both geologists and paleontologists view Earth and evolutionary history. It has even assumed an important role in explaining the distribution of plants and animals around the planet. So, before embarking on a detailed examination of the evidence for volcanism at the end of the Cretaceous Period, let us briefly introduce a simplified summary of how plate tetonics works.

EARTH STRUCTURE AND PLATE TECTONICS

As the Earth cooled after forming 4.5 billion years ago, it divided into several layers, including the core at the center, the mantle in the middle, and the crust at the surface. The rocky crust is the lightest layer, and it essentially floats on the lower denser layers. Basically, the crust is composed of minerals containing relatively light elements, such as silicon, aluminum, and potassium. It is actually extremely thin, averaging about 6 miles (10 km) under the oceans and about 25 miles (about 40 km) under the continents. The mantle extends from the base of the crust to the core, in other words, down to about 1,800 miles (2,900 km). The mantle is more dense than the crust because the minerals that make it up contain more heavy atoms. These include iron, magnesium, and even a little iridium (an element that will become very important to our discussion about possible K-T impacts). The core is even more dense than the mantle and is thought to be composed primarily of iron.

The mantle appears to circulate material in two layers of large convection cells—each cell moving material in a manner similar to that seen in a boiling pot of syrup.1 This convection, presumably fueled by the release of radioactive energy within the core, is thought to drive the motions of continents as they drift across the surface of the Earth as part of plate tectonic activity. Most of the movement within the mantle that drives the continents is thought to operate within the upper 400 miles (650 km).

In essence, one might think of plate tectonics in terms of a gigantic recycling process for the Earth's crust. As material in the mantle heats at the base of the upper layer of convection cells, it rises toward the surface of the Earth. There, it erupts through fractures or rifts, spreading the plates of the crust apart. New crust is created along such spreading centers, and older crust is pushed off to the side. One promi-

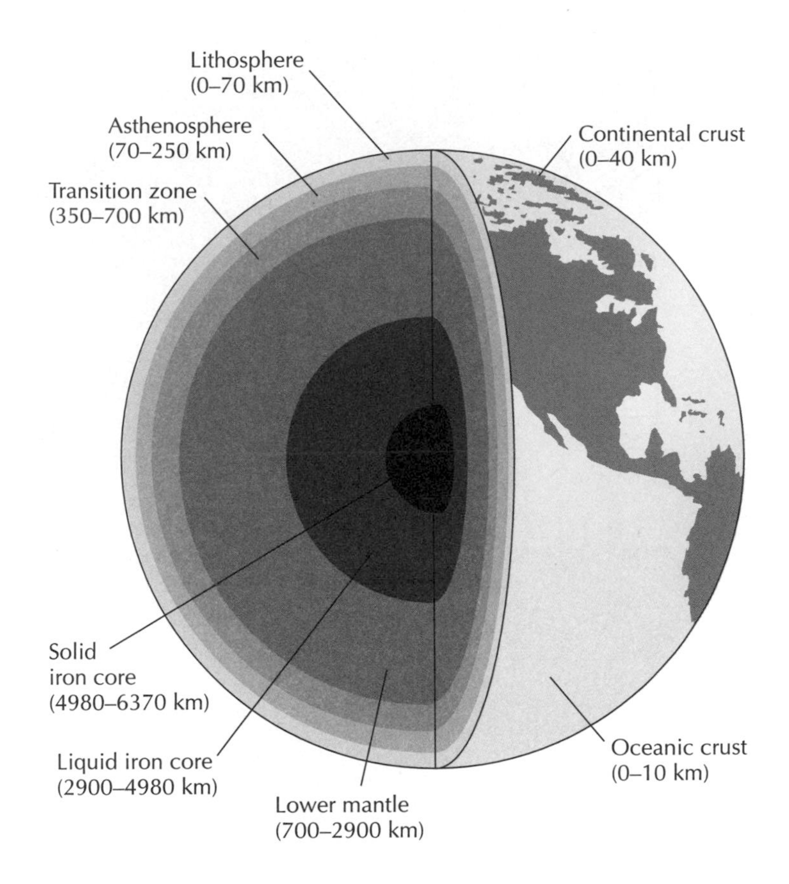

The Earth's interior is divided into several different layers. Both the continents and the ocean floors are plates, parts of the crust formed by the lithosphere. The asthenosphere, a ductile layer on which the plates float, lies below the lithosphere. The asthenosphere forms the lower part of the crust and upper part of the mantle. The lower part of the mantle lies above the Earth's core. The depth of each layer is shown.

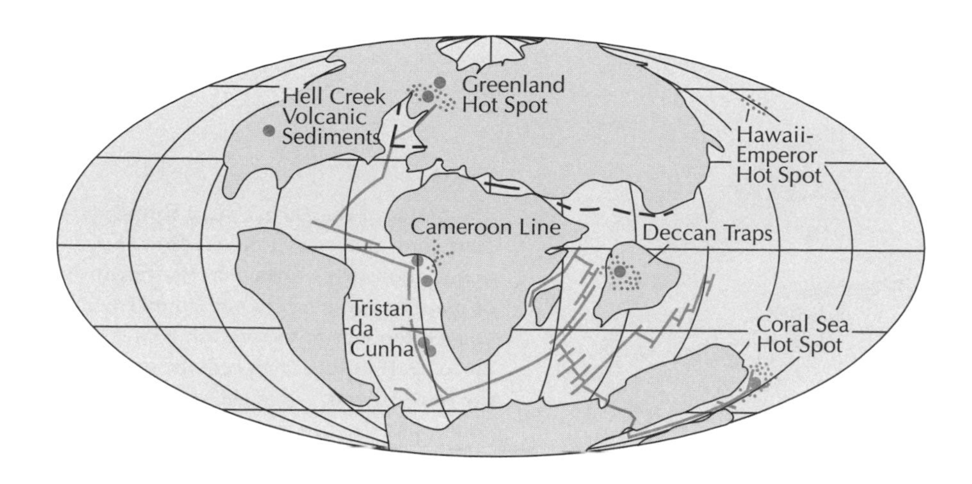

This surface map of the Earth's continents and ocean basins shows the areas of major volcanic activity at the time of the K-T transition.

nent spreading center is represented by the Mid-Atlantic Ridge, which runs from above Iceland all the way down below the southern tips of Africa and South America. When two plates collide, one dives beneath the other, and cooler mantle material descends back down toward the core. These areas are called subduction zones. One subduction zone borders the west coast of South America, where that continent overrides the floor of the Pacific Ocean. In essence, new crust is generally created at spreading centers, whereas old crust is melted and recycled at subduction zones. In addition to areas where crust is created and destroyed, plates can simply slide past one another at the Earth's surface. The San Andreas Rift, which cuts across the western edge of California, is an example of such a boundary between plates. Earthquakes are generated when the rocky plates fracture and rupture as they move past one another or where molten bodies of rock move toward the surface. Now, how does the volcanism at the K-T boundary reflect the actions of plate tectonics?

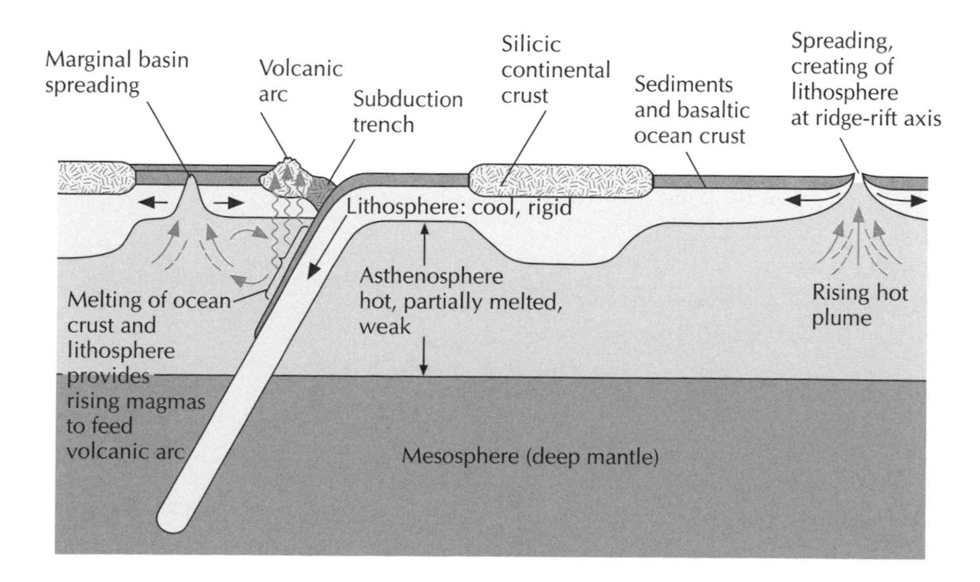

This cross section through the upper layers of the Earth illustrates how plates move apart at spreading centers and how one plate dives under another at subduction zones. Volcanic activity and earthquakes are most common at spreading centers and subduction zones.

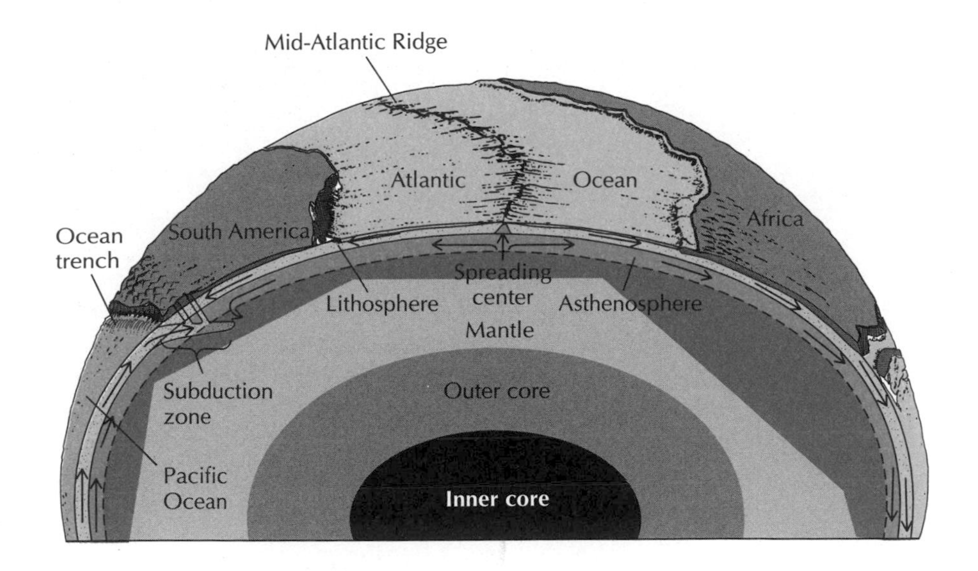

The plates of the Earth's crust form the continents and ocean floors. They move across the Earth's surface as the result of plate tectonic forces generated by convection within the mantle. Plates move apart at spreading centers, such as the Mid-Atlantic Ridge. At subduction zones, like the oceanic trench off the west coast of South America, one plate dives down under the adjacent one and melts. In this way, the Earth's crust is continually being recycled.

THE DECCAN TRAPS AND OTHER CRETACEOUS-TERTIARY VOLCANIC EVENTS

As early as 1972, a geologist named Peter Vogt of the Naval Research Laboratory in Washington, D.C., recognized that the extinctions at the end of the Age of Dinosaurs coincided with a tremendous increase in volcanic activity throughout the world.[2] Huge volumes of basaltic lava erupted and flooded over large areas on the floor of the Indian Ocean and India itself, as it drifted northward because of plate tectonic activity toward a collision with Asia. Basalt is the dark volcanic rock such as that generated by Hawaiian volcanoes in the Pacific. Vogt is now said to favor the impact hypothesis, but others have picked up on his original idea. Research relating these massive eruptions to dinosaur extinction has been spearheaded by Charles Officer and Charles Drake of Dartmouth College in the United States, as well as by Vincent Courtillot at the Institute of Physics of the Earth in France.[3–6]

Although about 15 percent of the Indian eruptions may have been catastrophically violent, most are thought not to have been terribly energetic. Nonetheless, the resulting lava flows document one of the largest single volcanic events in the history of the Earth. The scale of the eruptions is dumbfounding even to geologists. Over a 500,000 year interval between 66 million years ago and 65 million years ago, almost 480,000 cubic miles (2,000,000 km^3) of molten lava flowed out of cracks and fissures in the Earth's crust near the present-day Indian coast and flooded hundreds of miles inland.[6–9] As a result of the plate tectonic movements that have occurred since the end of the Age of Dinosaurs, some of these lava flows are now exposed along the western coast of India, while others form the seabed of the Indian Ocean around the present-day Seychelles Islands.

These flows make up a geologic feature called the Deccan Traps. The name carries a descriptive significance: In Sanskrit, *Deccan* means "southern," which is appropriate because the flows are now situated along the southwest coast of India. In Dutch, *trap* means "staircase," which refers to the steplike pattern that typifies the erosional surface of the flows piled one on top of the other. At some localities, the sequence of flows is over about 8,000 feet (2,400 m) thick.[6] This is more than 25 percent of the height of Mount Everest, which began to rise later as India slammed into Asia between 40 and 60 million years ago. The lava may have originally covered an area larger than California.[11] On average, it is estimated that the flows erupted at a rate of between 1/2 and 2 cubic miles (about 2 to 8 km^3) annually.[12] Individual lava flows in the Deccan Traps average between 33 and 164 feet thick (about 10 to 50 m), although the most extensive ones attained a thickness of almost 500 feet (150 m).

The question of how long individual eruptions lasted is controversial.[13] Volcanologists including Donald Swanson of the U.S. Geological Survey originally interpreted the geologic evidence to suggest that individual flows occurred over a matter of a few days. More recently, however, Stephen Self of the University of Hawaii and his colleagues have argued that the evidence suggests that the individual eruptions may have lasted a year or longer.

Much scientific interest and debate has surrounded the question of how such monumental volcanic events were generated. Recent research has provided strong evidence that the Deccan Traps represent the end product of a slowly ascending, gigantic plume of molten rock or magma from deep within the Earth. This plume may have originated as deep as the boundary between the core and the mantle.[14–17]

This rugged landscape shows some typical exposures of volcanic rock that outcrop as part of the Deccan Traps in India. The steplike nature of the exposures is created by erosion of the individual lava flows, which are piled one on top of the other.

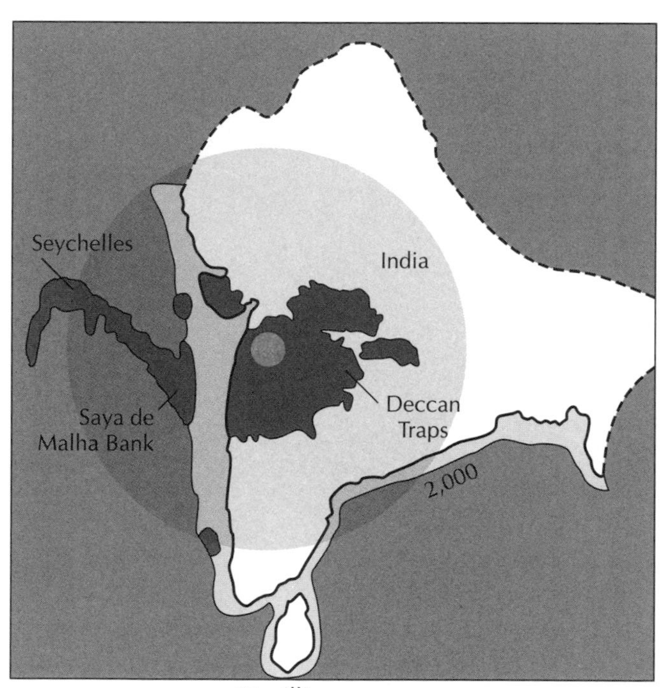

65 million years ago

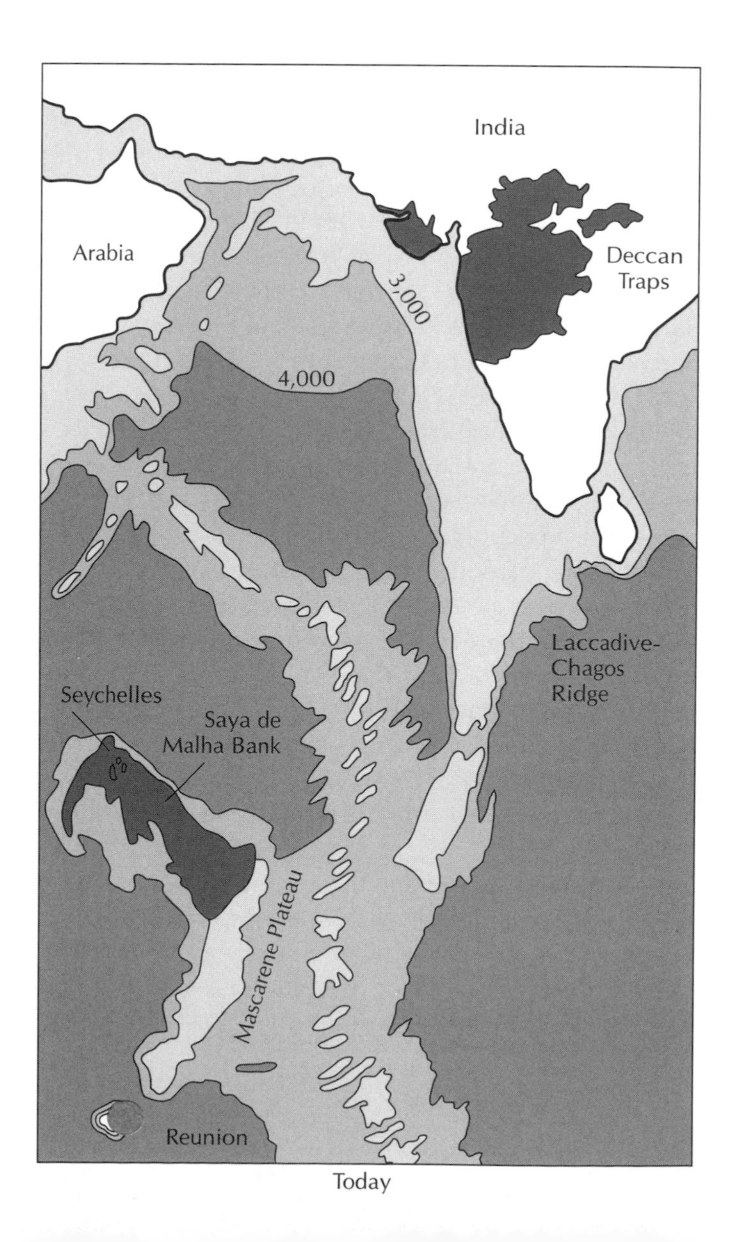

Today

Two maps contrasting the lava flows of the Deccan Traps as they existed 65 million years ago and as they exist today. The splitting of the flows occured as India drifted toward Asia as the result of the action of plate tectonics.

As early as the 1960s and 1970s geologists and geophysicists, such as W. Jason Morgan of Princeton University, proposed that flood basalts represented plumes of magma that arose from within the Earth's mantle.[18] The reason that this was suspected was because the chemical composition of the flood basalt that makes up the flows is similar to the composition of the material that geologists believe makes up the mantle. However, the evidence for the deep origins of this magma was not strong.

The scenario involving the origin of flood basalts goes like this. A tremendous amount of heat is released from the mantle through volcanism at spreading centers. In fact, this is generally how heat is released from inside the Earth. About one tenth of the heat escaping from the mantle, however, arises from deep-seated, thin "roots" that extend all the way down to the boundary between the mantle and the core.[19] Because these thin plumes are warmer than the surrounding mantle material, they rise until they reach the base of the crust. As they rise, a spherical head forms at the top of the plume as it forces its way up through the cooler surrounding material of the mantle. As the plume reaches the base of the crust, it spreads out, creating a mushroom-shaped body of material. The force of the rising plume with its mushroom-shaped head lifts a large area of the overlying crust, making it thinner and fracturing it as it is lifted.[20–22] Because the pressures at these shallower depths within the Earth are much less than those near the core, the plume decompresses and the rock partially melts. As it nears the surface of the crust, the volatile gases dissolved in the magma are released, somewhat like the effect generated by opening a bottle of soda or champagne. The upwelling lava then spews out from the fractures in the uplifted crust. The place where the lava erupts on the surface is called a hot spot. Once erupted, the lava floods downhill over the Earth's surface to form new crust.

Some geologists, including Vincent Courtillot, have argued that the record of magnetic reversals at the end of the Cretaceous also conforms to the scenario outlined above.[23] That argument proceeds as follows: the molten iron and nickel in the Earth's outer core is thought to circulate slowly by means of convection at a rate of about 6 miles (10 km) per year. In turn, these movements are thought by many earth scientists to produce the Earth's magnetic field. Unstable conditions that occasionally arise at the boundary between the mantle and core are believed to cause the reversals of the Earth's magnetic field, a phenomenon mentioned in Chapter 2. These unstable boundary conditions apparently result from the release of heat from the core into the mantle. As heat generated by radioactive decay radiates out of the core, the lower layer of the mantle heats up and becomes less dense. From time to time, this lowest layer of the mantle, called the "D-layer," becomes so thick and light that instability results and large volumes of material are released to rise as plumes of magma. As the D-layer grows thicker, reversals in the Earth's magnetic field are few and far between, such as the long interval of normal polarity that existed between about 120 million years ago and 80 million years ago. At the end of that period, which is called "The Cretaceous Long Normal," the D-layer apparently became unstable and broke up releasing enormous plumes of magma rising within the mantle. As the exchange of heat from the core to the mantle increased, reversals in the Earth's magnetic field resumed on a more frequent basis. Presumably it took millions of years for the largest of these plumes to reach the surface of the Earth, result-

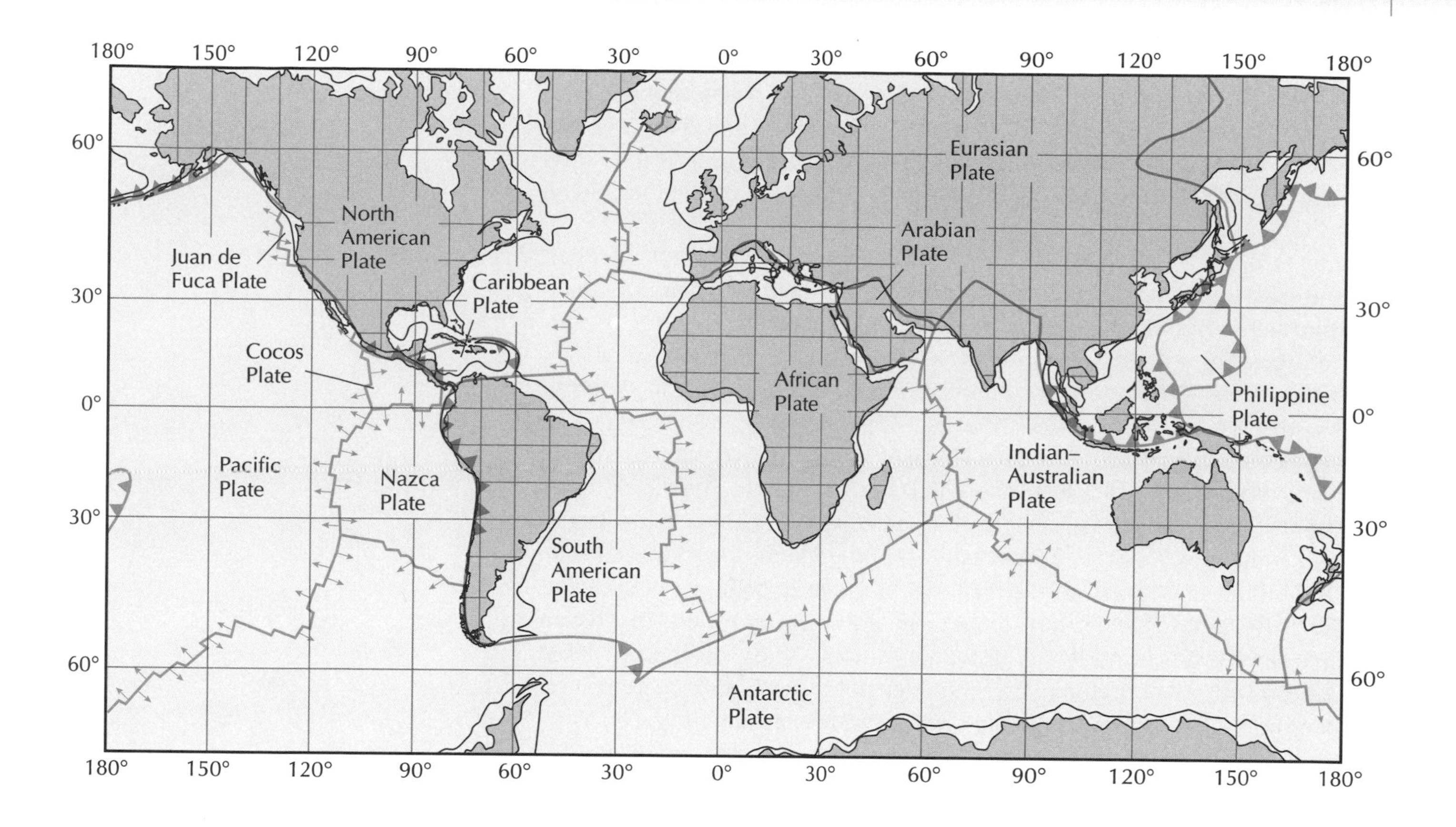

This surface map shows the present-day locations of the major oceanic and continental plates.

ing in the hot spot eruptions that formed the Deccan Traps and the undersea plateau around the Seychelles Islands.

Most of the lava flows of the Deccan Traps erupted over a period of about a half a million years between 66 and 65 million years ago in a series of massive pulses. According to Asish Basu from the Univeristy of Rochester and his colleagues, the earliest eruptions began about 68.5 million years ago, and the latest volcanic activity ended about 64.9 million years ago.[24,25] However, activity near the original point of eruptions continues even today in the form of a volcano called Piton de la Fournaise on Reunion Island in the Indian Ocean. In essence, the plume that reached the surface of the Earth's crust about 66 million years ago has created a hot spot for volcanic activity ever since.

The Deccan Traps, although the largest-known volcanic event at the end of the Age of Dinosaurs, was by no means the only one.[26] More volcanism is recorded in deposits at the bottom of the South Atlantic Ocean near Antarctica. The volcanic rocks of Cretaceous-Tertiary (K-T) age form part of the Walvis Ridge that ends at the volcanic island of Tristan da Cunha. In all, the ridge comprises a largely submerged mountain range that resulted from the eruption of another plume of magma that began before the end of the Cretaceous but remained active during the K-T transition. F. L. Sutherland of the Australian Museum identified as many as eight other centers of hot-spot volcanism in the southern Atlantic, as well as in southern Africa and eastern South America, which may have been active during the K-T transition. Precise radioisotopic dating has yet to pin down the ages of these eruptions very precisely, however.

More evidence for K-T volcanism is found in another part of Africa.[26] Crossing the west coast of central Africa is a linear deposit of basalt that began to erupt about 65 million years ago. It is called the Cameroon line, and it appears to be linked to nearby faulting that served to separate Africa from South America. These eruptions are thought to have been generated by another plume from the mantle.

Further north, the hot spot now associated with the volcanic activity in Iceland may have originated under Greenland and helped initiate the plate tectonic separation of Europe and North America from Greenland.[27] The latest recalculations of the age of this activity suggests that the volcanism began near the K-T transition, although much of the volcanism occurred later in the Tertiary.

In the central Pacific lies a familiar string of islands long recognized to have been created by hot-spot volcanism—the Hawaiian Islands. The oldest major eruptions that formed the Hawaiian-Emperor chain appear to have begun about 65 million years ago.[28] However, most of the islands are younger in age.

Off the northeast coast of Australia, an intensive episode of hot-spot volcanism is thought to have begun about 65 million years ago at rifts that began to open up the Coral and North Tasman seafloors. Other volcanic provinces associated with plate tectonic activity in Australia during the K-T transition include the Bass Basin and the main segment of the Tasman rift.[29]

Not all geologists agree that these volcanic plumes come from deep within the Earth, as proponents of volcanic extinction scenarios suggest. Don Anderson, a geophysicist at the California Institute of Technology, argues that flood basalts arise at depths of no more than 250 miles.[11] But what evidence is there to suggest that they originate from near the core-mantle boundary? Much of the work is based on analyzing different forms, or isotopes, of the atom helium.[30–32] This is a fascinating scientific detective story whose roots extend all the way back to the origin of the universe at the Big Bang about 15 billion years ago.

In some ways this apparent tie between the most momentous volcanic events on Earth and the most momentous event in the universe seems quite fitting. Cosmological models suggest that a particular kind of helium atom, termed helium-3, was created within the first two minutes after the Big Bang initiated our universe. Along with hydrogen, which was created at the same time, helium-3 comprised one of the two basic components from which galaxies and stars were eventually formed, including our solar system with its Sun and planets. Helium-3 was also incorporated into the Earth as it grew, or accreted, about 4.5 billion years ago by gravitationally attracting smaller objects, such as comets and asteroids. However, helium is not common in the rocks of the Earth's crust because it is so light, as evidenced by helium balloons which tend to float annoyingly away as any child can tell you. Similarly, helium escapes from the pores in rocks as they are eroded and broken during the process of weathering and remelting as igneous rocks are reconstituted into sedimentary and metamorphic rocks.

In the late 1960s and early 1970s, small traces of helium-3 were found in emissions from spreading centers under the ocean.[33] It was concluded that these traces were derived from material that had risen from within the mantle where it had been trapped since the early stages of the Earth's formation. In addition, relatively large amounts of helium-3 were discovered to be coming out of hot-spot volcanoes such as the one now located at Reunion Island, which represents the remnant of the source for the Deccan Traps. In the late 1980s, this led geochemist Asish Basu and his colleagues to carefully analyze the atomic composition of some of the earliest-known rocks erupted in the Deccan Traps. Amazingly, in some relatively unweathered samples of these rocks, relatively large traces of helium-3 were found to have been preserved in microscopic bubbles that formed inside the rock as it cooled and solidified before it could reach the surface and be released into the atmosphere. To Basu and many other geologists this represents strong evidence that the plume responsible for the Deccan eruptions originated at the base of the mantle near its boundary with the core.

Robert White and Dan McKenzie at the University of Cambridge have suggested that the rising plumes of magma that fed these huge volcanic eruptions served to lift the continents up and assist the process of continental breakup, or rifting.[34] According to the late Anthony Hallam of the University of Birmingham, such uplift may have had a role in causing the long-recognized retreat of shallow continental seaways back into the major ocean basins at the end of the Cretaceous.[35] To support the proposal that these plumes could have uplifted the continents and eventually caused the retreat of the shallow seaways, proponents like Hallam have noted that topographic high spots on the crust are associated with areas where plumes are rising to the surface today, such as around the Hawaiian Islands. This idea has yet to be thoroughly tested. It should be noted, however, that in some areas where seaways retreated, such as North Africa and Europe, no plumes appear to have been actively generating eruptions during the K-T transition. In addition, some of the K-T volcanic activity, including that at the Walvis Ridge and the part of the Deccan eruptions now located near the Seychelles Islands, occurred on the seafloor. Such seafloor eruptions would seem to have raised rather than lowered sea levels.

Finally, numerous layers of siltstone and mudstone, rich in altered volcanic ash derived from eruptions associated with the uplift of the ancestral Rocky Mountains, are preserved in ancient floodplain sediments now exposed in Wyoming, Montana, North Dakota, South Dakota, and Alberta. These were deposited by rivers along the shoreline of the Late Cretaceous seaway that bisected North America. Most beds are not pure volcanic ash. They have been washed around with other sediments on the floodplain adjacent to the seaway. Nonetheless, a good deal of this material was probably erupted between 70 and 65 million years ago.

Personally, this is one of our favorite field areas in the world. We have been making annual pilgrimages to study the rocks and fossils exposed in these rugged badlands ever since we became students at Berkeley. The ash-rich rocks are noted for the "popcornlike" texture found on the surface of these exposures. After the volcanic ash weathers to clay, it can expand by absorbing large quantities of rain water, the way a dry compressed sponge does when it is placed in water. This expansion creates the popcornlike surface texture. If you try to walk or drive across these exposures when they're wet, forget it. The rain transforms them right back into a Cretaceous swamp, and you'll either get stuck or fall flat on your behind before you can blink.

Exposures of the Hell Creek Formation in east-central Montana contain over 200 feet of bentonitic siltstones, mudstones, and sandstones. Many of these layers were derived from volcanic ash erupted out of the ancestral Rocky Mountains, which were rising about 100 miles to the west.

These beds form the rock units called the Hell Creek and Lance Formations, and they are world famous for the fossils that they contain. From these exposures, paleontologists and public alike got their first glimpses of the three-horned dinosaur, *Triceratops*, and the "king" of the dinosaurs, *Tyrannosaurus*. As we shall see in Chapter 7, fossils from these rock units provide one of the most complete pictures of biologic change on the continents during the K-T transition known from anywhere in the world.

Working with our former classmate, Carl Swisher of the Berkeley Geochronology Center, we collected samples from several layers of altered volcanic ash near the K-T boundary in this region. Previously, age estimates for the boundary had ranged between 64 and 66 million years ago. Based on our radioisotopic dating of minerals in these layers, the last dinosaurs and Cretaceous plants lived in this area slightly before 65.01 million + 30,000 years ago.[36] Our date coincides with the final phases of Deccan volcanism, which ended at 64.96 million + 110,000 years ago.[37] This suggests, but certainly does not prove, that the environmental effects of the Deccan eruptions could have been linked to the K-T extinctions. What would the environmental effects of the eruptions have been?

ENVIRONMENTAL EFFECTS ASSOCIATED WITH THE ERUPTIONS

As mentioned earlier, shallow continental seaways helped maintain the equable, subtropical environments in which the dinosaurs thrived. However, over a period of about 100,000 years at the end of the Cretaceous, as the seas rapidly retreated, climates became more extreme.

Vincent Courtillot, Charles Officer, Charles Drake, and Anthony Hallam argue that all this K-T volcanism could have spewed tremendous quantities of dust and sulfate aerosols into the atmosphere, generating both long- and short-term effects.[38,39] Shortly after major eruptions, the emissions of dust, sulfur dioxide, and other gases could have caused acid rain, acidic layers of water near the surface of the oceans, a cooling of air temperatures, depletion of the Earth's ozone layer, and possibly darkness.

The estimated amounts of pollutants that would have been injected into the atmosphere by the sequence of eruptions that formed the Deccan Traps are truly staggering: up to 33 trillion tons (30 trillion metric tons) of carbon dioxide, 6.6 trillion tons (6 trillion metric tons) of sulfur, and 66 billion tons (60 billion metric tons) of chlorine and fluorine molecules.[40] An initial cooling of the global climate of between 5 to 9 degrees F (3 to 5°C) would be triggered by the volcanic ash and sulfur emitted by a single Deccan eruption generating 240 cubic miles of lava (1,000 km^3). Sunlight to the Earth's surface would be cut substantially, inhibiting photosynthesis. Emissions of sulfur dioxide would turn the surface layers of the ocean into acidic baths, killing the microscopic organisms that normally process carbon dioxide in the atmosphere. Over the long term, the carbon dioxide rose to levels eight times that found in today's atmosphere, increasing the average global temperature as much as 9 degrees F (5°C) as the result of the greenhouse effect. As with earlier hypotheses involving the retreat of seaways, proponents of the volcanic scenario argue that the dinosaurs could not tolerate these toxic environmental conditions or climatic extremes and perished in the face of prolonged environmental stress.

However, critics respond: Why didn't the freezing winters and torrid summers extinguish turtles, snakes, lizards, and crocodiles—animals that are at the mercy of the climate to maintain a livable body temperature? They were all around at that time. It's hard to see why the physiology of those reptiles wasn't affected, whereas that of the dinosaurs was decimated. In addition, this was not the first time that large volcanic events had occurred during the Age of Dinosaurs. Why did dinosaurs survive earlier ones, but not this one?

COINCIDENCE BETWEEN OTHER LARGE VOLCANIC EVENTS AND MASS EXTINCTIONS

The K-T extinction episode is not the only mass extinction in evolutionary history that appears to be associated with an enormous volcanic event that resulted from plumes within the mantle.[41–43] The largest mass extinction in history occurred about 250 million years ago at the end of the Paleozoic Era and the beginning of the Mesozoic Era. This episode is believed to have extinguished as many as 95 percent of all the Earth's species living at that time.[44–47] Ironically, it was the extinction of early relatives of mammals, the dominant land animals of their time, that most paleontologists believe was responsible for creating the empty ecological niches which were filled by the dinosaurs after their origin about 230 to 240 million years ago.

The most enormous outpouring of continental flood basalts in the history of the Earth also occurred at that time. The Siberian Traps in Russia document that event. The amount of lava erupted to form the Siberian Traps was truly mind-boggling—up to 720,000 cubic miles (3 million km^3). If the lava had spread itself evenly across the entire surface of the Earth, it would have left a layer of basalt up to 10 feet (3 m) thick.[48]

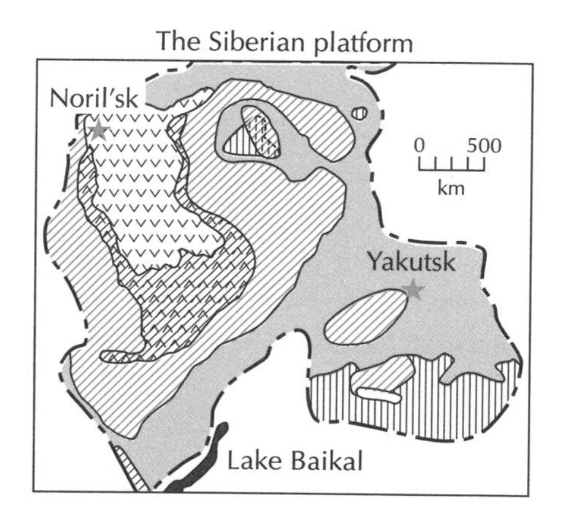

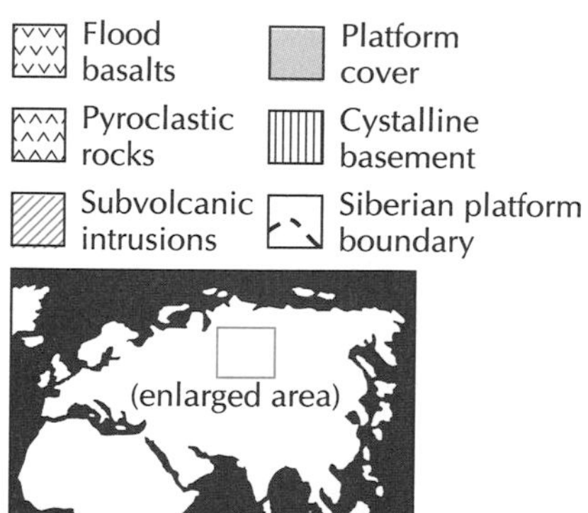

The Siberian Traps are the site of the largest volcanic event in the Earth's history. It happened at the end of the Paleozoic Era about 250 million years ago.

Paul Renne of the Berkeley Geochronology Center, along with Asish Basu and their colleagues, have established that the volcanism that generated the Siberian Traps began around 253 million years ago. However, the main pulse of eruptions began about 250.0 + 0.3 million years ago.[49,50] The time of the extinctions was 250.0 + 0.2 million years ago. Consequently, within our ability to tell time at the end of the Paleozoic, the dates are indistinguishable, suggesting that the eruptions could have caused the extinctions. Renne and his colleagues argue that the Siberian eruptions could have produced enough sulfur-based aerosols to cause acid rain and rapidly cool the global climate, generating polar ice caps. Ice cap formation could have initially led to a retreat of the seas, aided by the uplift generated by the enormous rising plume. Such environmental disruption is blamed for the massive extinctions. To date, no compelling evidence for an extraterrestrial impact at the Permian-Triassic boundary is known.

Although all the environmental effects associated with these large volcanic events are possible and may have caused the K-T extinction of many kinds of dinosaurs, this plethora of potential "killing mechanisms" presents us with a difficult problem. As the number of killing mechanisms increases, the scenario becomes more complicated, and it becomes even harder to test and pinpoint which mechanism, or combination of mechanisms, resulted in the extinction of a particular group of animals or plants. A wealth of ways to die doesn't necessarily lead to a clearer picture of why. Besides, with all these potential ways to become extinct, why were the plants and animals that survived not affected? The reasons for their physiological tolerance are difficult to understand. In addition, not all large continental episodes of basalt floods appear to be associated with large extinction events. One is the Columbia River Basalts in Oregon, which erupted about 16 million years ago.[51] Why are no extinctions associated with that event? Was it too small?

These problems and questions leave many scientists perplexed. The volcanic scenario for extinction just doesn't seem to explain all the available evidence in the rock and fossil record. So, in the late 1970s and early 1980s a new hypothesis was proposed while we were graduate students at Berkeley. This revolutionary proposal linked the K-T extinctions to the impact of an asteroid or comet. But how strong is the evidence documenting that K-T impact? Chapters 5 and 6 will examine the geologic record to evaluate that question.

The skeleton of *Moschops capensis,* one of the larger early relatives of mammals that lived near the end of the Paleozoic Era. The lineage of these early relatives of mammals was severely affected by the extinction event at the end of the Paleozoic, an event which killed off as many as 95 percent of the species living on the Earth at that time.

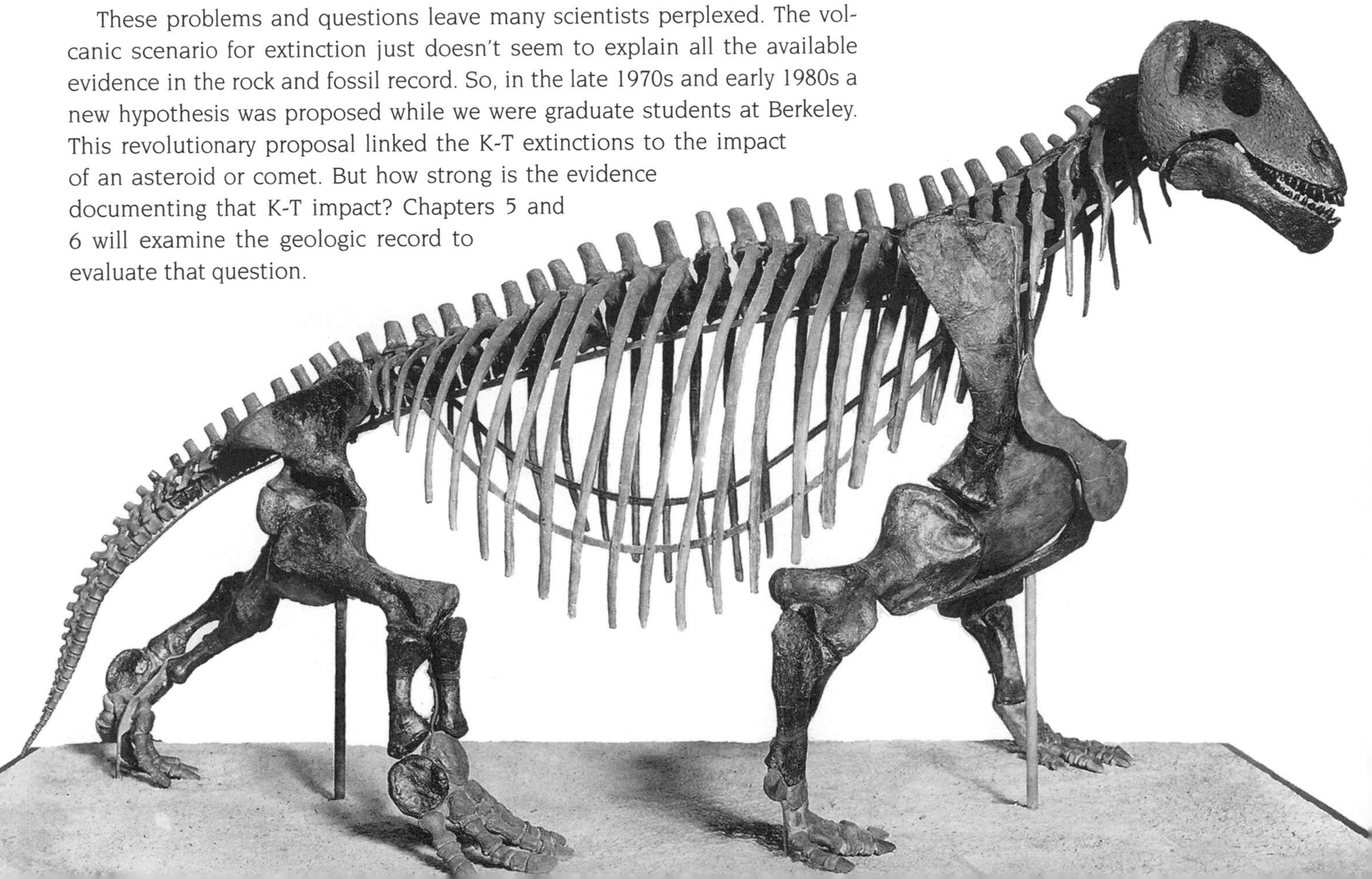

CHAPTER FIVE

The Fatal Impact

Dissatisfaction with both the conventional and fanciful explanations for the extinction of dinosaurs stimulated further research. This pursuit eventually led to a key observation that, in turn, fueled a decade-and-a-half-long vigorous, and often vitriolic, debate.

Actually, as early as the mid-1700s, it had been proposed that extraterrestrial events were responsible for extinctions on Earth.[1] In fact, one of these proposals by French scientist Pierre de Maupertuis in 1742 argued that impacts by comets on the Earth caused extinctions by modifying the atmosphere and oceans. Present-day scenarios are based on this same essential concept. In 1964, René Gallant wrote an essay on the geological and biological effects of meteorite impacts. In it he discussed the possibility that, as a result of associated changes in the Earth's rotation and environmental damage such meteorite impacts could be the cause of mass extinctions. In addition to impacts, Dale Russell of North Carolina State University and Wallace Tucker of the Center for Astrophysics at Harvard College Observatory proposed in 1971 that an explosion of a nearby supernova was the cause of dinosaur extinction.[2] In their scenario, organisms were exterminated by the lethal levels of radiation generated by the explosion. However, these early ideas involving extraterrestrial causes for extinctions on Earth had no tangible geologic evidence to back up the claims. They were just hypothetical scenarios, and as a result, none were taken very seriously.

In the early 1800s, Georges Cuvier interpreted the sequence fossil faunas in rocks of the Paris Basin to represent a series of catastrophic extinction events. This marked the beginnings of the "catastrophic" school of biological changes.

Since the end of the eighteenth century and beginning of the nineteenth century, the primary strategy that geologists have used to gain an understanding of Earth history has been to use the "present as a key to the past." In other words, those geological processes that we can see acting on the Earth today are most likely responsible for any evidence that we see preserved in the geologic record of the past. In general terms, this doctrine has become known as the principle of uniformitarianism.

The story of how uniformitarianism became the dominant school of geological thought is an interesting and informative one that continues to be debated and discussed two centuries after the fact. The story also has close ties to the contemporary debate between proponents of more catastrophic, extraterrestrially based scenarios and those favoring more gradual, terrestrially based ones. For it was at the end of the sixteenth century and the beginning of the seventeenth century that the geological schools of catastrophism and gradualism actually arose.[3–5]

The chief proponent of the catastrophic school was the influential French anatomist, Georges Cuvier of the Paris Museum of Natural History. Working in the early decades of the 1800s, his meticulous studies of the fossil faunas in rock layers of the Paris Basin led him to conclude that repeated catastrophes had been inflicted on the Earth and it's biota. He was not sure what caused these catastrophes, but he was sure that he could document the effects of them. The ideas and evidence are laid out in his *Essay on the Theory of the Earth*, the third edition of which was published in 1817.

Cuvier used a previously developed geological principle called the law of superposition. This principle generally states that rock layers are deposited sequentially one on top of the other, so that lower layers are older than higher ones. Cuvier noted that the kinds of fossil organisms preserved in the different layers of rock changed dramatically as one moved from a lower layer into the next higher layer of the sequence. He interpreted these changes to mean that each layer represented a distinct period of Earth history with its own distinctive fauna. The boundaries between the layers represented catastrophes that wiped out all the previously existing organisms. Subsequently in the next higher layer, a new set of organisms appeared.

Modern students of Cuvier's work disagree over how he viewed these catastrophes. Some, such as William Berry of UC Berkeley, interpret Cuvier's writings to suggest that he believed that during each catastrophe all of the previously existing organisms suddenly became extinct, and a complete new set of organisms was created.[6] Others, such as Anthony Hallam,[7] have suggested that Cuvier believed the new faunas might have migrated into the area from other continents or seas. Part of this interpretive problem appears to be tied up with understanding Cuvier's leanings in relation to the prevailing religious beliefs of the time. Across the Channel in England many people interested in geological issues were also clergymen and devout believers in the literal translation of the great biblical flood. In fact, in the earliest translation of Cuvier's seminal work into English from French, the most recent of Cuvier's catastrophic episodes was thought to represent the flood documented in Genesis. Cuvier's work was, therefore, of great interest to these clergymen because it could be easily rationalized with the religious teachings of the day. Whether Cuvier believed this himself is debated today.[8,9] Whether he did or not, he was convinced of the sudden and catastrophic nature of the biologic turnovers he saw in the geologic record.

English geologist James Hutton is commonly credited with establishing the principle of uniformitarianism. This idea argues that ancient geological deposits were created by the same geological processes operating on the Earth today.

During the same general time period, a distinctly different view of what the biologic changes between different rock layers meant was being developed in England and Scotland by other contemporary geologists. Beginning in 1780s, James Hutton, a Scottish farmer and dedicated collector of fossils, brought into focus the idea that natural geological laws could be derived from studying modern geological processes and the deposits that these processes generated.

In essence, he proposed that ancient geological deposits resulted from the same geological processes which we can see operating on the Earth today. His concepts were outlined in a talk to the Royal Society of Edinburgh in 1785 and eventually published as a book entitled *Theory of the Earth* in 1795. There, he stated, "In examining things present, we have data from which to reason with regard to what has been; and from what has actually been, we have data for concluding with regard to that which is to happen hereafter." Hutton traveled to many locations across the British Isles in search of fossils. As he did, he paid keen attention to the geologic processes that he could observe in operation. Although others had previously noted the same basic concept, Hutton is usually given credit for laying the foundation of this principle of uniformitarianism.

Hutton concluded that the geologic record had been generated by long episodes of rock formation and uplift, followed by erosion and the formation of younger layers of rock from previously existing rock. His observations were summed up in his statement that "The result, therefore, of our present inquiry is that we find no vestige of a beginning, no prospect of an end."[10] Hutton's concept of an immense expanse of geologic time and history flew in the face of the prevailing religious doctrine. According to the authority of Bishop Ussher's religious dictates, the Earth was only 6,000 years old, it having been created by God in the year 4004 B.C. Hutton died two years after his work was published in its most expanded form. Not suprisingly, his ideas were severely criticized by the religious establishment.

Nonetheless, Hutton's themes were carried on in the ensuing early decades of the 1800s by a friend and colleague, John Playfair, a Professor of Mathematics at the University of Edinburgh, as well as by Charles Lyell, a Fellow in the Geological Society of London. Lyell's *Principles of Geology*, first published in three volumes between 1830 and 1833, did much to influence the geological community's view of how on-going geologic processes could be used to interpret past geologic deposits and events.[11] In more detail than Hutton, Lyell extensively documented the processes of erosion and deposition to illustrate the uniformity of natural geological processes through time. He concurred with Hutton's conclusion that the expanse of geologic time was much longer than the prevailing religious doctrines allowed. Because of the popularity of his textbook, Lyell is often cited as the father of uniformitarianism.

Charles Lyell, often acknowledged as the father of modern geology, laid out the detailed geological evidence supporting uniformitarianism in his book entitled *Principles of Geology*.

In writing the *Principles of Geology*, Lyell appears to have had two primary items on his agenda. He wanted to counter the more catastrophic vision of Earth history embodied in the work of Cuvier, which had been adapted by some of his English colleagues for support of literal biblical rationalizations. Also, he sought to establish geology as a testable science based on empirical natural laws.

In attempting to counter Cuvier, Lyell argued that observable modern geologic processes operated at relatively constant or uniform rates rather than at varied rates that would generate catastrophic episodes of change. In essence, he argued that if the processes of volcanism, uplift, erosion, and deposition were allowed to operate constantly throughout the expanse of geologic time, they could easily explain the geologic structures and landforms found both on the Earth today and throughout the geologic past. No appeal to sudden catastrophes was required. By the time of his death in 1875, his goal had been fairly successfully accomplished.

Thus for two centuries, the principle of uniformitarianism has dominated our thinking, ever since the seminal work of James Hutton, Georges Cuvier, Charles Lyell, and others laid the foundation to transform geology into a testable science. Recent reviews of Lyell's brand of uniformitarianism have focused a critical light on some of his arguments. Steven Jay Gould of Harvard University played an instrumental role in this reevaluation when he recognized that the concept of uniformitarianism actually represented two different ideas.[12] Few geologists today agree with Lyell's conclusion, now termed substantive uniformitarianism, that geologic processes have operated at constant rates throughout geologic history. On the other hand, the other aspect of Lyell's uniformitarianism, termed methodological uniformitarianism, forms the cornerstone for all geological research. Basically, it states that natural geologic laws have applied throughout the Earth's history so that no hypothetical unobservable processes need to be invoked to explain the existing geologic record.

Because geologists have long focused almost exclusively on Earth-based observations and processes during their research, students have been cautioned to consider as bunk any catastrophic, extraterrestrial causes for geological phenomena observable on our planet. But the dawn of the Space Age in the latter half of the twentieth century called such Earth-centered approaches into question. Missions to the moon and planets produced a wealth of new data, including the undeniable effects of impacts on many objects in our solar system. In a very visceral and hyperreal sense, this exploration confirmed the Earth's relation to other bodies in the solar system and the universe. In effect, it became increasingly clear that impacts of asteroids and comets on the Earth could be legitmately considered to be methodologically uniformitarian.

Then, in the late 1970s when we were entering Berkeley as graduate students, several geological clues were discovered that provided the first compelling evidence for a plausible extraterrestrial culprit in the extinction of dinosaurs. In a sense, history had turned on itself, and the old arguments between the catastrophists and the uniformitarianists (or gradualists) were once again ignited into a cacophony of scientific conflict.

In a figurative sense, a meteorite had impacted just one floor above our offices in the Department of Paleontology, as a sensational new catastrophic extinction scenario was formulated and published by a group of researchers in the Department of Geology and Geophysics.[13] They were led by a Nobel-laureate physicist, Luis Alvarez, and his son Walter, a Professor in the Department of Geology and Geophysics.

In both marine and terrestrial rock layers, or sequences, many of the plants and animals that became extinct at or near the end of the Age of Dinosaurs drop out of the record abruptly as one moves from lower, older rocks documenting the end of the Age of Dinosaurs up into higher, younger rocks representing the beginning of the Age of Mammals. Such an abrupt disappearance of many forms of single-celled plankton in the marine limestones near Gubbio, Italy, led Walter Alvarez and his colleagues to wonder how long those extinctions really took.[14]

The first Ir-rich layer ever discovered was in marine limestones exposed near Gubbio, Italy. The coin is laid on the clay layer containing the fallout from the impact.

Between the white limestone representing the end of the Age of Dinosaurs and the pink limestone representing the start of the Age of Mammals lies a three-eighths of an inch thick (1 cm thick) bed of clay. Alvarez discussed this problem with his father and several of his father's associates. In the end, they felt that they might get some idea of how long it took to deposit this layer of clay by looking at the concentration of an element called iridium. Iridium (Ir) is one of the platinum group elements and has an atomic number of 77.

Iridium is not common in the rocks of the Earth's crust. It is a relatively heavy element that tended to sink toward the Earth's core as the planet cooled and differentiated into its aforementioned layers of different composition and density at an early stage in its history. Most of the iridium present at the formation of the Earth appears to be combined as an alloy with iron and is now concentrated in the core. However, iridium is found in higher concentrations in some meteorites and comets, where the original chemical composition of the solar system is preserved. These smaller bodies were not large enough to go through the same process of chemical differentiation that our planet did.

Nonetheless, small amounts of iridium continue to be deposited on the Earth's surface in microscopic meteorites which continually bombard the Earth. One estimate suggests that the mass of the earth is increased, on the average, by 12,100 tons (11,000 tonnes) per year as the result of this constant cosmic bombardment.[15] While falling through the atmosphere, most meteorites burn up. However, a small percentage survive the descent to impact on both land and sea. By measuring how many of these small meteorites fall to Earth over a given period of time in our modern world, scientists can estimate how much iridium is brought in because they know the concentration of iridium in the meteorites. Then, by measuring the amount of iridium in the boundary clay, the Alvarez team felt that it could estimate how long it took for the clay to be deposited, assuming that the rate at which small meteorites hit the Earth today is similar to the rate at which they hit the Earth at the end of the Age of Dinosaurs. In retrospect, it's extremely ironic to realize that this experiment represented a classic uniformitarian approach to solving the problem. The Alvarez team looked to the modern world for help in interpreting the ancient geologic record. Yet their experiment led to a very nonuniformitarian conclusion.

Based on the amount of iridium measured in the boundary clay exposed in the outcrops at Gubbio, Italy, the calculations made by the Alvarezes and their colleagues suggested a disturbing result. A period of about 1 million years would have been required to deposit the boundary clay. This contradicted the amount of time available based on a conflicting line of evidence suggested by reversals of the Earth's magnetic field. The interval of geologic history containing the last dinosaur fossils occurs within a period in which the Earth's geomagnetic poles were reversed. In other words, the poles were opposite the way they are oriented today. This interval, known as Chron 29r (for the twenty-ninth reversed interval before the present), was thought to have lasted only a half a million years. If so, the deposition of the boundary clay could not have lasted 1 million years.[16] Another estimate suggested that the amount of iridium present in the boundary clay was 30 times that which could be expected by the normal rate of gradual accumulation of small meteorites.[17]

Examining the same problem another way, the average, present-day concentration of iridium in seawater (11×10^{-12} grams per 10^{-3} cm^3) implies that it would have required the Earth's oceans to be over 435 miles (700 km) deep to have precipitated out the observed amount of iridium in the boundary clay under normal natural conditions.[18] However, the deepest known part of today's oceans is found in the Mariana Trench, a subduction zone between plates near Guam. There, the depth is less than 7 miles (11 km). So, the extraordinarily high concentration of iridium seemed to require a special explanation.

At first, the Alvarez team proposed that the explosion of a nearby supernova had generated the iridium. However, within a year, contradictory evidence had been found in the boundary clay.[19] Consequently, the Alvarez team regrouped and hypothesized that a large asteroid, about 6 to 9 miles

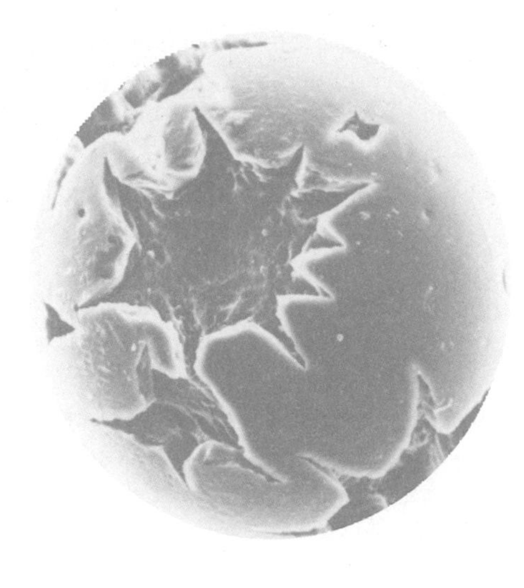

Photo taken through a microscope of a glassy tektite—one of millions of small meterorites that fall through the Earth's atmosphere to land on the ground and in the seas. Such objects are enriched in iridium.

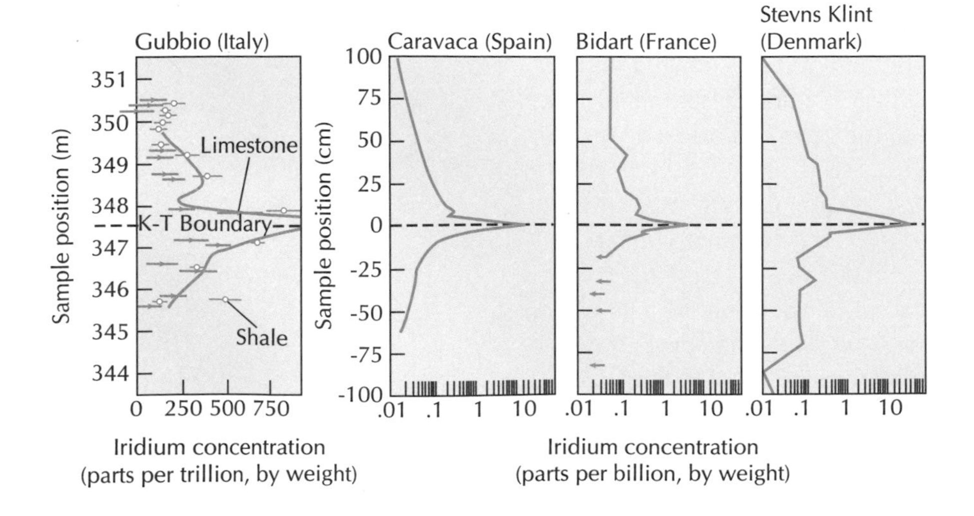

These charts show the elevated levels of iridium found at four different sites spanning the Cretaceous-Tertiary (K-T) boundary. Such "anomalous" concentrations of iridium have now been found in over 100 sites around the world.

This artist's rendering depicts how the Earth might have looked during the first few moments after impact, 65 million years ago.

(10 to 15 km) across, must have collided with the Earth, releasing the equivalent of about 10^{23} joules of kinetic energy. This is 100 times more than the amount of internal energy released each year from the Earth in the form of heat and earthquakes,[20] and between 1000 and 10,000 times the energy that would be released if all of the world's nuclear weapons were detonated simultaneously.[21] The fall-out resulting from that impact created the boundary clay.

As many people both inside and outside the ensuing debates have noted, this was a strikingly bold and professionally risky proposal, especially in view of geology's long history of excluding extraterrestrial events as one of its accepted uniformitarian processes. But the coincidence of the high iridium concentration with the biologic changes at the Cretaceous-Tertiary (K-T) boundary in Italy was difficult to dismiss as being purely coincidental. The most noteworthy effect of the hypothesis was to spawn a storm of multidisciplinary research around the world. What the Alvarez team had contributed to the debate over the cause of dinosaur extinction was a sensational hypothesis with predictive elements that could, at least to some degree, be tested both in the field and in the laboratory, and that's exactly what the scientific community set out to do.

The new theory contrasted markedly with the conventional wisdom long held in the paleontological community that the extinction of dinosaurs was brought about by more gradual, Earth-based processes. Consequently, the level of interest in both the Paleontology Department and the Geology and Geophysics Department at Berkeley was high. Interest quickly spread to other departments, especially as a rising wave of publicity in both the popular and scientific press began to win over members of what had been a deeply skeptical scientific audience.

In the interest of exploring this new hypothesis, a seminar was initiated involving the Alvarez group and a group in the Paleontology Department led by Professor William Clemens. For over a decade, Clemens had conducted a program of paleontological and geological field research in the most famous fossil locality documenting dinosaur extinction—the Hell Creek and Tullock Formations—which form the picturesque Missouri Breaks of east-central Montana. In 1980, Lowell was the senior graduate student helping to manage that research program.

In the beginning, this interdepartmental seminar seemed to symbolize the best that a multidisciplined academic environment could offer. We discussed the strengths and weaknesses of both the impact hypothesis and more gradualistic scenarios. A program was devised for how we might test the predictions of the impact hypothesis during our next field season in Montana. This initial spirit of cooperation led to the joint discovery of some of the first evidence for an impact ever identified in sediments that document the K-T transition in a land-based habitat that contained remains of dinosaurs. The evidence for impact consisted of an anomalously high concentration of iridium, similar to the one found between the marine limestones in Italy. This was a tremendously invigorating and intellectually intoxicating time. We were fully aware that our work was poised on the cutting edge of science.

However, as the months passed, the atmosphere of cooperation between some of the participants in the seminar began to deteriorate. A series of public debates between proponents of the impact and those favoring Earth-based scenarios served to polarize our positions. Some exchanges even took on a rather unpleasant personal pitch. From a personal point of view, we often found ourselves arguing toe-to-toe with a Nobel-laureate over issues involved in the debates. As new graduate students yet to pass our orals, this was an intimidating situation. At times, we wondered what effect it might have on our subsequent careers. In retrospect, these seminars set the tone for a decade and a half of dissension between proponents of catastrophic, extraterrestrially triggered extinction and proponents of generally more gradual, terrestrially

Participants in the extinction seminar at UC Berkeley posed for a picture early in their discussions. Seated *(left to right)* are Kevin Steward, Helen Michel, and Dale Russell. Standing *(left to right)* are Lowell Dingus, Alessandro Montanari, Luis Alvarez, Frank Asaro, William Clemens, Walter Alvarez, and Mike Greenwald.

The thin, light stripe of clay near the center of the photo represents the first evidence ever discovered in a terrestrial section of rocks for the Ir-rich fallout layer from the K-T impact. It was found at the boundary between the Hell Creek and Tullock Formations near Jordan, Montana.

based volcanic scenarios associated with the retreat of the seaways. In a very real sense, the debates took on the atmosphere and intolerance of wars—the "Iridium Wars."

At various times during this period, one side or the other claimed victory. Yet, still the polemics continue. Why is this? Why, after fifteen years of research and over 2,500 scientific publications, is the mystery not resolved to everyone's satisfaction? To understand, we need to examine the scenario thought to have caused the extinctions and principal lines of evidence in the debate.

ASSOCIATED ENVIRONMENTAL EFFECTS OF THE IMPACT

The initial calculations and scenario proposed by the Alvarez team suggested that the impact kicked up a globally distributed cloud of dust and water vapor that cut off light from the Sun for a period of several months to a few years. This cloud inhibited photosynthesis, which killed off many plants. Consequently, the herbivorous dinosaurs died out, and then the carnivorous dinosaurs perished.

The Alvarez team's original proposal eventually led to a contentious discussion of whether dinosaurs could have survived the several months of darkness posited by the original impact hypothesis. To try and test this scenario, paleontologists focused their attention on Cretaceous dinosaur sites located near the north and south poles.

One site, called Dinosaur Cove and located in southern Victoria, Australia, was the focus of work done by Pat Vickers-Rich of Monash Univeristy and Tom Rich of the Museum of Victoria—two of our predecessors at UC Berkeley. Dinosaur Cove preserved evidence of a thriving community of dinosaurs that lived about 100 million years ago as far as 80 to 85 degrees South latitude. Their home was well inside the Antarctic Circle, nuzzled up against the great cold continent of Antarctica.[22] The fossils were preserved in sediments laid down by rivers that ran through floodplains in a rift valley that formed as Australia was beginning to separate from Antarctica through the actions of plate

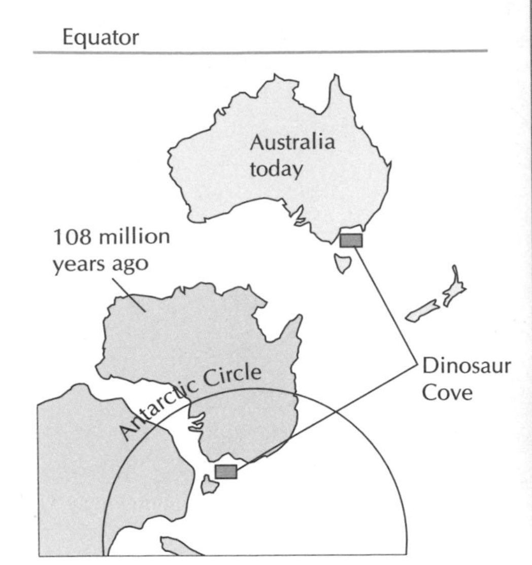

Dinosaur Cove in Australia has produced fossils of dinosaurs that lived on the continent about 100 million years ago when Australia occupied a position much nearer the South Pole than it does today.

tectonics. Based on data provided through an analysis of oxygen isotopes, the mean annual temperature was thought to have ranged between 32° F (0° C) and 46° F (8° C). However, an analysis of the flora, which included ferns, ginkgoes, conifers, and horsetails, suggests a higher mean annual temperature of about 50° F (10° C).

The kinds of animals present in the area around Dinosaur Cove incorporate a broad swath of dinosaurian diversity. These include a small meat-eating carnosaur that stood no more than 6 feet (2 meters) high; a member of the bird-mimic dinosaurs called ornithomimids; and a new, plant-eating hypsilophodontid, named *Leaellynasaura*. Because of the extreme southern positioning of this region 100 million years ago, it was concluded by paleontologists working on the fauna that these dinosaurs would have had to survive 3-month-long periods of darkness on an annual basis. This period would have constituted their winter, when the Earth's axis rotated into a position where the sun never illuminated this region. These conclusions left questions in the minds of the paleontologists as to why an impact-generated period of darkness would have extinguished groups of dinosaurs who were apparently well adapted to living in the cold and dark for periods of several months.

At the opposite end of the Earth on the north slope of Alaska, paleontologists, including Williams Clemens, focused their collecting efforts on a 76 to 66-million-year-old, Late Cretaceous dinosaur community. It included large, herbivorous duckbill dinosaurs, as well as carnivorous forms, such as a tyrannosaur and the diminutive *Troodon*.[23] Because of its younger age near the K-T boundary, an analysis of the conditions under which this fauna lived is more pertinent to the extinction debate than the situation at Dinosaur Cove. The north slope sites, located along the Coleville River, were deposited by streams and rivers that built a floodplain delta extending out into the Arctic Ocean. At the end of the Cretaceous, this area is thought to have been situated between about 70 and 85 degrees North latitude. These coordinates would have put the site within the Arctic Circle and subjected the fauna and flora to an annual winter darkness of about 3 months. Analysis of fossil pollen and spores indicates the presence of both ferns and deciduous conifers (similar to the dawn redwood *Metasequoia*) in a mild- to cold-temperate forest. Based on this flora and the absence of more temperate-adapted vertebrates, such as crocodiles and amphibians, the mean annual temperature is thought to have ranged between 35 to 46° F (2 to 8° C), similar to the temperature range of present-day Anchorage.

The resulting debate between proponents of volcanic and impact extinction scenarios centered on whether these high-latitude dinosaurs migrated south for the winter where the sun always rises throughout the winter months. However, Clemens and his colleagues argued that the presence of juveniles in the high latitude sites constituted evidence against this, because it would have been difficult for these young animals to survive such a long and rigorous trip. Most scientists involved in the extinction debate now concede that darkness alone was probably not responsible for causing dinosaur extinction, but many other potential "killing mechanisms" have been proposed for the impact.

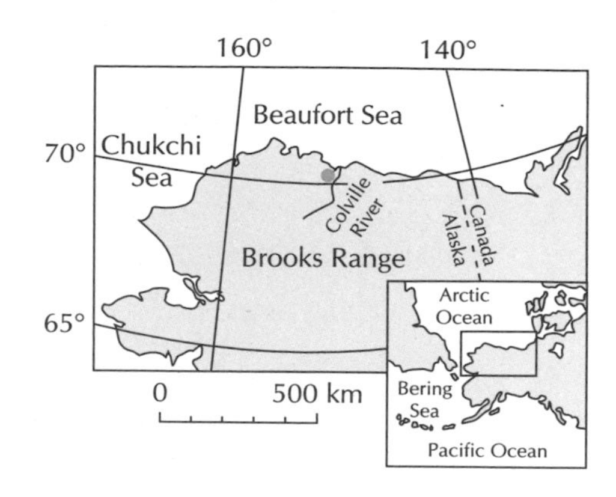

Near the North Pole, another asssemblage of dinosaurs was uncovered along exposures cut by the Colville River. Along with those found at Dinosaur Cove, these fossils raise questions about whether dinosaurs would have been decimated by long periods of darkness and cold.

Another killing mechanism posited that the impact caused extreme episodes of acid rain.[24] The intense, initial, atmospheric heating generated by the impact would have created temperatures of several thousand degrees at the sight of the impact and locally raised temperatures to broiling levels immediately after the impact. Such extreme heat would have caused the formation of nitric acid as nitrogen and oxygen reacted with water vapor. The resulting acid rain would have increased the acidity of the ocean's surface water, killing up to 90 percent of the microscopic organisms, such as planktonic foraminifera and coccoliths, that form the foundation of the food chain. Second, the impact may have set off wildfires around the globe as the result of the heat and falling debris generated by the impact.[25] Third, the impact was alleged to have dropped surface temperatures on continents to near or below freezing for a period of several months to a year when the dust and water vapor in the globally distributed cloud of debris cut off sunlight and killed plants needing sunlight to photosynthesize.[26] Fourth, the impact significantly raised global temperatures over the longer term, possibly tens of thousands of years, as a result of the "greenhouse effect." This rise in temperatue was driven by a two- to five fold increase in the atmospheric concentration of carbon dioxide.[27] Finally, the extensive extinctions of microscopic marine organisms would reduce the release of a chemical compound called dimethylsulfide, thereby inhibiting cloud formation and raising global temperatures more than 11° F (6° C).[28] This disruption of the food chain and climate was, in turn, hypothesized to have annihilated the dinosaurs and other organisms over a period of less than 50 years.[29]

EVALUATING THE EVIDENCE

There are really several separate questions involved in trying to scientifically test this scenario with the evidence available in the rock and fossil records. The first is, "Did an impact occur?" The second is, "If so, which killing mechanism extinguished each particular group of organisms?" And a third is, "Can we really tell time well enough to distinguish between short-term, catastrophic events and long-term, gradual events?"

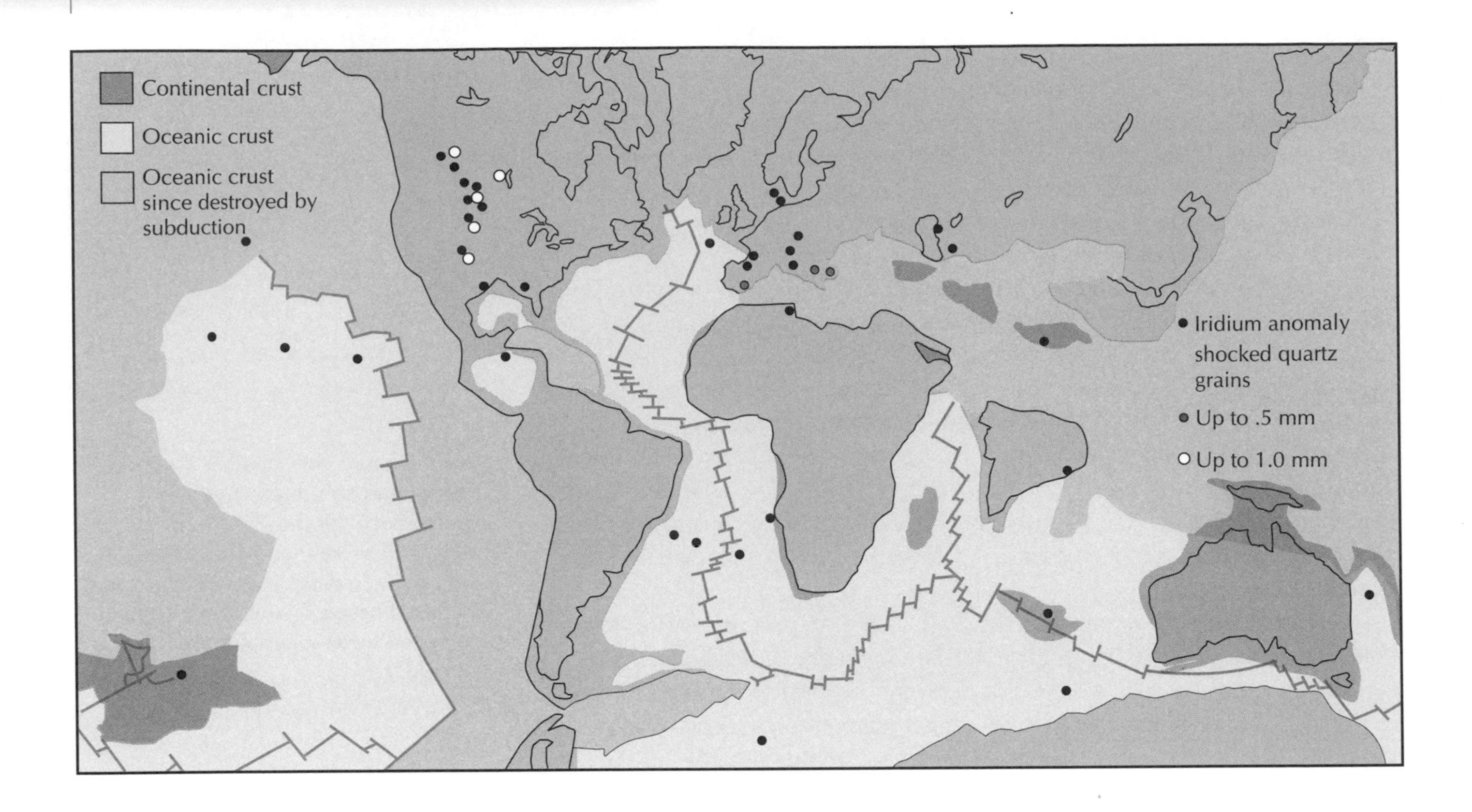

Map illustrating the global distribution of geological evidence used to support the impact hypothesis for dinosaur extinction.

Boundary clays with high concentrations of iridium have now been reported in over 100 continental and marine sequences around the world.[30] In terrestrial sequences, these clay beds are situated between the youngest plant fossils representing the Age of Dinosaurs and the oldest plant fossils representing the Age of Mammals. They're typically about 6 to 10 feet (2 to 3 m) above the highest or youngest-known dinosaur fossils that have not been reworked into younger layers.

To critics of the impact scenario, such as William Clemens and David Archibald of San Diego State University, [31,32] this "gap" between the youngest dinosaur fossils and the iridium-rich clay suggests that dinosaurs were extinct before the alleged impact occurred. Such an interpretation derives from a "strict translation" or literal reading of the fossil record. Proponents of the impact hypothesis, including Luis Alvarez, have responded that dinosaur fossils are not preserved as often as the pollen and spores representing the plants, so it is highly improbable that any dinosaur fossil will be found immediately under or at the same level as the boundary clay.[33] Others, including Archibald, have argued that fossils were originally in the several feet of sediment immediately underlying the boundary, but that these fossils could have been leached out by percolating ground water.[34] In any event, extensive searches for such fossil skeletons have thus far proved fruitless, although fossil footprints of dinosaurs have now been found within 15 inches (38 cm) of the boundary in New Mexico.[35] As we describe later on, rare reports of unreworked dinosaur fossils found above the level of the boundary clay remain controversial and only poorly substantiated by geologic evidence.

Advocates of the impact hypothesis argue that, in addition to the iridium, many other lines of geologic and chemical evidence suggest an impact-origin for the boundary clay. However, early on in the debate, critics proposed Earth-based mechanisms rather than an impact in an attempt to explain the same evidence. In an effort to determine whether an impact really occurred, let's now review the disputed scientific evidence involved in these arguments.

A HIGH CONCENTRATION OF IRIDIUM

As in any murder investigation, fingerprints of the potential culprits can play a major role. In the context of the K-T extinctions, the ratios of different atoms in the boundary can serve as geological "fingerprints," pointing to a probable source for the minerals in the boundary clay. Much of the geologic detective work done to establish whether rock layers resulted from impacts or volcanic events involves painstakingly detailed chemical analyses to measure the concentrations of rare elements contained within the rocks.

As mentioned earlier, the unusually high concentration of iridium found by the Alvarez team in the K-T boundary clay, estimated to total 550 million tons globally (500 million metric tons),[36] originally seemed to represent strong and unequivocal evidence in favor of the impact scenario. Within months, a number of iridium "anomalies" were found around the world at the paleontologically recognized boundary. Although such enrichments are termed "iridium anomalies," because the concentration of iridium in the boundary clay is 100 to 1000 times higher than that in the surrounding rock layers, these anomalous concentrations actually record only a few parts per billion of iridium. This means that for every billion atoms present in the boundary clay, less than ten are atoms of iridium.

In 1983, the idea that high concentrations of iridium represented definitive evidence for impacts was called into question by William Zoeller and his colleagues at the University of Maryland. They studied the gases emanating from a large volcano in Hawaii.[37] During their field work, these researchers inadvertently measured the amount of iridium contained in the airborne particles erupted from Kilauea. Although the amount of iridium in the dark basaltic rock formed from the lava of Kilauea is low, the iridium concentration in the airborne particles emitted by the volcano was 17,000 times higher. This raised the possibility that large eruptions of lava and gases could explain the high concentration of iridium in the K-T boundary clay. Zoeller's study noted that the volcanic gases generated by Kilauea contained abnormally high concentrations of another element, fluorine. The researchers noted that Kilauea was a hot spot volcano whose magma probably rose from deep within the Earth's mantle. The Deccan Traps were also generated by a plume from deep in the mantle, as we have seen in Chapter 4.

This previously mentioned extensive episode of volcanic activity at the end of the Age of Dinosaurs, especially the basaltic fissure floods in India, made it plausible that some or all of the iridium was generated by volcanism rather than impact. Most geologists agree that the tremendous amount of material erupted in the Deccan Traps is still not enough to explain the amount of iridium that would have been contained in the globally distributed fallout layer. However, as noted in previous chapters, there are several other

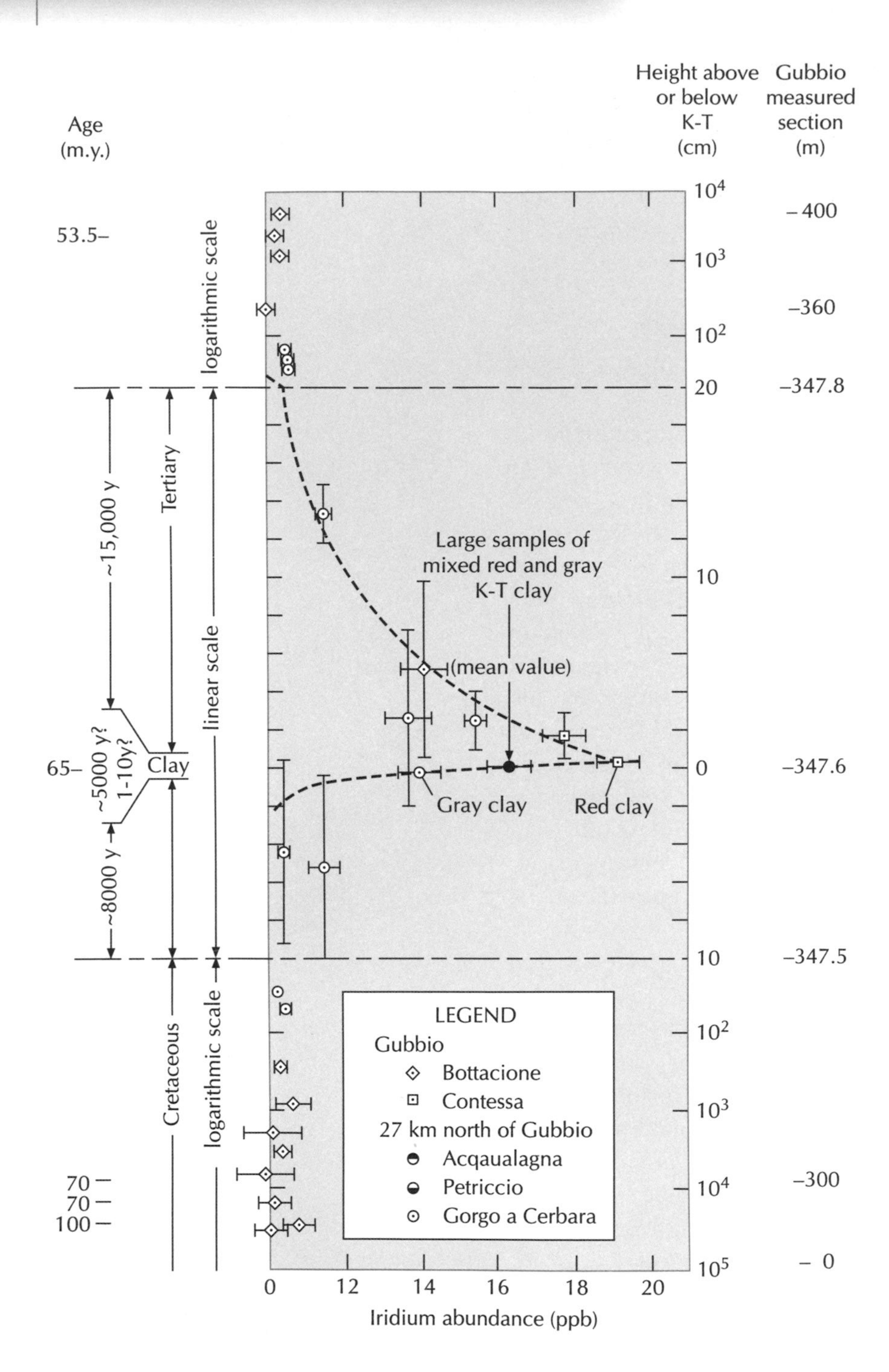

Concentration of iridium plotted against distance from the K-T boundary in marine limestones of Italy. Notice how the concentrations are low, less that one part per billion, in the Cretaceous limestone at the bottom of the section. The highest concentrations, about 8 parts per billion, are found in the boundary clay.

hot-spot eruptions that may have begun during the K-T transition. Proponents of the volcanic extinction scenario, such as Charles Officer, Charles Drake, and Anthony Hallam argued that these contributed some or all of the iridium in the boundary clays. [38–40]

This idea was reinforced by more measurements taken at Piton de la Fournais.[41] As mentioned in Chapter 4, this is a volcano on Reunion Island in the Indian Ocean. It represents the remnant of the hot-spot generated by the

A volcanic eruption rages at Kilauea on Hawaii.

plume of magma that was responsible for the eruptions that created the Deccan Traps and the undersea plateau around the Seychelles Islands between 68.5 and 64.9 million years ago. Geologists Jean-Paul Toutain of the Osservatorio Vesuviano in Italy and Georges Meyer of the Institut de Physique du Globe de Paris conducted chemical analyses of the gases and associated compounds emitted from that volcano. They documented iridium concentrations as high as 7 to 8 parts per billion. Calculations done by these researchers suggest that, given the amount of iridium emitted from Piton de la Fournaise (0.25 + 0.03 parts per billion), the earlier eruptions that produced the Deccan Traps would have produced more than enough iridium to account for the concentrations found in K-T boundary clays.

THE OVERALL CHEMICAL COMPOSITION

The overall chemical composition of the boundary clay was noted by the Alvarez team, and especially Miriam Kastner of the Scripps Institution of Oceanography, to be very similar to the composition of a certain type of meteorite called a type I carbonaceous chondrite.[42] These are among the most primitive known bodies in our solar system in terms of age and chemical makeup. This similarity in composition was based on the analysis of several elemental ratios present in the boundary clay, such as platinum/iridium and gold/iridium. Frank Asaro, one of the original members of the Alvarez team, working with George Bekov of the Institute of Spectroscopy in Moscow, discovered that the relative abundances of ruthenium, rhodium, and iridium were more similar to those found in meteroites than those found in terrestrial rocks. Atomic analyses eventually led to a debate within the community of impact proponents over whether the extraterrestrial body that hit the earth was an asteroid, as the Alvarez team had originally proposed, or a comet, as suggested by some other analyses.[43,44] The iridium concentration in comets is thought to be about ten times less than that for iron meteorites.

Critics, such as the late Anthony Hallam, pointed out that multiple impacts would probably need to have occurred in order to explain the observed iridium concentration.[45] This possibility and associated problems are discussed later. In addition, several contradicting points were raised by proponents of a terrestrial origin for the iridium, as noted in a paper by Vincent Courtillot.[46] A sampling follows.

Researchers, including W. Crawford Elliot of Case Western Reserve University, did chemical detective work on the boundary clay preserved at Stevns Klint in Denmark. They concluded that the clay was composed of an unusual kind of smectite (a type of clay mineral), which indicated that it represented a weathering product of volcanic ash. A study by Jean-Marc Luck of the Institut de Physique du Globe de Paris and Karl Turekian of Yale Univeristy indicated that the ratio of the elements rhenium and osmium in the clay resembled the ratio found both in meteorites and the Earth's mantle. Charles Officer and Charles Drake noted that the iridium concentration of boundary clays from different areas of the globe is rather variable, which suggested to them that the source was not a globally distributed dust cloud of constant composition. They also noted that the ratios of other elements differed from that of chondritic meteorites.[47] Finally, as Zoeller and his colleagues pointed out, enrichments of other rare atoms in the emissions from Kilauea, such as antimony, selenium, and arsenic, suggested that hot spot volcanism may have played a role in creating the elemental concentrations found in the boundary layers.[48]

Proponents of the impact hypothesis, such as Walter Alvarez and Frank Asaro, countered by arguing that the osmium concentration of the boundary layer seems to suggest a source exclusive of continental rocks.[49] In terrestrial rocks, the isotope of osmium that contains 187 protons and neutrons is much more abundant than the isotope of osmium that contain 186, but this is not what is seen in the boundary clay. In fact, the reverse is true, suggesting a source from extraterrestrial meteorites or volcanic rocks from deep within the mantle. The work of Turekian and Luck was critical in establishing this point. Also, an analysis of the concentration of rhenium in the boundary clay provides further support for this interpretation.

As Richard Grieve of the Geological Survey of Canada has summarized,[50] the chemical composition of some boundary layers suggested that the extraterrestrial body slammed into continental rocks. However, the chemical composition of a few layers yielded contradictory results, suggesting that the object may have slammed into a more basaltic material such as the floor of an ocean. A possible solution to this problem is discussed in Chapter 6.

All these factors led proponents of the impact hypothesis, such as Frank Kyte of the University of California at Los Angeles and his colleagues, to rationalize that any chemical inconsistencies between the boundary clay and extrterrestrial bodies were the result of natural variations.[51] Inconsistencies could have been caused when the fallout from the impact was mixed with the variety of materials found in different environments around the world in which the fallout was deposited and buried to form sedimentary rocks like the boundary clay. In addition, impact proponents argued that different geochemical processes could have altered the composition of the boundary layer after it was originally deposited. However, skeptics, such as Charles Officer and Charles Drake, argued that the compositional differences probably represented the result of natural variation in terrestrially based volcanic ejecta.[52]

Photograph taken under a microscope of a quartz crystal from the boundary clay in Wyoming. Although the crystal is less than 1 mm in length, it shows the multiple planes of fracturing that most scientists believe resulted from the impact near present-day Yucatan.

MINERALS FRACTURED BY THE SHOCK OF THE IMPACT

At more than 30 localities around the world, beds of the boundary clay have been found to contain microscopic grains of quartz with small fractures planes running in multiple directions through the crystal structure.[53] Under the microscope, these crystals appear to have a crosshatched pattern. The leader of these research efforts was Bruce Bohor of the U.S. Geological Survey. The first place that such "shocked" crystals were discovered was at Meteor Crater in Arizona. This was where a meteorite slammed into sandstone target rock between 50,000 and 25,000 years ago.

The sequence of events that create these fractured crystals during impact have been described by Richard Grieve.[54] Recall that we are dealing with a body traveling between 50,000 and 150,000 miles per hour (80,000 and 250,000 km/h)[34] and weighing as much as 1.1 trillion tons (1 trillion metric tons).[55] The energy of the speeding meteorite is translated into powerful, high-pressure shock waves that travel out from the point of impact. These shock waves are initially thought by some to be as strong as 100 billion pascals. To give you some idea of this force, a pascal is equal to applying a force of 1 newton per square centimeter. A newton represents the force required to constantly increase the velocity of a 1 kilogram object at a rate of 1 meter per second. In terms of miles per hour, this would be roughly equal to constantly increasing the velocity of a 2-pound object at a rate of about 2 miles per hour. The force of the impact was about 100 billion times greater. Another estimate of the impact force suggests that the shock would have been about 1 million times that exerted by the Earth's atmosphere.[56] These tremendous forces generated by the impact vaporize, melt, and fracture the rock near the point of impact. Fracturing occurs at the lowest temperatures and pressures, whereas melting and vaporization occur at the highest.

Meteor Crater in Arizona about $\frac{3}{4}$ mile (1.2 km) in diameter. It is thought to have been formed by the impact of an iron meteorite between 75 and 180 feet (25–60 m) in diameter, which created an explosion equivalent to between 4 and 60 megatons of TNT.

The impact excavates a crater in the shape of a bowl into which much of the damaged material falls. Consequently, much of the surface of the resulting crater is formed by shocked and fractured rock material, some of which is melted. In addition, smaller pieces of this pulverized material are ejected from the crater by the impact and fall as debris over a large surrounding area. In impacts as large as the one thought to have occurred at the K-T boundary, a central area of uplifted bedrock is surrounded by one or more rings of down-dropped material, as will be discussed in more detail in Chapter 6.

At many localities where the K-T boundary clay is present, it contains microscopic crystals of quartz that are fractured along planes that are oriented in several different directions. Many, including Bohor and the Alvarez group, feel that these multiplanar fracture patterns in the shocked quartz crystals represent unequivocal evidence in favor of the impact scenario. These proponents argue that the fractured texture could only have been created at the tremendous temperatures and pressures generated by an impact (between 10 and 40 GPa). In general, the number of different planar orientations of the fractures tends to increase with increasing shock forces. Some quartz crystals contained in exposures of the boundary clay have been claimed to have as many as seven different sets of planar fractures.

However, volcanic proponents, including Courtillot, Officer, Drake, and Hallam, noted that similar fracture patterns have been recognized in minerals of some rare volcanic rocks, like those from Toba, a volcano in Indonesia.[57] It must be pointed out, however, that, instead of having fractures oriented in more than one plane, the fracture pattern in these volcanic minerals is restricted to a single directional plane. Nonetheless, proponents of the volcanic extinction scenario maintain that theoretical modeling of the St. Helens eruption suggests that it might be possible to generate the necessary temperatures and pressures to create the multiplanar fracture pattern during large explosive volcanic eruptions. Until more definitive evidence is produced to demonstrate that multiplanar shocked quartz can be formed during volcanic eruptions, the presence of these crystals appears to be accepted by most geologists as strong geologic evidence for an impact.

Additionally, shocked quartz crystals have been analyzed through a technique called cathodo-luminescence by Michael Owen of St. Lawrence University and Mark Anders, one of Walter Alvarez's former students.[58] The technique has revealed that the shocked quartz present in the K-T boundary layers exhibits a range of luminescent colors not found in quartz erupted from volcanoes but more similar to the colors present in impact-altered quartz.

Along related lines, there are several different ways in which the atoms that make up quartz crystals, which contain one atom of silicon for every two atoms of oxygen (SiO_2), can be arranged into different mineralogical structures. Under intense pressures (beginning at 85 kilobars), the common crystal structure of quartz, in which the atoms are arranged in a four-sided tetrahedron, can be modified into a structure in which the atoms are arranged in an eight-sided octahedron. In the resulting eight-sided crystals, the atoms of silicon and oxygen are more densely packed together, so the mineral is given a new name—stishovite. Like iridium, stishovite is very rare in the Earth's crust. However, it is commonly found to be associated with debris in and around impact craters. Its presence has been documented by John McHone of Arizona State University and his colleagues in the boundary clay at a site in New Mexico.[59]

Finally, another mineral indicative of impact conditions, microscopic crystals of spinel with a high concentration of the atom nickel have also been found by Bohor in the boundary clay at Caravaca, Spain.[60] No known spinel crystals of volcanic origin have nickel content even approaching the concentration of those found in the boundary clay. However, the amounts and ratios of the elements iron, chromium, and nickel in these spinel crystals are similar to the concentrations of these atoms found in minerals from fractured rocks below impact craters located in Germany, France, and Russia.

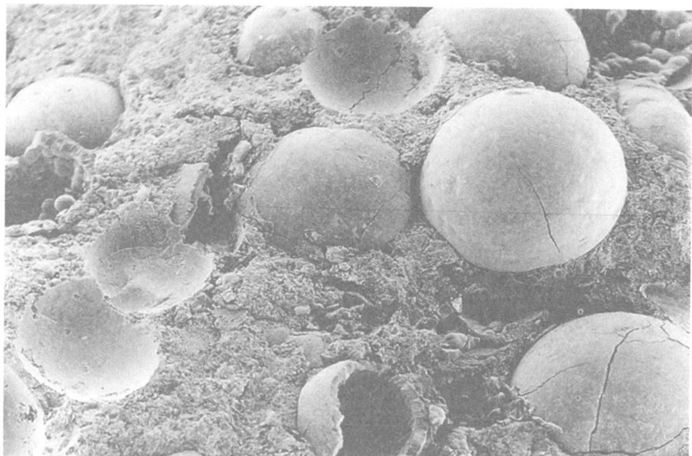

***(Left)* Scanning electron microscope photo of spinel crystals from the boundary clay at Caravaca, Spain, is thought by many scientists to represent evidence for impact.**

***(Right)* Microscopic spheres and globules, about 2 to 3 mm in diameter, thought by proponents of the impact hypothesis to represent the molten droplets of rock blasted out of the Chicxulub Crater by the impact.**

GLASSY SPHERES

Proponents of the impact scenario have noted the presence of extremely small spheres in the boundary clay at more than 60 sites.[61] Such spheres were discovered in the boundary clay at Caravaca, Spain, in 1981 by Jan Smit of the Free University in Amsterdam. More were separated from the boundary clay in Italy by Alessandro Montanari, another of Walter Alvarez's students at UC Berkeley.[62] They argue that these represent cooled drops of molten rock that were ejected into the atmosphere by the impact.

Critics, such as Officer, Hallam, Courtillot, and Drake, countered that these may instead represent fossilized algae, volcanic ejecta, or microscopic meteorites not associated with a large impact or organic remains. They noted that similar spheres were found in clay beds other than the boundary clay.[63]

More recently, some spheres with glassy cores contained within a rim of clay have been identified from the K-T boundary layer in Haiti. Several researchers, led by Haraldur Sigurdsson of the University of Rhode Island and Virgil Sharpton of the Lunar and Planetary Institute in Texas, have analyzed the chemical composition of the glass and reported that it was chemically distinct from the glass produced by volcanoes, especially in terms of its ratios of oxygen isotopes.[64] Additionally, the chemical composition of some of these spheres has been found to be very similar to the composition of a spot on the Earth's surface where an impact is thought to have occurred during the K-T transition, as will be discussed shortly.

SOOT

Chemical analyses have revealed a lot of soot and charcoal in the boundary clay at Denmark, New Zealand, and several other localities—concentrations of carbon thousands to tens of thousands of times higher than normal levels.[65] Some advocates of the impact theory suggest that this carbon enrichment came from massive global forest fires ignited by the fallout or the heat of the impact itself. Wendy Wolbach of the University of Chicago has been

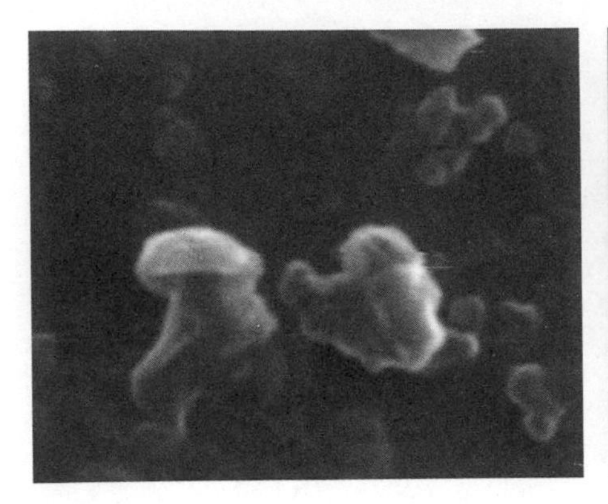
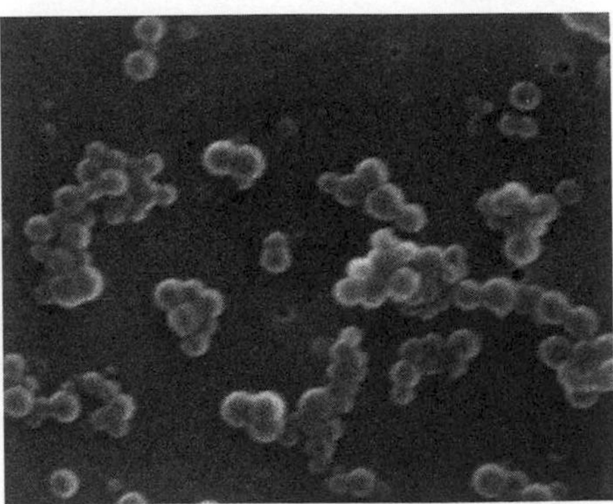
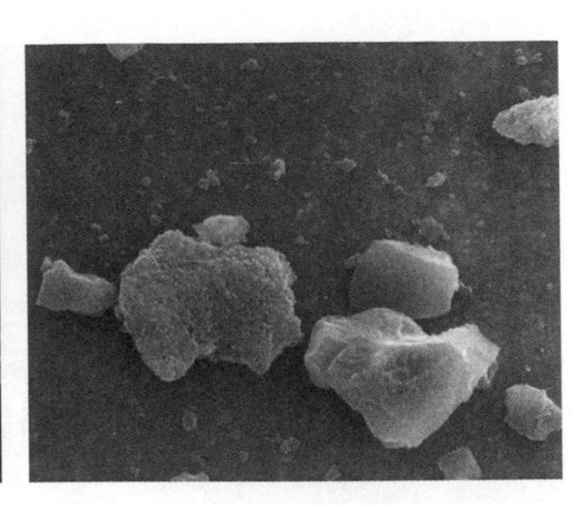

Photographs taken by a scanning electron microscope showing soot and carbon particles thought to have resulted from wildfires ignited by the impact and fallout.

instrumental in promoting and reporting on this evidence and hypothesis. Additionally, complex and elegantly constructed molecules called fullerines, comprised exclusively of carbon atoms bonded into a structure reminiscent of the structure found in geodesic domes and soccer balls, have been found in some exposures of the boundary clay.[66] These have also been interpreted by Dieter Heymann of Rice University and his collaborators to have formed as the result of wildfires caused by the impact and its fallout.

Overall, these impact-generated wildfires are estimated to have created 77 billion tons (7×10^{16} grams) of soot, which would have initially added to the opaque dust cloud that dramatically dropped temperatures. Eventually, these particles would have contributed to the "greenhouse effect" by raising global temperatures about 16° F (9° C).

However, David Archibald of San Diego State University argues that the soot could represent only the normal amount contributed by forest fires set by natural causes including volcanic activity.[67] If the clay layer took a long time to become deposited, as opposed to the several months or a year as called for by the single-impact hypothesis, soot from natural or volcanically triggered fires could be expected.

DIAMONDS IN THE BOUNDARY CLAY

Detailed analyses of the mineralogical composition of the iridium-bearing boundary clay in Alberta, Canada, during the early 1990s revealed some additional clues. After collecting less than 1 ounce (20 g) of the boundary clay at Knudsen's farm in Alberta, David Carlisle of Environment Canada and Dennis Braman from the Tyrrell Museum took it back to the lab and dissolved all the clay and coal out of the sample with hydrofluoric and hydrochloric acid.[68] After further chemical procedures and washing, they were left with a whitish sludge that based on X-ray analysis under a scanning electron microscope turned out to be 97 percent carbon, the only element that makes up the crystal structure of diamond. Of course, carbon is also the only component of graphite, the material that makes up pencil lead, so more definitive tests were needed to establish the identity of the mineral in the white residue. Using another X-ray procedure similar to the way one can shine sunlight through a prism and generate a rainbow of colors, the researchers discovered three lines in the material's spectrum that matched those which would be expected for diamonds. In addition, under a transmission electron microscope, the crystals in the white sludge were found to have an eight-sided octahedral shape—again, the same as would be expected for diamonds. So, they concluded that the material they had separated from the boundary clay was indeed minute diamond crystals.

Now, before you run out to stake your claim on the closest exposure of the K-T boundary clay, please realize that these diamonds are only nanometers (a millionth of a millimeter) in size. Nonetheless, when the concentration of the diamonds was compared to that of iridium, the results were very close to the concentration of diamonds to iridium found in meteorites called carbonaceous chondrites. Furthermore, the ratios of the carbon isotopes making up these diamonds were found to be very close to that of interstellar dust, suggesting that the diamonds were made from star dust and not made from material on the Earth.[69]

AMINO ACIDS

Two scientists at the NASA Ames Research Center, Kevin Zahnle and David Grinspoon, did additional geochemical analyses on the boundary clay from Denmark. These resulted in the identification of two amino acids common in the same kind of meteorites but very rare on Earth.[70] Critics pointed out, however, that these two amino acids were in fact present on Earth, so that a terrestrial origin could not be ruled out.[71] Further research on the boundary clay in Alberta by Carlisle revealed the presence of 18 amino acids present in meteorites but not found on Earth. Furthermore, the ratios of these amino acids to iridium are about the same as that found in meteorites, suggesting an extraterrestrial origin for the amino acids in the boundary layer.[72]

TSUNAMI DEPOSITS

In the late 1980s, Joanne Bourgeois from the University of Washington and her colleagues conducted field work along the Brazos River in Texas. That sequence of rocks, which contained an iridium anomaly and the characteristic paleontological signature of the K-T boundary, also contained an unusual layer of sandstone that appeared to represent a deposit generated by a "tidal wave," or more accurately, a tsunami.[73] The sandstone directly underlies the iridium anomaly, as well as the disappearance of marine fossils used to mark the end of the Cretaceous Period. Most of the layers of rock near the boundary are composed of mudstone that was deposited along the middle to the outer edge of the then-existing continental shelf in the Gulf of Mexico. The site is thought to have been about 62 miles (100 km) from the shoreline, and the water is thought to have been between 164 and 328 feet (50 and 100 m) deep.

There is only one layer of sandstone in the sequence near the boundary, and it is slightly over 3 feet (1 m) thick. The base of the sandstone layer contains a chaotic mixture of coarse sand grains jumbled together with shell fragments, fish teeth, pieces of fossilized wood, rounded chunks of limy mudstone, and angular hunks of mudstone. The latter are commonly as large as 2 inches (5 cm) across and occasionally as large as 3 feet (1 m) across.

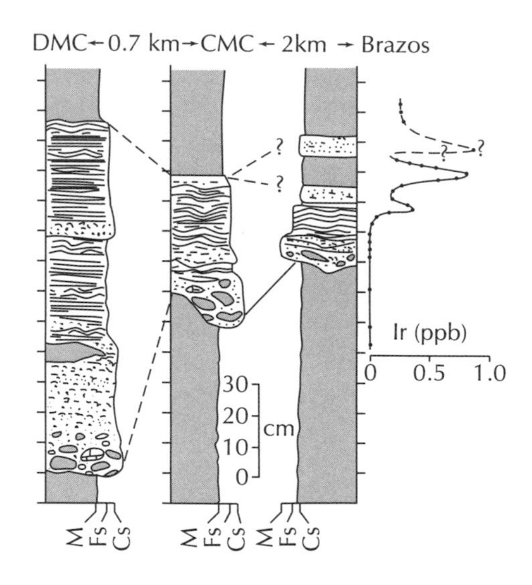

Diagram illustrating the layers of rock spanning the K-T boundary along the Brazos River in Texas. Note the large blocks at the base of each section. These are interpreted to have been ripped up by a tsunami generated by the Chicxulub impact.

Further up in the layer, smaller sand grains, silt, and mudstone show evidence of current action in the form of rippled bedding surfaces.

Calculations based on the size of the larger chunks of mudstone suggest that current velocities between 6 and 39 inches (15 and 100 cm) per second would have been required to rip the chunks of mudstone out of their original beds and redeposit them as part of the sandstone layer. The researchers concluded that a tsunami was the only kind of depositional event that could have been responsible for creating this deposit on the continental shelf. Because these deposits were located immediately below the iridium anomaly and the K-T paleontological signature, they proposed that the tsunami deposits were the result of gigantic waves generated by the impact. Based on the size of the ripped-up chunks and the estimated depth of the water, they suggested that the wave was 164 to 328 feet (50 to 100 m) high when it reached this area—about as high as a 16 to 33 story skyscraper. It is thought that the whole layer of sandstone may have settled out of the disturbed waters of the Gulf of Mexico in as little as one day.

Earlier estimates of the tsunamis generated near the point of impact suggested that they could have been as high as 13,000 feet (4000 m) if the impact had occurred in a deeper part of a major ocean basin. Mount Whitney, the highest peak in the continental United States, is slightly over 14,000 feet in elevation. That would have been as high as a stable water wave could have been. However, water in the Gulf region was much more shallow.

Scientists led by Florentin Maurrasse from Florida International University found similarly suspicious deposits in northeastern Mexico, Haiti, and Arkansas, where blocks as large as 16 feet (5 m) were incorporated into a tsunami deposit that totaled 65 feet (20 m) in thickness.[74] In sediments deposited in troughs on the bottom of the Caribbean near what is now Cuba, a chaotic layer as much as 1500 feet (457 m) in thickness contains jumbled blocks as large as 5 feet (1.5 m) across that were apparently swept off the lowlands. In addition to having sedimentary structures that were interpreted to be the result of deposition and scouring by gigantic waves, some of these rock layers also contained particles interpreted to represent altered droplets of melted rock from the impact.

Not all scientists agree with these interpretations, however. Some of the proponents of volcanically based extinction hypotheses, including Charles Officer, argue that these distinctive beds represent the normal deposition and scouring of large underwater debris slides, called turbidity currents. Officer argues that they began on the continental slope of the Caribbean and traveled on out into deeper parts of the basin.[75] Also, it has been argued that if these beds represent tsunami deposits they should contain no evidence of burrowing by bottom dwelling marine organisms.[76] Some, at least, appear to contain burrows that are cut off at the same stratigraphic level. This suggests that in these instances an episode of erosion removed the upper part of the burrows before sediment was once again deposited. One would not expect such gaps during the rapid deposition of sediment after a tsunami.

In the 1980s thousands of scientists from numerous disciplines joined the debate on one side or the other. As a result of this interdisciplinary cross-fertilization, a plethora of new ideas spun off from the original impact hypothesis. However, our ability to successfully test these ideas scientifically has been limited, as the following example illustrates.

PERIODIC EXTINCTIONS

Paleontologists David Raup of the Field Museum of Natural History and Jack Sepkoski of the University of Chicago proposed that episodes of mass extinction occur every 26 million years.[77] Their effort to sort through the fossil record of the last 500 million years was used to identify five or six episodes of "mass extinction" at the end of the Cambrian, Ordovician, Devonian, Permian, Triassic, and of course the Cretaceous Period. Walter Alvarez and Richard Muller of UC Berkeley reported that impact craters appeared to be produced in a 28.4 million year cycle, within the error margins of the 26 million year mass extinction cycle.[78]

There are three hypothesized causes that have been proposed to link such cycles to extraterrestrial impacts. One was put forward by Marc Davis of UC Berkeley, Piet Hut of the Institute for Advanced Study in Princeton, and Richard Muller. Their idea is that the Sun either has a dwarf twin star, appropriately christened Nemesis, orbiting around it.[79] Every 26 million years, this twin disturbs a cloud of comets in the distant reaches of the solar system called the Oort Cloud, sending a swarm of them speeding on collision courses toward the Earth. A similar proposal by Daniel Whitmire from the Univeristy of Southwestern Louisiana and his colleagues suggests that a yet to be discovered planet, also appropriately identified as Planet X, resides in the outer reaches of the solar system beyond Pluto and disturbs the Oort Cloud in the same way.[80] These ideas are very controversial in themselves, and astronomers are actively searching for such a star or planet—so far without success. A third idea is that, as the solar system periodically moves through the plane of our Milky Way Galaxy, dense gas clouds that reside near the galactic plane disturb the comets in the Oort Cloud, altering their orbits and sending them toward the Earth.[81] This idea was proposed by Michael Rampino of New York Univeristy and Richard Strothers of the NASA Goddard Institute for Space Studies, as well as by Richard Schwartz and Philip James of the University of Missouri.

Critics have offered several arguments to counter the proposed idea of periodic extinctions. One is that, since iridium concentration of comets is

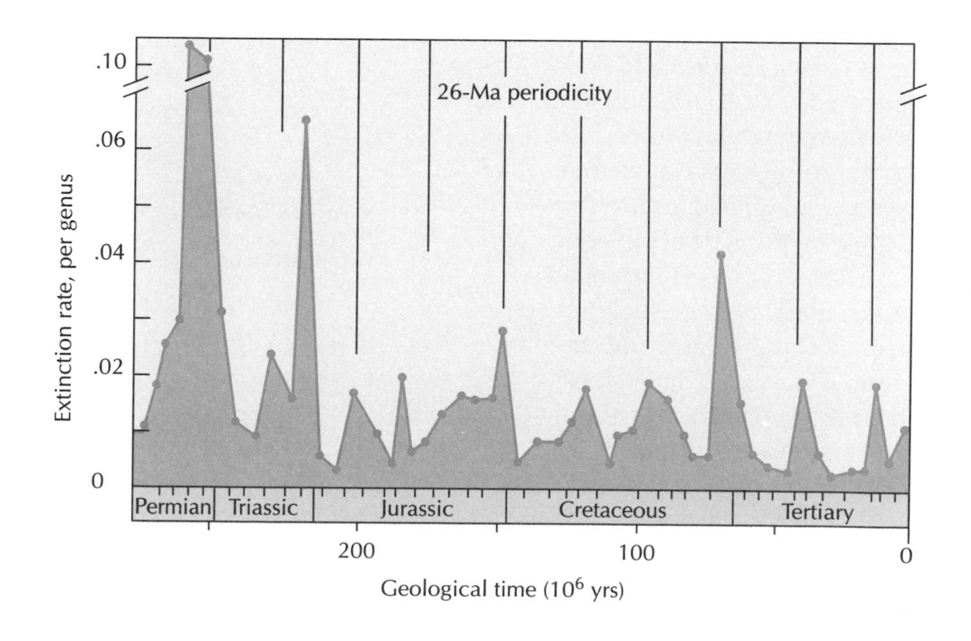

These peaks illustrate how major episodes of extinction in the fossil record might correlate with a periodicity of 26 million years.

about ten times less than that of iron meteorites, as many as 50 impacts may have been required to provide the observed amount of iridium in the boundary clay at Italy.[82] To the critics, like Anthony Hallam, this seems rather improbable. He suggests that a series of volcanic eruptions might be more plausible in explaining the high concentration of iridium.

Also, the body of statistics and paleontologic data used to argue for periodic extinctions has been vigorously questioned. Is the fossil record really complete enough to document all the episodes of mass extinction? Do we know the age of the extinctions precisely enough to demonstrate that they are happening every 26 million years? Many scientists do not believe that we do. For example, opponents such as Antoni Hoffman of the University of Warsaw[83] have argued that the ages of the period boundaries were arbitrarily chosen to fit the desired temporal pattern. In the process, some conflicting paleontologic evidence was simply eliminated from consideration in the studies purporting to document periodic extinction. The conclusion of some of these critics is that there is no periodicity among these extinction episodes. Instead, the extinction events are argued to exhibit a pattern of random occurrence. Hoffman's criticism was, in turn, rebutted by a response from J. J. Seposki and D. M. Raup, among others.[84] Other pessimists, including Stephen Stigler and Melissa Wagner from the University of Chicago, contend that the periodicity coming out of the recent research is simply a statistical artifact.[85] In essence, these skeptics argue, particular kinds of measurement errors can artificially generate a cyclic pattern in the data or force a false periodic signal to emerge from data that is not truly cyclic. Jennifer Kitchell and Daniel Pena of the University of Wisconsin argue that the spacing and magnitude of mass extinctions over the last 250 million years is best explained by a random model.[86]

Finally, paleontologists Colin Patterson and Andrew Smith of the British Museum of Natural History have noted the dependence of the periodic extinction results on the way new species, genera, and families have been established and described in the literature.[87] These paleontologists limited their study to animals with which they are especially familiar and utilized a method of systematically organizing organisms called cladistics (see Part II for a detailed explanation). Their results showed that only about 25 percent of the alleged extinctions of fish and echinoderms (sea urchins, starfish, etc.) were confirmed to have really happened. In fact, a few paleontologists, like John Briggs at the University of South Florida, have recently taken an even more extreme position. Briggs argues that, because of such problems in interpreting the fossil record, no mass extinctions have actually ever happened.[88] This represents a return to an extremely gradualistic view of the mode of evolution.

In all, a lot of geologic evidence has been cited by advocates of the impact hypothesis as documentation for an extraterrestrial impact, but most of these lines of evidence have been countered by proponents of the volcanic extinction hypothesis. In our role as geologic detectives, it would be nice to have *unequivocal* evidence in order for us to decide once and for all whether an impact really occurred during the K-T transition. To provide that, impact advocates would have to find the crater left by the impact itself—the "smoking gun" as it has been called. The search for that evidence is the subject of the next chapter.

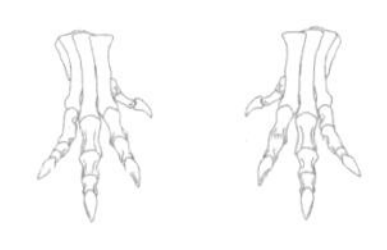

CHAPTER SIX

Direct Evidence of Catastrophe

Throughout the 1980s, the proponents of the impact scenario searched feverishly through geologic data and satellite imagery for the "smoking gun" or, more appropriately, for traces of the "festering wound" inflicted on the Earth's crust by such an impact.

The Earth has long been bombarded by extraterrestrial objects. Although the rate of impacts appears to have decreased in the last 4.5 billion years since the Earth first formed, extraterrestrial objects (large enough to pass through the atmosphere without burning up) continue to

Artist's conception of what the Chicxulub Crater might have looked like soon after it was formed by the impact of a comet or asteroid about 65 million years ago.

hit the Earth with tremendous force. Scars from some recent impacts remain clearly visible. For example, an iron meteorite between 80 and 200 feet (25 and 60 m) across is believed to have been responsible for Arizona's Meteor Crater. It formed when a meteorite hit the Earth between 50,000 and 25,000 years ago, creating a bowl-shaped depression about 3/4 mile (1.2 km) across.[1] The impact released between 10^{16} and 10^{17} joules of kinetic energy—about the same amount as would be released by the explosion of 4 to 60 megatons of dynamite. Estimates of the frequency of crater impact suggest that a meteor large enough to leave a crater 6 miles (10 km) across hits the Earth about once every 100,000 years.[2]

Even in the twentieth century, an apparent explosion of a comet or meteorite devastated a large area of Siberia. This near-impact, called the Tunguska event, leveled 1,000 square miles (2,600 km^2) of forested land on June 30, 1908.[3] The devastated region was equivalent to a square about 32 miles on each side. The explosion, which released energy equivalent to a 15-megaton nuclear bomb, was heard as far away as Moscow.[4] The Tunguska event is thought to have been the result of a object that exploded between 4 and 6 miles (6 and 10 km) above the Siberian landscape. Debate has ensued as to whether the explosion was caused by a comet between 295 and 623 feet (90 to 190 m) across, or by a carbon-rich asteroid between 164 and 328 feet (50 to 100 m) across. It is estimated that the body was traveling between 7.5 and 12.5 miles per second (12 and 20 km/s), which is equivalent to between 27,000 and 45,000 miles per hour. The impact would have generated shock layer temperatures of somewhat less than 36,000° F (20,000° C). In 1956, one student of that event, M. W. De Laubenfels, went so far as to suggest that a similar but larger impact could have been responsible for the extinction of dinosaurs.[5]

But these relatively recent events are dwarfed by what was proposed to have happened at the end of the Cretaceous. An asteroid or comet large enough to have caused the Cretaceous-Tertiary (K-T) extinctions would have left a truly enormous crater. One early estimate suggested that the collision would have generated a crater about 31 miles (50 km) deep that extended across an area of almost 5,800 square miles (15,000 km^2).[6] This would represent a crater with a diameter of almost 90 miles (140 km). But the question remained, where is that crater?

A K-T IMPACT CRATER

For several years, no one could find a crater of the right size and age. Many potential explanations were offered by impact proponents, including Walter Alvarez and Frank Asaro.[7] They pointed out that the crater might have been eroded away by subsequent geologic weathering or even destroyed by the recycling of crust at subduction zones. Alternatively, it might now be too deeply buried under subsequently deposited sediment to be easily recognizable. The opposition favoring volcanic extinction scenarios countered that we should have found at least one crater, especially if multiple impacts had occurred, as some impact advocates had begun to suggest by the mid-1980s.

One proposed candidate for the K-T impact crater was the Manson Crater, now buried under the ground by sediments that form the surface of the land in Iowa. However, it is only about 22 miles (35 km) across. This is much too small to have caused the observed global thickness of debris and iridium concentration found in the boundary layer.[8] One sure test of whether the Manson Crater was a legitimate candidate would be to date the rock that had been melted at the time of its impact. Such a test should yield an age of about 65 million years. Initial radiosiotopic evidence developed by Michael Kunk of the U.S. Geological Survey suggested that it was about the right age—65.7 million years old. In 1993, however, further tests conducted by Kunk in conjunction with Glen Izett and others from the U.S. Geological Survey established that the Manson Crater is about 74 million years old—much too old to have had any role in the K-T extinctions.[9]

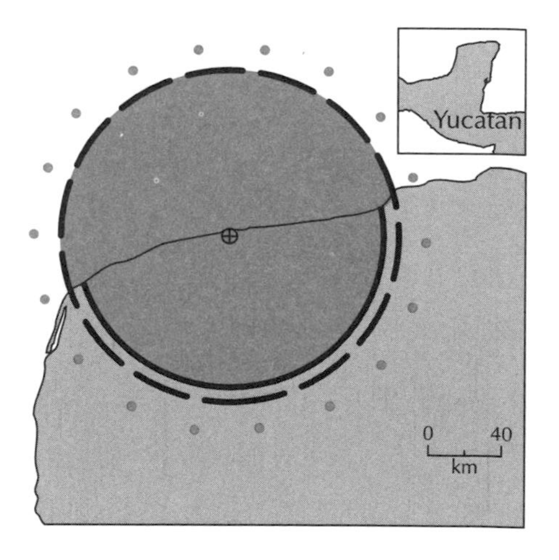

The Chicxulub crater occupies a corner of the Yucatan Penninsula in Mexico and forms some of the bed for the adjacent Gulf of Mexico. The solid semicircle represents a ring of sinkholes found on the ground surface. The dashed circle represents the approximate location of a negative gravity anomaly. The dotted circle represents the approximate outer limit of a concentric positive magnetic anomaly.

Another serious candidate was actually discovered as the result of exploration for oil in 1981.[10] However, little attention was paid to the announcement by Glen Penfield and his colleagues because it was before the search for the K-T impact crater had actually begun. In an ironic twist of fate, most of the drilling-core samples that documented the geologic evidence at this site were destroyed by a fire in the warehouse where they were stored. Consequently, the crater's possible existence and relationship to the scenarios involving the K-T extinctions only came to light in the early 1990s. Studies by Alan Hildebrand of the Geological Survey of Canada and his colleagues, as well as research by Kevin Pope of Geo Eco Arc Research and his colleagues, brought this geologic structure to the attention of the scientific community.[11]

The site is located near the town of Chicxulub. It lies partly on the northwest corner of the Yucatan Peninsula and extends into the adjacent Gulf of Mexico. On land, the form of the buried crater can be seen, to some degree, on topographic maps and aerial photos.[12] They reveal a semicircular ring of sink holes that correlate well in position with the margin of the crater that was identified at first by anomalous readings of gravity. These sink holes range from about 164 to 1640 feet (50 to 500 m) in diameter and are from 6 to almost 400 feet (2 to 120 m) deep. Apparently, the edge of the crater formed a barrier to the later migration of groundwater under the surface of the Earth. This caused an increased flow of subsurface water, which dissolved the limestone layers underneath the ground. Then, a collapse of the limestone around the boundary of the crater formed the sink holes.

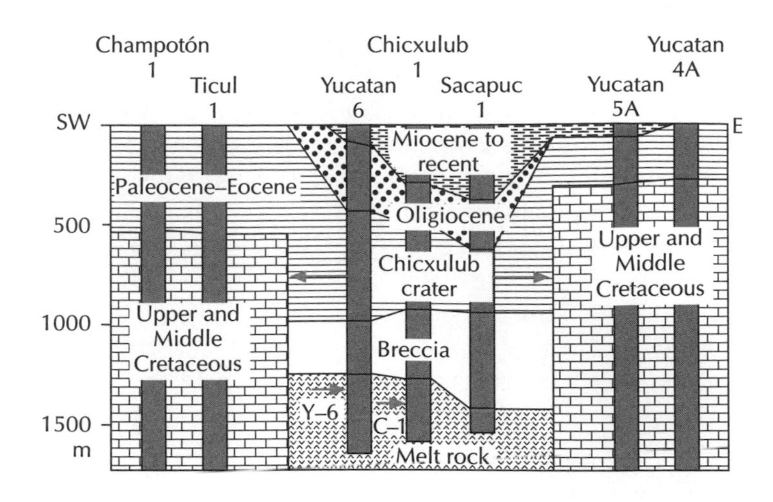

Generalized cross section of the Chicxulub crater illustrating both the rock units that form the crater and its contents, as well as the younger units on top that have buried it since it was formed.

Chicxulub is truly enormous, whether you believe scientists who advocate a mind-boggling diameter of about 175 miles (280 km) or those who propose a more modest, but still daunting, figure of about 110 miles (180 km). The conflicting dimensions were reported in different studies published in 1993 and 1995. Virgil Sharpton of the Lunar Planetary Institute and his co-authors published the earlier study, which argued for the larger diameter.[13] Alan Hildebrand and his co-authors published the later study.[14] In both papers, the estimated diameter of the crater was based on interpretations derived from gravity measurements taken at the site.

Although it would seem to be an easy job to determine the crater's size by measuring the diameter of the crater rim, the job is made more difficult by the fact that the crater's full topography is no longer clearly expressed at the surface of the Earth. Since the actual crater formed, it has been buried by between 1,000 and 3,300 feet (300 and 1,000 m) of sediments. But the gravity measurements allow geophysicists to "see" through the overlying sediments deposited on the land and on the seafloor to interpret the underlying structure of the crater.

The process of crater formation is fairly well understood as the result of ballistic experiments in the laboratory and observations of craters on the Earth, moon, and planets.[15,16] Upon impact, shock waves compress and accelerate the target rocks downward and outward to velocities of a couple of miles per second. Then rebound changes the direction of movement to upward and outward for target material near the surface. These two phases of movement excavate a transient crater by moving target material downward and out from the point of impact, then upward and out of the crater, to create a layer of debris ejected outside the crater in the form of fallout. Calculations indicate that the great majority of material in the impacting object is blasted upward at high velocities as ejecta. In addition, a mass of target rock equal to about 10 to 100 times the mass of the meteorite is exploded out of the crater along similar paths.

Relatively large craters on Earth have a central peak surrounded by a series of concentric terraces. These formed as large, unstable blocks of surrounding rock, propelled by gravity, slid down into the crater's cavity soon after it formed. The floor of the crater is formed by a layer of rock that was melted by the impact. Some of this melted material was blasted out of the crater. In terms of the Chicxulub impact scenario, this material is thought to have formed the glassy, spherical, and globular droplets of melted impact debris found in some boundary clays. These are called tektites. The layer of melted rock on the crater floor overlies a layer of fractured rock including fragments of the target rock. Some of the fractured target rock is thought to have been blasted into orbit as part of the dust cloud. Smaller fractured fragments eventually settled out of the atmosphere as the badly fractured crystals of quartz and other minerals discussed in Chapter 5.

Within the largest craters on Earth the large central peak is represented instead by a tall ring of mountains. These are formed by uplifting movements after the initial downward and outward forces reverse in the center of the crater to generate upward and inward motions as a result of rebound and the collapse of the crater rim. The process seems to be kind of like watching a slow-motion movie of a drop of water hitting the surface of a bowl of water, but obviously, water is much less viscous. Such large craters are more easily seen on the Moon where erosion and subsequent deposition have been minimal due to the lack of rain and wind. An example is the crater that forms the Orientale Basin, which has five stupendous concentric rings that range in diameter from 200 to 800 miles (320 to 1,300 km).

Some of the crater rings were generated by gravitational collapse after an initial crater, shaped like a hemisphere, was created by the impact. The depth of the crater continues to grow until the pull of gravity becomes too weak to pull the material forming the floor further down. The resulting hole is termed the transient crater, and it has a ratio of depth/diameter of about 1 : 3. The fractured walls of the transient crater are not stable because their slopes are too steep. Consequently, they collapse under the force of gravity to form additional rings, as well as the layer of fractured material in the bottom of the crater. So the game is to identify the diameter of transient crater so that the force of the impact can be estimated.

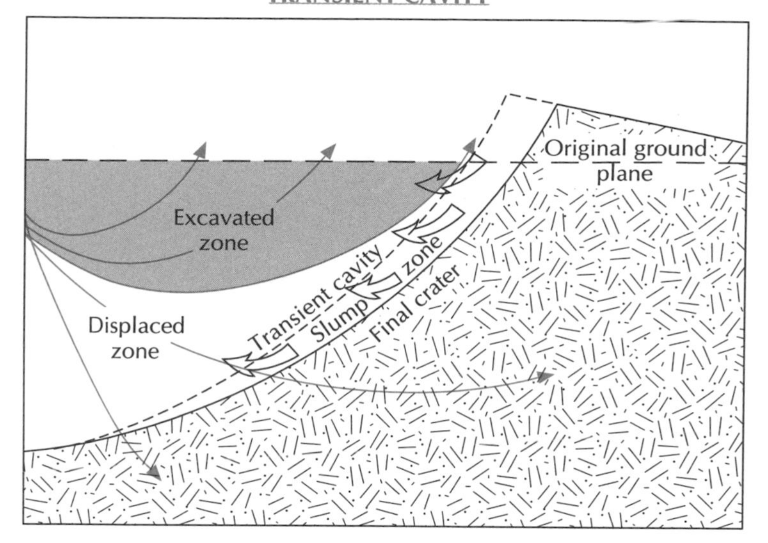

This schematic diagram illustrates how the transient crater is formed by the impact, as well as the subsequent rebound and collapse of the crater rim.

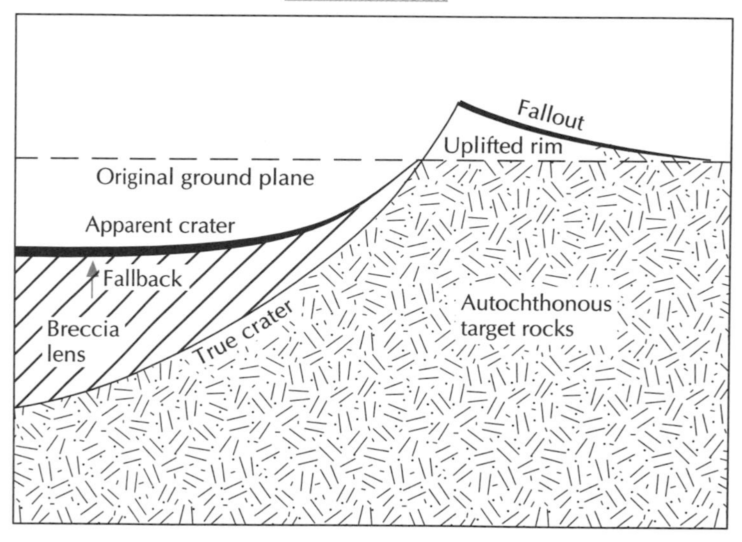

The final form of the crater takes shape after the collapse of the unstable walls of the transient crater.

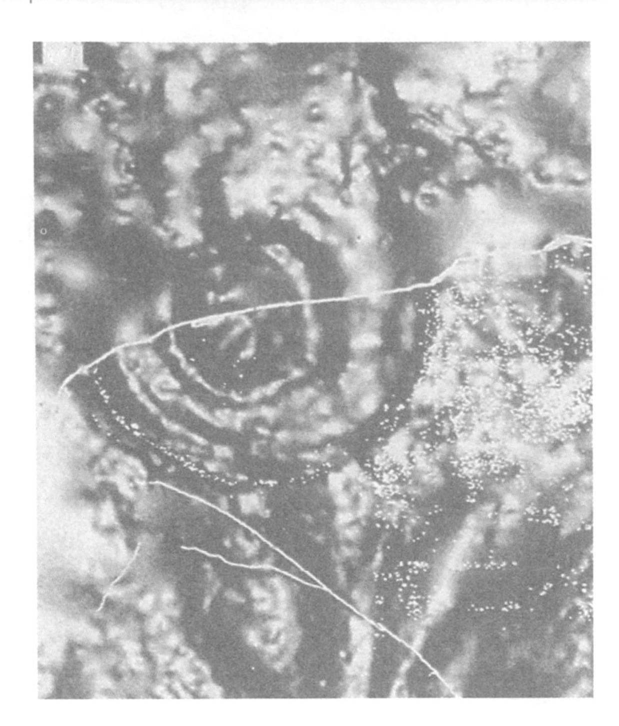

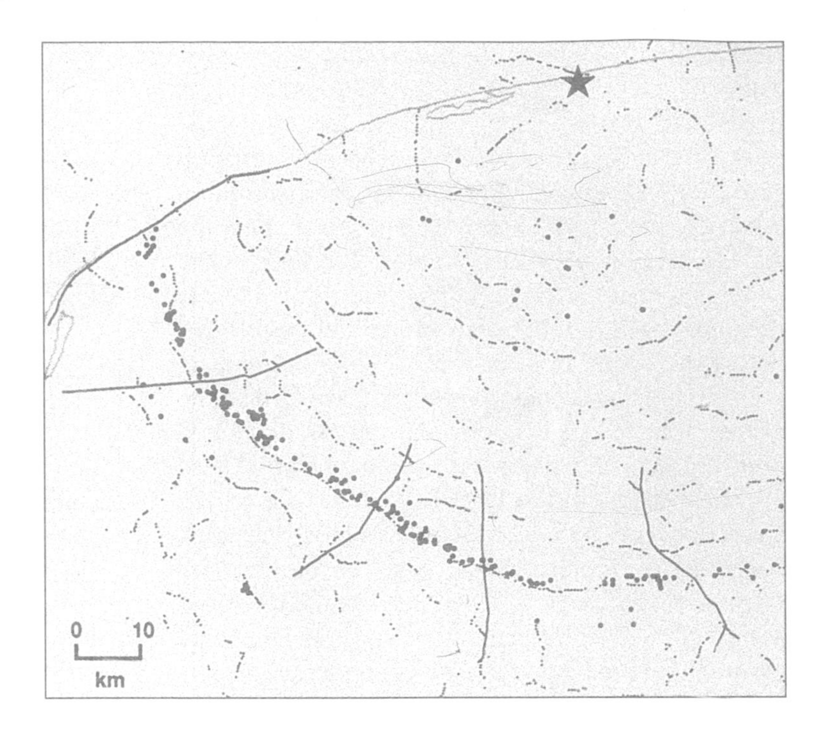

The image on the left has been developed from studies analyzing the force of gravity around the impact site at Chicxulub. The image shows the rough outline of the crater and its concentic rigns. The map on the right documents the curved line of sinkholes (green dots) delineating the circular outline of the crater.

On gravity surveys, the Chicxulub crater shows up as a basin almost 125 miles (200 km) in diameter that contains two ring-shaped structures with diameters of 65 and 96 miles (104 and 154 km). One of these may correspond to the central ring of mountains seen in large lunar craters. The 1993 study by Sharpton and his colleagues argued for the presence of a fourth ring with a diameter of about 175 miles (280 km). This suggested to those researchers that the transient crater was about 106 miles (170 km) across.[17] However, using the same gravitational data supplemented with other new data and a slightly different method of analysis, the 1995 study by Hildebrand and his colleagues failed to confirm the presence of the outside ring. This later study concluded that the 125-mile-diameter ring represented the rim of the collapsed crater and that the transient crater was only 50 to 56 miles (80 to 90 km) in diameter.[18]

If the interpretation of the 1993 Sharpton paper is correct, the estimated force of the impact may have been one of the largest in the last 4 billion years within the inner part of the solar system. The impact—judged to be equivalent to simultaneously detonating between 1,000 and 10,000 times the number of nuclear weapons present in all the world's arsenals—would have excavated a crater about 35 miles (56 km) deep. This results from the calculating the depth based on the depth to diameter ratio of about 1 : 3. If the 1995 Hildebrand study is correct, the depth of the crater would have been about 18 miles (28 km). No matter which estimate is considered valid, the impact was obviously an event of staggering proportions.

But what geologic and paleontologic evidence is there to suggest that Chicxulub is, in fact, the crater left by the impact at the K-T boundary? One test would be to date the meltrock in the crater radioisotopically to see if it yields the same age as the tektites preserved in K-T boundary sections in other areas.

This is what Carl Swisher and his team of colleagues did. Their radioisotopic dating of the melted rock within the Chicxulub crater formed at the time of the impact has established its age at 64.98 million ± 50,000 years.[19] The microscopic glassy tektites at the boundary found in Haiti and at Arroyo El Mimbral, Mexico have been dated by the same laboratory to be 65.01 million years ± 80,000 years and 65.07 million ± 10,000 years old. These radioisotopic ages represent strong geologic evidence that these tektites, found at the K-T boundary where the change in microscopic marine organisms occurs, were at least formed at the same time as the Chicxulub impact.

The geographic location of this impact crater near the coastline of the ancient Gulf of Mexico is also significant. Some chemical analyses of boundary clay suggested that the impact had occurred in the crust of the ocean because the tektites and other boundary material were of similar composition. One such study was conducted by Don DePaolo and Frank Kyte of the University of California at Los Angeles and their colleagues.[20] Other analyses by Michael Owens and Mark Anders suggested that the impact had occurred in the crust of a continent.[21] Since the Chicxulub crater contained both continental and marine rocks, these seemingly contradictory results could be explained by the fact that the impact debris contained remnants of both.

More definitive geologic evidence comes from geochemical "fingerprinting"—a method whose name sounds especially appropriate for solving mysteries related to a scientific detective story. As mentioned in Chapter 5, geochemical fingerprinting involves analyzing the chemical composition of elements in different rock material at different sites to see if they might represent rocks formed by the same event. In this case, scientists compared the chemical composition of tektites from the boundary layer near Beloc, Haiti, with the chemical composition of the rocks in the crater at Chicxulub.

Haraldur Sigurdsson and his colleagues found that the boundary layer at Beloc contains two principal kinds of spherules thought to represent tektites that were blasted out of the crater during the impact.[22] One is composed of black glass and the other is composed of black glass covered by a coating of yellow glass. The black glass is rich in silica and has a chemical composition similar to that of a particular kind of volcanic rock called andesite, which is commonly erupted along the edges of continental plates. More impressively, the chemical composition of this black glass is almost exactly like that of some andesites found in Mexico, and andesite is present in the Chicxulub area.

Analyses of the yellow glass coatings, however, reveal a distinctly different chemistry. The yellow glass is depleted in rare earth elements, but it is enriched in atoms of calcium and magnesium. To try and recreate the composition of the yellow glass, Sigurdsson and a second group of colleagues melted minerals such as gypsum and anhydrite—common in the magnesium-rich, limy sediments along the Mexican coast near Chicxulub—with andesite.[23] The temperature in the experimental melting tubes made of platinum was raised to between 2200 and 2500° F (1200 and 1400° C). The result, upon cooling, was the formation of a yellow glass enriched in calcium, similar to that found coating the Haitian tektites.

In 1993, another piece of mineralogical "fingerprinting" evidence was discovered by Thomas Krogh of the Royal Ontario Museum and his colleagues. In the Raton Basin of Colorado, the boundary layer was found to contain microscopic crystals of the mineral zircon weighing from 1 to 4 one-millionths of a gram (1 lb = 454 g).[24] Zircon is very durable and often contains in its crystal structure atoms of uranium that naturally decay to atoms of lead. Zircon can, therefore, be used for radioisotopic dating. Such age analyses on these crystals established that they had originally been formed 544.5 ± 4.7 million years ago, then reheated 65 million years ago. At Chicxulub, fragmented target rock from the crater created by the impact also contains crystals of zircon. When they were analyzed, they were found to have an original age of formation of 544 ± 5 million years. Similar age relationships have been found for microscopic zircon crystals separated from the boundary clay in Haiti. What makes this evidence compelling is that 545-million-year-old rocks are very rare in North America. Consequently, the presence of such zircon crystals at the impact site and at these two other geographically widespread localities is very difficult to explain without invoking the Chicxulub impact and its globally distributed fallout layer.

OTHER POTENTIAL K-T IMPACT CRATERS

There is now no shortage of other large impact craters that potentially could have been created at the same time as the Chicxulub crater. They range in location from Siberia, to the Black Sea region, to northern Africa, to Alaska.[25] One example is the 60-mile-diameter (96 km) Popigay crater in eastern Siberia. Based on its size in relation to the Chicxulub crater, only about one-tenth the amount of material is thought to have been ejected from this smaller crater. Yet, based on modeling, the Popigay impact alone would have generated a dust cloud thick enough to cut off light from the Sun to the Earth for three months. However, radioisotopic dates indicate that it is around 39 ± 9 million years old. It is much too young to have been involved in the K-T extinctions.

Some scientists have argued that several potential K-T impact craters tend to line up on a single great arc that extends across the surface of the Earth.[26] This implies that they may represent the break-up of a single large extraterrestrial body into several smaller bodies that impacted in a machine gun–like barrage. One example of such a barrage was witnessed in the summer of 1994 when the comet Shoemaker-Levy broke up into several pieces before pummeling Jupiter.

At this point, however, such claims that more than one K-T impact occurred represent a significant leap of faith. The age of these other craters is not tightly constrained by published radioisotopic dating in the way that the age of the Chicxulub crater is.

Although not all proponents of the volcanic extinction scenario would probably agree, it appears to us that there is now good geologic evidence to demonstrate that an extraterrestrial impact of huge proportions did occur at the end of the Cretaceous. But can effects of the potential killing mechanisms from the impact at Chicxulub be distinguished from the potential killing mechanisms of the volcanic activity? To see, we must closely examine clues present in the fossil record spanning the K-T boundary, paying special attention to the kinds of organisms that perished and those that survived. This exercise is the subject of the next chapter.

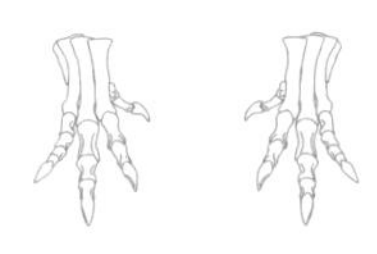

CHAPTER SEVEN

Patterns of Extinction and Survival

By the mid-1980s, the impact hypothesis had secured a strong foothold in both scientific circles and in the public press. Proponents of the volcanic hypothesis were quick to vociferously raise their objections. Throughout this debate, however, there was a rather counterintuitive aspect to the problem of what decimated the dinosaurs and other groups at the Cretaceous-Tertiary (K-T) boundary—how did the survivors survive?

Soon after the impact scenario was proposed, an "arms race" to model its effects and generate enough potentially lethal mechanisms to do the job began. Indeed, that list of alleged killing mechanisms was truly incredible: an earthquake 1 million times larger than any recorded in human history, a global inferno following the impact, a subsequent period of darkness lasting months, an interval of months on the continents when temperatures dropped below freezing, episodes of searing acid rain, tsunamis hundreds of meters high. Now, add to those presumably lethal effects the enhancements contributed by the volcanic scenarios and one cannot help but wonder how anything survived. As Anthony Hallam noted in 1989:

> Even granted the evidence for impact, it is still by no means clear how many of the end-Cretaceous extinctions relate to it, and no really plausible, exclusively impact-based, 'killing scenario' has yet been put forward. Whether involving dust clouds blocking out sunlight, dramatic rise or fall of temperature, acid rain on a massive scale or spectacular wildfires on the continents, they are all too drastic in their environmental effects to account for the selectivity of the extinctions and the high rate of survival of many groups.[1]

Hallam and several other scientists began to focus on this important point. If we were to gain a better understanding of how the K-T extinctions occurred, we would have to come to grips with the pattern of extinction and survival among different organisms. We would have to try to understand why some organisms perished while others survived—not just locally, but globally.

Explaining patterns of survival and extinction requires not only fairly complete fossil records but also an understanding of the physiology of these ancient organisms. We must specify which "killing mechanism" eradicated a particular group of animals or plants. As the number of potential killing mechanisms increases and the scenario for extinction becomes more complex, it becomes that much more difficult to test and pinpoint which mechanism, or combination of mechanisms, was responsible for a particular group's extinction. Further, since the impact and volcanism could have generated many of the same "killing mechanisms" (acid rain, short-term cooling, long-term greenhouse heating), it becomes even more difficult to pinpoint which of these events was responsible.

EXTINCTION AND SURVIVAL IN THE OCEANS

Despite the complexity of the problem, a number of ideas regarding the selectivity of extinctions have been put forward. In marine environments, for example, many forms of microscopic organisms that live near the ocean's surface, such as planktonic Foraminifera and Coccolithophora, were apparently decimated at the boundary. Jan Smit has presented documentation on marine sections from around the world.[2]

Perhaps the organisms perished because of acid rain. The more acidic waters could have dissolved their calcium carbonate shells.[3] The shells of clams and snails are made of calcium carbonate and if you drop acid on these shells, it triggers a bubbling reaction. This reaction represents the acid dissolving the shell material. It takes a lot of strong acid to dissolve a large clam shell. But because the shells of microscopic plankton are so much smaller, it is that much easier to dissolve their shells and kill the tiny organisms within.

These microscopic organisms formed the foundation of the food chain in ancient oceans, just as they do today. Their extinction could have set off an ecological chain reaction. Organisms that fed on these kinds of plankton would have been the most vulnerable, especially those organisms that lived in the tropics.

Jennifer Kitchell and her colleagues suggested that those kinds of plankton that have a method of reproduction that involves a dormant spore stage, such as dinoflagellates and diatoms, survived at higher rates than the forams and coccoliths which do not.[4] However, these statements do not really conclusively resolve the issue of which "killing mechanism" from which event was responsible.

Some workers, such as Gerta Keller of Princeton University and Norman MacLeod of the British Natural History Museum, have argued that many of the sequences of rock layers used by proponents of impact-generated extinction do not contain a complete fossil record of the events during the K-T transition.[5] Consequently, what appears to be an abrupt and catastrophic extinction in the rock record may have actually happened more gradually. To give you a sense of how contentious this issue is, we need only take a look at the debate swirling around one relatively complete section at El Kef, Tunisia. Smit argues that almost all the Late Cretaceous forams disappeared at the K-T boundary, whereas Keller argues that only about one-third did.[6] Independent tests are now underway to try and resolve the dispute. However, disagreements still exist about the species identification of many specimens. Furthermore, arguments abound about whether or not some of these microscopic fossils were eroded out of Cretaceous sediments and then redeposited in younger Tertiary sediments.[7] This process of reworking greatly complicates our attempts to interpret the sequence and pace of events in the fossil record.

Among other marine organisms, estimates for extinction of the nautilus-like ammonites reach 100 percent, whereas clams exhibit about a 55 percent extinction, and snails about 35 percent.[8] Especially with ammonites, the issues of reworking and the frequency at which specimens have been found at different levels in the sequences of rock layers have been used to raise questions about whether these organisms actually went extinct before the K-T boundary. However, it is now generally accepted that ammonite extinction was part of the K-T event.

ONE SYNTHETIC EXPLANATION FOR THE VICTIMS AND SURVIVORS

There have been few attempts to generate an all-encompassing explanation for the patterns of survival and extinction on both land and sea. One such attempt was put forward by Anthony Hallam.[9] Hallam argued that the selectivity of the extinctions could be most logically explained by a combination of climatic and other environmental consequences related to falling sea levels and volcanism. He even attempted to attribute the extinction of particular groups to particular killing mechanisms.

First, Hallam compiled his list of victims. On land, dinosaurs perished, as did pterosaurs—a group of flying reptiles. Some groups of mammals, such as the rodent-like multituberculates and the marsupials, suffered high levels of extinction, along with some groups of land plants. Yet many other land plants, freshwater fish, amphibians, snakes, and placental mammals were not severely affected. In the oceans the Late Cretaceous extinctions completely eliminated ichthyosaurs, plesiosaurs, and mosasaurs—three groups of marine reptiles. Ammonites also became extinct. The microscopic marine organisms that lived near the surface of the ocean, such as coccoliths and forams, were almost completely decimated, as were relatives of modern squid called belemnites. Numerous groups of corals, clams, snails, and echinoderms (starfish, sea urchins, and their relatives) were hard hit, along with bottom-dwelling forams and the microscopic radiolarians. But many kinds of marine invertebrates and microscopic organisms, called dinoflagellates, were not badly affected.

Examining this list of victims, Hallam correctly tried to use the environmental limits of the closest living relatives of these victims as a guide for interpreting what killing mechanism was responsible for their extinction. Much of his discussion of extinctions was based on the acidity of the seawater and freshwater. Scientists use the pH scale to standardize these measurements. A pH of 1 is very acidic. A pH of 7 is perfectly balanced between acidic and alkaline. A pH of 7 is about the value for normal drinking water. A pH of 14 is highly alkaline. Slightly alkaline water is commonly called "hard" water, whereas slightly acidic water is commonly called "soft" water.

Modern surface-dwelling forams have difficulty living in seawater with acid levels below a pH of 7.6–7.8, and extant coccoliths die off in water with levels below 7.0–7.3. Accordingly, Hallam reasoned that volcanic eruptions, such as those that formed the Deccan Traps, released almost 21 trillion tons (19 trillion metric tons) of hydrogen sulfate (H_2SO_4) and 298 billion tons (270 billion metric tons) of hydrochloric acid (HCl). As a result, tremendous amounts of acid rain would be dumped into the surface waters of the world's oceans—14 times the amount of acid rain generated by the burning of fossil fuels in Europe and the United States in 1976. In turn, the acidity levels of the surface waters in the ocean would drop to pH levels of 7.4 or lower, resulting in massive extinctions of planktonic forams and coccoliths. Since acidity levels in deeper levels of the ocean were less affected, the bottom-dwelling forams were reasoned to have fared much better. Similarly, living dinoflagellates can survive in water with pH levels as low as 4.0–5.0, and this group of microorganisms was one of the few to sail through the K-T crisis virtually unscathed.

In freshwater environments, the effects of acid rain were also implicated in the selective extinction of fishes. Hallam noted that in modern Canadian lakes, different kinds of fish can survive different levels of acidity. Walleye and lake trout, for example, disappear when pH levels reach 5.8–5.2, but yellow perch and lake herring can survive at a pH of less than 4.7. If freshwater fish living during the K-T transition were similarly diverse in their tolerance of acidity, selectivity in extinction could be expected. Similarly, in the Canadian forests, red cedar and sugar maple trees are more tolerant of acid rain than white pine and white birch. Therefore, acid rain might be responsible for the selectivity seen in land plant extinctions at the end of the Cretaceous.

Many of the larger marine invertebrates, such as corals, inoceramid clams, rudistid clams, some snails, and various echinoderms, that went extinct lived in reefs and other near-shore environments. Hallam argued that these shallow-water animals were essentially left high and dry at the end of the Cretaceous by the drop in sea level associated with the retreat of the shallow continental seaways.

In attributing the extinction of dinosaurs to the increased climatic extremes that resulted from this regression of the seaways, Hallam also invoked an increase in ultraviolet radiation. This effect, he argued, was the result of the tremendous amount of hydrochloric acid injected into the stratosphere by the K-T volcanism, which depleted the ozone layer in the atmosphere. He asserted that the deleterious effects of this increased radiation would be especially acute on terrestrial animals that were unprotected or unable to hide during either their adult or reproductive stages. However, he was unable to explain precisely why dinosaurs were so affected while other terrestrial vertebrates, such as placental mammals, birds, and freshwater amphibians, were not. This is especially puzzling given the recent reports that an increase in ultraviolet radiation, attributed to decreased ozone levels in our modern atmosphere, may be wreaking havoc on modern species of frogs and other amphibians.

Hallam's was a noble attempt to solve the murder mystery. But once again, given the fact that many of the same killing mechanisms that Hallam invoked as the result of volcanism have also been associated with the impact scenario, it becomes impossible to tease apart which event was responsible.

In addition, the role of acid rain in these extinctions has been called into question by Steven D'Hondt of the University of Rhode Island and his colleagues. They argue that acid rain is not a viable explanation for the selective extinction of plankton. Their calculations suggest that acid yields from the Chicxulub impact were not high enough to significantly alter the acidity of the ocean's surface waters. Further, living plankton with calcium carbonate shells exhibit a range of sensitivity to acid, and plankton with shells made of other materials are often acid sensitive.[10] D'Hondt also raised the point that, based on his calculations, about 20 percent of the entire sequence of Deccan eruptions would have to have occurred in less than 20 years in order to have significantly affected the pH of the ocean's surface waters. This is because a 20-year period is needed in order to thoroughly mix the surface waters with deeper ocean waters. Such mixing would have diluted the acid in the surface waters.[11]

WERE EXTINCTIONS MORE SEVERE NEAR THE EQUATOR THAN AT THE POLES?

In the original formulation of the impact hypothesis, a dust cloud that cut off photosynthesis was implicated in the extinction of many species of plants. The role of seed, spore, and pollen reproduction has also been invoked to explain the pattern of extinction and survival among plants.[12] The idea being that species that had their reproductive structures sheltered underneath the soil would have been more protected from extinction. However, the big picture of plant extinction and survival on a global scale documents some interesting variations and complications.

Kirk Johnson of the Denver Museum of Natural History, Leo Hickey of Yale University, Doug Nichols of the U.S. Geological Survey, and Carol Hotton formerly of the University of California at Davis have conducted studies on species of fossil pollen and fossil leaves that occur near the K-T boundary at midlatitude sites in the Western Interior of North America that reveal relatively high extinction rates for plants at midlatitude sites (roughly half way between the equator and the poles).[13] As many as 30 to 50 percent of the fossil pollen species drop out at or near the boundary clay marked by the iridium anomaly. Among other kinds of fossil leaves, the extinction rate is even higher (see fossil leaf photos in Chapter 3). Almost 80 percent of the latest Cretaceous leaf flora disappeared at the boundary. Most researchers conclude that this pattern of change is consistent with a sudden and catastrophic event such as the Chicxulub impact.

One specialist on fossil plants, Jack Wolfe of the U.S. Geological Survey, has even gone so far as to conclude that the impact occurred in early June. Wolfe based his interpretation on the reproductive stages exhibited by aquatic fossil plants preserved in the sediments and fallout debris at the bottom of an ancient pond now located at Teapot Dome, Wyoming.[14] However, Nichols and Hickey have seriously questioned Wolfe's identifications of the species present and his interpretive methodologies.[15]

Near the south pole, the rates of extinction for fossil pollen species seem to be much lower than those in the midlatitudes of the northern hemisphere. On Seymour Island near the Antarctic Peninsula, only a few species disappear at the iridium anomaly according to Rosemary Askin of the University of California at Riverside and her colleagues.[16] Paleontologic studies on Seymour Island's marine invertebrate fossils by workers such as William Zinsmeister of Purdue University and his colleagues are also interesting.[17] In terms of typical marine invertebrates, such as clams, snails, lobsters, and ammonites, there is no abrupt change at the K-T boundary as marked by the iridium anomaly. Instead, there is a pattern of gradual extinction. The researchers originally argued that this is because this section is more complete than other marine K-T sections and thus preserves a more detailed picture of the sequence of extinctions. This pattern of extinctions has been used to argue against a single catastrophic extinction event caused by an impact. In addition, studies of forams by Gerta Keller in sediments preserved on the sea bottom off the Antarctic coast are also thought to exhibit a gradual pattern of extinctions.[18]

Not surprisingly, proponents of the impact hypothesis have argued that the extinction was abrupt but that because of preservational gaps, sampling problems, and reworking, the pattern of extinctions appears gradual. Recently, Zinsmeister revised his opinion; he now believes that the section at Seymour Island reflects a catastrophic rather than a gradual faunal turnover.[19]

A gradual pattern of extinctions is also found for pollen species of plants in sections spanning the K-T boundary in New Zealand, which was much farther south 65 million years ago than it is today. The rate of extinction is reported by Kirk Johnson to be much lower than that found in the Western Interior of North America.[20] In addition, the fossil record of radiolarians, microscopic marine organisms, has been studied in five rock sequences by Chris Hollis of Utsunomiya University. They show that all 44 of the Late Cretaceous species survived the boundary events, then gradually became extinct above the K-T boundary.[21]

These contrasts in extinction rates and patterns have led some paleontologists, including David Archibald, Norman MacLeod, and Gerta Keller, to argue that the effects of the impact, if any, were greatly reduced at higher latitudes than at midlatitudes. This would make good intuitive sense, given that the impact occurred at a midlatitude site near present-day Yucatan.

This idea also squares with a proposal made by Peter Schultz of Brown University and Steven D'Hondt. Based on the Chicxulub crater's asymmetry, they hypothesized that the impacting body came in through the atmosphere at a low angle—not straight down.[22] The gravity maps of the crater appear to document

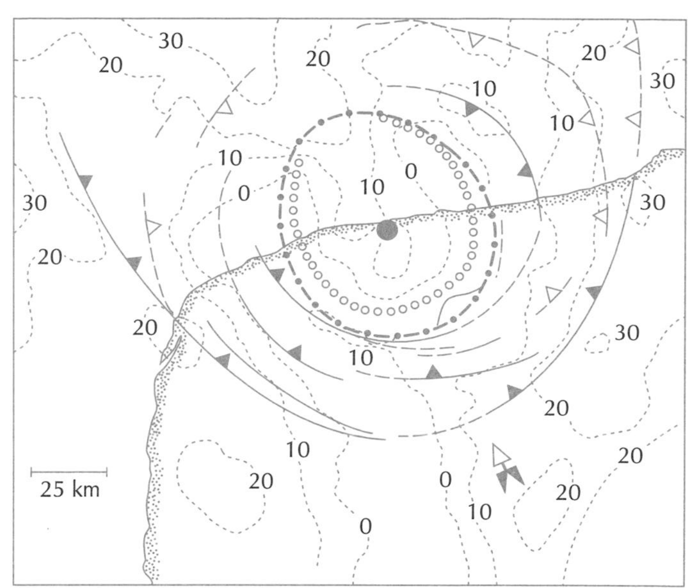

The map of the Chicxulub crater is based on contours representing gravity data. Note the asymmetrical, horseshoe-shaped depression defined by the gravity data.

a horseshoe-shaped crater with a steep southeast side and a shallower northwest side. Such a configuration seems to indicate that the impacting body came in at an angle of only 20 to 30 degrees from the southeast. If so, the fireball resulting from the impact would spread directionally to the northwest over North America. The evidence from both the fossil and rock records is consistent with this scenario. Extinctions over the midcontinent of North America were especially severe. At some North American sites, there are two fallout layers, characteristic of oblique impacts simulated in laboratory experiments. Also, the fragments of shocked quartz are larger in North America than in K-T sections from other parts of the world. This hypothesis will undoubtedly catalyze further tests to generate evidence that either confirms or refutes the scenario.

This scenario also seems to be consistent with the research of David Jablonski and David Raup, who noted an apparent correlation between the geographic range of a group and its survival rate.[23] Those organisms that were geographically widespread fared better than those whose ranges were more restricted.

Nonetheless, for plants and animals, it's still almost impossible to pinpoint one "killing mechanism" from any one event. But what if we narrowed the scope a bit and focused just on vertebrates living on the continents, including the dinosaurs?

THE FOSSIL RECORD OF DINOSAURS AROUND THE WORLD

To document a global extinction of dinosaurs clearly requires having a late Cretaceous fossil record of dinosaurs from many areas around the world. Does such a fossil record exist?

Two vertebrate paleontologists have looked closely at this question, Peter Dodson of the University of Pennsylvania and David Archibald.[24] In assessing the late Cretaceous record, these workers have noted that only twenty-six of latest Cretaceous dinosaur localities have been discovered throughout the

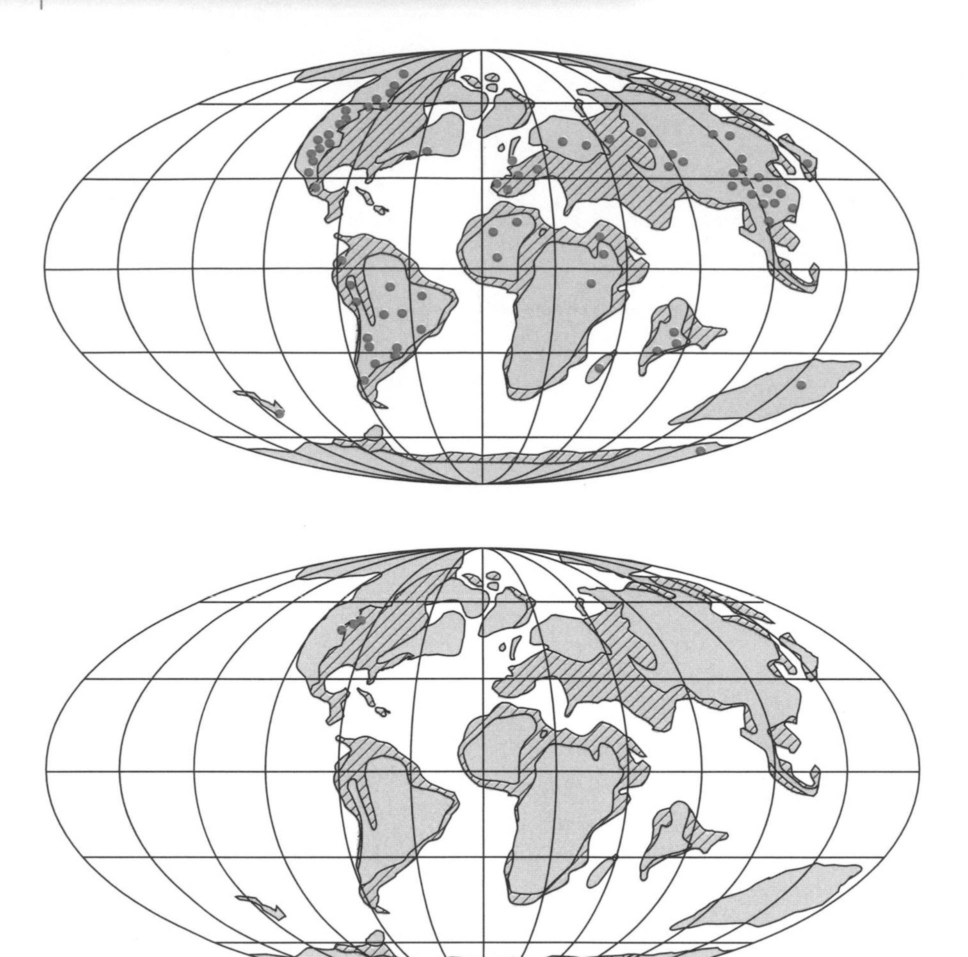

Maps showing locations of latest Cretaceous and earliest Tertiary fossil localities throughout the world. In the map *(at left)*, the dots indicate the latest Cretaceous dinosaur localities. In the map *(below)*, the dots indicate localities which span the K-T boundary. These localities can be used to document the changes that occurred during the extinction event.

whole world. All but six of these localities are found in the central interior of North America. These twenty-six localities preserve remains of only twenty different genera of dinosaurs, and fourteen of these twenty are restricted to North American localities. So, although we probably have a fairly good idea of what was happening in the interior of North America, we are almost totally ignorant of how dinosaurs were faring at the end of the Cretaceous in other parts of the world.

It's certainly possible that dinosaurs might have gone extinct in one geographic area but not in another area. Thus, the geographic incompleteness of the fossil record greatly hinders our ability to evaluate exactly what effect the volcanic eruptions and impact had on dinosaur faunas on different continents.

PATTERNS OF EXTINCTION AND SURVIVAL OF VERTEBRATES ON THE CONTINENTS

On land, reptiles larger than about 44 pounds (20 kg) suffered extinction at higher rates than smaller reptiles.[25] This accounts for most of the dinosaurs that lived at the end of the Cretaceous, including the North American forms *Tyrannosaurus*, *Triceratops*, *Anatotitan*, *Edmontosaurus*, and *Pachycephalosaurus*.

However, size is not a foolproof guide. The crocodilian lineage, for example, apparently sailed right through, although a number of reptiles smaller than 44 pounds became extinct, including the dinosaurs *Dromaeosaurus* and *Troodon*.

In addition to body size, an animal's position in the food chain has also been raised as a factor influencing its survival or extinction. Some researchers, including Peter Sheehan at the Milwaukee Public Museum, have noted that those animals dependent on living plants, or on animals that ate living plants, were more susceptible to extinction than those which fed on dead plant or animal material.[26] Again, it's not possible know the diet of an extinct animal, making this hypothesis difficult to test rigorously. Also, these supposed factors do not specify which "killing mechanism" was responsible.

By far the most detailed look at what happened to vertebrates on land during the K-T boundary has been compiled by a group of paleontologists who began working out of UC Berkeley under the direction of William Clemens, including David Archibald, Howard Hutchison, Laurie Bryant, and Donald Lofgren.[27] Their work focuses on the continental vertebrate fauna in the Western Interior of North America. Archibald's 1996 book summarizes this massive body of research, explaining the extinction patterns group by group and contrasting how many species in the latest Cretaceous rocks of the Western Interior had representatives in the early Tertiary rocks of the region. Before attempting to provide an explanation for the pattern of victims and survivors, Archibald first lists the species that form the pattern.

Sharks and rays were hard hit. Of the five species present in streams or ponds below the K-T boundary, none are found above, resulting in an extinction rate of 100 percent. Among ray-finned fishes, the most common and diverse group of modern bony fish, fifteen species are recorded, including seven representatives of relatively primitive groups, such as paddlefish, sturgeons, bowfins, and gars. All of these groups have members living today. Five of the seven late Cretaceous species are represented by survivors found above the boundary. Among the more advanced groups of ray-finned fish (the teleosts), four of eight species have descendants that survived events at the boundary. Some of the surviving groups are related to living pike, perch, and mudminnows. Overall, nine of fifteen ray-finned fish species are found above the boundary, constituting a survival rate of 60 percent.

Seven species of salamanders and one species of frog are found in the late Cretaceous. Interestingly, all of these species survived the events at the K-T boundary, yielding a survival rate of 100 percent.

Of seventeen species of turtles present in the late Cretaceous, fifteen survived the extinction event. This results in an impressive survival rate of 88 percent. Among lizards and snakes the results were more grim. Only three of ten species survived the boundary events. However, the crocodiles and alligators did better: four out of five species continued to exist in the Tertiary.

In terms of mammals, different groups suffered vastly different rates of extinction. Late Cretaceous faunas in North America are known to have contained ten species of rodentlike multituberculates. Five of these are represented by descendants above the K-T boundary, resulting in an extinction rate of 50 percent for the group. Among marsupial mammals, only one out of eleven species survived. However, among placental mammals, all six species are found in rocks above the boundary. Clearly, the events at the K-T boundary affected these groups of small mammals quite differently.

SURVIVAL OR EXTINCTION OF VERTEBRATE SPECIES ACROSS THE K-T BOUNDARY

Species from the Upper Cretaceous Hell Creek Formation, Montana		Survivors of K/T
ELASMOBRANCHII		
RHINOBATOIDEI		
Family Indeterminate		
Myledaphus bipartitus		O
SCLERORHYNCHOIDEI		
SCLERORHYNCHIDAE		
Ischyrhiza avonicola		O
?SCLERORHYNCHIDAE		
"Squatirhina" americana		O
?ORECTOLOBIFORMES		
Family indeterminate		
"Brachaelurus" estesi		O
POLYACRODONTIDAE		
Lissodus selachos		O
Number and % survival		0/5 (0%)
ACTINOPTERYGII		
CHONDROSTEI		
ACIPENSERIDAE		
"Acipenser" albertensis		X
"Acipenser" eruciferus		X
Protoscaphirhynchus squamosus	r	O
POLYODONTIDAE		
undescribed Polyodontidae	r	X
NEOPTERYGII		
(HOLOSTEANS)		
AMIDAE		
Kindleia fragosa		X
Melvius thornasi		
LEPISOSTEIDAE		
Lepisosteus occidentalis		X
(TELEOSTS)		
"APIDORHYNCHIDAE"		
Belonostomus longirostris	r	O
Belonostomus sp.	r	X
ESOCIDAE		
Estesesox foxi		O
Family indeterminate		
undescribed Esocoidei	r	X
New, unpublished family		
Platacodon nanus		O
PACHYRHIZODONTOIDEI, indet.		
species indeterminate	r	O
PALAEOLABRIDAE		
Palaeolabrus montanensis	r	X
PHYLLODONTIDAE		
Phyllodus paulkatoi	r	X
Number and % survival		9/15 (60%)
LISSAMPHIBIA		
ANURA		
DISCOGLOSSIDAE		
Scotiophryne pustulosa		X
CAUDATA		
BATRACHOSAUROIDIDAE		
Opisthotriton kayi		X
Prodesmodon copei	r	X

Species from the Upper Cretaceous Hell Creek Formation, Montana		Survivors of K/T
LISSAMPHIBIA		
PROSIRENIDAE		
Albanerpeton nexuosus	r	X
SCAPHERPETONTIDAE		
*Lisserpeton bairdi*X		
cf. *Piceoerpeton* sp.	r	X
Scapherpeton tectum		X
SIRENIDAE		
Habrosaurus dilatus		
Number and % survival		8/8 (100%)
MAMMALIA		
MULTITUBERCULATA		
CIMOLODONTIDAE		
Cimolodon nitidus		X
CIMOLOMYIDAE		
Cimolomys gracilis	r	O
Meniscoessus robustus		O
Family indeterminate		
Cimexomys minor	r	X
Essonodon browni		O
Paracimexomys priscus		O
NEOPLAGIAULACIDAE		
Mesodma formosa		X
Mesodma hensleighi		O
Mesodma thompsoni		X
? *Neoplagiaulax burgessi*	r	X
Number and % survival		5/10 (50%)
EUTHERIA		
GYPSONICTOPIDAE		
Gypsonictops illuminatus		X
PALAEORYCTIDAE		
Batodon tenuis	r	X
Cimolestes cerberoides	r	X
Cimolestes incisus	r	X
Cimolestes magnus	r	X
Cimolestes propalaeoryctes	r	X
Number and % survival		6/6 (100%)
METATHERIA		
DIDELPHODONTIDAE		
Didelphodon vorax		O
Family indeterminate		
Glasbius twitchelli		O
PEDIOMYDAE		
Pediomys cooki	r	O
Pediomys elegans	r	O
Pediomys florencae		O
Pediomys hatcheri	r	O
Pediomys krejcii	r	O
PERADECTIDAE		
Alphadon marshi		X
Alphadon wilsoni		O
Protaiphadon lulli	r	O
Turgidodon rhaister	r	O
Number and % survival		1/11 (9%)

SURVIVAL OR EXTINCTION OF VERTEBRATE SPECIES ACROSS THE K-T BOUNDARY *(continued)*

Species from the Upper Cretaceous Hell Creek Formation, Montana		Survivors of K/T
REPTILIA		
TESTUDINES		
ADOCIDAE		
Adocus sp.		X
BAENIDAE		
Eubaena cephalica		X
Neurankylus cf. *N. eximius*		X
Palatobaena bairdi		X
Plesiobaena antiqua		X
Stygiochelys estesi		X
CHELYDRIDAE		
Chelydridae indet.		X
Emarginochelys cretacea		X
KINOSTERNIDAE		
Kinosternidae indet.		X
MACROBAENIDAE		
"Clemmys" backmani		X
NANHSIUNGCHELYDIDAE		
Basilemys sinuosa		O
PLEUROSTERNIDAE		
Compsemys victa		X
TRIONYCHIDAE		
Heloplanoplia distincta		O
"*Plastomenus*" sp. A		X
"*Plastomenus*" sp. C		X
Trionyx (*Aspideretes*) sp.		X
Trionyx (*Trionyx*) sp.	r	X
Number and % survival		15/17 (88%)
SQUAMATA		
ANGUIDAE		
Odaxosaurus piger		X
?HELODERMATIDAE		
Paraderma bogerti	r	O
NECROSAURIDAE		
Parasaniwa wyomingensis		O
SCINCIDAE		
Contogenys sloani		X
TEIIDAE		
Chamops segnis		O
Haptosphenus placodon		O
Leptochamops denticulatus		O
Peneteius aquilonius	r	O
?VARANIDAE		
Palaeosaniwa canadensis	r	O
XENOSAURIDAE		
Exostinus lancensis		X
Number and % survival		3/10 (30%)
CHORISTODERA		
CHAMPSOSAURIDAE		
Champsosaurus sp. indet.		X
Number and % survival		1/1 (100%)

Species from the Upper Cretaceous Hell Creek Formation, Montana		Survivors of K/T
CROCODILIA		
ALLIGATOROIDEA		
Brachychampsa montana		O
undescribed alligatoroid(?) A		X
undescribed alligatoroid(?) B		X
CROCODYLIDAE		
Leidyosuchus sternbergi		X
THORACOSAURIDAE		
Thoracosaurus neocesariensis	r	X
Number and % survival		4/5 (80%)
DINOSAURIA		
ORNITHISCHIA		
ANKYLOSAURIDAE		
Ankylosaurus magniventris	r	O
CERATOPSIDAE		
Torosaurus ? *latus*	r	O
Triceratops horridus		O
HADROSAURIDAE		
Anatotitan copei		O
Edmontosaurus annectens		O
NODOSAURIDAE		
? *Edmontonia* sp.	r	O
PACHYCEPHALOSAURIDAE		
Pachycephalosaurus wyomingensis	r	O
Stegoceras validus	r	O
Stygimoloch spinifer	r	O
Number & % survival		0/10 (0%)
SAURISCHIA		
DROMAEOSAURIDAE		
Dromaeosaurus sp.		O
? *Velociraptor* sp.		O
ELMISAURIDAE		
? *Chirostenotes* sp.		O
ORNOTHOMIMIDAE		
Ornothomimus sp.		O
TROODONTIDAE		
Paronychodon lacustris		O
Troodon formosus		O
TYRANNOSAURIDAE		
Albertosaurus lancensis		O
Aublysodon cf. *A. mirandus*	r	O
Tyrannosaurus rex		O
Number & % survival		0/9 (0%)
TOTAL NUMBER AND % SURVIVAL		52/107 (49%)

SOURCE: Data from J. D. Archibald, 1996, *Dinosaur Extinction and the End of an Era*, Columbia University Press, New York, pp. 84–85.

NOTE: X by scientific name means the species is represented by a survivor in the Tertiary. O by scientific name means the species went extinct. r means the species is rare in fossil collections.

The fossil record of birds is very poor in the Hell Creek Formation, although its implications for dinosaur extinction and survival are quite important, as will be discussed in great detail in Part II of this book.

Finally, none of the nineteen species of conventionally recognized dinosaurs survived the boundary events. The decimated genera included: *Ankylosaurus*, *Torosaurus*, *Triceratops*, *Anatotitan*, *Edmontosaurus*, *Thescelosaurus*, *Edmontonia*, *Pachycephalosaurus*, *Stegoceras*, *Stygimoloch*, *Dromaeosaurus*, *Velociraptor*, *Chirostenotes*, *Ornithomimus*, *Paronychodon*, *Troodon*, *Albertosaurus*, *Aublysodon*, and of course, *Tyrannosaurus*. Regarding this list, *Albertosaurus* might be replaced with *Nanotyrannus*, and *Leptoceratops* and *Alamosaurus* might be added.[28]

Overall, of 107 species of terrestrial vertebrates present in the Hell Creek deposits, 49 percent appear to have been survivors of the extinction event in the Western Interior of North America. These are puzzling numbers that make up a complex pattern of survival and extinction. Archibald notes that, among terrestrial vertebrates, extinctions in just five of the twelve major groups (sharks and rays, marsupials, lizards and snakes, along with ornithischian dinosaurs and saurischian dinosaurs) account for 75 percent of the extinctions that occurred in the Western Interior during the K-T boundary. He argues, quite correctly, that this pattern must be explained by any mechanism invoked as a cause for the extinctions.

Using this extensive paleontologic census as a foundation, Archibald has offered a thoughtful analysis of why some groups of vertebrates, such a dinosaurs, suffered more than others.[29] In essence, he considered three potential causes for the extinctions, the impact, volcanic events, and the retreat of shallow continental seas. In relation to the impact hypothesis, he argued that the known record of continental vertebrates does not jibe well with the killing mechanisms alleged to have resulted from the impact. For example, one would expect the cold-blooded vertebrates to have been affected most by the subfreezing temperatures. For lizards, this seems to be true, as evidenced by a 70 percent rate of extinction across the boundary. However, other cold-blooded groups—frogs (0 percent), salamanders (0 percent), turtles (12 percent), and crocodiles and alligators (20 percent)—were minimally affected. In all, only four of the twelve groups of vertebrates studied (bony fish, multituberculate mammals, placental mammals, and lizards) fit the extinction pattern predicted by sudden cooling. (Steven D'Hondt has pointed out that many of these groups can survive subfreezing temperatures by estivating.)[30]

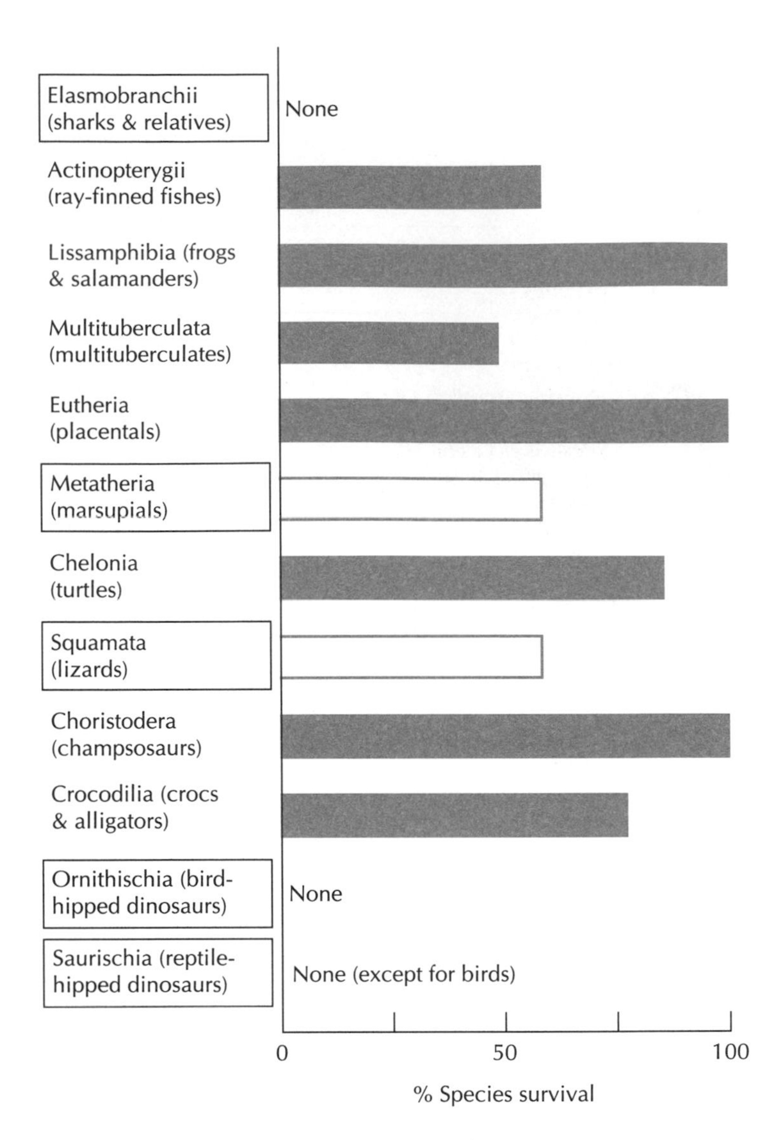

Chart documenting the rate of survival for late Cretaceous vertebrates during the K-T transition in the western part of North America. Overall, 49 percent (52 of 107 species) survived the K-T transition.

Questions remain about which, if any, of the extinct dinosaurs were actually warm-blooded, so it is difficult to be certain how freezing temperatures might have affected them. Archibald argues that the large body size of most dinosaurs would have, in essence, insulated them from the effects of lower temperatures. However, not all the dinosaurs that went extinct, including the bone-headed pachycephalosaur *Stegoceras*, and the small carnivores *Dromaeosaurus and Troodon*, had large bodies.

In terms of acid rain, Archibald noted that impact modelers had argued that both nitric acid (HNO_3) and sulfuric acid (H_2SO_4) would be produced as a result of the impact. These acids would then fall on the continents and oceans during rainstorms. The pH of such an acid rain has been estimated by modelers to have been as low as 0 to 3. Archibald added that today rain below a pH of 5 is considered unnaturally acidic, although rains as low as 3.8 and fogs as low as 2.1 have been recorded. The eggs and adults of aquatic vertebrates are damaged at a pH lower than 3. If the pH of rains dropped to 0, the results on terrestrial vertebrates would have been devastating. However, among aquatic vertebrates in the Western Interior, only sharks and rays suffered drastically (with an extinction rate of 100 percent). Bony fish (40 percent), amphibians (0 percent), turtles (about 12 percent), and crocodiles and alligators (20 percent) each fared pretty well. Overall, only three of the twelve major vertebrate taxa (sharks, multituberculate mammals, and placental mammals) fit the expected extinction pattern for acid rain.

Again, it's not certain how such acidic rains might have affected the dinosaurs that went extinct. However, Archibald notes that all life forms would die from exposure to water of pH less than 1.5. However, his table does not reflect that these groups would suffer extinction as a result of acid rain. This might well be debatable.

In terms of the global wildfires caused by the impact, researchers have estimated that the infernos would have burned the equivalent of half of all the modern forests—possibly 25 percent of all the terrestrial biomass that existed at the end of the Cretaceous. Archibald argues that such a massive apocalypse would have been devastating to both plants and animals, generating a fallout of organic and inorganic debris beyond any in our human experience. To him, such a horrific event could not have produced the selective pattern of survival and extinction reflected in the known vertebrate fossil record. Of the twelve groups Archibald studied, only five (sharks, marsupials, lizards, ornithischians, and saurischians) exhibit near total rates of extinction as would be expected by him if a global wildfire had actually occurred.

In all, Archibald concludes that only five of the twelve groups of vertebrates he studied exhibit extinction patterns consistent with the predicted patterns for impact-related effects (sharks and rays, marsupials, lizards, and the two groups of dinosaurs).

Turning to volcanic scenarios, Archibald notes that Deccan researchers have estimated that enough lava was produced to cover both Alaska and Texas to a depth of 2000 feet (600 m). In Archibald's view, the primary consequence would have been higher levels of debris in the atmosphere, especially carbon dioxide. This resulted in global warming due to the greenhouse effect, and it may have been a boon for plants, which utilize carbon dioxide in photosynthesis. Yet, Archibald notes that some evidence based on measuring different

TESTING POSSIBLE CAUSES OF K-T EXTINCTIONS AGAINST SURVIVORSHIP PREDICTIONS AND OBSERVED PATTERNS

	Sharks and relatives	Bony fish	Lissamphibians	Multituberculates	Placentals	Marsupials	Turtles	Champsosaurs	Lizards	Crocodilians	Bird-hipped dinosaurs	Reptile-hipped dinosaurs	Number of correct predictions
OBSERVATIONS													
Number L K vert. species	5	15	8	10	6	11	17	1	10	5	10	9	
Number of K/T survivals	0	9	8	5	6	1	15	1	3	4	0	0	
Significant extinction	YES	NO	NO	NO	NO	YES	NO	NO	YES	NO	YES	YES	
ULTIMATE CAUSE													
Impact (and volcanism?)	YES	yes	yes	yes	yes	YES	yes	yes	YES	yes	YES	YES	5
Proximate corollaries													
Sudden cooling	no	NO	yes	NO	NO	no	yes	yes	YES	yes	no	no	4
Acid rain	YES	yes	yes	NO	NO	no	yes	yes	no	yes	yes	no	3
Global wildfire	YES	yes	yes	yes	yes	YES	yes	yes	YES	yes	YES	YES	5
ULTIMATE CAUSE													
Marine regression	YES	NO	NO	NO	NO	YES	NO	NO	no	NO	YES	YES	11
Proximate corollaries													
Habitat fragmentation	no	NO	NO	NO	NO	no	NO	NO	no	NO	YES	YES	9
Lengthening of streams	YES	NO	NO	NO	NO	no	NO	NO	no	NO	no	no	8
Competition	no	NO	NO	NO	NO	YES	NO	NO	no	NO	no	no	8
Corollaries not ultimate-cause specific													
Local wildfire	no	NO	NO	yes	yes	YES	NO	NO	YES	NO	YES	YES	9
Detrital influx	no	NO	NO	yes	yes	YES	NO	NO	YES	NO	YES	YES	9

SOURCE: J. D. Archibald, 1996, *Dinosaur Extinction and the End of an Era*, Columbia University Press, New York, p. 128.

NOTE: This table attempts to correlate the potential killing mechanisms resulting from impact, volcanic activity, and seaway retreat with the pattern of extinctions and survival in vertebrates of the Western Interior in North America. Major groups of vertebrates are listed across the top. Along the left-hand side at the top, an analysis of extinction and survival rates for those groups is shown. The third row down shows which groups suffered major extinction (less than 30 percent survival = YES; greater than 50 percent survival = NO). An analysis of how the patterns of extinction and survival for the major groups corresponds to the patterns predicted by the different extinction scenarios follows. Capitalized "YES" and "NO" designations indicate agreement between the predictions of the scenario and the observed pattern. Lowercase "yes" and "no" indicate a lack of agreement. Numbers in the right-hand column indicate how many of the major groups conform to the predictions of the particular killing mechanism.

forms of oxygen atoms in marine sediments off of Africa suggest that the temperature of the ocean actually decreased by about 8°C (14° F) during the K-T transition. He suggests that because of the long-term nature of this temperature change, most organisms, especially those with short generation times, would have been able to adapt successfully. Those with longer generation times, such as dinosaurs, might not have been able to.

Curiously, Archibald does not address potential effects of acidic rain generated by volcanic eruptions, an effect prominently discussed by proponents of the volcanic scenario for extinction. But, in his table summarizing his results, he lumps the effects of impacts and volcanism together, so presumably his conclusions would mirror those argued for impact-based acid rains. If volcanism is assumed to have caused sudden cooling and acid rain, but not global wildfires, then the table can be used to estimate how many of the twelve groups fit the expected pattern of extinction. Four of the twelve groups fit the expected pattern (sharks and rays, multituberculate mammals, placental mammals, and lizards).

Finally, Archibald discusses how the fossil record of vertebrates fits the potential effects resulting from the retreat of continental seas back into the major ocean basins at the end of the Cretaceous. As noted previously, the retreat of these seaways is probably related to plate tectonic motions and may even be related to the plumes of magma that created the large episode of volcanic activity at the end of the Cretaceous.

Archibald, citing work by Alan Smith of Cambridge University and his colleagues, first presents the case that the retreat of the continental seas at the end of the Cretaceous was among the most extensive in the history of the Earth.[31] As the seaways pulled back, the land surface of the planet is thought to have increased by more than 25 percent—an increase equal to adding a continent the size of modern-day Africa. He notes three resulting potential causes for extinction: habitat fragmentation, lengthening of streams, and competition from invading species across land bridges.

Archibald argues that the retreat of the seaways generated a reduction in the area of coastal plains, where most of our evidence of Late Cretaceous vertebrates is found. He reasons that such a reduction of habitat would affect large vertebrates, such as most dinosaurs, first because these require the most habitat area per animal to satisfy their needs. As evidence for such an effect, Archibald cited the deleterious effects of habitat loss on today's large mammals in the Rift Valley System of East Africa. He goes on to discuss another effect of the marine regressions: habitat fragmentation. He notes that some paleontologists have argued that the concept of habitat fragmentation is not testable with geographically incomplete fossil records. But citing the work of modern ecologists, Archibald states that, if a habitat is broken up into many pieces, the ability of animals to move from one area of the habitat to another is impeded. Consequently, it may not be possible to maintain populations of animals large enough to assure a species' reproductive viability, especially for large organisms like dinosaurs.

Comparing this prediction to the fossil record, Archibald notes that only eight of thirty large vertebrate species present in the Late Cretaceous survived the extinction events. All eight survivors were at least partially aquatic, including two fish, one turtle, and a number of crocodiles and alligators. The extinguished species include one turtle, one lizard, one crocodilian, and nineteen dinosaurs. He argues that nine of the twelve groups he studied fit the expected extinction pattern for habitat fragmentation (all but sharks, marsupial mammals, and lizards).

Critics such as Steven D'Hondt have raised questions about the validity of Archibald's analysis.[32] One is that not all dinosaurs lived along the coastal plain. With greater continental areas exposed as the result of seaway retreat, other environments must have expanded. Why didn't dinosaurs living in those environments prosper?

Archibald also notes that with the retreat of seaways, river systems must have become longer, increasing the area of these freshwater aquatic habitats. Freshwater vertebrates, including ray-finned fish, salamanders, aquatic turtles, crocodiles, and alligators, did pretty well. However, the sharks and rays, with their ties to marine environments, disappeared from the area. In all, Archibald feels that eight of the twelve vertebrate groups studied fit the extinction pattern predicted by lengthened rivers and reduced coastal marine environments (all but marsupial mammals, lizards, and the two groups of dinosaurs).

Finally, Archibald argues that the drastic reduction of marsupial mammals across the K-T boundary was due to the immigration and resulting competition of archaic hooved mammals from Asia. The fossil record shows that about fifteen species of marsupials were present in the Western Interior just before the K-T boundary. That number dropped precipitously to one after the extinction events. As the seaways retreated, the Bering Land Bridge between Alaska and Asia became exposed above sea level. Within a million years of the K-T boundary, thirty species of archaic hooved mammals, thought to have originated in Asia between 80 and 85 million years ago, reached North America. It's this coincidence that led Archibald to suggest that the archaic hooved mammals out-competed the marsupials. This is certainly possible, but competition in the fossil record is a very difficult proposition to test because we cannot observe the populations interacting. Despite this, Archibald argues that eight of the twelve groups have extinction patterns consistent with the expected effects of competition (all but sharks, lizards, and the two dinosaur groups).

In all, Archibald argues that the fossil records for eleven of the twelve groups of vertebrates studied fit the anticipated patterns for the three extinction mechanisms related to marine regression. Only lizards do not. Accordingly, Archibald concludes that marine regression explains the most evidence available in the fossil and rock records.

At the time of this writing, Archibald's study has just been published. It is clearly the most comprehensive effort of its kind and will undoubtedly generate a great deal of further discussion from advocates of all the different extinction hypotheses.

Thus, there is good geological evidence for a monumental impact, a massive volcanic event, and the retreat of shallow continental seaways at the end of the Cretaceous. The "killing mechanisms" ascribed to each event could have taken their respective tolls on the Earth's biota. Since many of those killing mechanisms were associated with more than one event and since different killing mechanisms could have killed off a particular group of organisms, it is difficult to associate the extinction of particular groups with a particular event and a particular mechanism. This is especially true since the period of volcanism and seaway retreat overlapped in time with the Chicxulub impact.

If we could assure ourselves that the extinctions occurred very quickly at the end of the Cretaceous, rather than over a longer period of time, that might help us decide which event was responsible. This is because the effects of the impact are thought to have operated over a period of only months to a few thousand years, whereas those associated with volcanism and seaway retreat are generally thought to have operated over tens or hundreds of thousands of years. How well can we really distinguish the durations of these kinds of events in the geologic record at the end of the Cretaceous? That question is the focus of the next chapter.

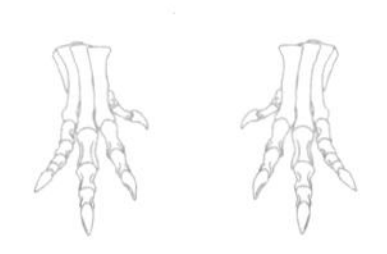

CHAPTER EIGHT

Our Hazy View of Time at the K-T Boundary

The Cretaceous-Tertiary (K-T) boundary lies 65 million years in the past. While this is a fairly recent event in relation to the origin of the Earth (4.5 billion years ago) or the origin of dinosaurs (around 240 million years ago), trying to tell geologic time 65 million years ago is a significant challenge for paleontologists and geologists. Our ability to tell time precisely has a direct effect on our ability to establish whether the K-T extinction happened very quickly, as predicted by the impact hypothesis, or more gradually, as suggested by most scenarios involving volcanic activity and the retreat of shallow seaways.

There are two common ways to tell geologic time. In the first, "relative" geologic time, all one is trying to do is decide whether a particular event happened earlier or later than another event. No judgment is made about how much earlier or later that event occurred. As mentioned in earlier chapters, this concept is based on the law of superposition, which generally states that layers of rocks and fossils that lie lower in a sequence are older than layers and fossils that lie higher in the sequence. This is because layers are laid down one on top of the other.

The second way to tell geologic time is termed "scaled" or "absolute" geologic time. Here, instead of simply trying to determine whether one event occurred earlier or later than another, we try to estimate how much earlier or later it occurred in terms of years. The only way to reliably scale time in terms of "years before the present" is through radioisotopic dating. It is through radioisotopic dating that the age of the K-T impact has been estimated to be 64.98 million ± 50,000 years. Other geologic timescales exist, such as the previously mentioned timescale based on the reversals of the Earth's magnetic field, but almost all of these timescales ultimately rely on radioisotopic dates for calibration in terms of years before present.

In terms of the debate between proponents of the impact and volcanic extinction scenarios, how does our ability to tell time through these relative and scaled methods affect our ability to assign responsibility for the extinctions to the two competing hypotheses?

STEPWISE EXTINCTIONS AND MULTIPLE IRIDIUM ANOMALIES

In its original form, the impact hypothesis posited a single catastrophic blast as the cause of the terminal Cretaceous extinctions. However, in the last half of the 1980s a modified hypothesis began to take shape. The large iridium enrichment associated with the boundary clay and the last appearances of typical Cretaceous organisms could not explain the detailed pattern of the last occurrences of extinguished organisms seen in the rock layers leading up to the K-T boundary. Instead, a series of steplike extinction events appear to have occurred in the layers just below the boundary clay containing the major iridium enrichment.[1] In some rock sequences spanning the K-T boundary, the high concentrations of iridium are not restricted to the boundary clay itself. The iridium enrichment appears to more gradually diminish in intensity as one moves either up or down from the boundary clay over distances of a few meters. These are termed "smeared" anomalies. Occasionally, subsidiary peaks of iridium concentration are associated with these smeared enrichments and are interpreted by some to represent multiple iridium anomalies.

Correlated with the apparent stepwise extinction events below the boundary are significant changes in the ratios of different forms of carbon, oxygen, and osmium atoms. These are interpreted by Jeffrey Mount of the University of California at Davis and his colleagues to represent significant environmental and/or climatic changes during an extended period of the K-T transition. Taken as a whole, the picture seems to suggest that a series of small extinction events occurred over a period of several hundreds of thousands or even a few million years before the major impact.

Not surprisingly, proponents of impact scenarios have distinctly different interpretations of what these stepwise extinctions represent than do proponents of volcanically-based scenarios. To impact proponents, the stepwise extinctions and the multiple or smeared iridium enrichments are documentation of multiple impacts, supporting the belief that the extinctions were caused by a swarm of cometary impacts that lasted less that a couple of million years. To proponents of volcanic scenarios, the stepwise extinctions and iridium anomalies document the major pulses of volcanic activity and associated environmental havoc resulting from the eruptions that formed the Deccan Traps and other terminal Cretaceous volcanic activity.

In large part, these differences of opinion result from the approaches used to interpreting the relative positions of the fossils present in the sequence of rock layers spanning the K-T boundary. One problem, pointed out by Phil Signor of the University of California at Davis and Jere Lipps of UC Berkeley, is that organisms that were relatively rare elements in these faunas are not likely to be preserved as often as more abundant members of the faunas.[2] Consequently, the less common organisms may *appear* to drop out of the fossil record well below the boundary, even though they may have lived right up to the last moment of the Cretaceous. They appear to disappear because we have yet to find them preserved as fossils close to the boundary. This sampling or preservational phenomena might make one large extinction event look more gradual or "stepped" in the geologic record.

On the other hand, because gaps in time may be present between different layers of rocks containing the fossils, artificially large numbers of organisms can appear to disappear from the record abruptly. Some of these animals that appear to drop out might have actually lived longer, but there are no layers of sediment to document their existence.

Similarly, as David Jablonski of the University of Chicago and Karl Flessa of the University of Arizona point out, there are instances in which a fossil animal appears to drop out of the record temporarily, only to reappear several layers higher up in the sequence.[3] Such a pattern might be interpreted to mean that the organism was temporarily not living in the local area but survived in other regions before returning to its original habitat.

Finally, because layers of sediment and enclosed fossils can be eroded from their original position in the sequence and redeposited later into younger layers

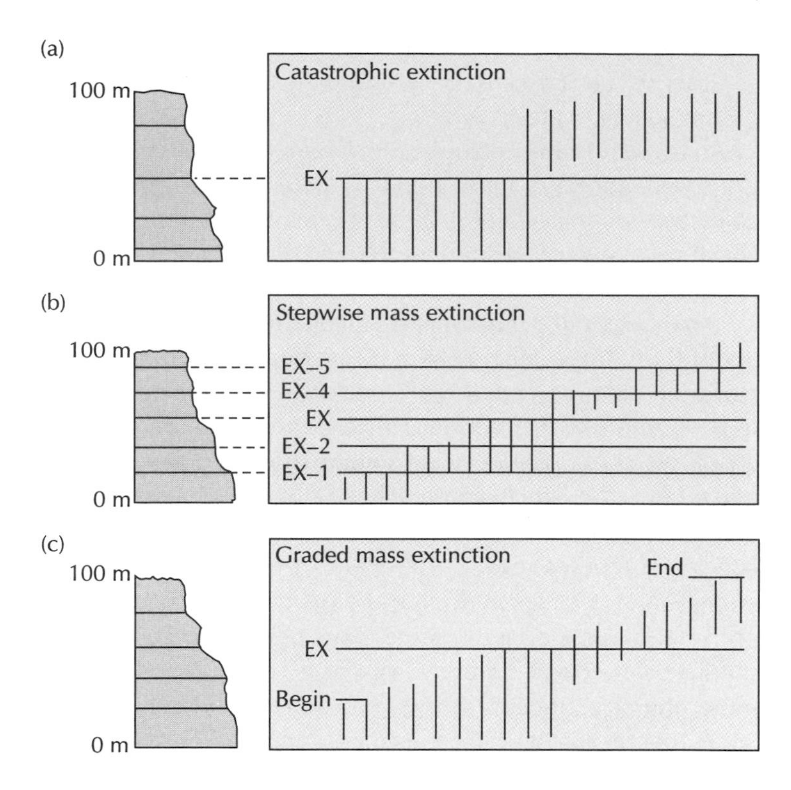

These diagrams illustrate how different patterns of extinction might look in the geologic record. Three different hypothetical stratigraphic sections are shown on the left. Each section is 100 meters thick and contains several layers of rock. To the right, the vertical ranges of several kinds of fossil animals are plotted. In other words, each kind of animal is found throughout the interval of rock layers corresponding to the height of the vertical line. In (a), many kinds of animals disappear at the same level in the sequence of rocks, suggesting a catastrophic extinction. In (b), different kind of animals disappear at five different levels within the sequence of layers, suggesting a stepwise pattern of extinction. In (c), almost every kind of animal disappears at a different level in the sequence, suggesting a more gradual pattern of extinction.

higher in the sequence, it is dangerous to read the fossil record too literally. In essence, nature can subtly fool with the law of superposition, as David Archibald has noted.[4]

All these possibile complications make a consensus about whether extinctions were abrupt or more gradually stepped difficult to achieve. This is especially true when paleontologists are dealing with the fossil record on either very fine or very broad stratigraphic and temporal scales. Take the fossil record of dinosaurs in North America, for example. A literal reading of the record shows that, about 8 million years before the end of the Cretaceous, there were about thirty-three genera of dinosaurs recorded in the Judith River Formation. However, near the end of the Cretaceous, only nineteen genera were present in the Hell Creek Formation. This appears to document a gradual decline in the kinds of dinosaurs living before the terminal Creatceous events. This agrees with the conclusion reached by Robert Sullivan of the State Museum of Pennsylvania. His global census of Late Cretaceous and Early Tertiary species of reptiles reveals that many of the major dinosaur lineages experienced major declines in diversity before the latest stage of the Cretaceous. All of these lineages either became extinct or suffered a severe reduction in the number of species before the last phases of the Cretaceous.[5] Proponents of a more abrupt extinction at the end of the Cretaceous, such as Dale Russell, argue that the latest Cretaceous dinosaur fauna from the Hell Creek Formation is not as thoroughly sampled as that from the older Judith River Formation.[6] However, after fairly intensive collecting in both areas for the last 100 years, this seems like a fairly weak argument.

Another study has been conducted within the Hell Creek Formation itself to estimate whether the diversity of dinosaurs was declining before the terminal Createceous events. Peter Sheehan along with David Fastovsky of the University of Rhode Island and their colleagues divided the 300-foot-thick (90 m.) formation into three parts—the lowest 100 feet, the middle 100 feet, and the upper 100 feet.[7] Then they recorded where they found fossils of different kinds of dinosaurs. They tracked fourteen different genera of dinosaurs that belonged to eight different families and concluded that there was no statistically significant change in diversity from the bottom to the top of the formation.

Unfortunately, the Hell Creek Formation is widely noted for its ancient river channels that are capable of eroding and redepositing fossils from the original position in which they were preserved. A simple assignment of fossils to one of the three levels within the formation ignores this possibility of reworking. Other paleontologists, including Archibald, have argued that the statistical approach used by Sheehan and his colleagues is not truly able to measure whether the diversity of dinosaurs actually declined or remained constant.[8] Consequently, the conclusion that dinosaur diversity did not decline until the last moments of the Cretaceous is still very controversial.

HOW PRECISELY CAN WE REALLY TELL TIME AT THE K-T BOUNDARY?

Regardless of whether one favors a more gradual, terrestrial scenario or a more catastrophic extraterrestrial scenario, the question of how well we can "tell time" at the end of the Age of Dinosaurs directly affects our ability to test these competing hypotheses.

tieth and twenty-first centuries. Most records detailing deaths during the twentieth and twenty-first centuries were forever destroyed during global wars at the end of the twenty-first century. However, some sketchy records remain of these events. It's known that World War II happened during a five- to seven-year period in the middle of the twentieth century, and that environmental pollution caused numerous deaths during both centuries. In your research, you are trying to interpret whether a large cemetery—essentially a bone bed—that you've excavated in Europe represents the effects of war casualties that occurred during World War II or deaths that resulted from environmental pollution. If you could find a way to tell time on the scale of decades, you could at least decide whether the bones in the cemetery represented individuals that died during the decade in which World War II was fought. However, if your ability to tell time was limited to establishing that these deaths occurred sometime during the twentieth or twenty-first centuries, you could not be sure whether the deaths were the result of casualties during World War II or the result of that era's environmental pollution.

Now let's take that concept and apply it to extinction events that happened 65 million years ago during the K-T transition.[13] First, we'll look at the completeness of the fossil record in deep marine sections containing the fossils used to document the extinction of microscopic plankton. The latest estimates suggest that we can expect many 100,000-year intervals spanning the K-T transition to be complete in these sections. In other words, many of the 100,000-year intervals during the transition can be expected to be represented by sediment. For example, in the Spanish K-T boundary sequence at Caravaca, all the 100,000-year intervals are thought to be represented. Thus, it is said to be "complete" at the 100,000-year timescale. This is very encouraging because this estimate is telling us that we can expect to find fossils and sediment deposited during the last 100,000-year period before the extinction event and from the first 100,000-year period after the extinction. However, in the Italian sequence at Gubbio, where the iridium enrichment was first discovered, slightly more than half of the 100,000-year intervals are expected to be represented by fossils or sediment. So, we have about a 50 percent chance of not having fossils from the last 100,000-year interval before the extinction event and a 50 percent chance of not having a fossil record from the first 100,000-year interval after the extinction

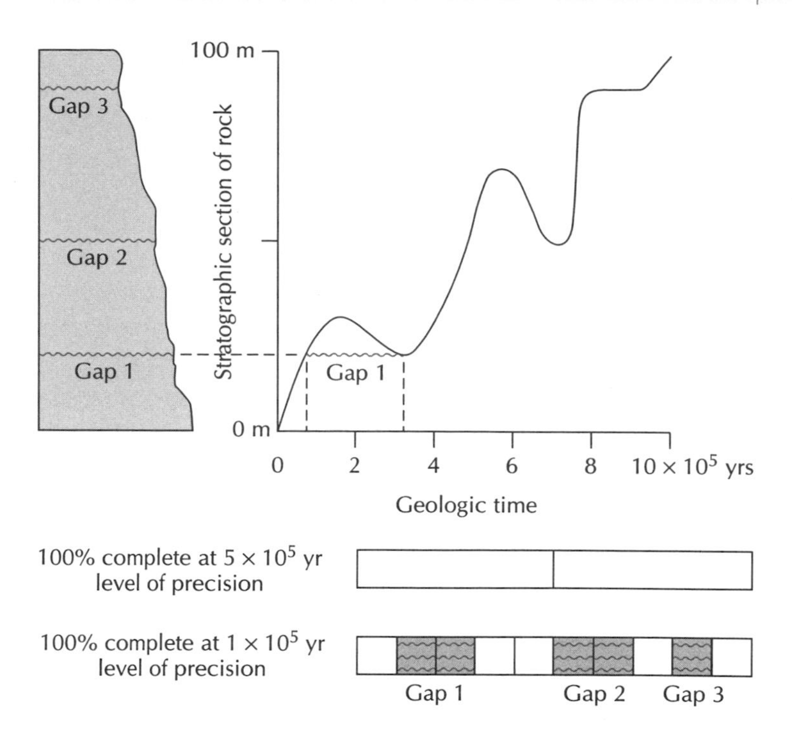

The graph illustrates how gaps not represented by sediment in a section of rocks can affects our ability to read the fossil record at short time scales. Basically, the record is revealed to be less and less complete as one tries to read the fossil record at increasingly finer timescales.

A stratigraphic section of rock on the left (vertical axis) is graphed as a function of its history in geologic time (horizontal axis). The line on the graph represents a history of when layers of sediment were preserved and when they were not. For example, between 0 and about 100,000 years, as the line rises form the origin of the graph, about 25 meters of sediment was deposited in the section. However, between about 100,000 and 300,000 years, a period of erosion, (represented by the downward segment of the line) removed about 15 meters of sediment from the section, creating gap 1 in the section. In other words, there is no sediment preserved in the section representing the interval between 100,000 and 300,000 years. A period of rapid deposition followed between about 300,000 and 500,000 years, before another period of erosion between 500,000 and 700,000 years removed about 20 meters of sediment to create gap 2. Between about 800,000 and 900,000 years no sediment was either being deposited or eroded, as evidenced by the horizontal line segment. This period of nondeposition generated gap 3 in the section.

Below the horizontal time axis, the time span represented by the section is divided first into two 500,000-year long intervals. Both of these intervals are represented by sediment preserved in the section. Consequently, the section is complete at the 500,000 year level of precision. Below, the time span represented by the section is civided into ten 100,000-year-long intervals. Only five of these ten intervals are represented by sediment preserved in the section. Thus the section is only five-tenths, or 50 percent, complete at the 100,000-year level of precision.

event. The picture is even worse at Stevns Klint in Denmark, where only one in four 100,000-year intervals is expected to be represented by sediment layers spanning the K-T boundary. In other words, we have about a 25 percent chance of finding fossils of organisms that lived during the last 100,000-year interval of the Cretaceous and a 25 percent chance of finding fossils from the first 100,000-year interval of the Tertiary.

In continental sections containing dinosaur and pollen fossils documenting the extinctions, the story is similar. Estimates based on this method suggest that most complete sequence is found in the San Juan Basin of New Mexico, where all the 100,000-year intervals are expected to be represented by fossils and sediment. In the Bug Creek sequence of Montana slightly over half are probably represented, while in the Red Deer River Valley in Canada the estimate is slightly less than half.

For testing questions concerning more catastrophic, impact-triggered killing mechanisms that operated over a period of 100 years or less, the estimates are more pessimistic. This makes good intuitive sense when one considers that sediment is only occasionally deposited and preserved in rock layers. It is much more likely that some rocks will be deposited and preserved during a 100,000-year period than during a 100-year period. Nonetheless, to rigorously test predictions of catastrophic impact scenarios, we would need to have had sediment preserved during each 100-year period at the end of the Age of Dinosaurs and the beginning of the Age of Mammals. What are our chances?

In the same marine sections we examined before, we can expect only three out of four 100-year intervals to be represented by sediment at Caravaca. So, we have a 75 percent chance of finding fossils that lived during the last century of the Cretaceous and the first century of the Tertiary. Only about one out of ten 100-year intervals can be expected to be represented at Stevns Klint and Gubbio. So our chances of actually documenting the effects of a catastrophic impact in these sections is quite low. In the continental sections where the last large dinosaurs lived, we can expect about one out of seventy 100-year intervals to be represented at the San Juan Basin, one out of two hundred at Bug Creek, and one out of three hundred at Red Deer River Valley. So, our chances of actually being able to test the predictions of the effects generated by a single impact are pretty discouraging—roughly 1 percent or less.

This does not mean that an impact did not cause the extinctions, nor that volcanic activity associated with the retreat of seaways did not cause the extinctions. It means that given our limited ability to tell time at the end of the Age of Dinosaurs, we probably can't distinguish between the effects of the two processes, at least for now.

Critics, including Mark Anders of Columbia University and his colleagues, have noted that the statistical base for making these estimates of completeness in geological sequences spanning the K-T boundary was constructed in a way that tends to underestimate the completeness of deep sea rock sequences.[14] Their recalculations suggested that Caravaca and Gubbio might preserve sediment during every 100-year interval during the K-T transition. Their work did not address the completeness of continental sections where dinosaurs lived, however, which are generally agreed to be much less complete. Anders and his colleagues concluded that this method "should not be used to assess completeness for individual sections." This is because compaction of sediment, reworking by burrowing organisms, and uncertainties in radioisotopic ages make estimating short-term sedimentation rates and completeness risky, if not unreliable. This conclusion corresponds closely with the warnings discussed in the original study.[15] In the end, the conclusions of both these studies only reinforce the point that we can't really test the predictions of short-term, catastrophic scenarios in a truly definitive scientific sense. At this point, we can *believe* what we *want* to believe, but we can't rigorously test catastrophic hypotheses about dinosaurian extinction to establish that the extinctions occurred as fast as the impact hypothesis predicts.

COULD A COMBINATION OF THE VOLCANIC/MARINE REGRESSION AND IMPACT SCENARIOS BE POSSIBLE?

Not surprisingly, some scientists are now beginning to argue for the possibility that a combination of the volcanic/marine regression and impact scenarios caused the mass extinctions at the K-T boundary. For example, F. L. Sutherland has suggested that the extinctions were caused by a combination of environmental effects generated by the monumental K-T impact at Chicxulub and the hot-spot volcanism originating deep within the Earth's mantle that created the Deccan Traps and other K-T volcanic deposits.[16]

Since the early 1980s, even some of the staunchest proponents of the impact and volcanic secnarios have softened their stances a bit. In 1990 Walter Alvarez and Frank Asaro,[17] original members of the team that proposed the impact hypothesis, wrote:

> In the past few years the debate between supporters of each scenario has become polarized: impact proponents have tended to ignore the Deccan Traps as irrelevant, while volcano backers have tried to explain away evidence for the impact by suggesting that it is also compatible with volcanism. Our sense is that the argument is a Hegelian one, with an impact thesis and a volcanic antithesis in search of a synthesis whose outlines are yet unclear.

In 1997, Alvarez elaborated on this point.

> Impact as a geologic process . . . must be recognized as a rare but significant kind of event, and evidently the cause of at least the K-T mass extinction . . . can volcanism be dismissed from the list of catastophic events with global effects? Not yet, . . . I would have dismissed the apparent age match between the Deccan Traps and the K-T impact-extinction event as a strange coincidence, if it were not that a second such coincidence has turned up. . . . Recently, Paul Renne . . . has obtained reliable dates on both the Siberian Traps and the Permian-Triassic boundary, . . . they are indistinguishable. A good detective shouldn't ignore even a single coincidence like the K-T-Deccan match in timing, and when it is bolstered by a second coincidence . . . , it just has to be significant. . . . Right now . . . there is an intriguing mystery, some obviously significant clues, and nobody has any idea what the explanation will be.[18]

On the other side, Vincent Courtillot now beleives that between one-third and one-half of Late Cretaceous extinctions resulted from the asteroid impact.[19] David Archibald, based on the analysis described in the last chapter, concluded that the extinctions were probably a combination of causes related to marine regression and impact. He was less convinced that volcanism had much to do with it, however.[20] Nonetheless, given that four of the twelve groups of vertebrates that he studied fit the expected pattern for extinctions resulting from sudden cooling and acid rain, it's clear that volcanism could have played some role.

Some scientists, including Mark Boslough of Sandia National Laboratory and John Hagstrum of the U.S. Geological Survey, have even gone so far as to propose that the impact served as the trigger for the plume volcanism that catalyzed the eruptions that formed the Deccan Traps.[21] This scenario suggests that the impact at Chicxulub generated a force equivalent to the simultaneous explosion of millions of hydrogen bombs, which, in turn, is calculated to have generated an earthquake of an unprecedented magnitude—13 on the Richter scale. The shock waves from the impact would have traveled through the Earth, which would have acted like a lens to refocus the energy of the shock waves at the point on the Earth directly opposite the place of impact—the antipode. This focusing effect is somewhat like the way sound travels through a domed chamber such as those found in many state capitol buildings. As opposed to the smoke-filled, rectangular caucus chambers, a whisper on one side of a relatively quiet rotunda can often be clearly heard on the opposite side of the room because the sound waves reflect off the dome to be focused on the area directly opposite the point of origin. Based on one computer simulation of the impact, the shock waves focused at the antipode would have caused the ground to flex as much as 60 feet (18 m), opening a network of huge cracks in the Earth's surface, heating the rocks in the upper 100 miles (about 150 km) of the crust and mantle, and catalyzing massive floods of basalt. To provide some scale for comparison, the earth-

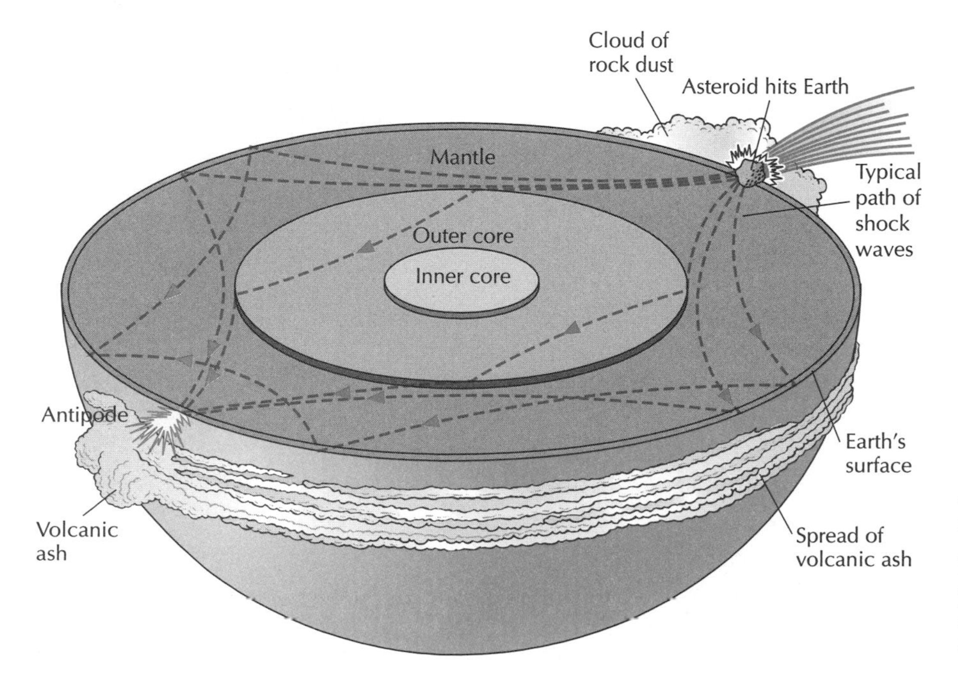

A massive impact might trigger extensive volcanic eruptions on the opposite side of the Earth. Such a mechanism has been advocated by some scientists to be responsible for both the Chicxulub crater and the Deccan Traps.

quake that rocked San Francisco in 1906 is though to have generated shock waves that lifted the ground surface only about 3 feet (1 m) or so.

Although not well documented on the Earth, evidence for antipodal volcanism appears to exist on other rocky planets in our solar system, as noted by David Williams and Ronald Greeley of Arizona State University. The largest known impact crater visible on the surface of Mars, named Hellas Plenitia, is located at the antipode for the lava flows, christened Alba Patera. These flows extend across about 930 miles (1,500 km) of the planet's surface and were erupted from the largest known volcano in the solar system. It is estimated that the cracks opened by the shock waves generated by the impact that formed Hellas Plenitia may have been as much as 9 miles (15 km) deep.

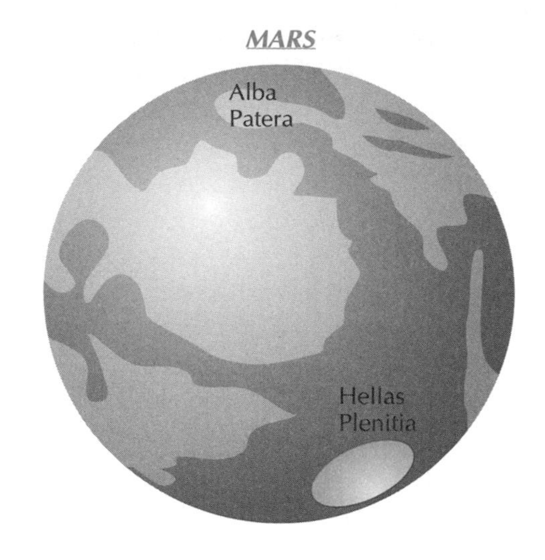

The manifestations of antipodal volcanism seem to be clearly demonstrated on Mars. The largest known volcano in the solar system is located at the opposite point on the planet from where a large impact occurred, creating the crater Hellas Plenitia.

However, within the context of the Deccan Traps and the one well documented K-T crater at Chicxulub, they did not appear to have been at opposite points on the Earth 65 million years ago. Thus proponents of antipodal volcanism must search for the impact source of the Deccan Traps volcanics in the eastern Pacific. Although John Hagstrum of the U.S. Geological Survey feels that there may be geologic evidence in that area for an impact, no clear evidence has yet been documented and published. At this point, therefore, the existence of impact-generated antipodal volcanism on Earth and its possible connection to K-T events must be considered to be rather speculative.

In addition, most, but not all, experts in flood basalt volcanism argue that the amount of energy required to have produced the half-million-year-long flows of magma that characterized the most copious eruptions of the Deccan Traps would have greatly exceeded that generated by the Chicxulub impact. Thus, the Deccan eruptions must have arisen from a more powerful source of energy deep within the Earth itself. Furthermore, the Deccan Trap volcanism began about 68 million years ago and continued to erupt material in several major pulses until about 65 million years ago. However, the impact at Chicxulub did not occur until right at the end of the Deccan volcanism, not right at the start. So any antipodal relationship between the two most well documented geological features associated with these two scenarios seems completely unfounded.

SO, WHERE DO WE STAND?

The murder mystery at the end of the Cretaceous remains a fascination for scientists and public alike. Despite an abundance of clues, the case of whether the extinction of large dinosaurs was caused by volcanic eruptions and seaway retreats or by impact is not yet closed. In essence, the jury composed of members of the scientific community is still out because a clear consensus has yet to emerge. Many researchers have concluded that an impact was responsible. Others maintain that the volcanic acitivity was the cause. Still others, as noted above, think that a combination of events was to blame.

To us, it seems that the coincidence of the volcanic activity, the retreat of the seaways, and the impact would have put tremendous environmental stress on the Earth's biota. The available geologic evidence suggests that each event was among the largest of its kind ever to be inflicted on the Earth. Disagreement about which was primarily responsible reflects the fact that none of these hypotheses unequivocally and exclusively explains all of the available evidence in the rock and fossil records. In part, this appears to be because many of the proposed killing mechanisms ascribed to these events are the same. In addition, both the impact and volcanic scenarios appear to be consistent with the pattern of extinctions predicted by the law of superposition which is used to tell relative geologic time. In other words, one does not find unequivocal evidence of dinosaurs above the boundary clay or above the Deccan lava flows. Where radiosiotopic dates are available, the last dinosaurs appear to have died out either at or slightly before both the end of the volcanic activity and the impact. One important and unresolved problem is that we cannot tell time precisely enough in continental records spanning the K-T boundary to determine whether large dinosaurs perished over a period of a few decades or millennia, as predicted by the impact hypothesis, or over a longer period of hundreds of thousands of years, as predicted by the scenarios invoking volcanism and seaway retreat.

An improved sense of timing would be most helpful in resolving these questions. As in many modern murder cases, the solution depends on figuring out not only what happened but also when it happened. Radiosotopic dates have now established the time of the impact as precisely as our current techniques will allow. As previously stated, the melted rock in the Chicxulub crater yields an age of 64.98 million ± 50,000 years.[22] The timing of the Deccan eruptions remains a bit more vague. The initial stages of the volcanic acitivity associated with the plume began about 68.57 million ± 80,000 years ago, and the last phases of activity near the K-T boundary ended about 64.96 million ± 110,000 years ago.[23] But it is much less certain when most of the lava flows that form the Deccan traps occurred within that interval. The most recent statement suggests that most of the Deccan Traps were emplaced in less than 1 million years, although the uncertainty in the absolute age, about 2 million years, remains larger than that often quoted in published work.[24] More precise resolution of when the main pulse of Deccan eruptions occurred could help us decide which animals and plants went extinct due to the impact and which went extinct due to the volcanic eruptions and seaway retreat.

In the marine realm, establishing that a catastrophic impact was responsible will require us to identify which marine organisms went extinct within 100 years after the impact. That is not presently possible. However, if the method based on cycles in the precession of the Earth's axis can be used to establish that certain extinctions occurred within one cycle, we would know that the extinctions took place over less than 21,000 years. That would represent a vast improvement over the precision of our best current estimates.[25]

In the continental realm, it will be important to find a rock layer containing minerals that can be radiosiotopically dated just below the highest dinosaur fossils. That will help us constrain the period in which the extinction of Late Cretaceous dinosaurs occurred.

At this point, however, the situation within the scientific community is analogous to having a hung jury. So to us as authors, the cause for the extinction of large dinosaurs at the end of the Cretaceous remains unresolved, but we have tried to lay out the evidence so that you have the opportunity to act as a juror in the case and decide for yourself.

Regardless of what caused it, events surrounding those moments of evolutonary history during the K-T transition radically changed the course of life on this planet. No longer did dinosaurs, as we commonly recognize them, play the dominant role on land that they had for over 165 million years. No longer did beasts like *Tyrannosaurus* and *Triceratops* roam the continents.

Although a consensus verdict of the scientific community has not been rendered, we have progressed in terms of the historical debate between catastrophists and uniformitarianists or gradualists. As a result of the contemporary extinction debates, the concept of uniformitarianism has now been expanded to encompass two kinds of natural geologic events that were originally deemed to be suspect. Large extraterrestrial impacts and massive eruptions of flood basalt are now clearly established as normal, if relatively rare, geological processes. As such, they have been incorporated into the methodological uniformitarian repertory of natural processes that can be considered in explaining Earth-based events, including extinctions. It's the question of how these new processes in the uniformitarian repertory affect the extinction and evolution of life on Earth that is still at issue. In retrospect, even the volcanic scenario based on plumes of magma rising from the mantle is rather catastrophic in terms of geologic time in relation to the uniform, gradual, stately rate of evolutionary change originally envisioned by proponents of substantive uniformitarianism.

It really should not be too surprising to us that a consensus has not developed among scientists on what extinguished all the large dinosaurs at the end of the Cretaceous. We have similar problems in modern murder cases. Take the assassination of President John F. Kennedy, for example. There was a lot of evidence to help us decide exactly how it happened, including photos and even home movies taken at the moment of the shooting. However, controversy still abounds about many aspects of the case. Was there just one gunman or more? Was there a conspiracy involved? This assassination even happened within our lifetime in front of dozens of eye witnesses, yet we still can't all agree about the exact sequence of events and their causes. Extrapolate these kinds of problems back 65 million years into the past and you get some sense of how difficult it is to be certain about what caused the extinction of *Tyrannosaurus*—the king of the dinosaurs.

To amplify this point, one need only look back to the extinctions at the end of the ice ages about 11,000 or 12,000 years ago. North America was then populated with a remarkable fauna of large mammals, including sabertooth cats, giant ground sloths, camels, mammoths and mastodons. But at the end of the ice ages, they all went extinct. Although the primary suspects in this prehistoric case seem to be either climatic change or human predation, rather than volcanic activity and extraterrestrial impact, a similar debate between researchers over the cause characterizes that more recent mass extinction.

Given the lack of consensus over what caused the extinction of large dinosaurs at the end of the Cretaceous, is there another way to solve the mystery of dinosaur extinction? In fact, there is. Recall that our charge in this book is to examine all the evidence in the rock and fossil records involving dinosaur extinction. This requires us to investigate another basic question. Ironically, despite several decades of debate concerning what caused the extinction of dinosaurs 65 million years ago, another line of evolutionary research has rendered the main point of that debate moot. It is true that neither *Tyrannosaurus* nor *Triceratops* roams our world anymore. Yet in a very important sense, the extinction debate has been focused on the wrong question. Until recently, scientists participating in that debate have been asking, "What caused the extinction of all dinosaurs about 65 million years ago?" But amazingly, an equally appropriate question to ask is, "Did all the dinosaurs really go extinct 65 million years ago?" That investigation, which revolves around the search for living descendants of the dinosaurs, is the focus of the second part of this book.

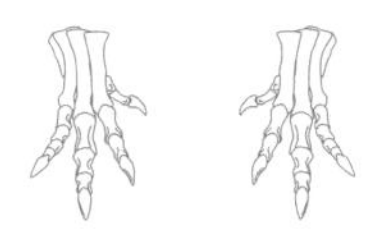

PART TWO

Dead or Alive?

CHAPTER NINE

Living Dinosaurs?

Now, let us be quick to clarify what we mean by "living" dinosaurs. We're not talking about dinosaurs hiding in the black waters of Loch Nesse or the deepest jungles of Africa. Only a few crackpots believe that sauropods and ceratopsians aren't extinct. But what about the dinosaurs living in our own back yard? The notion seemed ridiculous. Or so we thought as we packed up our books and moved to California.

When we arrived at Berkeley we encountered the prospect of living dinosaurs, and it seemed surprising to be confronted with this argument by members of our own department, instead of the street life of Telegraph Avenue.

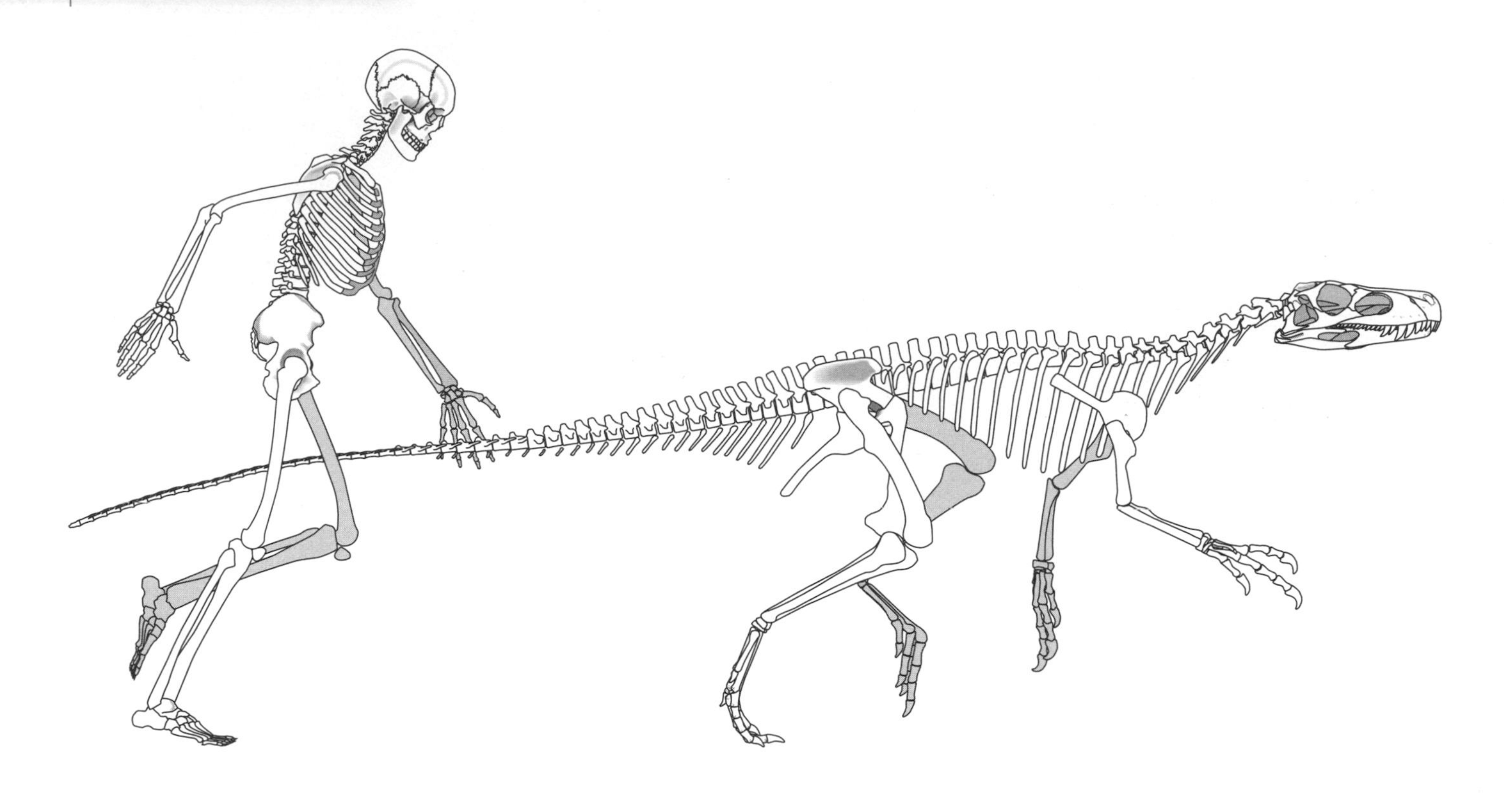

Obviously *Tyrannosaurus* and *Triceratops* disappeared at the end of the Cretaceous, our new colleagues conceded. Nonetheless, one lineage that descended from the ancestral dinosaur survived the impact and eruptions at the Cretaceous-Tertiary (K-T) boundary. Today, that lineage is represented by over 9,000 species of birds. Furthermore, if birds are the living descendants of dinosaurs, how can we claim dinosaurs are extinct?

What does it mean to say that birds are dinosaurs? Proponents of this theory argued that newly discovered fossils demonstrated that dinosaurs were the ancestors of birds. In addition, new methods for reconstructing evolutionary history and establishing evolutionary relationships were being developed. Applying these methods to the fossil record not only resurrected dinosaurs from extinction but, in effect, changed the course of the history of life.

Others at Berkeley argued that the issue of living dinosaurs was trivially semantic and of no scientific significance. Even if birds did evolve from dinosaurs, they are totally different from the extinct Mesozoic monsters. With the evolution of flight and warm-bloodedness, birds entered a new adaptive zone, hence their placement in a separate class of vertebrates—Aves. Any genealogical connection must be dwarfed by the shift to a new adaptational way of life.

If science were democratic the notion of living dinosaurs would have quickly gone extinct. The new methods for reconstructing evolutionary history that produced this perspective were attacked by some of the world's most influential evolutionary biologists. They argued that the procedures were flawed and that they produced flawed results. Moreover, most scientists believed that reconstructing ancient genealogies was virtually impossible to do with any accuracy. They believed that scientists should stick to biological issues that could be directly observed in the modern world. As a result, most scientists adamantly maintained that dinosaurs were extinct.

Dr. John Ostrom and *Deinonychus*.

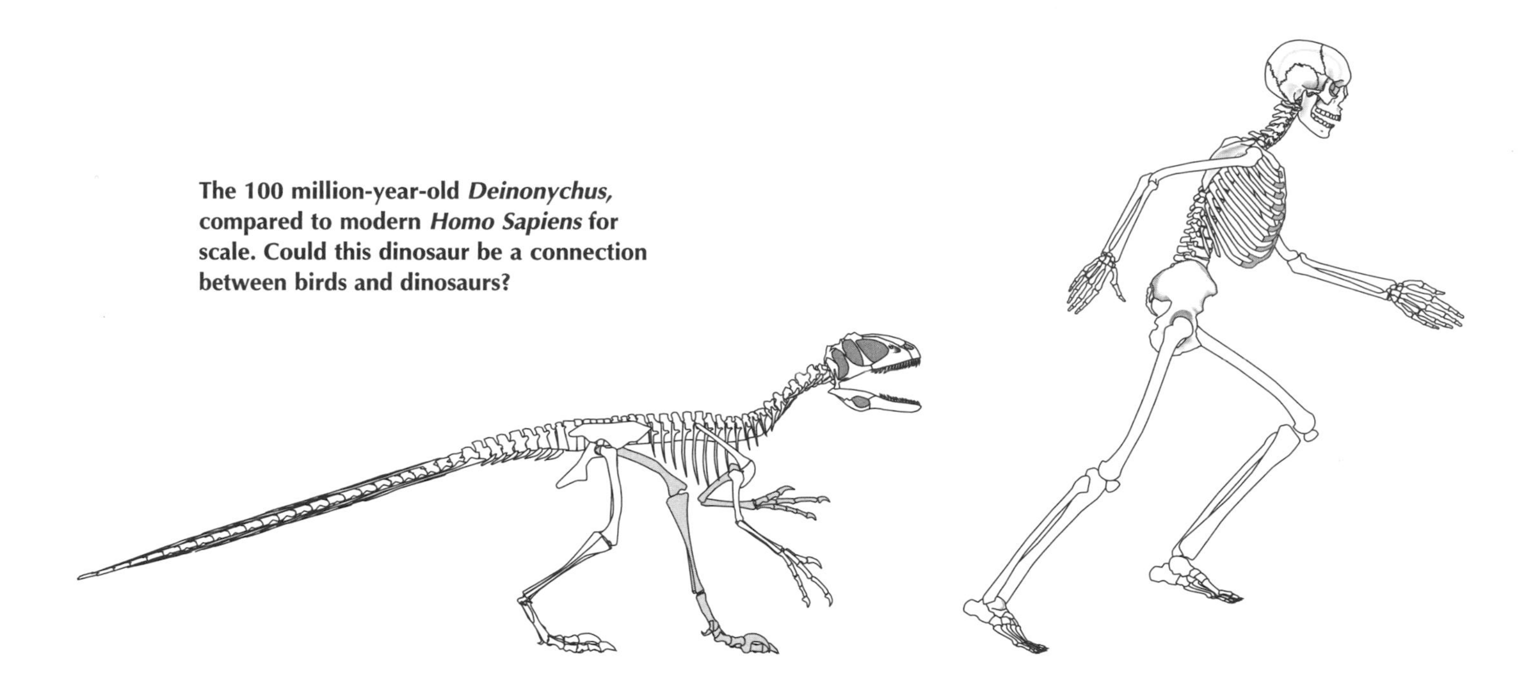

The 100 million-year-old *Deinonychus*, compared to modern *Homo Sapiens* for scale. Could this dinosaur be a connection between birds and dinosaurs?

But scientific arguments are decided on the weight of the evidence rather than the weight of opinion or appeal to authority. With new fossil discoveries, the evidence for a connection between birds and dinosaurs grew steadily in the shadow of the more public brawl over extraterrestrial events and the extinctions at the K-T boundary. What is this evidence, and how do paleontologists view it today?

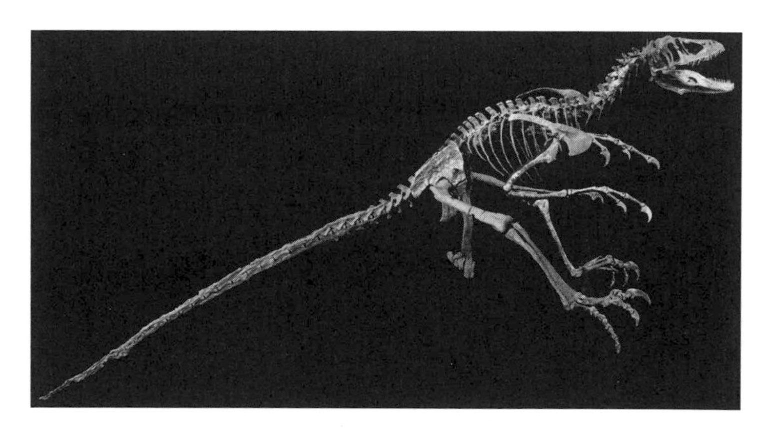

***Deionychus antirrhopus*, leaping to attack.**

JOHN OSTROM AND *DEINONYCHUS*

The seeds of this debate were planted many years before we arrived at Berkeley by Yale University's celebrated paleontologist John Ostrom. His 1964 discovery of the birdlike dinosaur *Deinonychus antirrhopus* was already a famous story. By our arrival, the discovery's significance had grown far beyond Ostrom's expectations when he first gazed upon the bones in the ground southeast of Bridger, Montana. As he studied them over the next two decades, he became increasingly convinced that the bones might hold the key to solving one of the oldest and most vexing problems of evolutionary history—the origin of birds.

Deinonychus was both one of Ostrom's best discoveries and one of his best rediscoveries.[1] Actually, the first bones of *Deinonychus* were collected in 1931 and 1932 by Barnum Brown, the legendary dinosaur hunter from the American Museum of Natural History. The bones came from a ranch on the Crow Indian Reservation, about 35 miles (56 km) northeast of Ostrom's Yale locality. During the 1931 season, Brown discovered a poorly preserved skull and several dozen additional fragments, including partial hands and feet. Brown knew

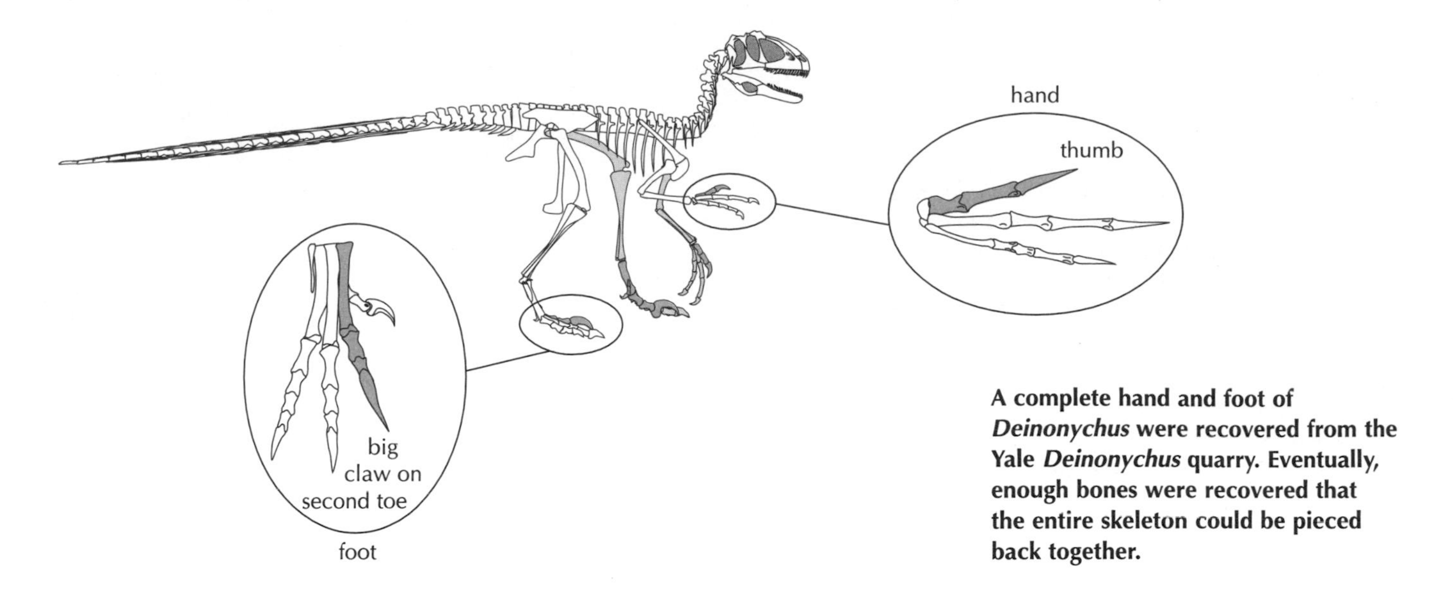

A complete hand and foot of *Deinonychus* were recovered from the Yale *Deinonychus* quarry. Eventually, enough bones were recovered that the entire skeleton could be pieced back together.

that small carnivorous dinosaurs are rare and that he had discovered something new and important. Back in New York, he had illustrations prepared, developed plans to mount a skeleton for display, and even came up with a new name for his find—"Daptosaurus." But Brown never finished the manuscript, so the name and illustrations were never published, and the bones never went on display. Thirty-two years after it was made, the first discovery of *Deinonychus* died with Barnum Brown.

Shortly before Barnum Brown's death in 1963, he spoke to John Ostrom, who was then a graduate student at Columbia University looking for a dissertation topic in the American Museum's vast fossil collections. Ostrom met Brown to discuss potential topics, including the geology and paleontology of the beds that Brown had prospected thirty years earlier. While Ostrom did not choose this as a dissertation topic, shortly after completing his Ph.D. he picked up the lead again and led a party from Yale to Montana to rediscover *Deinonychus*. His crew found an entirely new site, the Yale *Deinonychus* Quarry. The fossils were hard, black, shiny, and beautifully preserved. The isolated bones of at least three individuals were discovered at this site.[2] A complete hand and foot were collected, as well as bones of the skull, so that a whole skeleton could be pieced together.

Ostrom's rediscovery of *Deinonychus* proved to be more significant than Brown's, because with a more complete skeleton, Ostrom was able to study *Deinonychus* in much more detail. It had teeth like steak-knives, whose pointed tips curved backwards, with sharp, serrated edges for sawing flesh. It also had long, recurved claws on its hands and feet. *Deinonychus* was predaceous, like its cousin *Tyrannosaurus rex*. But it was much smaller, weighing in at roughly 100 lbs. The hands on its long, slender arms made a peculiar swivel motion at the wrist. It ran at fairly high speeds. Ostrom recognized *Deinonychus* as "an animal so unusual in its adaptations that it will undoubtedly be a subject of great interest and debate for many years among students of organic evolution."[3] He was right. *Deinonychus* focused the problem of avian ancestry squarely on dinosaurs, and in the 1970s this was a wrenching shift in perspective.

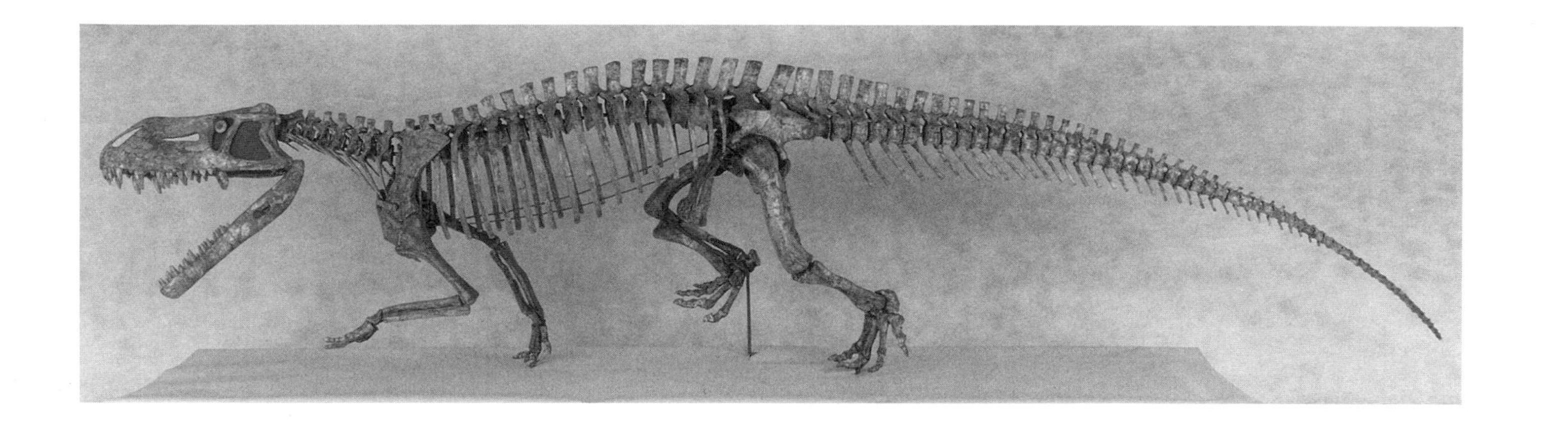

Prestosuchus chiniquensis is not a dinosaur. Instead, it is closely related to crocodylomorphs. In the proportions of its limbs and body, it reflects the condition from which birds, dinosaurs, and several other archosaur lineages evolved.

ROBERT BROOM AND *EUPARKERIA*

Before the rediscovery of *Deinonychus*, most paleontologists agreed that birds descended from primitive archosaurian reptiles, often referred to as thecodonts, which died out 100 million years before *Deinonychus* walked across southern Montana. Dinosaurs were also traced back to the same ancestor, making birds sort of siblings of dinosaurs. Modern crocodylians, the extinct flying pterosaurs, and other extinct reptilian lines were also thought to have evolved from primitive archosaurs. From then on, however, each lineage split off onto its own separate evolutionary path. Accordingly, dinosaurs, crocodylians, and pterosaurs were classified together in a reptilian group named Archosauria, the "ruling reptiles." Although the common ancestry of birds and archosaurs was acknowledged, birds were placed in a totally separate class—Aves—to signify that, by evolving feathers, flight, and warm-bloodedness, they had traveled across a far greater distance of evolutionary change than the others.

But putting birds in a separate class obscured the answer to the principle evolutionary mystery: From which particular lineage of archosaurs did birds descend? This is a difficult problem because when the different lineages are compared with birds, special resemblances are evident with each. For example, birds and pterosaurs both have wings for flight, suggesting to some paleontologists that birds are most closely related to pterosaurs. But a mosaic of resemblances suggests other possibilities. The feet of birds look much like the feet of dinosaurs such as *Deinonychus*, having three principal toes that point forward. In fact, the nineteenth century naturalists who discovered the first dinosaur trackways thought they were the tracks of giant birds. Some paleontologists, like John Ostrom, argue that foot structure provides evidence of close relationship between birds and dinosaurs. Still other features have been identified that seem to link birds to crocodylians, turtles, and mammals.

The explanation offered for this confusing mosaic of similarities is that the special resemblances, like the wings and three-toed feet, evolved separately in the different lineages, a phenomenon known as convergent evolution. According to this hypothesis, any special resemblances that birds share with any of the other archosaurian lineages are the results of convergence, and not evidence of close relationship.[4] This solved many of the problems that arose from conflicting combinations of similarity—they were all independently evolved.

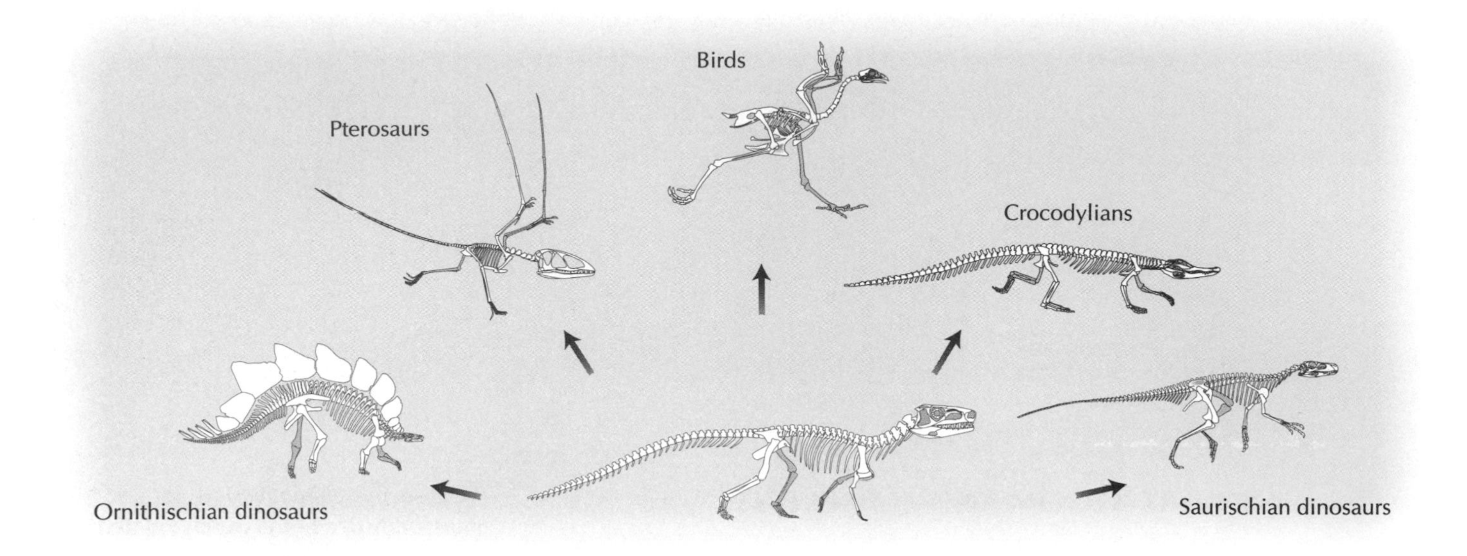

According to the popular twentieth-century "thecodont hypothesis," *Euparkeria,* or something very much like it, was the ancestor to all these lineages (skeletons not to scale).

Logically, of course, only one archosaur lineage could contain the real ancestor of birds, just like only one person can be your genealogical father or mother, and we'll investigate the "paternity" of birds in upcoming chapters.

This view of avian origins, called the "thecodont hypothesis," sprouted over a century before the name *Deinonychus* was coined. It has been advocated ever since by a sizable constituency of paleontologists and is the view championed by most ornithologists. At the time, there were several competing views about the relationships among birds and the various archosaurs. But the idea that birds, pterosaurs, and dinosaurs all diverged onto separate evolutionary pathways, grew in popularity toward the end of the nineteenth and through the first half of the twentieth century as more and more fossils came to light from different parts of the world. Around 1910, the discovery of *Euparkeria capensis* by Mr. Alfred Brown (no relation to Barnum Brown) of Aliwal North, a small town along the Orange River of South Africa, solidified the perspective that *Deinonychus* would later threaten.[5]

Euparkeria is small, just under a meter from its snout to the tip of its tail. It is still among the oldest known archosaurs, dating back to the Early

The skull of *Euparkeria capensis.* This specimen is housed in the South African Museum, Cape Town.

Triassic, about 240 million years ago. Both its generalized structure and its antiquity implicate *Euparkeria* in archosaur ancestry. Its teeth are bladelike, serrated, and curved, leaving little doubt that *Euparkeria* was predaceous. Its limbs indicate that it usually walked around on all fours—an habitual quadruped. However, the forelimbs are slightly shorter than the hindlimbs, implying that it occasionally ran on its hindlimbs—a facultative biped. Facultative bipedal running is the fastest gait in modern reptiles of like proportions, and this was probably the case in *Euparkeria*.

A dozen or so individuals were collected near Aliwal North, but the exact location of their burial site is unknown today. Many paleontologists have searched, but none has found more specimens. Our effective knowledge of this important early archosaur resides in Alfred Brown's original collection, which today is distributed among museums in South Africa, Germany, England, and the United States.[6]

Alfred Brown eventually showed the fossils to Robert Broom, South Africa's preeminent paleontologist. Broom immediately recognized that *Euparkeria* was ". . . very near to the ancestor of the Dinosaurs, Pterodactyles [sic.], Birds and Crocodiles, [and] its extreme importance will at once be manifest."[7] So it was. *Euparkeria*, the primal archosaur, long reigned as the most important discovery of the twentieth century in terms of archosaur evolution and the origin of birds. In a highly influential book published in 1927, entitled *The Origin of Birds*, Gerhard Heilmann further solidified the view.[8] Since that time, most students have learned that from an ancestor such as *Euparkeria*, over the vastness of Mesozoic time, birds, crocodilians, pterosaurs, and dinosaurs slowly, independently, and divergently evolved.

The great South African paleontologist Robert Broom, circa 1950, during a visit to the University of Texas at Austin.

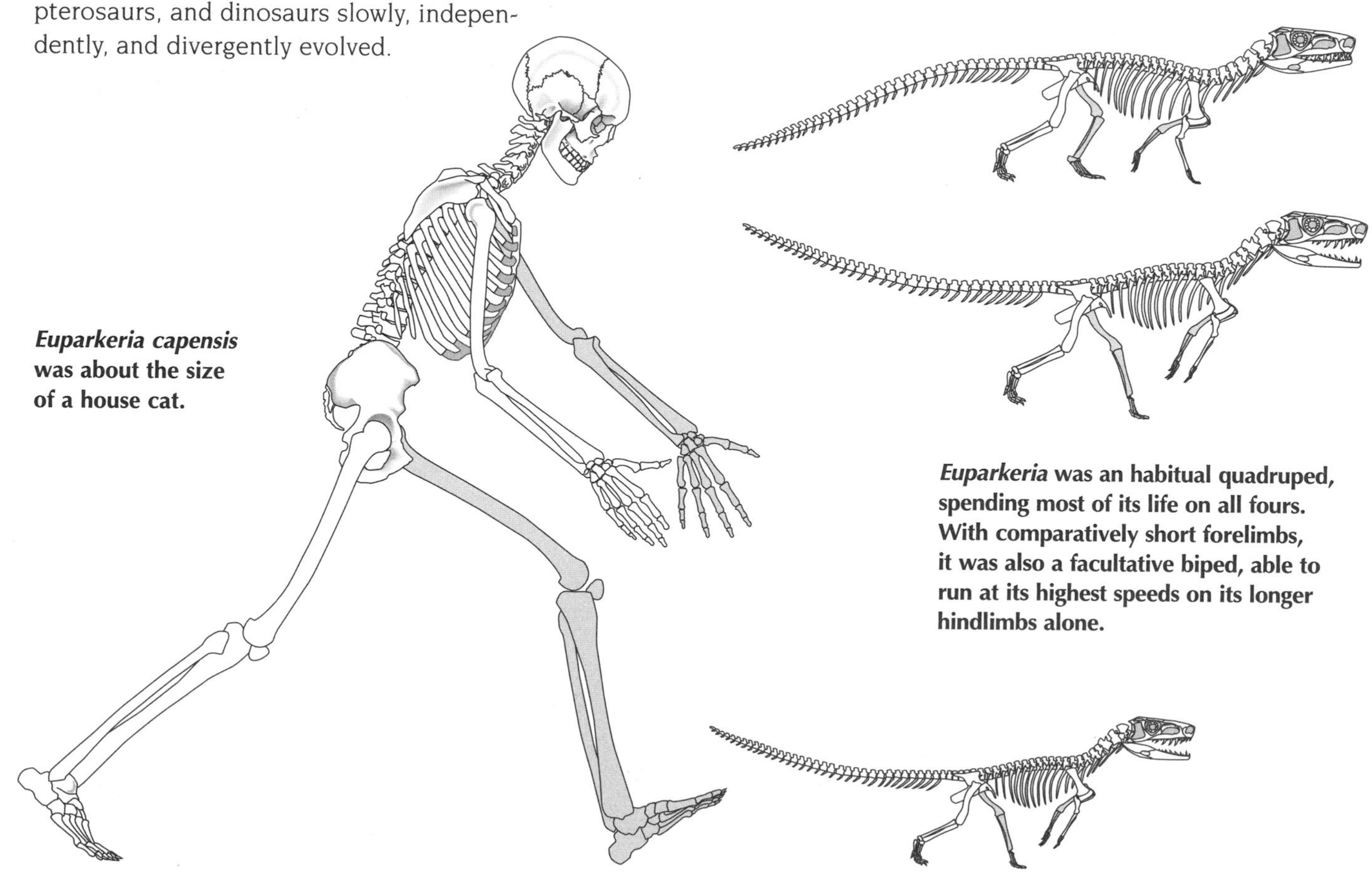

***Euparkeria capensis* was about the size of a house cat.**

***Euparkeria* was an habitual quadruped, spending most of its life on all fours. With comparatively short forelimbs, it was also a facultative biped, able to run at its highest speeds on its longer hindlimbs alone.**

FROM THE TREES DOWN

Identifying *Euparkeria* as a possible ancestor to birds and other archosaurs presented one problem, however. The oldest known bird is 100 million years younger than *Euparkeria*, leaving a long, dark gap of pre-avian history. Paleontologists wondered what the intermediate "protobirds" might have looked like, how they functioned, and behaved. Without fossils, only speculation was possible. Most of the controversy in early avian history revolved around bird's revolutionary new form of locomotion—flight. How did birds evolve flight from ground-dwelling, four-legged ancestors?

Most scientists, Broom included, believed that the ground-living ancestors first moved into the trees. These "protobirds" scrambled up tree trunks and along branches using all four limbs and jumping branch-to-branch much like today's squirrels. To reach the ground to feed, they scampered down the trunks or leapt from low branches. With evolutionary change, "protobirds" began to parachute and then glide down from greater heights. At first, gravity provided most of the power for airborne locomotion, which at this stage was limited to parachuting and gliding. Minor increases in the body's surface area and positioning of that surface improved their gliding ability. To embryologists, feathers are merely modified scales, so the reptilian *Euparkeria* probably possessed the evolutionary forerunners of feathers. It only required a selective environment that might lead to their elaboration into feathers, and the trees would seem to provide just that. Gliding eventually led to powered, flapping flight as "protobirds" learned to use their arms more effectively while in the air. Then, the gliders became extinct whereas true birds survived. In essence, flight evolved from the trees down.[9]

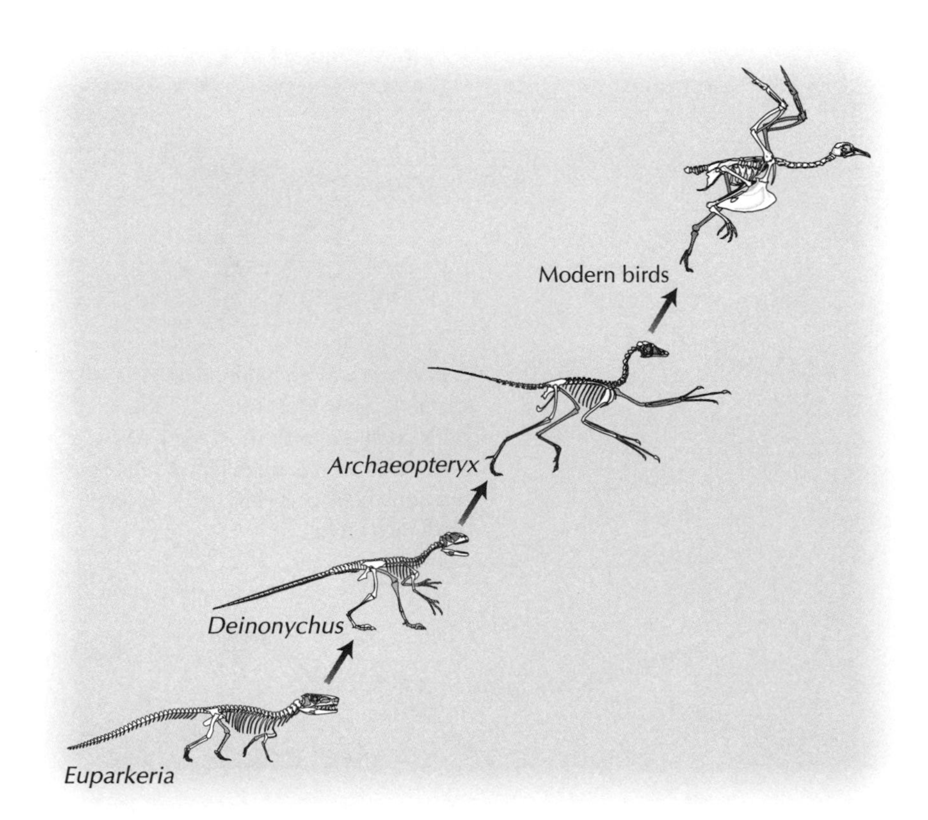

According to the "bird-dinosaur hypothesis," flight evolved from the ground up, as birds evolved from extinct dinosaurs resembling *Deinonychus*.

Although this argument had been around several decades before the discovery of *Euparkeria*, no known fossil could reasonably be interpreted as ancestral to the various archosaurs and birds as well. With Alfred Brown's discovery near the Orange River, the last piece of evidence seemed to be in place.

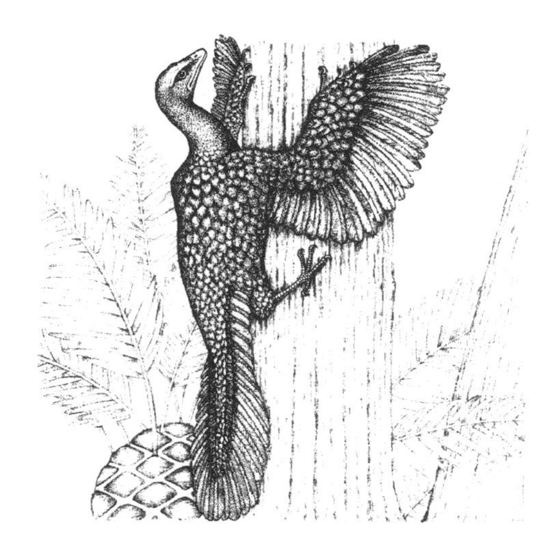

Under the thedocont hypothesis, the ancestors birds are thought to have first moved into the trees and then learned to fly from the trees down.

FROM THE GROUND UP

The "trees down" hypothesis has been compelling to generations of scientists. Intuitively, it would seem that birds *must* have gone through an arboreal stage, and flight *must* have evolved from the trees down. *Deinonychus* came as a rude jolt. It is larger than any modern tree-dwelling bird. Moreover, its skeleton is built for running and taking down prey. *Deinonychus* has long, curved raptorial claws on its hands and feet, including a huge sickle-shaped claw on the end of its second toe for rending flesh, not climbing. Ostrom described the etymology of the name he had coined: *Deinos* (Greek), "terrible," and *onyx* (Greek, masculine), "claw or talon."[10] This name does not smack of squirrels scampering through the trees after nuts. If birds are most closely related to dinosaurs like *Deinonychus*, flight must have evolved from the ground up.

According to the "ground up" hypothesis, the predatory dinosaurs ancestral to birds were fast-running animals that moved on their hindlimbs alone—obligate bipeds—like modern birds. These early dinosaurs also had very powerful forelimbs, like birds do today. The running speed they achieved chasing prey eventually came into play helping them reach speeds necessary to lift off the ground. Throwing their hands forward to rake or grasp for their prey, the ground-dwelling dinosaurs performed an action that was the precursor of the flight stroke. The elaboration of scales into feathers—even small feathers—may convey greater maneuverability in chasing prey over broken ground. Even a small increase in the surface area of the forelimb might provide some lift, and greater degrees of lift could be generated as the powerful limbs swiveled rapidly forward through the air. The act of catching fleeting prey eventually led to the evolution of powered, flapping flight, as the hands and arms of bipedal dinosaurs became elongated and as flight feathers evolved.

For most people, it was hard to imagine how flight could have evolved through intermediate stages from the ground up, where at some point, the forelimbs must have been transitional between the arms of predatory dinosaurs and the wings of a fully flying bird. The thought of a partly volant bird was like the thought of being partly pregnant. Nevertheless, Ostrom argued that the discovery of *Deinonychus* provided evidence that flight evolved from the ground up.[11]

ARCHAEOPTERYX: THE OLDEST KNOWN BIRD

Before even attempting to explain how flight might have evolved from a dinosaur, John Ostrom first had to test whether a close evolutionary relationship really existed between *Deinonychus* and birds. If not, then there was no point in trying to explain how flight evolved from the ground up. The logical place to start was by comparing *Deinonychus* to the oldest known bird, *Archaeopteryx lithographica*. In the process, Owen made several more significant rediscoveries.

To museum curators the name *Archaeopteryx* rings like that of Rembrandt, Stradivarius, or Michelangelo. *Archaeopteryx* is known from the world's most celebrated, controversial, valuable, and rare fossils. The importance of *Archaeopteryx* rests upon its antiquity of 150 million years, the transitional nature of its skeleton, and the timing of its discovery, which occurred only two years after Darwin's *On the Origin of Species* was published. The peerless preservation of the specimens, one a single, perfect feather, has further enhanced their importance and mystique.

In addition to the exquisite but rather uninformative feather, only seven specimens of *Archaeopteryx* are known. Each preserves some or all of the skeleton, along with feather impressions that are more or less obvious. Each is named according to its proprietor or place of discovery. The first skeleton to be discovered is now in England, the London *Archaeopteryx*, arriving in 1862 to become the center of historic debates about evolution that erupted into scientific and public forums with the publication of *On the Origin of Species*.[12] Consequently, it is considered a "crown jewel." Four specimens are in Germany and one is in the Netherlands, where they are similarly acclaimed. The seventh specimen has vanished.

SOLNHOFEN

All known specimens of *Archaeopteryx* were collected from stone quarries in the Solnhofen limestone, located in Bavaria's Franconian Alb region of southern Germany. The limestone outcroppings were exposed by erosion of the Altmühl River. The quarried stones were famous for their marblelike beauty centuries before their fossils were appreciated by the scientific community. Romans first used them for buildings and road paving. Since then, they have been widely valued as ornamental building materials, and they adorn centuries-old palaces and modern state buildings throughout Europe and the world.

The unusual setting in which these rocks formed is responsible for both their fine building attributes and their exquisite, rare fossils.[13] During the Late Jurassic, Bavaria was covered by a shallow sea. Across the arid Altmühl region was a tropical or subtropical lagoon. The lagoon's floor was composed of a chain of basins divided by reefs and shell mounds that restricted water circulation. The bottom water was probably stagnant, very salty and devoid of oxygen, except when flushed out by occasional storm surges. A steady rain of microscopic shells produced by plankton living in the surface waters filtered down onto the quiet basin floors, burying whatever else had drifted in. That limy ooze later turned to rock, petrifying the carcasses of *Archaeopteryx*, pterosaurs, crabs, insects, shrimp, crinoids, and a diversity of fishes. Because the bottom waters lacked oxygen, the usual diversity of scavenging and bottom-feeding creatures was absent. Cadavers deposited on the floor of the lagoon remained undisturbed, producing some of the world's most spectacular fossils with intricate impressions of skin, feathers, and other soft tissues.

An emerging local technology in the eighteenth century catalyzed the discovery of the first *Archaeopteryx* specimens. In the town of Solnhofen, for which the limestone beds were named, Alois Senefelder invented the process known as lithography, using slabs of rock taken from nearby quarries in the Solnhofen limestone.[14] Lithography enabled the first mass publication of illustrations. For many decades it was the principal method for reproducing imagery, fostering a thriving business in the Altmühl district. In the lithography process, an image is rendered onto a flat surface and treated so that it will retain ink, while the nonimage areas are treated in a fashion that repels ink. Once inked, detailed images can be printed by pressing or rolling paper onto the stone. Any flaw, such as a fossilized shell or bone, mars the printing surface, rendering the stone unsuitable for lithography. The Solnhofen stone quarries produce some of the world's finest lithographic stones and command high prices even today, although lithography may be a dying art.

Discovery of the first *Archaeopteryx* specimens occurred near the peak of demand for lithography stones. Looking for stones with perfect surfaces, each slab of limestone was chiseled from the quarry by hand and checked for blemishes. The stones were then set aside for lithography or trimmed into shingles, floor tiles, and so on. For two centuries, workers have also set aside slabs with fossils, owing to their growing commercial value. As quarrymen inspected untold thousands of Solnhofen slabs, eight specimens of *Archaeopteryx* have been found.

THE EIGHT SPECIMENS

The first *Archaeopteryx* specimen recognized to be a bird was the perfectly preserved impression of a single feather found in 1860. Early the next year, Dr. Hermann von Meyer, a preeminent paleontologist from the Senckenberg Natural History Museum, published a short report announcing the unprecedented discovery of a Mesozoic bird. He quickly followed with another publication that included an illustration—a lithograph—showing the unmistakable resemblance of this ancient fossil to the flight feathers of modern birds. His announcements raised a rumble of controversy. The authenticity and age of the fossil were immediately questioned. But a few months later the discovery of the first skeleton temporarily put to rest the question of authenticity. Most of the skeleton was preserved, along with impressions of feathers radiating fanwise from each of the forelimbs and along each side of a long bony tail. In that same year in which he had

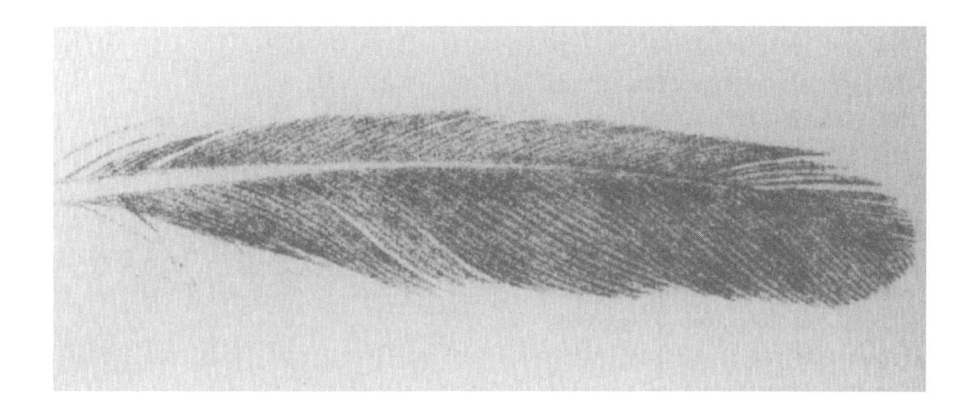

Hermann von Meyer's lithograph of the single feather of *Archaeopteryx*.

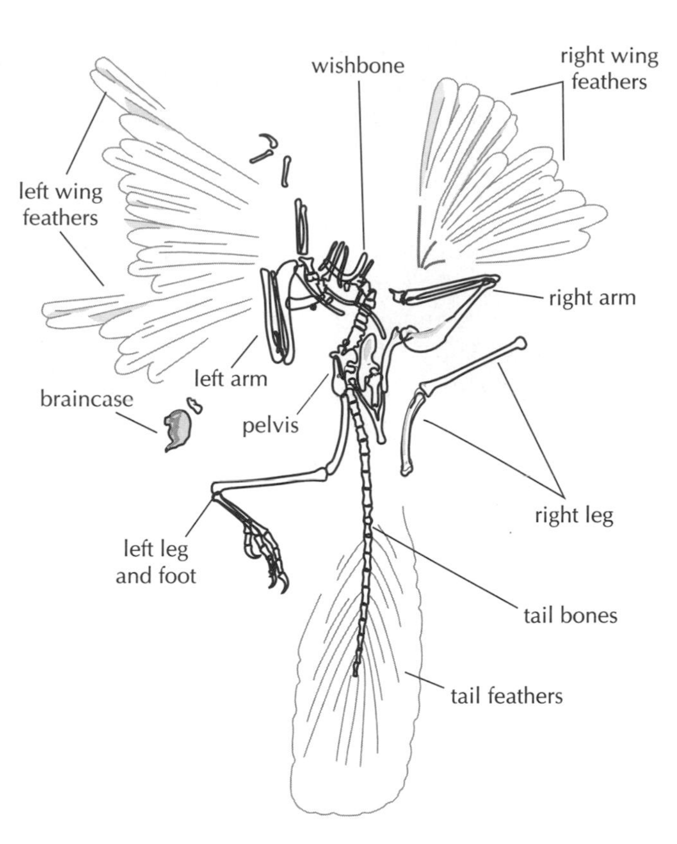

The London specimen of *Archaeopteryx* is housed in the British Museum (Natural History). A photograph *(at left)* of the main slab is shown accompanied by a drawing of the specimen as it was preserved.

announced the feather, von Meyer published a two-page announcement of the first skeleton of a Mesozoic bird. For this specimen, he coined the name *Archaeopteryx lithographica*, the root *archaios* meaning "ancient" and *pteryx* meaning "wing." The species name, *lithographica*, refers to the lithographic limestone in which it was buried and the state of its preservation. By the end of the year, von Meyer's short announcements had grabbed the attention of the scientific community.[15]

The hype surrounding this skeleton was promoted by its first owner, Dr. Karl Häberlein, a medical officer for the district of Pappenheim. Häberlein was a collector who accepted the fossil in payment for his medical services. He appreciated the potential value of such an ancient and well-preserved bird. He let several people inspect it, but no one could make drawings or take photographs. One visitor did make a drawing from memory shortly after viewing the specimen, fueling speculation about the specimen's importance and raising its value. Häberlein offered the specimen for sale at, probably, the highest price yet asked for a fossil, some £750.[16] A scramble for the fossil bird ensued. The German court tried to secure it for the State Collection in Munich, but the British Museum succeeded in negotiating Häberlein's unprecedented final price of £700. The Museum's payment for this and other fossils in Häberlein's collection had to be spread across two fiscal years. The slab and counter slab became known as the London *Archaeopteryx*.

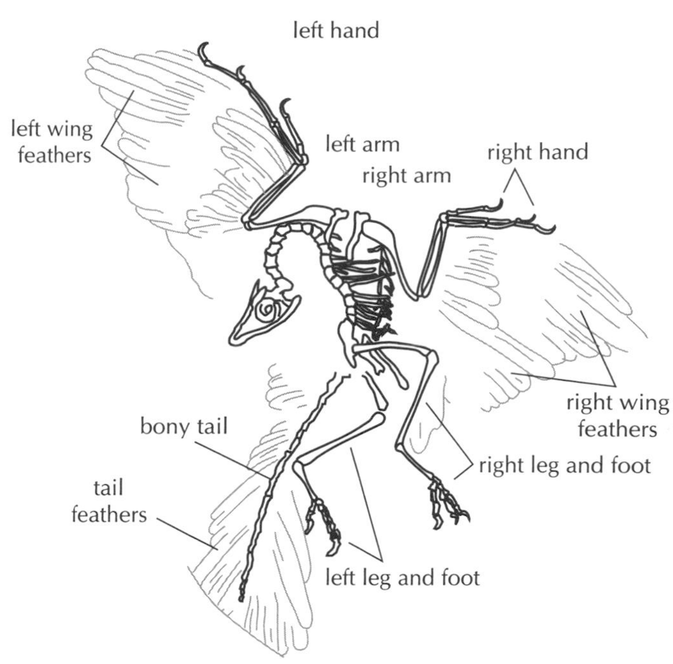

The Berlin specimen of *Archaeopteryx* is now housed in the Humboldt University Museum of Natural History. A photograph of the main slab *(at left)* is shown accompanied by a drawing of the specimen as it was preserved.

A century later, the London *Archaeopteryx* is still regarded as the world's most valuable fossil. Although Britain's Natural History Museum usually displays real specimens, only a cast of *Archaeopteryx* is exhibited for safety, and the museum avoids publicly divulging exactly where the original is stored in the museum.[17]

A third, even more extraordinary *Archaeopteryx* specimen was discovered in the fall of 1876.[18] This skeleton is virtually complete and posed in a natural death posture, with feather impressions spreading out from the forelimbs and tail. It came from a quarry near Eichstätt about 9 miles (15 km) from where the London specimen was found and was acquired by Dr. Ernst Häberlein, the son of Karl. Ernst sold the specimen in 1881, this time to The Museum for Natural History of Humboldt University in Berlin, for £1000.[19] The Berlin *Archaeopteryx* is the most complete and exquisitely positioned of all the specimens found to date. Unlike the London specimen, the Berlin skull remains attached to the neck. It has teeth in the jaws instead of a beak. Parts of the head and teeth were later recognized on the London specimen, but the Berlin *Archaeopteryx* is still the most complete.

Eight decades passed before the fourth *Archaeopteryx* specimen was recovered in 1956, from a quarry near the discovery site for the London specimen. It was quickly identified and described as a new specimen of *Archaeopteryx* in 1959 by Florian Heller,[20] a paleontologist from Erlangen University. It was privately owned by Mr. Eduard Opitsch and exhibited for two decades at the Maxberg Museum near Solnhofen. For a time, the

Maxberg *Archaeopteryx* was accessible to researchers, including John Ostrom.

When Ostrom began his comparison of *Deinonychus* and *Archaeopteryx*, only the London, Berlin, and Maxberg skeletons, along with the single feather, were known. While touring European museums to study them, Ostrom looked at other fossils collected from the Solnhofen limestone. Naturally, he was also interested in pterosaurs, another group of extinct flying vertebrates. This interest led him to the Teylers Museum in Haarlem, the Netherlands, to examine a specimen that had been collected in 1855 and described in 1857 as a new species of pterosaur.[21] The author was none other than Hermann von Meyer, who would four years later coin the name *Archaeopteryx lithographica*. Pterosaurs were well known from the Solnhofen limestone by the mid-nineteenth century. Perhaps because it was incomplete, the true identity of this specimen went unrecognized until 1970, when Ostrom peered into its storage cabinet.[22] Ostrom realized that von Meyer had actually described the first *three* specimens of *Archaeopteryx*.

The sixth and seventh specimens also sat unrecognized for many years after they were first discovered. The sixth is a complete, articulated skeleton that was misidentified as the contemporary dinosaur *Compsognathus*, which is about the same size as *Archaeopteryx*. It was collected in 1951 but not recognized to be *Archaeopteryx* until 1970. Franz Mayr of the University of Eichstätt illuminated the specimen with oblique lighting and was the first to see faint feather impressions along the tail and arms.[23] The Eichstätt *Archaeopteryx* is exquisite and has the best skull of the lot. The seventh specimen was also misidentified as *Compsognathus* for many years after it was collected. Its amateur collector failed to keep record of exactly where near Eichstätt he found it. In 1987 the curator of the Jura Museum, Gunter Viohl, recognized this seventh *Archaeopteryx* in the private collection of Freidrich Müller, a former mayor of Solnhofen. It now belongs to the village and is on display in the Burgermeister Müller Museum. The Solnhofen *Archaeopteryx* is quite complete, and its bones remain in their natural positions, although some critical parts of the skull were lost during excavation or preparation.

The most recent and eighth discovery of an *Archaeopteryx* specimen was made in 1992 from near the sites of the London and Maxberg specimens.[24]

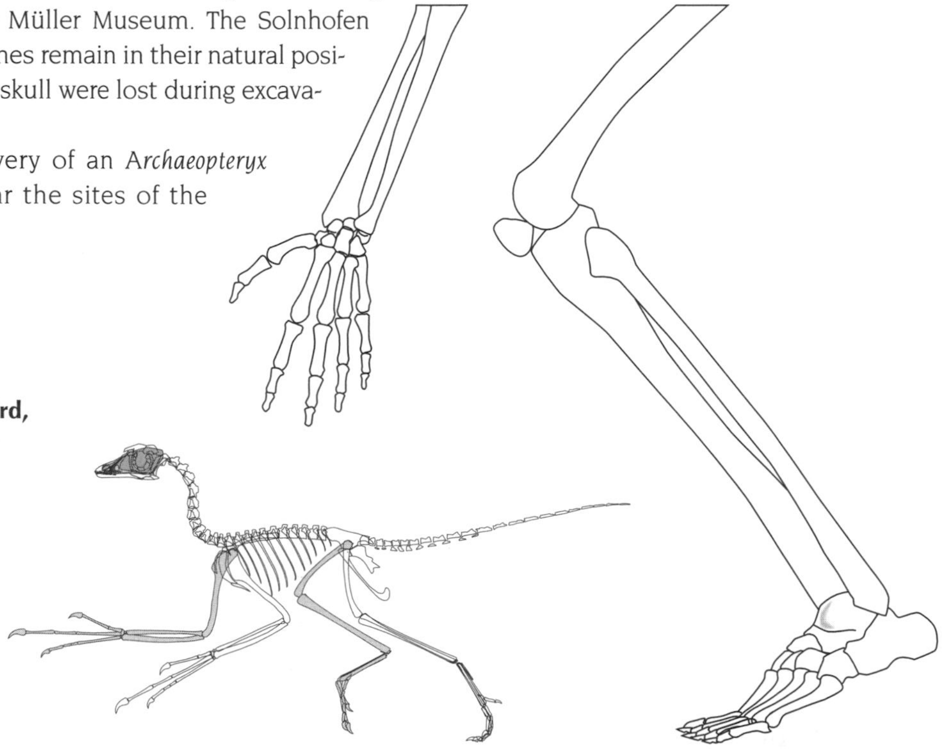

The skeleton of the 140 million-year-old bird, *Archaeopteryx lithographica*, compared to the arm and leg of modern *Homo sapiens* for scale.

It is larger than the other skeletons and preserves a number of important anatomical details not previously seen. As long as the Solnhofen quarries are worked with current techniques, a new specimen of *Archaeopteryx* will probably be unearthed every decade or so.

THEFT?

Tragically, in 1982, Eduard Opitsch, owner of the Maxberg *Archaeopteryx*, removed it from display and took it to his home. Until his death in 1991, he refused scientists access to the specimen. Opitsch even ignored a request to borrow the specimen briefly for the 1984 International *Archaeopteryx* Conference in nearby Eichstätt organized by Dr. Peter Wellnhofer, a preeminent authority on *Archaeopteryx* and director of the nearby Bayerische Staatssammlung fur Paläontologie in Munich. Wellnhofer wanted to gather together all of the *Archaeopteryx* specimens and all of the *Archaeopteryx* experts in one place. Sadly this unprecedented and distinguished conference went off without the Maxberg specimen. Opitsch died a bachelor at the age of 91. Immediately after his death, his nephew and only heir tried to locate the specimen, but failed. There is no evidence that Opitsch sold the specimen. German authorities handling the case believe that someone carried off the Maxberg *Archaeopteryx* in the commotion of activity that had attended Eduard Opitsch's death. Wellnhofer has notified the scientific community of the disappearance, but its whereabouts remain unknown.[25]

FRAUD?

An unexpected challenge to *Archaeopteryx* came from a famous physicist in 1986. Sir Fred Hoyle and Chandra Wickramasinghe, otherwise respected for their works on astronomy and mathematics, published a book entitled *Archaeopteryx*, the Primordial Bird: A Case of Fossil Forgery.[26] Their writings and public presentations attracted considerable interest. With one author knighted, considerable clout accompanied their challenge.

Hoyle and Wickramasinghe claimed that the London specimen was an authentic *Compsognathus*, but that the feather impressions were faked to perpetrate a hoax, in support of Darwin's false theory of evolution. They claimed that *Archaeopteryx* was like the famous Piltdown Man hoax in which altered modern ape bones were planted together with some authentic fossils, fooling paleontologists for about 50 years. Hoyle and Wickramasinghe asserted that the scientific world has been fooled by *Archaeopteryx* for even longer.

These authors alleged that limestone, gouged from around the bones of a genuine *Compsognathus* skeleton, was later poured back into the gouges as a limestone cement and, while still wet, modern feathers were pressed into it. Voilà, a dinosaur with feathers. The forger was the original owner, Dr. Karl Häberlein, who sought to augment the specimen's value. Richard Owen, who orchestrated the specimen's purchase from Häberlein, knew of the forgery and in fact had probably commissioned it himself. Owen may even have arranged the untimely death of Andreas Wagner who published a notice shortly before his death questioning the authenticity of the specimen. Owen, a life-long opponent of evolution, planed to reveal the fraud and discredit the evolutionists.

Inexplicably, the trap was never sprung. Nevertheless, Hoyle and Wickramasinghe allege that subsequent generations of British paleontologists have conspired to maintain the hoax, in a desperate attempt to prop up Darwin's flawed theory of evolution. These authors' preferred evolutionary theory is that viral invasions from outer space periodically introduce new genes into the chromosomes of living organisms. Thus, evolution proceeds in sudden bursts, and transitional forms, like *Archaeopteryx*, must be bogus.

The late Alan Charig, curator of the British Natural History Museum and custodian of the London *Archaeopteryx*, must have relished his chance to respond.[27] Working with a crew of technicians, he detailed the many properties of the specimen that would have had to have been faked. Indeed, the specimen preserves so much detail that the accomplishment of producing a convincing fake would be far more remarkable than the conspiracy itself. Most important would be to match hairline fractures that crosscut both slabs and which were, therefore, made before the slab and counterslab were split. In addition, the impressions of bones and feathers match exactly on the slab and counterslab, so they could only have been made before the split. There is no evidence of a secondary infilling of cement. Charig summarized the charges as "based on a plethora of faulty observations, incorrect data, wrong interpretations, untrue statements and misleading arguments; which, in turn, are due to sheer carelessness, lack of knowl-

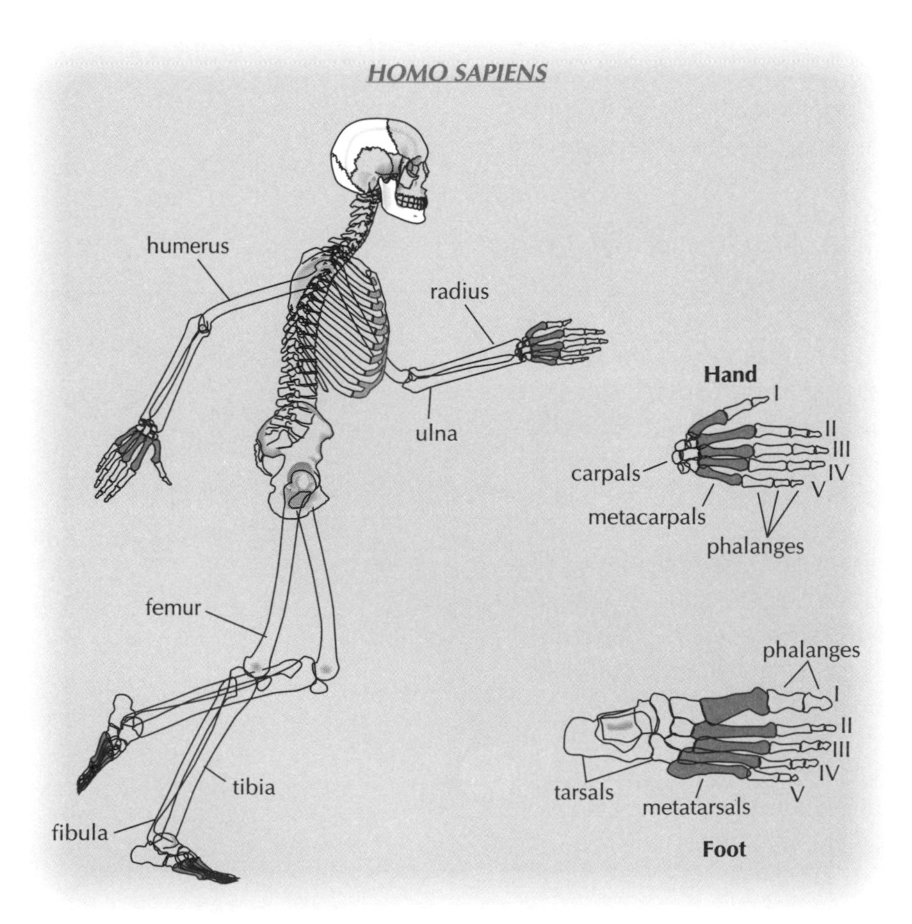

In virtually all tetrapods, the forelimb contains a consistent pattern of bones that extend from the shoulder to the fingers.

edge of the relevant subjects, false logic and a fertile imagination."[28] The charge that *Archaeopteryx* is a fraud is as believable as the "constipation cause" for extinction. The authenticity of the eight *Archaeopteryx* specimens is one of the very few points that everyone on both sides of the bird-dinosaur debate can agree upon.

DEINONYCHUS AND *ARCHAEOPTERYX* COMPARED

During his European trip, Ostrom noted that the limbs of *Deinonychus* and *Archaeopteryx* exhibited many unique resemblances.[29] Animals with four limbs are called tetrapods (*tetra*, "four", *pod*, "feet"); these include birds, dinosaurs, humans, lizards, crocodilians, turtles, and many others. As we will see, evolution has produced many variations on a central pattern of bones that underlies the skin in all tetrapod limbs.

The tetrapod forelimb contains a string of bones that extend from the shoulder to the fingers. Between the shoulder and the elbow is a single bone, the humerus. Between the elbow and the wrist are two bones, the radius, which lie on the inside or thumb side of the forearm, and the ulna, which lies on the outside or "pinkie" side. The arrangement is the same in all tetrapods, except some, like snakes, who lose their forelimbs altogether. The bones of the wrist called carpals, however, vary in number and shape. There were originally probably eight, but they have been variously lost, fused, and otherwise transformed in different tetrapod lineages. Next, the bones forming the palm of the hand are referred to as metacarpals, and there are usually five of them—one for each finger. The fingers are made from numerous phalanges (phalanx is the singular).

Ostrom noted that both *Archaeopteryx* and *Deinonychus* have an unusual three-fingered hand, in which the thumb, digit I, is the shortest and the index finger, digit II, is the longest. Like humans, most other reptiles have five fingers. In *Deinonychus*, there is also an unusual half-moon-shaped bone in the wrist—the semilunate carpal. This bone's shape helps to direct the peculiar swivel path the hand of a living bird takes during the flight stroke, and it is present in *Archaeopteryx*. Might this swivel-wrist be evidence for a close evolutionary relationship?

There are also unique resemblances between them in the hindlimb. The hindlimb is built much like the forelimb in tetrapods. The femur or thigh bone extends between the hip socket and the knee. Between the knee and ankle, the tibia or shin bone runs along the inside, whereas a more slender bone, the fibula, lies on the outside. In birds, the fibula tapers to a point a short distance below the knee and fails to reach the ankle, a very distinctive configuration. The ankle bones, or tarsals, like the bones of the wrist, are numerous and complex. The foot is made up of the metatarsals, which extend between the tarsals and each toe, and there is a variable number of phalanges in each toe.

In both *Archaeopteryx* and *Deinonychus* the outer toe, digit V, has been lost, and the inner toe or first digit, the hallux, is shortened. The principal toes, digits II, III, and IV, are arranged symmetrically about digit III, which is the longest. Ostrom argued that these resemblances are unique, and that they point to a close evolutionary relationship between birds and dinosaurs.

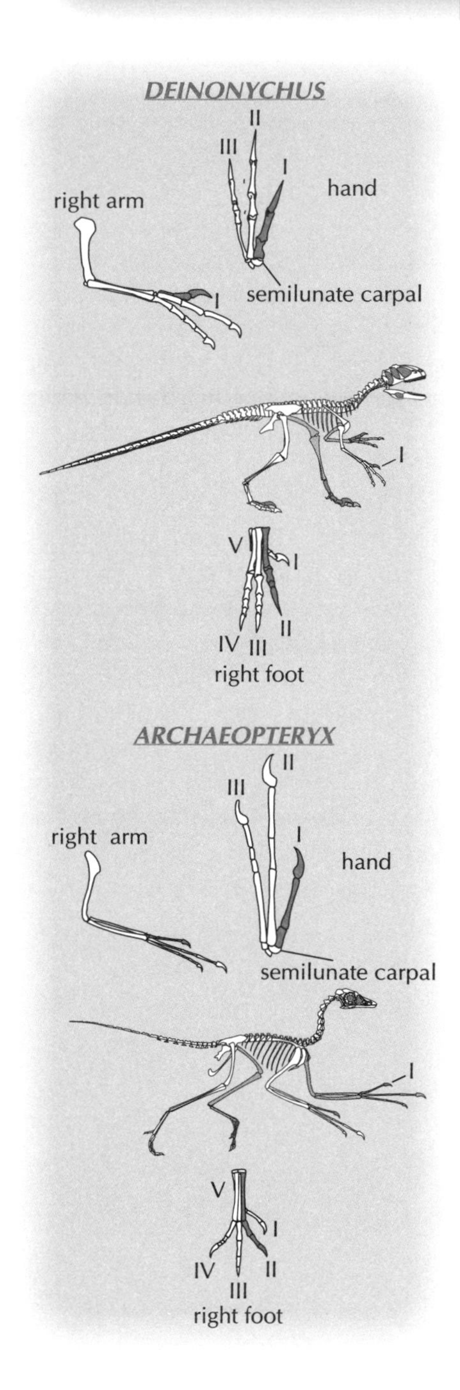

The limbs of *Deinonychus* and *Archaeopteryx* compared.

STILL NO CONSENSUS

Still, there were problems that Ostrom's bold hypothesis did not explain. *Deinonychus*, the birdlike dinosaur thought to resemble the ancestor of birds, lived tens of millions of years *later* than *Archaeopteryx*. So, *Deinonychus* could not be ancestral to birds, and no other dinosaur was thought to share all the unique features pointing to a resemblance with birds. In addition, critics pointed to unique resemblances between birds and other reptiles. For instance, both birds and pterosaurs have tubular skeletons made up of thin-walled, hollow bones, and the braincase of *Archaeopteryx* resembles crocodilians. And what about the origin of flight—how could it have evolved from the ground up?

By the time our first semester at Berkeley ended, there was still no clear resolution to the argument about "trees down" vs. "ground up" origin of birds. Who were the ancestors of birds? Were they dinosaurs or undiscovered tree-dwelling reptiles? Nonetheless, something else was becoming clear. The debate that Ostrom carried to our generation was not new. Virtually the same battle had been fought a century and a half ago in Victorian England, between Darwin and the critics of his theory of evolution. Like the rebirth of catastrophism that occurred with the discovery of iridium in the K-T boundary clay, we watched as a nineteenth-century drama replayed in Berkeley's Paleontology Department. Ostrom's proposed connection between birds and Mesozoic dinosaurs was yet another rediscovery. As we will see, this storm of controversy, like the first one, violently tore at some large branches on the tree of Life.

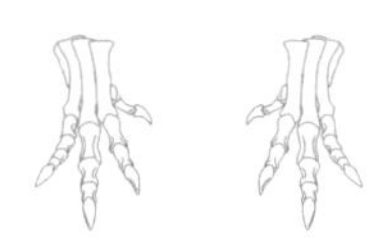

CHAPTER TEN

Dinosaurs Challenge Evolution

More than 150 years ago, the great British naturalist Richard Owen ignited the controversy that the discovery of *Deinonychus* would eventually inflame. The word "dinosaur" was probably first uttered by Owen in a lecture delivered at Plymouth, England, in July of 1841. He had coined the name for the giant fossil reptiles that were discovered in England earlier in the century. The root, *Deinos*, is usually translated as "terrible" but in his first report on dinosaurs, published in 1842, Owen chose the words "fearfully great."[1] To Owen, dinosaurs were the fearfully great saurian reptiles, known only from the fossilized skeletons of huge extinct animals, unlike anything alive today.

Richard Owen as (*left*) a young man at about the time he named Dinosauria, (*middle*) in middle age, near the time he described *Archaeopteryx*, and (*right*) in old age. Today, Owen is recognized as one of the greatest naturalists of all time.

Dinosaur bones were discovered long before Owen first spoke their name, but no one understood what they represented. The first scientific report on a dinosaur bone was printed in 1677 by Reverend Robert Plott in his work, *The Natural History of Oxfordshire*. The broken end of a thigh bone had come to Plott's attention during his research. It was nearly 23 inches (60 cm) in circumference—greater than the same bone in an elephant. We now suspect that it belonged to *Megalosaurus bucklandii*, a carnivorous dinosaur now known from Oxfordshire. But Plott concluded that it "must have been a real Bone, now petrified" and that it resembled "exactly the figure of the lowermost part of the Thigh-Bone of a Man, or at least of some other Animal . . ."[2] Plott did not believe it belonged to a horse, an ox, or even an elephant, but rather to a giant man, evidently relying on biblical accounts of "giants in the earth."

In 1763, Robert Brooke published an even more amazing interpretation of the same specimen. In his *Natural History of Waters, Earths, Stones, Fossils and Minerals, With their Virtues, Properties and Medicinal Uses: To which is added, The method in which* LINNAEUS *has treated these subjects*, Brooke labeled Plott's original illustration *Scrotum humanum*.[3] This ribald name might have been forgotten were it not for Carolus Linnaeus. The great Swedish botanist established a system of taxonomic classification and nomenclature that was already in widespread use in 1763. The Linnaean system uses a binomial or two-name system to describe species, for example, our name *Homo sapiens*.

To correct previous errors and to accommodate reinterpretations, taxonomic names are constantly changing and evolving. Linnaeus set out rules governing the coining and use of taxonomic names, and a commission of scientists now oversees amendments to the Linnaean rules. The rule of priority states that when one species (or specimen) has been given different names (usually by different naturalists), the valid name will be the first given. Brooke's ignominious name for Plott's specimen used a proper Linnaean binomial, the first such name ever applied to a dinosaur bone and, under the Linnaean rules, the proper name for what we now call *Megalosaurus*.

(*At left*) Robert Plott's original figure (1677) of what was probably the first dinosaur bone known to science. (*Above*) Richard Brooke's figure (1763) of the same bone, now christened with the first proper Linnaean name for a dinosaur.

***Megalosaurus,* as restored by Richard Owen in 1854.**

It is perhaps fortunate, then, that Plott's specimen has been lost, so it is impossible to confirm that it belongs to what we call *Megalosaurus*. Citing technicalities of the rules, two paleontologists[4] formally recommended abandoning Brooke's terminology. The Linnaean Commission agreed, laying to rest this false start to the scientific study of dinosaurs.

The early nineteenth century saw the beginnings of our modern understanding of dinosaurs. Numerous discoveries of dinosaur bones were made in Britain, and they provoked more insightful interpretations. In 1824, British naturalist William Buckland described a broken jaw with a single, recurved, serrated tooth of *Megalosaurus*, also from Oxfordshire. Buckland's interpretation is evident from his title "Notice on the *Megalosaurus* or great Fossil Lizard of Stonesfield."[5] Apart from size, Buckland found nothing remarkable about *Megalosaurus*. The next year Gideon Mantell, a British physician and naturalist, described *Iguanodon*[6]—now recognized as a giant herbivorous dinosaur. Mantell thought that "*Iguana* tooth", as its name translates, was a giant lizard resembling today's green *Iguana*. In 1833, Mantell described a second dinosaur, *Hylaeosaurus armatus*, an armored herbivore, which he thought to be another extinct lizard. Early nineteenth-century collections made by the British Geological Survey contain additional dinosaur specimens that went unrecognized until seen through Owen's eyes. Buckland and Mantell came a lot closer than Plott, but as Owen showed, they still missed the mark.[7]

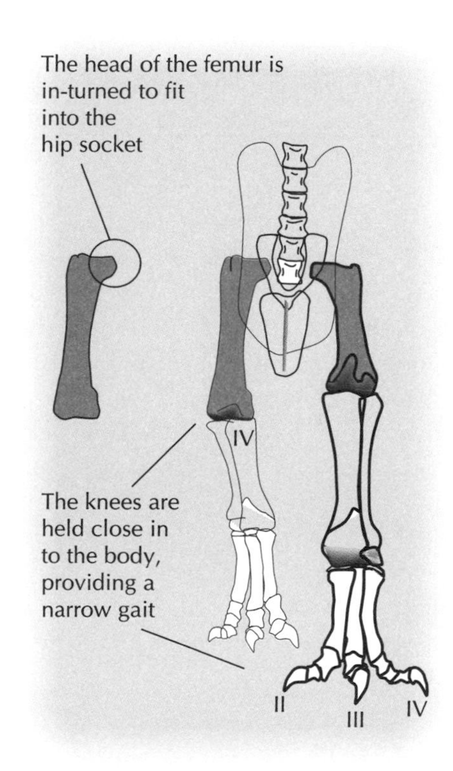

NOT GIANT LIZARDS?

When he first recognized dinosaurs, Owen observed that *Megalosaurus*, *Iguanodon*, and *Hylaeosaurus* shared distinctive anatomical features, so they could be grouped together using the Linnaean system. In addition to their huge size were several characters involving the hip. The thigh bone of dinosaurs has a head that bends at an angle of about 90 degrees, permitting it to fit into the hip socket or acetabulum. Consequently, the femur is held vertically alongside the body, and the hip socket has a hole where, in other reptiles, it is solid. Also, the entire pelvis, which attaches to the backbone at the sacrum, is elongated and reinforced to withstand great forces. So the knees of dinosaurs are turned forward and held against the body, and the head of the thigh bone presses upward against the top of the hip socket. Dinosaurs had a narrow gait, their feet striking the ground beneath the body instead of sprawling out to the side as in lizards. Mesozoic trackways confirm this.

Owen pointed out that dinosaurs more closely approach the limb structure found in mammals and birds than in other reptiles. This was a striking observation because mammals were then viewed as representing a higher level of creation than reptiles, and yet here were reptiles that approached that high plane. Moreover, these long-extinct reptiles appeared far more advanced than any living reptiles. Weighing this evidence, Owen wrote, "The combination of such characters, some, as the sacral ones, altogether peculiar among reptiles, others borrowed, as it were, from groups now distinct from each other, and all manifested by creatures far surpassing in size the largest of existing reptiles, will, it is presumed, be deemed sufficient ground for establishing a distinct tribe or sub-order of Saurian Reptiles, for which I would propose the name of *Dinosauria*."[8]

Major parts of the dinosaur skeleton remained unknown to Owen in 1841. Consequently, his reconstructions of the entire animals were based on some guesswork that later proved false. Owen knew nothing of the dinosaur hand and at first reconstructed them walking on all fours like giant bears. Nearly two decades later more complete skeletons were discovered, indicating that *Iguanodon*, *Megalosaurus*, and many other dinosaurs habitually walked on their hindlimbs alone. But despite incomplete evidence, Owen drew remarkable insights about dinosaurs right from the start. On the last page of his 1842 manuscript, he wrote, "The Dinosaurs, having the same thoracic structure as the Crocodiles, may be concluded to have possessed a four-chambered heart; and, from their superior adaptation to terrestrial life, to have enjoyed the function of such a highly-organized centre of circulation in a degree more nearly approaching that which now characterizes the warm-blooded Vertebrata."[9] To Owen, these skeletons were not merely giant lizards. They bore distinctive anatomical resemblances to modern warm-blooded mammals and birds.

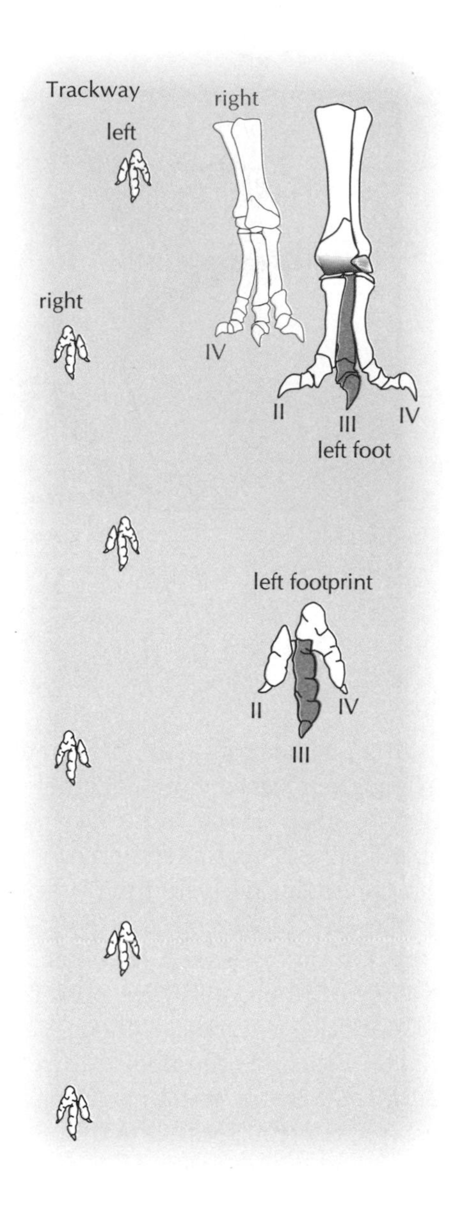

The feet of early dinosaurs left narrow three-toed trackways, indicating that their limbs were held close to their sides.

DINOSAURS AND EVOLUTION

Owen's remarkable conception of dinosaurs alone might have ensured the group's popularity. But another thrust of his 1842 report ensured that dinosaurs would always remain at the forefront of scientific thought. Owen tied dinosaurs to the theory of evolution.

By the 1840s, the scientific community was reacting to the influential work of James Hutton, George Cuvier, and Charles Lyell as their logic and observations transformed geology into a testable science (see Chapter 5). Naturalists had already been discussing various theories of evolution or transmutation of species for several decades. Darwin's grandfather Erasmus had written a book on the subject and several theories had been advanced by the great French biologists Etienne Geoffroy Saint-Hilaire and Jean-Baptiste Lamarck. Darwin's ideas were as yet only sketches in one of his notebooks and remained unpublished.

Like Darwin's, early theories of evolution tried to explain a variety of patterns in Nature. One compelling pattern was the orderly succession of groups in the fossil record. As we saw earlier, George Cuvier interpreted the succession of fossils in rocks around Paris as evidence of successive catastrophes. By Owen's time much of Britain had been geologically mapped by Hutton and his successors. Knowing the stratigraphic sequence of rock layers, geologists could determine which of two fossils was younger and which was older. And there was a consistent pattern. The oldest vertebrate fossils, from the stratigraphically lowest rocks, mostly Paleozoic in age, were fishes. Only in younger Mesozoic rocks did the first land-dwelling tetrapods appear, and mammals and birds were confined to still younger Tertiary rocks. Early evolutionary theorists attempted to account for this orderly succession by invoking a process of progress. Evolution proceeded as a progressive ascent on the "scale of nature." Later organisms exhibited what were deemed to be evolutionary "improvements" over earlier organisms. Owen's first study on dinosaurs would refute this view.

Owen observed that "many races of extinct reptiles have succeeded each other as inhabitants of the portion of the earth now forming Great Britain; their abundant remains, through strata of immense thickness, show that they have existed in great numbers, and probably for many successive generations." He then asked, "To what natural or secondary cause . . . can the successive genera and species of Reptiles be attributed? Does the hypothesis of transmutation of species, by a march of development occasioning a progressive ascent in the organic scale, afford any explanation of these surprising phenomena? Do the speculations of Maillet, Lamarck, and Geoffroy derive any support or meet with additional disproof from the facts already determined in the reptilian department of Palaeontology?"[10]

Ideas about progressive evolution were rejected by later scientists because they made testable predictions about the distribution of organisms in time and space that were refuted by observations of Nature. In this way, Owen used dinosaurs to test and reject the hypothesis. Progressive evolution predicted that modern reptiles should be more advanced than their extinct counterparts. However, Owen observed that "The period when the class of Reptiles flourished under the widest modifications, in the greatest number and the highest grade of organization, is past; and, since the extinction of the Dinosaurian order, it has been declining. The Reptilia are now in great part superseded by higher classes: Pterodactyles have given way to birds; Megalosaurs and Iguanodons to carnivorous and herbivorous mammalia; but the sudden extinction of the one, and the abrupt appearance of the other, are alike inexplicable on any known natural causes or analogies."[11] If evolution occurred at all, it was not via the march of progress envisioned by Owen's French colleagues.

Owen was right, but history has treated him harshly because he advocated a view with even less merit than progressive evolution. His 1842 report concluded that "The evidence . . . permits of no other conclusion than that the different species of Reptiles were suddenly introduced upon the earth's surface, although it demonstrates a certain systematic regularity in the order of their appearance. . . . Thus, though a general progression may be discerned, the interruptions and faults, to use a geological phrase, negative the notion that the progression has been the result of self-developing energies adequate to a transmutation of specific characters; but on the contrary, support the conclusion that the modifications of osteological structure which characterize the extinct Reptiles, were impressed upon them at their creation, and have been neither derived from improvement of a lower, nor lost by progressive development into a higher type."[12] And with that, Owen's lifelong campaign against evolutionists began.

Charles Darwin in 1854, at the time he began full-time research on species.

DARWINIAN EVOLUTION

Darwin's theory of evolution basically states that descent with modification is the process that has yielded today's diverse biota. All species, however divergent, are related, more or less closely, to each other through a long chain of ancestors and descendants. Speciation, the process of diversification, is influenced by a variety of mechanisms.

The principle mechanism of evolutionary change, according to Darwin, is natural selection that results from a struggle for survival faced by all organisms. He described it this way:

> A struggle for existence inevitably follows from the high rate at which all organic beings tend to increase. Every being, which during its natural lifetime produces several eggs or seeds, must suffer destruction during some period of its life, and during some season or occasional year, otherwise, on the principle of geometrical increase, its numbers would quickly become so inordinately great that no country could support the product. Hence, as more individuals are produced than can possibly survive, there must in every case be a struggle for existence, either one individual with another of the same species, or with the individuals of distinct species, or with the physical conditions of life. It is the doctrine of Malthus applied with manifold force to the whole animal and vegetable kingdoms; for in this case there can be no artificial increase in food, and no prudential restraint from marriage. Although some species may be now increasing, more or less rapidly, in numbers, all cannot do so, for the world would not hold them.[13]

A pool of natural variation is present among individuals in any population or species, and natural selection favors traits in the pool that enhance the survival of some individuals over others. Over the vastness of geological time, survival of the fittest leads to gradual change of characteristics from one generation to the next, so descendants differ markedly from their distant ancestors. In a similar way, humans have artificially selected the traits of domestic animals and crops possessing desirable traits, whereas individuals lacking those traits are prevented from breeding. So, domestic organisms are quite different from their closest wild relatives. With sufficient time, natural selection may produce new races and species or lead to extinction:

> As man can . . . and . . . has produced a great result by his methodical and unconscious means of selection, what may not nature effect? . . . natural selection is daily and hourly scrutinising, throughout the world, every variation, even the slightest; rejecting that which is bad, preserving and adding up all that is good; silently and insensibly working, whenever and wherever opportunity offers, at the improvement of each organic being in relation to its organic and inorganic conditions of life. We see nothing of these slow changes in progress, until the hand of time has marked the long lapses of ages, and then so imperfect is our view into long lost geological ages, that we only see that the forms of life are now different from what they formerly were.[14]

Charles Darwin, 20 years later, in 1874.

Darwin's theory of natural selection is often thought to be synonymous with his theory of evolution, but they are really two distinct theories. The theory of evolution relates the process of descent to the pattern of diversity that exists today. Darwin's theory of evolution differs from other theories by saying that descent leads only to divergence, and not necessarily to progress. Both progressive and degenerative changes could occur, depending on whatever particular mechanism of change was operating on a particular lineage. Descendants will simply be different from their ancestors to greater or lesser degrees, not necessarily better or worse.

Natural selection, on the other hand, is a separate theory about one of several mechanisms that might drive an episode of diversification. Darwin identified several other mechanisms as well, such as sexual selection and what he called blending inheritance. Because genetics and genes had yet to be discovered, some of Darwin's ideas about evolutionary mechanisms, like blending inheritance, have been refuted by later researchers. Identifying which mechanisms have operated in particular evolutionary cases has always been the most problematic and controversial aspect of Darwinian evolution. Even Darwin's strongest supporters, including Thomas Huxley, had a hard time accepting the slow, gradual mechanism of natural selection, and with good reason as we will see in a later chapter. But Darwin's basic tenet of evolution, descent with modification, remains the grand unifying theory of biology. Of course in the mid-nineteenth century these issues were far less clear. By the time Darwin published his theory in 1859, numerous theories had been advanced and refuted. After all, Owen had been correct in refuting the progressive evolution of Lamarck. Any new theory about evolution should be met with healthy skepticism.

One of the earliest and greatest tests of Darwinian evolution involved the fossil record. All debaters agreed that today's diverse species are not arrayed into a complete continuum of form, and that Nature is instead divided into distinctive clumps of species separated by wide gaps. Green plants are unmistakably different from animals, turtles from birds, and so on. Within larger groups, smaller groups and gaps are also discernible. Among birds, hummingbirds, woodpeckers, and ducks are readily distinguished. The critics of evolution asked, if today's diversity emerged by gradual modification from some common ancestral form, why do these gaps exist?

In response, Darwin's ally, Thomas Huxley argued that "We, who believe in evolution, reply that these gaps were once non-existent; that the connecting

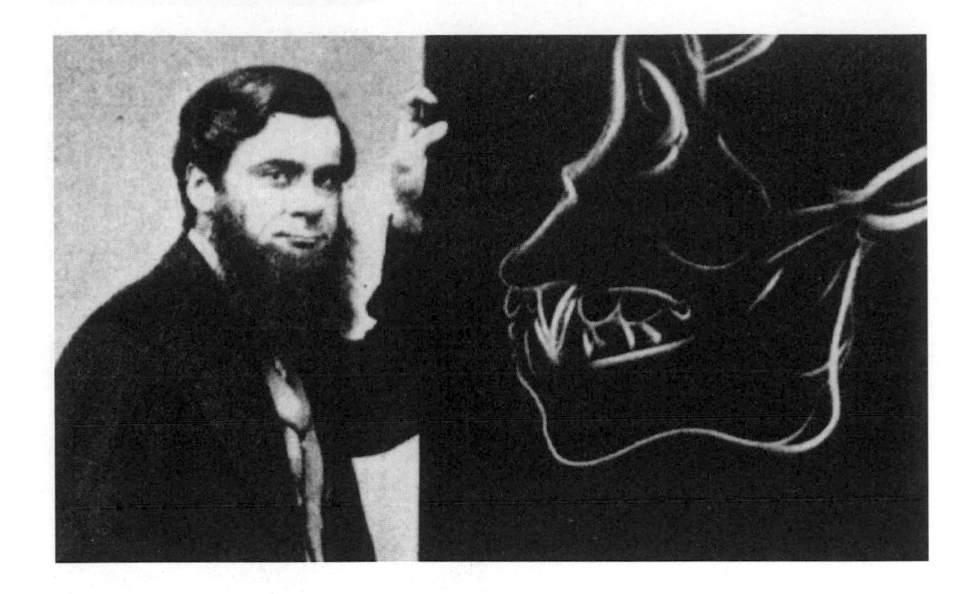

Thomas Huxley, one of Darwin's greatest supporters, was the first great promoter of the dinosaur-bird hypothesis.

forms existed in previous epochs of the world's history, but that they have died out."[15] Living lineages, however distinctive and different from one another, should be connected in the past through their fossil records.

Testability was the standard for Darwin's scientific theory, as it had become for the geological sciences in the decades preceding publication of the *Origin*. Naturally, evolutionists were challenged to produce evidence of the extinct transitional forms that Darwin's theory predicted. But just as incompleteness of the rock record compromises our ability to test both catastrophic and gradualistic explanations for the K-T extinction episode, incompleteness also compromises our ability to connect lineages back through time. Darwin appreciated this and explained that we should expect such gaps.[16] Fossilization is an exceptional process. Only a minute fraction of the organisms that have lived on Earth are preserved as fossils; far fewer yet have been discovered by naturalists. So our record of ancient life is highly incomplete.

Darwin's critics labeled this as an excuse to cover another fundamentally flawed theory of evolution. And without any transitional fossils to counter this claim, they might just as well have been right.

THE PROBLEM WITH BIRDS

When *On The Origin of Species* hit the newsstands, one of the most glaring gaps in the fossil record was the early history of birds. Birds are highly distinctive. Nobody would mistake a bird for any other animal. But in the mid-nineteenth century their fossil record appeared to begin abruptly during the Tertiary. Some naturalists argued that some three-toed Mesozoic trackways were made by more ancient birds. But some of these were much too large for known birds to make, so most naturalists agreed with Richard Owen that these "Foot-prints alone . . . are insufficient to support the inference of the possession of the highly developed organization of a bird of flight by the creatures which have left them."[17]

The earliest fossil birds known at that time were already very highly distinctive, having wings and the ability to fly. And at their first appearance in the fossil record, there were already a number of different lineages. To Owen, this reflected the pattern of creation. Darwinians, however, viewed this as an artifact of an incomplete fossil record. They predicted that a long pre-Tertiary history of fossil birds linking the various bird lineages together through common ancestors would eventually be found, as well as fossil "protobirds" linking birds with other tetrapods. But if this is true, where were the fossils?

Compounding the incompleteness problem, birds are so distinctive that evolutionists weren't sure which other tetrapod represented their closest living relative. With their warm-bloodedness, feathers, and flight, birds seemed as different from other tetrapods as tetrapods are from fish. All tetrapods, including birds, mammals, crocodylians, lizards, turtles, and amphibians, could be distinguished from fish by their four limbs. Turtles, lizards, and crocodylians were all cold-blooded and had scales. So they belonged to a group within tetrapods called reptiles. But that was as far as general agreement went.

This limited understanding was reflected in the classification system used at that time. Owen and Darwin were basically still using Linnaeus' system. Birds, mammals, reptiles, amphibians, and fishes were each placed in a separate class, a Linnaean rank denoting a large group that looks very distinct form other animals. Class Aves, for birds, was equal to but entirely separate from Class Reptilia, Class Mammalia, and Class Amphibia. While this allowed all animals to be properly classified in the Linnaean system, it ignored the critical genealogical question: Are birds more closely related to mammals, reptiles, or amphibians?

Working in 1763, Linnaeus had no idea that species were related, so he never attempted to design a system of classification that reflected genealogy. But as the Darwinian Revolution swept across the scientific community, naturalists began to search for the closest relative of birds. Most naturalists speculated that birds were most closely related to reptiles, but which reptiles? Some allied birds with turtles, based on the shared presence of a toothless, horn-covered bill or beak. Other naturalists allied birds with crocodylians or lizards. Richard Owen classified animals based on "unity of type" not genealogy. His solution to the problem was to classify birds and mammals together based on warm-bloodedness. But if Darwin was right about all species being related, these possibilities could not all be true. There could be only one true, historic set of relationships. Who were the closest living relatives of birds? Who were the ancestors of birds among extinct organisms?

ARCHAEOPTERYX AND EVOLUTION

Two years after publishing *On The Origin of Species*, Darwin was under attack in both the scientific and popular press. And then *Archaeopteryx* was discovered. It was an instant scientific sensation—both evolutionists and antievolutionists claimed it as their prize and used it to open a new battlefront in the escalating war over evolution.

It is difficult to imagine a more highly charged atmosphere than the one existing when *Archaeopteryx* was discovered. Its Late Jurassic age fitted Darwinian predictions perfectly. The feather impressions looked like modern bird feathers down to their microscopic detail. But *Archaeopteryx* also had a long bony tail, more like the cold-blooded reptiles than any living birds. In addition, the three fingers of its hand (or wing) were separate, with claws on the end. Modern birds have three fingers that fuse into a single bone to help strengthen the wing. And who ever heard of a bird with claws on its hands? Were these the primitive features expected in a Mesozoic intermediate between birds and other tetrapods?

On November 9, 1861, Professor Andreas Wagner, a distinguished German paleontologist, announced the discovery of what would become known as the London specimen at a meeting of the Royal Academy of Sciences of Munich. However, the real issue of Wagner's address was evolution. Wagner had never actually seen the skeleton, so he relied upon a report by a lawyer named Witte who had seen it in Karl Häberlein's possession. As Owen later recounted the story, "Upon the report thus furnished to him, Professor Wagner proposed for the remarkable fossil the generic name *Griphosaurus*, conceiving it to be a long-tailed Pterodactyle with feathers. His state of health prevented his visiting Pappenheim for a personal inspection of the fossil; and, unfortunately for palaeontological science, which is indebted to him for many valuable contributions, Professor Wagner shortly after expired."[18]

Wagner's terminal scientific thought that night was a shot at evolution: "In conclusion, I must add a few words to ward off Darwinian misinterpretation of our new saurian. At the first glance of the *Griphosaurus* [Wagner's proposed name for *Archaeopteryx*] we might certainly form a notion that we have before us an intermediate creature, engaged in the transition from the saurian to the bird. Darwin and his adherents will probably employ the new discovery as an exceedingly welcome occurrence for the justification of their strange views upon the transformation of animals but in this they will be wrong."[19]

ARCHAEOPTERYX: BIRD OR FEATHERED REPTILE?

Was *Archaeopteryx* a feathered pterosaur or some other kind of feathered reptile? Was it really a reptilelike bird—a transitional fossil like Darwin's theory predicted? Grasping instantly its pivotal importance to the debate over evolution, Owen snatched up the specimen for the British Museum. It arrived in London in 1862.

Now, with both the specimen and one of the world's premiere skeleton collections to compare it with, Owen wasted little time in responding to the speculations of evolutionists, who had never even seen *Archaeopteryx*. In 1863 he published a thorough description, masterfully crafted to rebut the publicity *Archaeopteryx* had already received as a potential link between birds and reptiles.

The skeleton is divided between two halves of the split slab. Prior to burial, the skeleton became partly dismembered, and the skull and jaws were apparently lost. But most of the skeleton and many feathers were preserved. As Owen described it, "The remains of *Archaeopteryx*, as preserved in the present split slab of lithographic stone, recalled to mind the condition in which I have seen the carcase of a Gull or other sea-bird left on an estuary sand after having been a prey to some carnivorous assailant. The viscera and chief masses of flesh, with the cavity containing and giving attachment to them, are gone, with the muscular neck and perhaps the head, while the indigestible quill-feathers of the wings and tail, with more or less of the limbs, held together by parts of the skin, and with such an amount of dislocation as the bones of the present specimen exhibit, remain to indicate what once had been a bird."[20]

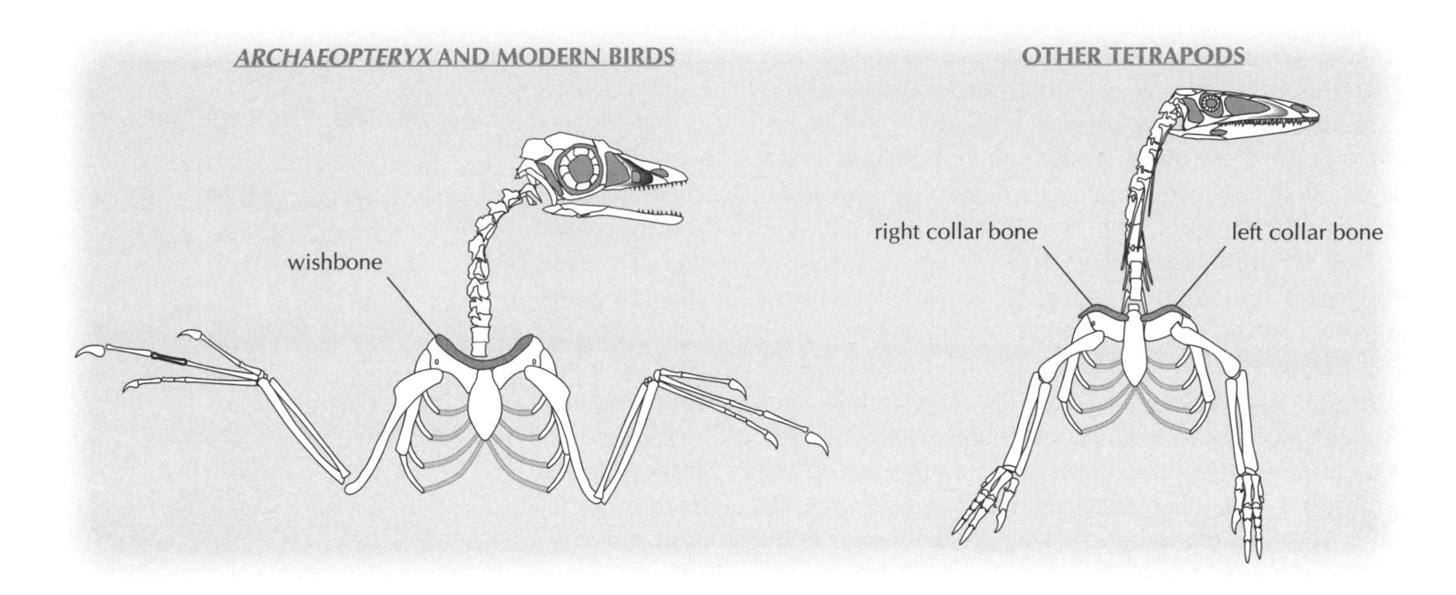

Along with its size, proportions, and feathers, the skeleton showed many other resemblances to modern birds, including a wishbone or furcula. Among living species, the furcula is known only in birds and had never before been discovered in a fossil reptile. It represents a fusion of the right and left collar bones or clavicles. As a young bird embryo develops, the collar bones form separately, reflecting the condition found in most adult tetrapods. But by hatching time, the avian clavicles fuse to form the highly characteristic U-shaped furcula. Once fused, the wishbone acts as a spring to help power

***Archaeopteryx* has what Owen considered to be definitive avian features, like the wishbone or furcula, which forms when the two collar bones become fused together.**

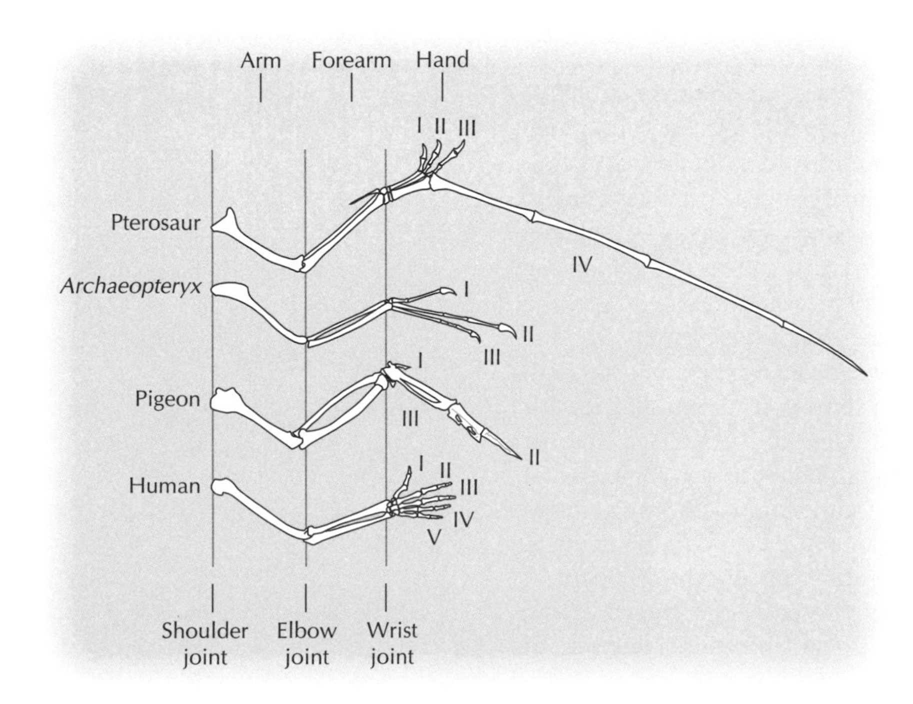

The arms of a pterosaur, *Archaeopteryx*, a modern bird, and a human are compared.

the flight stroke and may also facilitate breathing during flight. Because the furcula is absent in flightless birds like ostriches, some naturalists have argued that it must be an essential component of flight. Who else but a fully volant bird would have a furcula?

Also birdlike are the shoulder and arm of *Archaeopteryx*. The wrist has the half-moon-shaped semilunate carpal, just like *Deinonychus*. The hand proportions, the three fingers, and the relationships of the wing feathers to the fingers are also birdlike. Owen went to great length to show that pterosaur wings found around Solnhofen are different from *Archaeopteryx*. The pterosaur wing's membrane of skin often forms an impression on the lithographic stone. The membrane is supported by the fourth finger, a digit that is entirely absent in *Archaeopteryx* and adult living birds. This was clearly no feathered pterosaur, as Andreas Wagner had claimed.

Owen reported that the pelvis and hindlimb are especially birdlike. One of the most important resemblances is in the sacrum, and the pelvis that attaches to it. The sacrum is composed of separate vertebrae whose short and massive ribs attach to the inside of the pelvis, anchoring the backbone to the hindlimb. The elongated sacrum in *Archaeopteryx* and modern birds is massive and reinforced to withstand the forces generated in landing. The inner wall of the hip socket is perforated. The head of the thighbone is bent sharply to fit deeply into the hip socket, in characteristically avian fashion. Even more birdlike, the bones of the ankle are fused together, some to the end of the shin, others to the top of the foot, producing a hingelike ankle. The upper foot bones, the metatarsals, are partially fused as well for both strength and stability. Three toes point forward, while the first toe is turned backward so that it can grasp against the other three. Does any of this sound familiar?

Owen noted that the limb bones in *Archaeopteryx* are hollow, thin-walled tubes. Modern birds have an expanded respiratory system that extends throughout much of the body as a series of air sacs. They enhance the efficiency of oxygen extraction by the lungs and lighten the skeleton. Some air sacs actually extend into the hollows of the bones. The first clues to this remarkable lung system were discovered a century before Owen set eyes on *Archaeopteryx*. In 1758, John Hunter, a leading British physician and naturalist, discovered the unique connection. After tying closed the windpipe of a domestic fowl, he cut through the humerus, the bone between shoulder and elbow, and "found that the air passed to and from the lungs by the canal in this bone. The same experiment was made with the os femoris [the femur] of a young hawk, and was attended with a similar result."[21] Hunter had demonstrated that birds' bones contain air, not marrow, and that the air spaces are connected to the respiratory system. The hollow skeleton of *Archaeopteryx* was further evidence that it is a bird.

But Owen noted striking differences between *Archaeopteryx* and modern birds, and it is these features that caused the sensation surrounding its discovery. Its peculiar hand had three separate fingers, two of which were equipped with claws. The unusual, 8-inch (20-cm) bony tail was composed of sixteen vertebrae, and impressions indicated that the quills of the tail feathers were attached to the vertebrae by thin ligaments. In modern birds, there are only a few vertebrae at the base of their stubby tail, and the tail feathers attach to a plow-shaped bone at the end of the backbone known as the pygostyle. Were the evolutionists right that these are transitional features, or was *Archaeopteryx* just another bird?

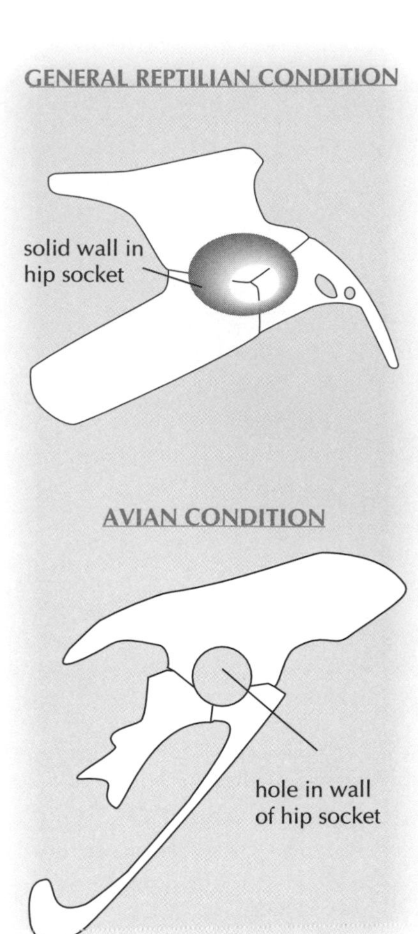

The hip socket, or acetabulum, is perforated in *Archaeopteryx* and other birds, contrasting with the general reptilian condition in which there is a solid wall inside the socket.

Utilizing his understanding of embryology, Owen downplayed the importance of these potentially transitional features. He pointed out that the embryonic pygostyle forms from separate cartilages, like the separate vertebrae in *Archaeopteryx*, before they fuse to form the adult pygostyle. Similarly, the avian hand forms from the fusion of the second and third fingers, which are separate in early development. To Owen, both the hand and tail of *Archaeopteryx* were minor developmental variants of the basic avian plan; similar variants are seen in the tails of bony fishes and other vertebrate groups. And the presence of claws, Owen argued, was highly variable in living groups. Even some modern birds, like the Kiwi, have a claw on the end of one finger in the hand.

In summation, Owen counterattacked against Darwin, "Thus we discern, in the main differential character of the by-fossil-oldest known feathered Vertebrate, a retention of a structure embryonal and transitory in the modern representatives of the class, and a closer adhesion to the general vertebrate type. . . . The best-determinable parts of its preserved structure declare it unequivocally to be a bird, with rare peculiarities indicative of a distinct order in that class."[22] The differences between *Archaeopteryx* and other birds were minor compared to the overall similarities in the skeleton and the presence of feathers. In Owen's view, *Archaeopteryx* fell squarely in the Class Aves; it was not transitional between birds and reptiles.

In embryonic birds and young hatchlings, the tail is made up of several separate bones or cartilages, like the adult tail of *Archaeopteryx*. In modern birds, these separate elements eventually fuse to become the adult pygostyle.

BRIDGING THE GAP

But to an equally famous evolutionist like Thomas Huxley, *Archaeopteryx* looked like a perfect transition between birds and reptiles, suggesting that one class could "transmute" into another.[23] Furthermore, Huxley loathed Owen, and relished the opportunity for a fight. Their personal acrimony spilled out into the tabloids, and the debate became highly adversarial.

Unlike Owen, Huxley compared *Archaeopteryx* to fossil reptiles as well as to living birds, noting that today "no two groups of beings can appear to be more entirely dissimilar than reptiles and birds. Placed side by side, a Humming-bird and a Tortoise, an Ostrich and a Crocodile offer the strongest contrast, and a Stork seems to have little but animality in common with the Snake it swallows."[24] He then asked, "How far can this gap be filled up by the fossil records of the life of past ages? This question resolves itself into two:—1. Are any fossil birds more reptilian than any of those now living? 2. Are any fossil reptiles more bird-like than living reptiles?"[25]

Huxley answered both in the affirmative. For a fossil bird more reptilian than any modern bird, Huxley of course pointed to *Archaeopteryx*. Like reptiles, the digits of the hand are unfused, and two of them have claws, whereas in living birds there is never more than one claw. The long tail made up of separate vertebrae is also reptilian. Owen's objection that these were merely embryonic and transitory structures notwithstanding, Huxley concluded "it is a matter of fact that, in certain particulars, the oldest known bird does exhibit a closer approximation to reptilian structure than any modern bird."[26]

In answering the second question, Huxley slapped Owen's face by turning to Dinosauria, the very group Owen had discovered. In the two decades since Owen named Dinosauria, great new discoveries of dinosaurs had been made in Britain, Germany, and North America. They showed that some dinosaurs walked on their hindlimbs like birds. Mesozoic trackways confirmed this and showed, moreover, that both huge animals and tiny ones were walking bipedally. In Bavaria a nearly complete skeleton as old as *Archaeopteryx* was described and named in 1861 by Andreas Wagner shortly before his death.[27] *Compsognathus longipes* is only about two-thirds of a meter long. Wagner recognized it as a reptile, but failed to see what kind.

The brilliant German anatomist and evolutionist Carl Gegenbaur studied *Compsognathus* soon after its discovery.[28] He observed that its ankle closely resembled the ankle of birds, one of the features Owen had used to place *Archaeopteryx* within birds. Huxley instantly concurred with Gegenbaur, adding that "there can be no doubt that the hind quarters of the Dinosauria wonderfully approached those of birds in their general structure, and therefore that these extinct reptiles were more closely allied to birds than any which now live."[29] Huxley also recognized that *Compsognathus* must be "placed among, or close to, the Dinosauria; but it is still more bird-like than any of the animals which are ordinarily included in that group."[30]

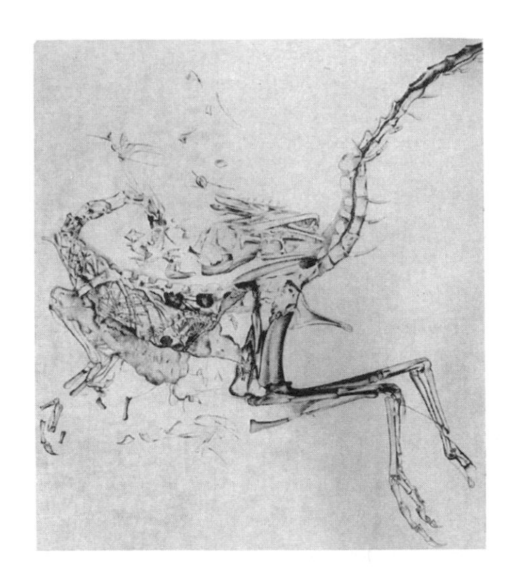

The skeleton of *Compsognathus* was found in the Solnhofen limestones at about the same time as *Archaeopteryx*. It led the great German anatomist Carl Gegenbaur and developmental biologist to discover the connection between Mesozoic dinosaurs and modern birds.

Huxley next turned to embryology, highlighting a special similarity between dinosaurs and young embryonic birds. Expanding on Gegenbaur's observations, he described the leg and ankle of a young Dorking fowl, concluding that if "found in the fossil state, I know not by what test they could be distinguished from the bones of a Dinosaurian. And if the whole hindquarters, from the ilium to the toes, of a half-hatched chicken could be suddenly enlarged, ossified, and fossilized as they are, they would furnish us with the last step of the transition between Birds and Reptiles; for there would be nothing in their characters to prevent us from referring them to the Dinosauria."[31]

Many characters that Huxley used to identify *Compsognathus* as a dinosaur were the same ones Owen had used to distinguish Dinosauria, and Huxley must have appreciated the irony of it. The fact that two specimens of *Archaeopteryx* were long mistaken for *Compsognathus* further highlights the degree of resemblance. *Compsognathus* has the sacrum and hip of a dinosaur but the foot more like that of a bird. The major difference is that the metatarsal bones remain separate in *Compsognathus*, whereas they become fused in birds. It was surprising to find these features in such a small animal, however. Dinosaurs were supposed to be huge. But other paleontologists quickly saw the resemblance, and the idea of a close connection between birds and dinosaurs found a wide following over the next few years.

Huxley even inferred that the physiology of dinosaurs might have achieved a level found in modern birds. "Birds have hot blood, a muscular valve in the right ventricle, a single aortic arch, and remarkably modified respiratory organs; but it is, to say the least, highly probable that the Pterosauria, if not the Dinosauria, shared some of these characters with them."[32] Huxley and Owen had both seen birdlike features in extinct dinosaurs.

THE THEORY OF HOMOPLASY OR INDEPENDENT EVOLUTION

Showing that tiny, birdlike dinosaurs existed closed the gap between birds and reptiles even more than *Archaeopteryx* had, by linking birds with a particular kind of reptile. It was also the critical step, for both Huxley and John Ostrom a century later, in connecting birds to dinosaurs.

So what happened to the connection? Why did *Deinonychus* resurrect the point Huxley apparently won a century before? What changed the scientific world's mind in the last quarter of the nineteenth century?

Harry Seeley, Owen's protégé, proposed the argument at one of Huxley's lectures before the Geological Society of London. Huxley cited the similarities in the hindlimbs of birds and dinosaurs as evidence of a close evolutionary relationship. At the end, as recorded by the secretary of the society, Seeley rose and noted "that the peculiar structure of the hinder limbs of the Dinosauria was due to the functions they performed rather than to any actual affinity with birds."[33] Later, he enlarged on the idea: "All the characters whereon are based the claim of dinosaurs to be regarded as the ancestors of birds are only related to the power of keeping an upright position upon the hind feet."

Seeley claimed that the resemblances between birds and dinosaurs, like running on the hindlimbs, could have evolved independently. No close evolutionary connection was necessary. Natural selection might possibly shape comparable evolutionary solutions in distant relatives, owing to comparable demands of the environment. Seeley also believed that the linking together of extinct lineages, as Huxley was trying to do with birds and dinosaurs, was virtually impossible. Lastly, he noted that only some, not all, dinosaurs had the birdlike resemblances. There was a mosaic of characters distributed among these fossils that he argued were best interpreted as independent acquisitions.

This notion was eventually termed the "theory of homoplasy."[34]Homoplastic features are corresponding and similar parts in different organisms whose similarity is not inherited from a common ancestor. Independent acquisitions like the wing of a bird and a bat are said to be homoplastic because, based on what we know about their relationships, we can conclude that their wings evolved independently. The last common ancestor shared by birds and bats did not fly.

The theory of homoplasy was widely accepted by naturalists. Those who supported dinosaurian ancestry complicated the problem by arguing over which dinosaur was the ancestor. Some would derive birds from the carnivorous dinosaurs while others argued that it was the herbivorous forms like *Iguanodon* that held the ancestry of birds. Even Huxley later straddled the fence. Although he derived birds from reptiles, he did not subscribe to the direct derivation of birds from dinosaurs, stating "It may be regarded as certain that we have no knowledge of the animals which linked reptiles and birds genetically, and that the Dinosauria, with *Compsognathus*, *Archaeopteryx*, and the struthious birds only help us to form a reasonable conception of what these intermediate forms may have been."[35]

THE $64 QUESTION

What Huxley had originally resolved into two questions, paleontologists now resolved into one: Are birds directly descended from dinosaurs or not? Huxley had answered the question with a strong affirmative to an initially warm reception. But his following in the scientific community on this point steadily diminished toward the end of the century. One reason is that no known dinosaur presented a suitable ancestor. The particular mosaic of characters observed in known fossils seemed to disqualify each one of them. One seemingly insurmountable problem was that no dinosaur was known to have clavicles—the collar bones—for without the clavicles where could the furcula have come from? Several influential paleontologists argued that complex features like the clavicles, once lost, could never reappear. "Dollo's law," as it was called, seemed to put the last nail in the coffin for dinosaurs as the ancestors of birds. According to Dollo's law, once complex features like a clavicle have evolved and then been lost, they were unlikely to ever reappear. As we will see, there are many exceptions to Dollo's law, which has been discarded by twentieth-century biologists. But for a time it seemed to offer a more accurate explanation for the actual patterns observed in the paleontological record. Birds and dinosaurs evolved as separate lineages arising from some earlier, common ancestor. Any advanced similarities they might share must have arisen numerous times.

Several additional discoveries were taken to support this view. The discovery of part of the head and jaws of *Archaeopteryx* on the London slabs must have embarrassed Owen, who had seen the bones but dismissed them as belonging to a fish. The jaws had teeth, but the braincase was indicative of a big brain, like living birds. The Berlin *Archaeopteryx* came to light at about this time, confirming the presence of teeth in a fossil bird. The discovery of additional toothed Mesozoic birds by O. C. Marsh in the Cretaceous chalks of Kansas seemed to add support to the homoplasy theory. His two most complete finds were *Hesperornis* and *Ichthyornis*, both of which were aquatic and had a mosaic of characters best explained as having evolved independently from the comparable features in dinosaurs. The common ancestor for the two groups must have existed in the early Mesozoic, or before.

In Robert Broom's hands, the Early Triassic *Euparkeria* became the perfect candidate for the ancestor from which birds and dinosaurs split off. It was almost uniformly primitive. Crocodylians, birds, dinosaurs, and pterosaurs could all have evolved directly from *Euparkeria* or from something very much like it. There were several other primitive archosaurs, all poorly known, of Triassic age, that became grouped together with *Euparkeria*. Somewhere within this assemblage of primitive archosaurs, named Thecodontia, was the ancestor. The thecodont ancestry of birds was the most popular view from then on, or at least until *Deinonychus* raised its distinctive birdlike head.

By the time Ostrom began to suspect that *Deinonychus* was more than just another dinosaur, the homoplasy theory had even been carried to Dinosauria itself. Some paleontologists were arguing that the resemblances Owen had originally noted between the various dinosaurs were also homoplastic, and the things we call dinosaurs had evolved these features several times. Not only were birds not derived from dinosaurs, but the dinosaurs themselves did not form a natural group and were instead separately evolved from different thecodont ancestors.

The many new discoveries since 1841 only seem to complicate our picture of the past. Did the discovery of *Deinonychus* take us a step backwards or a step closer to answering whether birds and dinosaurs are related? By the time our first year of graduate school at Berkeley had ended, it seemed that the only thing paleontologists all agreed on is that dinosaurs are extinct.

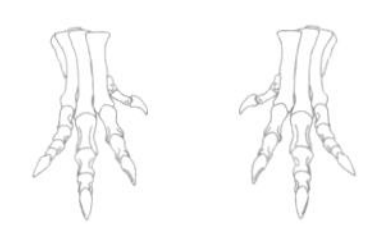

CHAPTER ELEVEN

Dinosaurs and the Hierarchy of Life

Are birds related to dinosaurs or not? *Deinonychus* carried the question full circle. First Huxley, and then Ostrom, argued that small dinosaurs are closely related to birds. And the same charges of homoplasy launched at *Compsognathus* in Huxley's time were thrown at *Deinonychus* a century later. Are the resemblances of modern birds to extinct dinosaurs genealogical, or merely a coincidence—the result of convergent evolution?

As we watched the debate between Ostrom and his critics unfold, each side emphatically asserted that it was correct. But they couldn't both be right. Birds could have only one true set of relationships, one historic line of descent. The trick was to figure out how to test between the alternatives: How do we distinguish genealogical similarities from those that reflect homoplasy?

To answer this question, we must first come to grips with what a dinosaur is from a modern scientific viewpoint. What features must an animal have to be a dinosaur? While this seems like a simple question, when we arrived at Berkeley we encountered an intense debate over how to answer questions like this. The debate eventually grew across departmental lines to involve many of the faculty and students studying evolutionary biology. It commanded wide attention because of the more fundamental issue at stake involved in reconstructing the past—how can we testably reconstruct the relationships among living and extinct organisms?

RECONSTRUCTING RELATIONSHIPS: MAPPING THE PHYLOGENY OF LIFE

Evolutionary relationship, a shared common ancestry, is what makes each of the various groups of living organisms distinctive and is what provides their biological identities today. Before there was a theory of evolution to suggest that species are in fact related, there was no reason to search for a way to map their relationships. But in the wake of Darwin's theory, many different methods have been advanced for what is known as phylogeny reconstruction—the mapping of evolutionary relationships.

Reconstructing and mapping ancient phylogenetic relationships has become a highly sophisticated science; as a result, these maps can provide answers to a vast range of fundamental biological questions. Biologists now study organisms in microscopic detail as they search for information pertaining to phylogeny, using advanced technologies like high resolution X-ray CT scanners and scanning electron microscopes. Modern computers perform sophisticated image analyses, a wide array of statistical tests, a variety of phylogenetic analyses, and simulations aimed at understanding the past. Just as remote sensing technologies have revolutionized our mapping of the Earth's surface, our ability to map evolutionary relationships has made enormous strides with the advent of microprocessor computing. Thousands of biologists and paleontologists are now involved in mapping the phylogenetic relationships among the myriad branches of life.

But even with modern technology there remain daunting obstacles. Chief among these is the fragmentary record of the past. For most living species there simply is no fossil record. Preservation is the exception to the rule—far more species have come and gone than were ever captured by the fossil record. Of those that did leave fossils, the records are at best incomplete. In general, the older the event, the less information is preserved. Even at best, fossils are mere fragments of a once living, breathing organism. How much do they really tell us about the distant past of the Mesozoic?

Several prominent biologists recently argued that fossils are in fact worthless for reconstructing ancient relationships, and that only modern species need be studied. Colin Patterson of the British Museum of Natural History, a preeminent paleontologist whom we met in an earlier chapter, argued that fossils had no effect whatsoever on modern conclusions about relationship.[1] After all, in living species we can directly observe molecules, soft tissues, coloration patterns, and behaviors, whereas in fossils we can only speculate about these features. Living species present so much more information that they will simply swamp any signal preserved in fossils. Coming from an enormously influential paleontologist like Patterson, this was a powerful argument. Other biologists carried the argument further, claiming that our modern capability to measure the sequences of nucleic acids in the DNA of living species means that we don't even need to keep museum specimens—a drop of fluid is all it takes to reconstruct their relationships.

But many modern species are so radically altered from the appearance of their ancestors, that little of their past is preserved. All systems, including bones, soft tissues, molecules, and behaviors can transform; none of these systems is immune to evolution. And as they change, they overwrite and gradually erase their past like a palimpsest. The problem using modern species alone becomes increasingly severe in reconstructing progressively more ancient patterns. For example, the modern amniotes—birds, crocodylians, lizards, turtles, and mammals—diverged from each other about 300 million years ago. Over that expanse of time, they evolved in such differing directions that it is difficult today to see any clues as to their relationships. Bird feathers, mammal hair, the shells of turtles, and lizard scales all seem equally different from each other. At first glance, their skeletons look equally different as well. Still, if Patterson were correct, the fossil record had no bearing on our understanding of these relationships.

Thomas Huxley took a different position while defending Darwinian evolution a century earlier. Arguing that the gaps separating species today were less marked in the past, he predicted that the fossil record would provide critical intermediary stages that are clues to the evolutionary linkages between modern species. The new tools and computer programs for phylogeny reconstruction gave us a chance to test the importance of fossils in a collaborative study with Jacques Gauthier and Arnold Kluge (University of Michigan, Ann Arbor) on amniote phylogeny.[2] A series of computer analyses that alternatively included and deleted fossils from a phylogenetic analysis of the major groups of living and extinct amniotes found that different genealogies arose when fossils were added or deleted from the analysis. Fossils unquestionably made a difference to mapping phylogeny, so the assertion that fossils were irrelevant to phylogeny reconstruction was refuted. The details of the tests were revealing. Analyzing just the most primitive amniote fossils failed to reveal very strong information on relationships among the major lineages, because those lineages had not yet become markedly differentiated. Analyzing modern species and some of their closest fossil relatives also produced poor results, because those particular fossils had already taken on most of the distinctive patterns of their living relatives. It was the intermediate fossils, which documented the history of evolutionary transformation over several hundred million years, that provided the key to understanding amniote relationships. This might seem like a case of scientists discovering the obvious, but a growing number of biologists had begun to operate under the assumption that the modern biota alone could tell the same story. Our tests showed that by looking at *both* fossils and recent species, we have the best chance of accurately reconstructing the past.

HOW TO READ A PHYLOGENETIC MAP

Life is arrayed in a hierarchy, in which lineages are arranged within more inclusive lineages. There are distinguishing features at each level. The relationships among species can be diagrammed in several ways. These diagrams represent the same hierarchy of relationships, in which B *and* C *are each other's closest relatives. Together, they form the sister lineage to* D. *Similarly,* B–C *and* D *together form the sister lineage to* A.

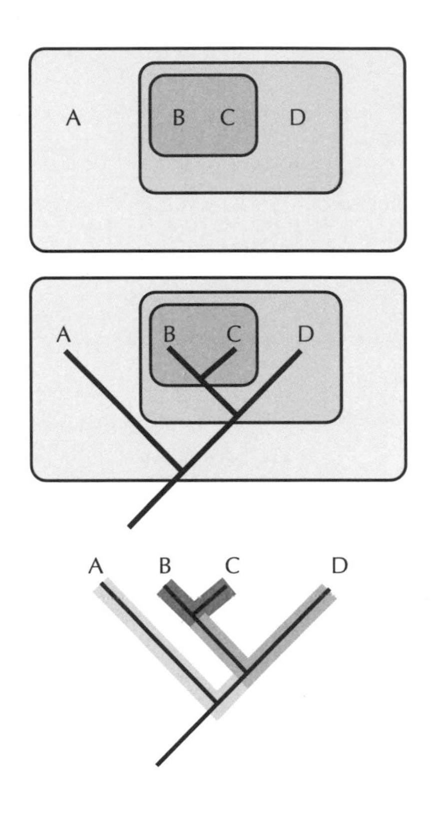

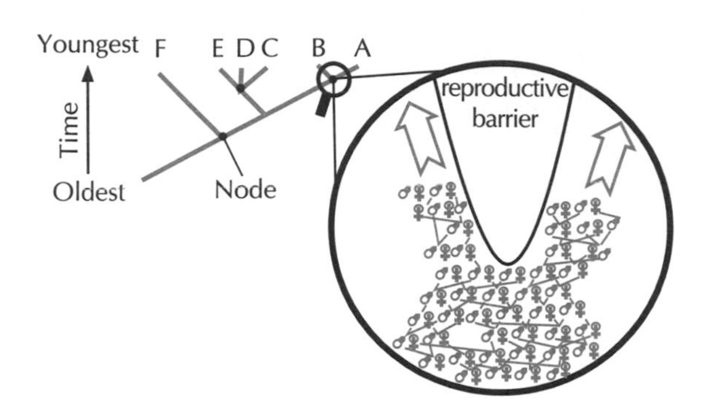

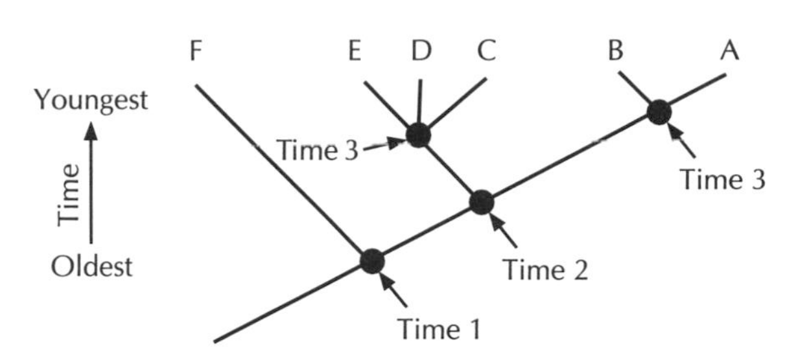

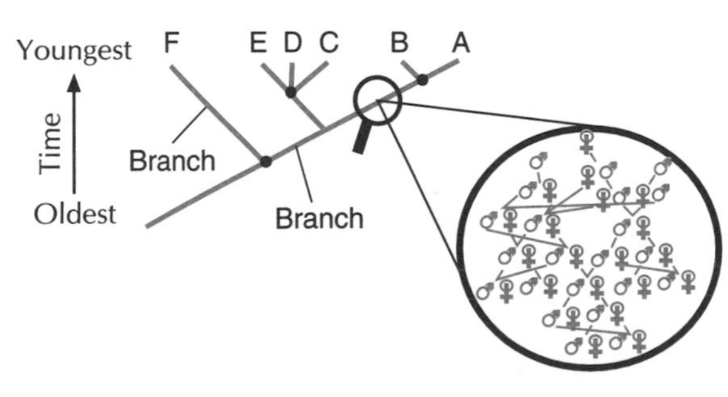

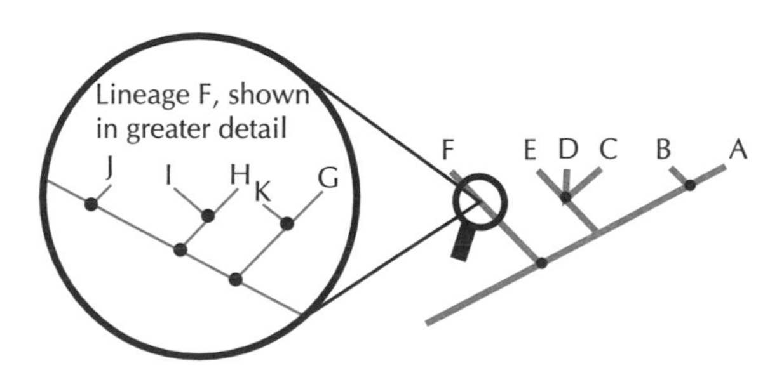

Phylogenetic maps, otherwise known as cladograms, use solid lines to represent lineages. At the lowest levels in the hierarchy of life, these are lineages of interbreeding organisms *(bottom left)*. Environmental factors sometimes create reproductive barriers that split a population into separate, diverging lines of ancestry and descent. These are speciation events. On a cladogram they are represented by nodes—where two or more branches split apart *(top right)*. The relative timing of speciation events is represented vertically, from oldest to youngest branchings. Like geographic maps of different scales, branches on a phylogenetic map, like those labeled A to F, may consist of a single species, or thousands of related species. Branch F, for example, might represent hundreds of species that are simply depicted as a single lineage *(bottom right)*.

As we saw in Chapter 10, a second factor that complicates phylogeny reconstruction is homoplasy, otherwise known as convergent or independent evolution. This was the basis for Harry Seeley's argument against a close relationship between birds and dinosaurs. The problem for Seeley's generation, and for John Ostrom's as well, was that there were no clear methods to determine which similarities reflected homoplasy and which reflected evidence of genealogy.

Given the complications of incompleteness and homoplasy, how do modern biologists reconstruct phylogeny? While we were students, a method called "cladistics" was developed for reconstructing or mapping phylogenetic patterns. Although the methods of cladistics are still evolving, there are several basic ideas that underlie most of this work. One is to assemble all the data one can possibly find with any bearing on a given problem. Like most detectives and most juries, scientists generally prefer explanations that address all the evidence, not just some of it. An explanation of relationships among modern species that rejects fossils is weak compared to one that accounts for all the information. A single explanation for *all* the data is more powerful than a series of special arguments that address different parts of the data.[3] Underlying this idea is the same principle of parsimony, often called Occam's Razor—seeking the simplest explanation for all the data—that we introduced earlier.

A second basic idea in cladistics is to measure and map patterns of unique similarity among the intrinsic, heritable characteristics of organisms. Heritable features include molecular structures, DNA sequences,

Vertebrata forms a distinct lineage whose members are distinguished by their common possession of a backbone or vertebral column (dark green).

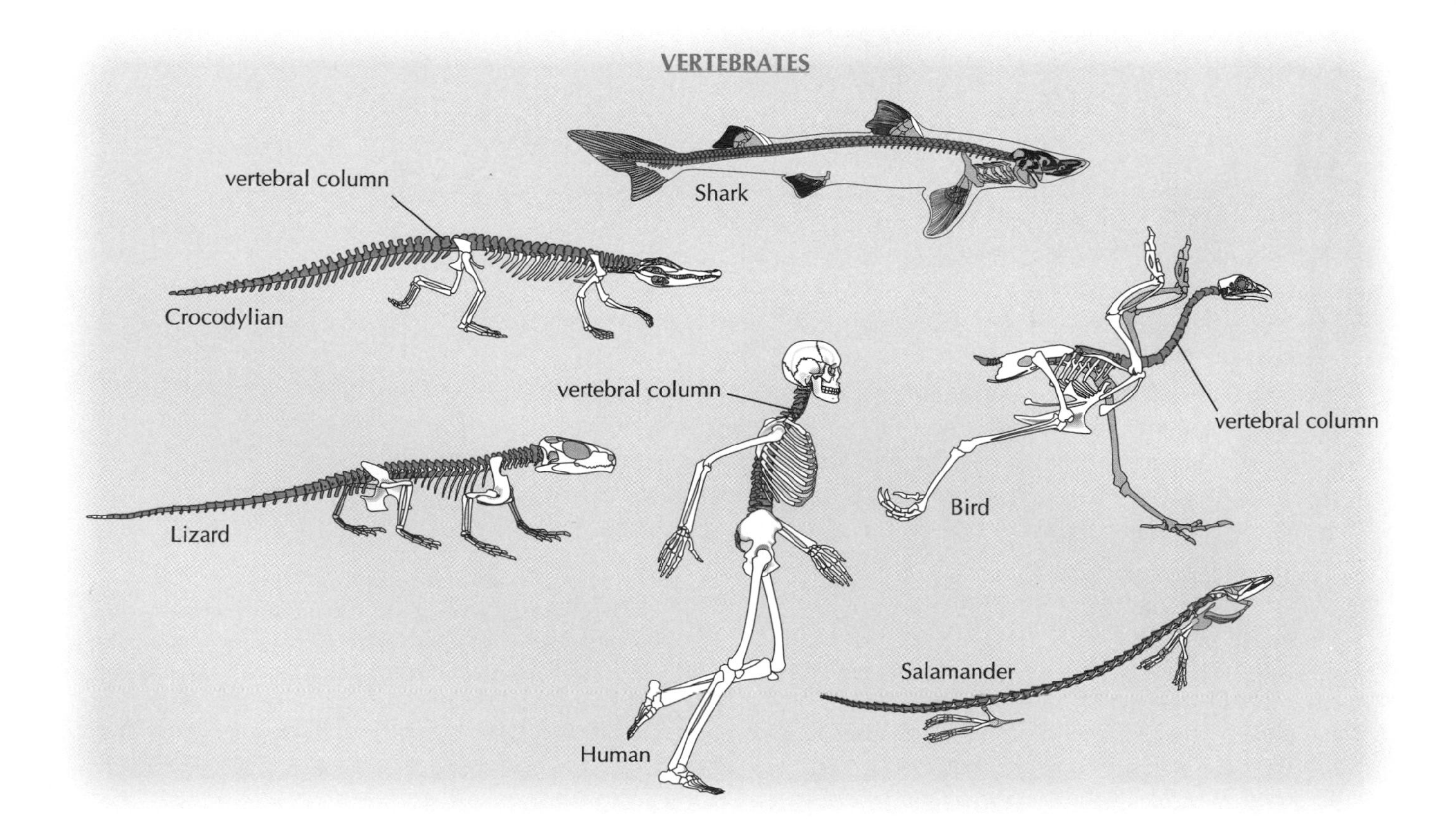

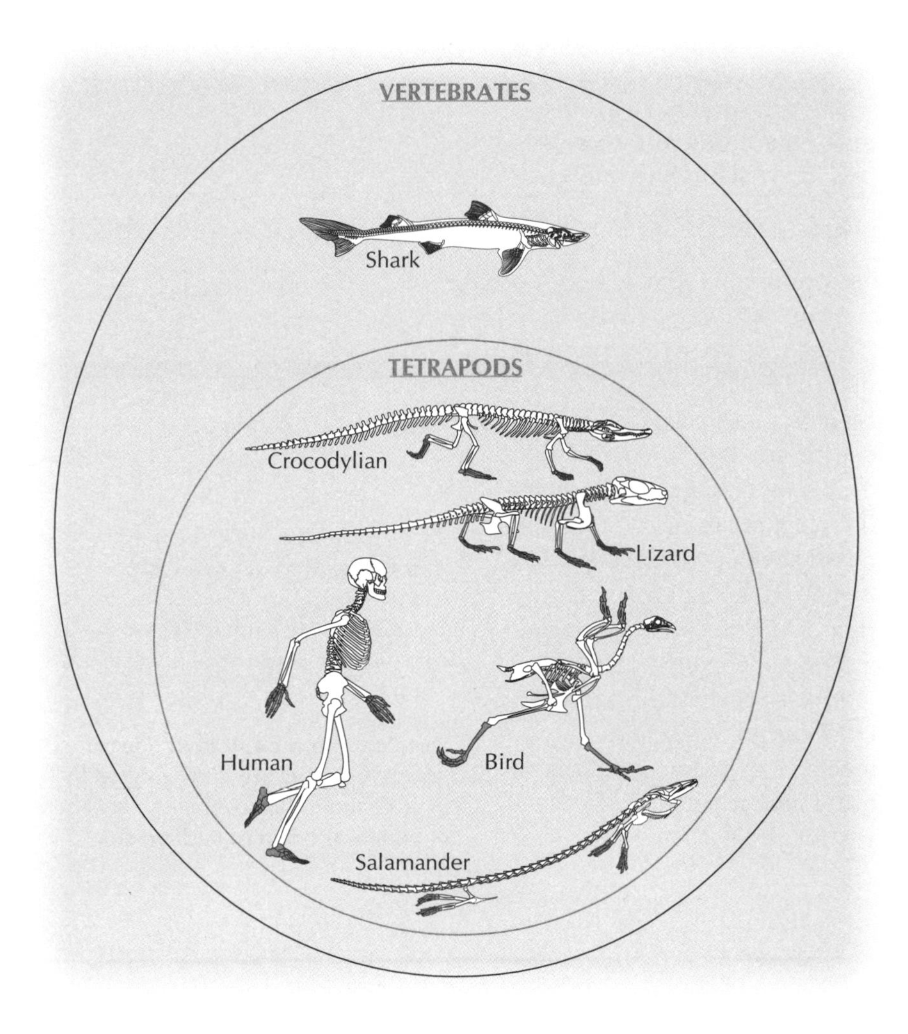

Among vertebrates, members of the tetrapod lineage are distinguished by limbs that develop into hands and feet (dark green).

bone structure, muscle anatomy, coloration patterns, instinctive behaviors, and so on. Only shared heritable characteristics bear on genealogy, and only shared evolutionary novelties provide evidence of shared genealogical history. Just as children share unique resemblances to their parents, closely related species reveal their relationship in the shared possession of unique features.

To reconstruct patterns of evolutionary relationship, one of the challenges is to distinguish features that arose in a particular ancestor of interest, such as the ancestral dinosaur or the ancestral bird, from older features that had simply been inherited from even more distant ancestors in the lineage, like the ancestral reptile or the ancestral vertebrate. For instance, both birds and humans have a vertebral column, but this is not necessarily evidence of close relationship because many other organisms, like lizards, salamaners, and sharks, also possess a vertebral column. The vertebral column distinguishes vertebrates as a group separate from other organisms, but it provides no information on whether humans and birds are more closely related to each other, or to lizards, salamanders, sharks, or some other vertebrate.

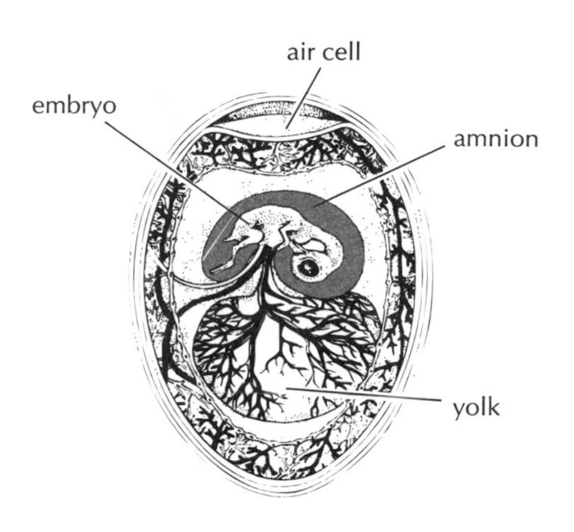

Birds, mammals, lizards, crocodylians, and turtles have an egg with an amnion—a fluid-filled sack in which the embryo develops. This distinctive egg distinguishes members of Amniota from all other vertebrates.

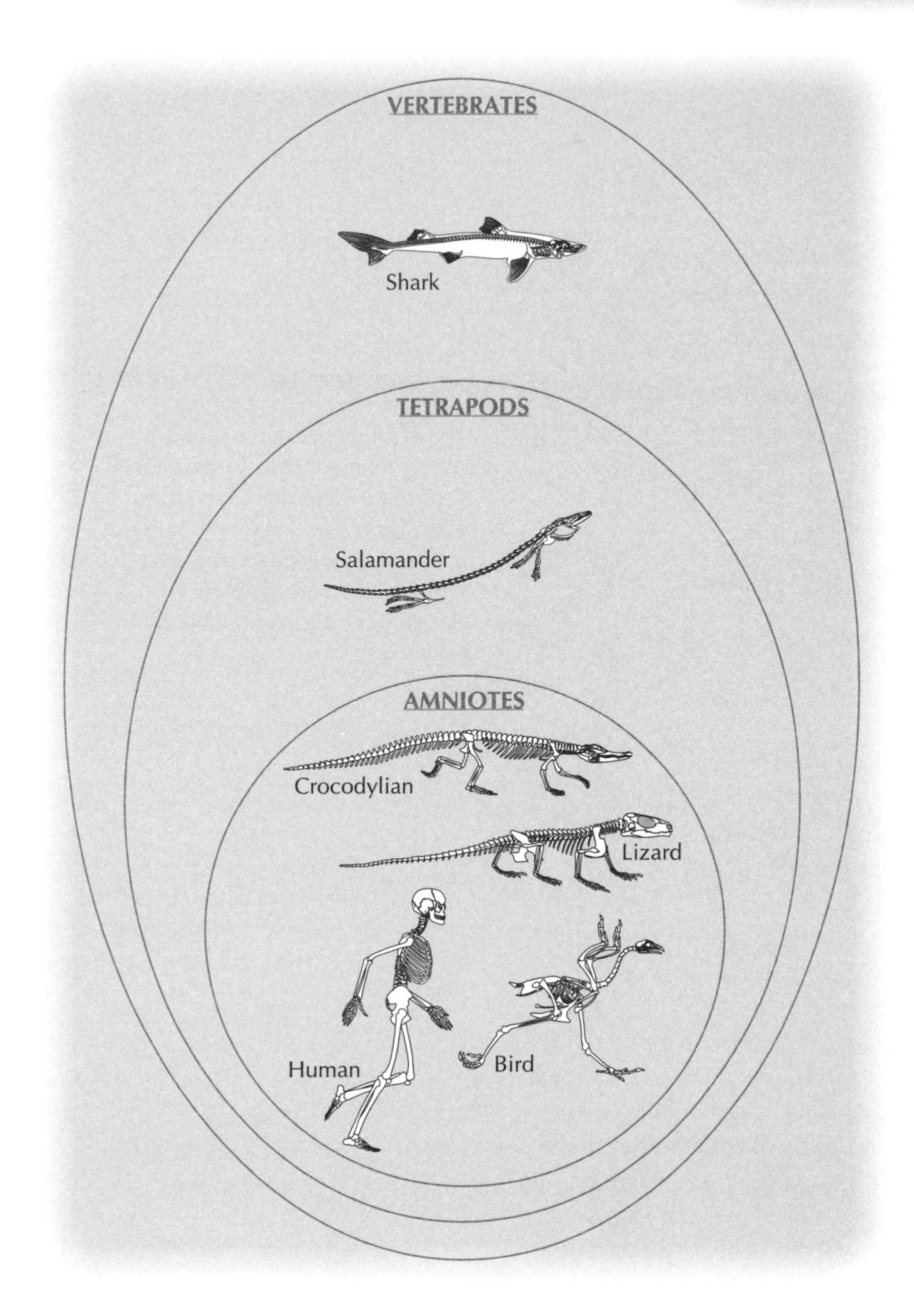

Amniota is a lineage contained within the more inclusive lineages Tetrapoda and Vertebrata.

The structure of the limbs is a different story. Birds and humans both possess limbs equipped with hands and feet. Sharks and the various "fishes" have fins, instead. Whereas all vertebrates possess a backbone, only some called tetrapods possess limbs that develop into hands and feet. Thus, from the structure of their limbs, we can distinguish members of the tetrapod lineage. Within tetrapods, we can discern smaller groups. One lineage is marked by the amniotic egg, which provides evidence that birds and humans are more closely related to each other than they are to salamanders or sharks. By comparing the many different features of organisms and mapping which features are found in what particular species, patterns of relationship among species emerge and the evolutionary histories of

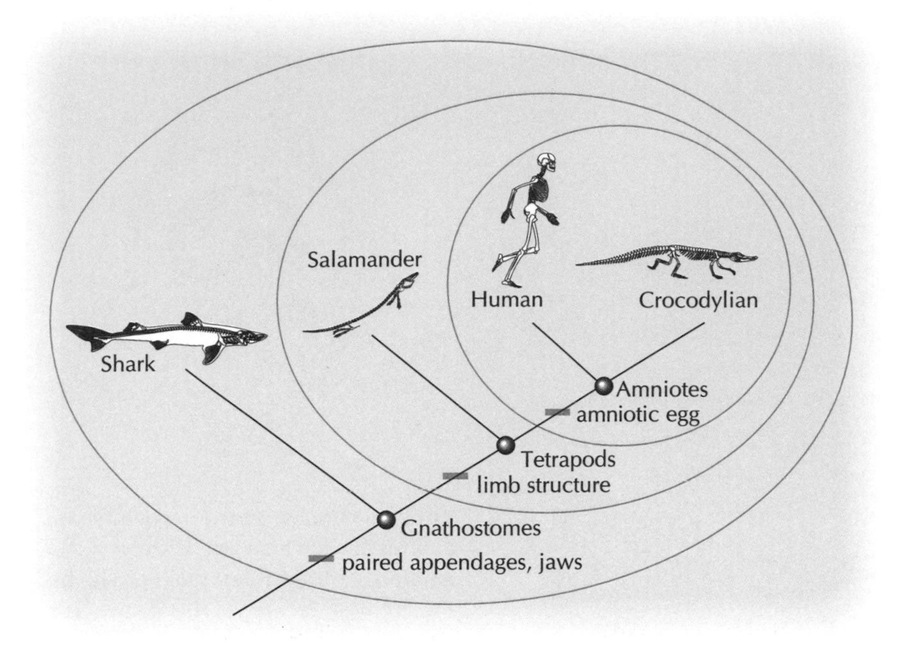

The hierarchy of vertebrate relationships can be mapped by tracing the hierarchical arrangement of shared evolutionary novelties, otherwise known as synapomorphies. First paired appendages and jaws appeared, then the distinctive tetrapod limb structure, then the amniote egg.

different features can be viewed in an intelligible fashion. Comparisons like this can get complicated if a lot of species and characters are analyzed, which is where the computers come in.

This brings us to a third idea behind reconstructing evolutionary relationships. Evolutionary relationships form hierarchical patterns, in which large, encompassing lineages, like Vertebrata, include smaller lineages, like Tetrapoda, which in turn includes even smaller lineages like Amniota. The metaphor of a family tree is apt, because from one trunk sprout many branches, which in turn sprout twigs, and the hierarchy of relationships is obvious. To discover these hierarchies of relationships, we map patterns of characteristics that are hierarchical, like the pattern just noted. All tetrapods, species with hands and feet, have a vertebral column but not all vertebrates have hands and feet. Tetrapod species with an amnion all have hands and feet, but not all tetrapods have an amniotic egg. By following the pattern of shared characteristics from larger groups into smaller groups, one can map out the sequence of evolution for life's diverse structures. This is like following an evolutionary path on the map of life. First the vertebral column evolved, then hands and feet, and later the amniote egg appeared in a descendant lineage. Amniotes are members of Tetrapoda and Vertebrata, which is the more-inclusive level in the hierarchy. There are about 40,000 living vertebrate species, while Tetrapoda includes only about half of the living vertebrate species.

These ideas can be applied together to tell resemblances that are due to homoplasy from resemblances that are genuine clues to common ancestry, and this is one of the great strengths of cladistics. Evolutionary resemblances are arrayed in a hierarchical pattern, while homoplasy is indicated by points of resemblance. Consider insects and birds. Both have wings, so one might argue that they are close relatives. But when the characteristics of all of their anatomical systems are considered, the preponderance of evidence places insects deeply within the hierarchy of arthropods—the crabs, lobsters, and other organisms with jointed limbs and a rigid external skeleton. This hierarchy is built from thousands of detailed anatomical observations, and it indicates that the insect wing evolved from nonflying arthropod ancestors. Birds, on the other hand, lie deeply within the hierarchy of vertebrates, organisms with an internal skeleton made of bones, as we will see below. Only the shared possession of wings tends to pull birds and insects away from their respective hierarchies and into a separate group by themselves. But this point of resemblance between birds and insects reflects the common mechanical demands of flight, not common ancestry.

This conclusion is testable by comparing in minute anatomical detail the structures of the two wings. If the two wings evolved from a common ancestral pattern, we might expect some degree of resemblance to persist. In fact, the two wings are radically different in virtually all details of anatomy and development. A second test is to add new information to the analysis, and to recalculate the hierarchical pattern that their characters support. For example, each new fossil discovery provides unique combinations of characteristics for the analysis and can potentially change the map of relationships.

Does the evolutionary path to birds lead through dinosaurs as Huxley and Ostrom argued? If birds and dinosaurs are closely related, then the various characteristics of their skeletons should be arrayed into a single hierarchical pattern. If the resemblances are homoplastic, then there should be only isolated points of resemblance between the two, and we should be able to map dinosaurs and birds into separate evolutionary hierarchies. And with each new fossil discovery, we can test and retest older conclusions.

BIRDS AND DINOSAURS: ONE HIERARCHY OR TWO?

Now we can settle into our role as evolutionary detectives, tracking a trail of evolutionary clues spread across 3.5 billion years and millions of diverging pathways on the map of life. Our first quest is to follow and find the evolutionary path leading to dinosaurs. Then we will determine whether any of the paths within dinosaurs lead to birds. Dinosaurs and birds are undoubtedly members of many of the same levels in the hierarchy of life. For example, like all life forms, both share an organization based fundamentally on cells and a reproductive mechanism that uses DNA to pass parental traits on to descendants. Both are multicellular organisms with differentiated tissues, such as muscles and nerves, and with tissues that in turn are organized into organs and organ systems. Further along the map, they belong to a lineage with an elongate body, a head at one end and a tail at the other, a mouth in front and an anus behind, and bilateral symmetry in which the right and left sides of the body are mirror images of each other. Dinosaurs and birds have a vertebral column, the central support of their internal skeletons, and so they lie on the vertebrate path of the hierarchy of life. Humans also fall along this path of the hierarchy.

But just how far into the vertebrate hierarchy do these lineages travel together? To best appreciate the controversy embroiling birds and dinosaurs, it is helpful to follow the map of relationships forward from a level that is not

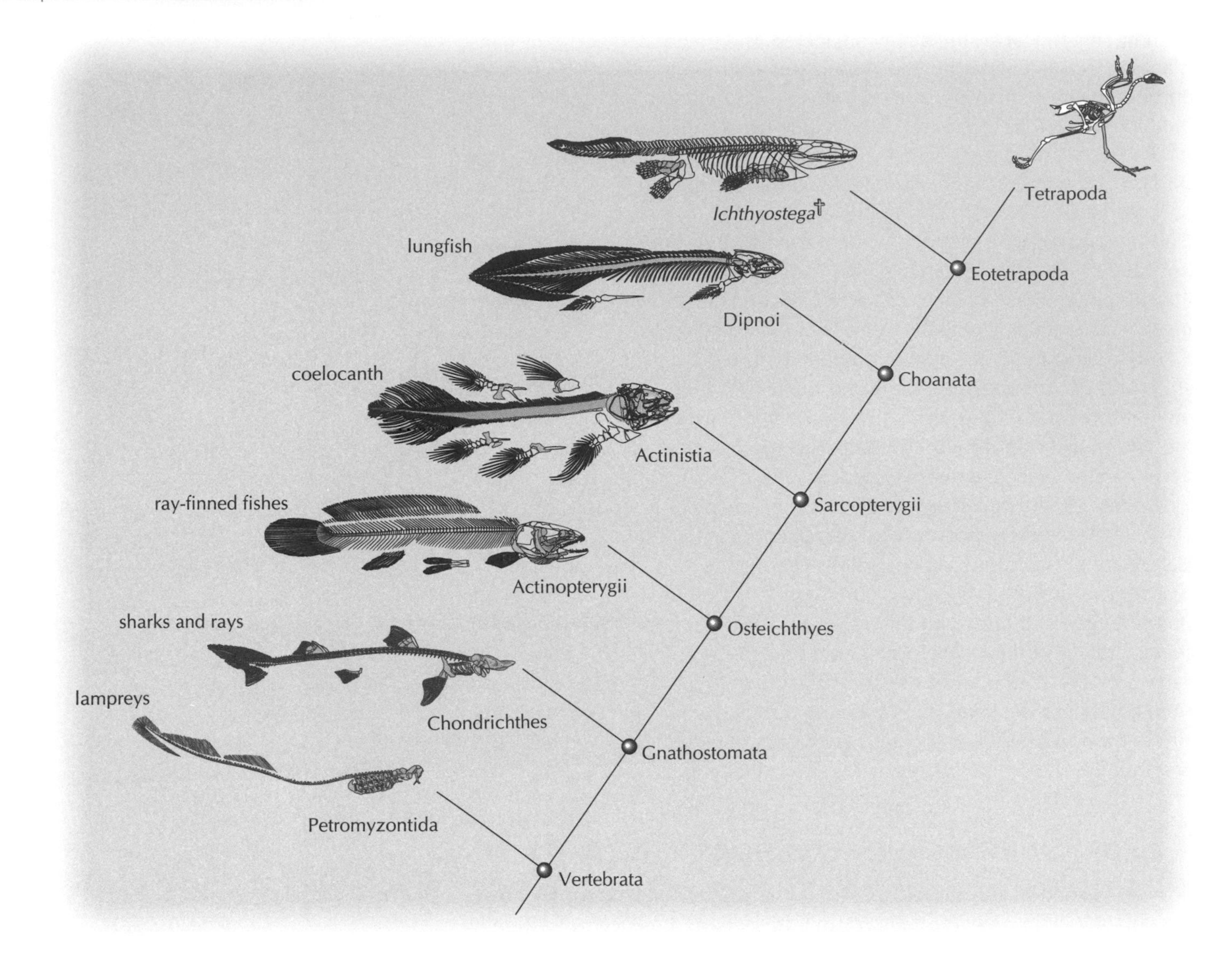

Phylogenetic map, or cladogram, of vertebrate relatinships. (Cross denotes extinct species.)

controversial. In order to provide an evolutionary context in which to evaluate the controversy, we will pick up the trail to birds from the beginning of vertebrate history. From there we will follow the phylogenetic map forward in time to see where these lineages split and diverge from one another onto their own evolutionary trails.

The first cladistic maps of vertebrate phylogeny were generated by Gareth Nelson[4] and Donn Rosen[5] of the American Museum of Natural History, and their work has been extended by Colin Patterson,[6] Bobb Schaeffer,[7] and John Maisey[8] also of the American Museum of Natural History, and many others. All agree that at the most general level, the map of vertebrate relationships is split into two major paths which can be regarded as "sister lineages" because they sprang from the same common ancestor. Each sister lineage forms a hierarchy within the more inclusive hierarchy of vertebrate relationships. These sister lineages are of unequal size today. The less diverse of the two has been christened Petromyzontida and includes the modern parasitic lampreys. All other vertebrates are members of a lineage known as Gnathostomata. Among their many distinctive characters, gnathostomes have jaws that are lined with teeth, improving the grasping capability of the mouth. They also have fins on either side of the body, one pair in front and one in back. Paired fins offer greater maneuverability up and down in the water column and more rapid turning than was the case in the earliest vertebrates. The oldest fossil evidence of gnathostomes indicates that the lineage extended back in time at least 500 million years. Dinosaurs, birds, and humans all have a vertebral column, jaws, and paired appendages, indicating that they belong to the gnathostome path in the vertebrate hierarchy.

Gnathostomata includes two great sister lineages. Chondrichthyes includes the sharks and rays while its sister lineage, Osteichthyes, includes all the rest. Chondrichthians have lost virtually all of the bone in the skeleton, while in contrast Osteichthyes have increased the extent of their internal bony skeleton by adding bony ribs that articulate movably with the vertebral column, and a bony shoulder girdle that firmly anchored the front fins to the body. The ancestral osteichthyan lived in the ocean, and many of its living descendants have remained in their ancestral environment. Chief among them is the actinopterygian lineage, which contains about 18,000 species of ray-finned fishes, including all that are native to North American fresh and coastal waters. However, its sister lineage, known as Sarcopterygii, has fins that were modified in a highly characteristic fashion, paving the way toward life on land. In the front fins there appeared a single bone, corresponding to our upper arm or humerus, in the part of the fin closest to the body. Following the humerus are two more bones, which correspond to the radius and ulna,

the two bones

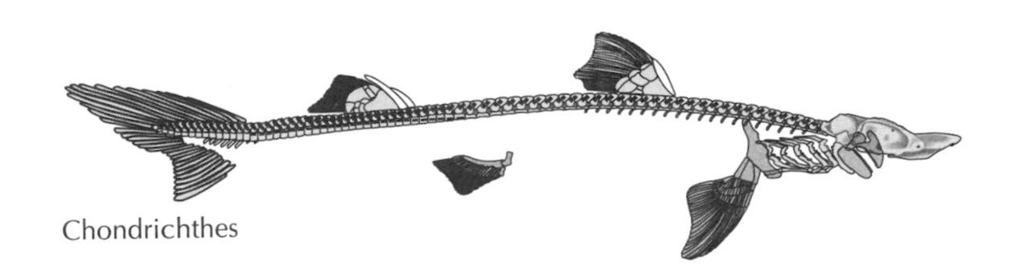

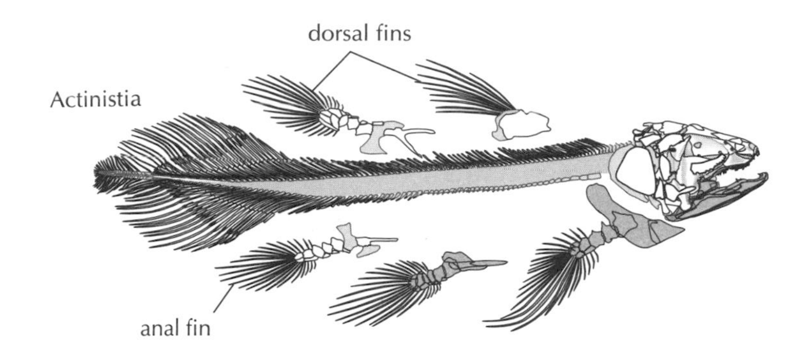

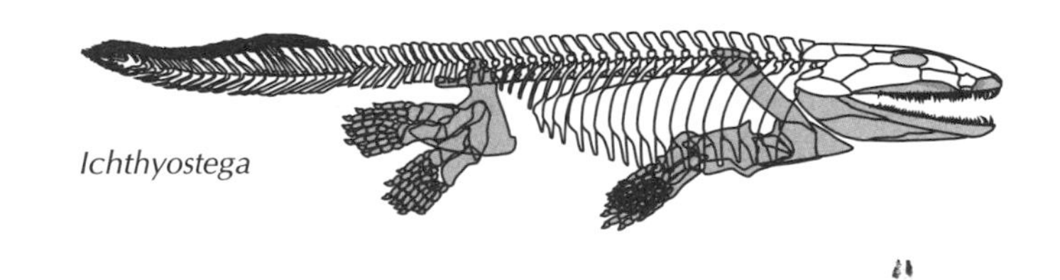

Three gnathostomes; their characteristic jaws and paired appendages are highlighted in green.

***Xiphactinus audax* belongs to an extinct family of primitive actinopterygian called ichthyodectids.**

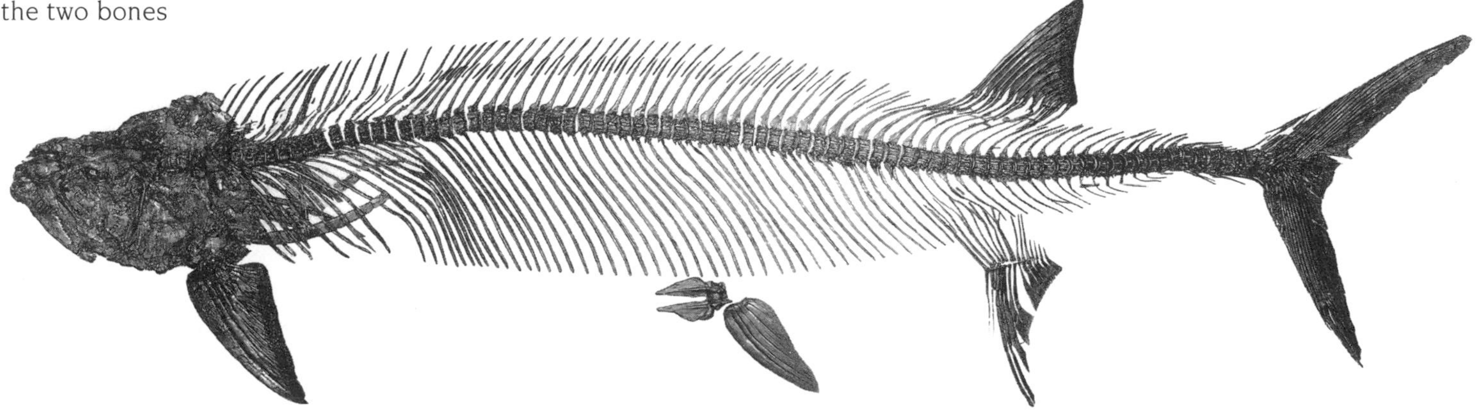

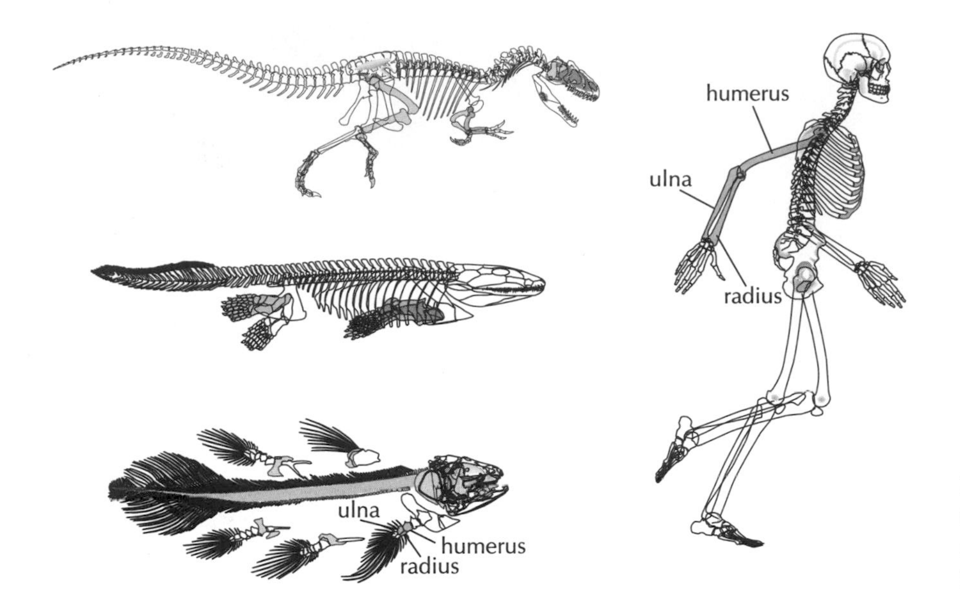

Representative members of the sarcopterygian lineage, showing the common pattern of organization in the front limb.

making up our forearms. As we saw in Chapter 9, both *Deinonychus* and *Archaeopteryx* have the same bones in their skeleton. Birds, dinosaurs, and humans all inherited this pattern from the ancestral sarcopterygian so the evolutionary trail leading to these groups runs through sarcopterygii.

Sarcopterygii includes two surviving sister lineages. One of these is known as Actinistia and is represented by a single species, which today lives only in deep waters around the Comoros Islands of the Indian Ocean. Its sister taxon is Choanata, which has nearly 20,000 living species. It is through the Choanata lineage that the evolutionary path to dinosaurs can

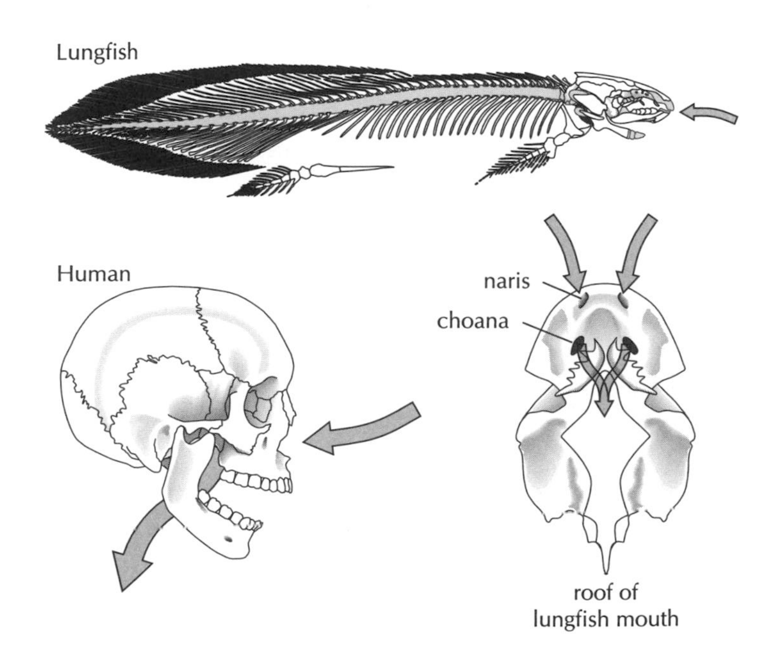

A pathway for water or air from the naris to the choana is diagnostic of the Choanata lineage.

be traced. Members of Choanata are distinguished by a continuous passage from the nose through to the roof of the mouth known as the choana.

There are two major living lineages of Choanata. One, named Dipnoi, includes the three living species of lungfish, which live today in freshwater streams and water holes in Africa, South America, and Australia. Its sister lineage is Tetrapoda.[9] Tetrapods are the vertebrates who moved onto land, transforming profoundly from the habitat and appearance of their fishlike ancestors. This momentous relocation occurred in several stages, beginning about 350 million years ago. The first tetrapods were helped along by the transformation into limbs of their more distant ancestor's two pairs of fins. The first tetrapods added a series of interlocking bones that form the wrists and ankles. Following these is a series of bones forming the hand and foot, which in turn

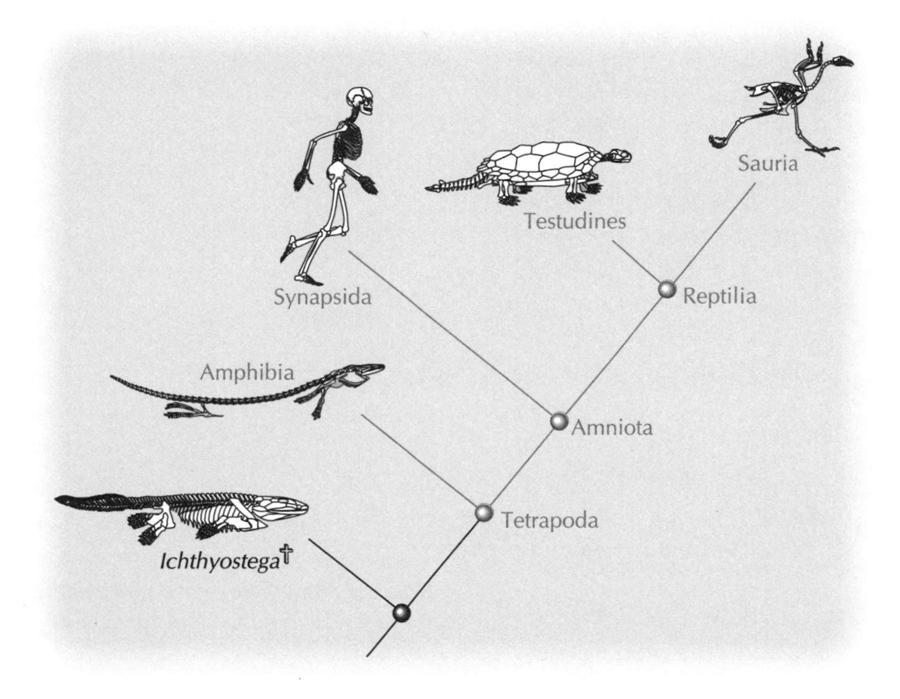

Phylogenetic map of relationships of the major lineages of Tetrapoda. (Cross denotes extinct species.)

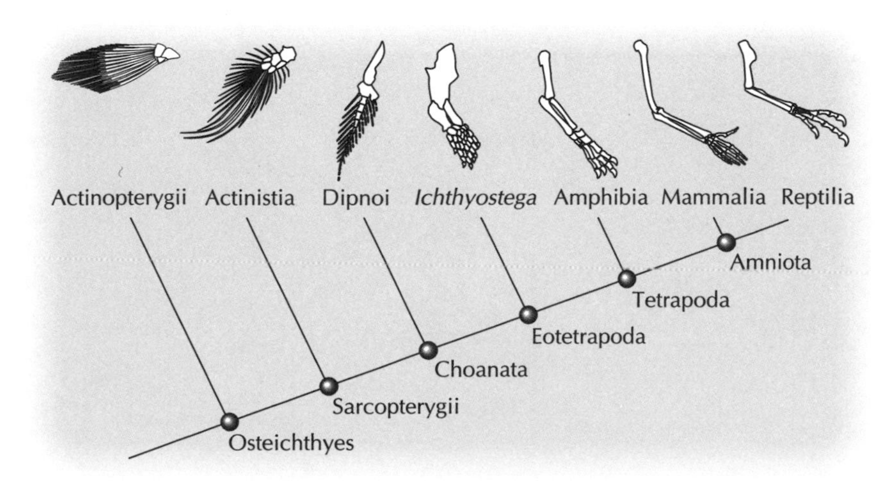

The evolution of the tetrapod forelimb occurred through successive modifications of the fin.

are followed by additional rows of bones forming the fingers and toes, producing the basic pattern of bones found in the hands and feet of living tetrapods. One glance at a dinosaur skeleton confirms that dinosaurs are part of this lineage. One glance in a mirror should convince you that you are too.

Locomotion in early tetrapods was only slightly different from the motion through the water column of their fishlike ancestors. Sigmoidal side-to-side undulation of the vertebral column provided the basic thrust, as was the case in Vertebrata ancestrally, but the body was propped against the ground instead of against a water column. Even with this new ability, the earliest tetrapods probably spent most of their time in the water, feeding there and laying their eggs in the water as well. Their tails, moreover, were still designed for swimming. If we could go back in time to hunt for early tetrapods, it would take a fishing pole to catch one.

Two major tetrapod lineages survive today, namely, Amphibia and Amniota. Amphibians living today include the frogs, salamanders, and caecilians. As the name of this group implies, most amphibians have ties to both the land and the water. Modern amphibians were once looked upon by naturalists as being uniformly primitive. However, in the 350 million years since branching out on their own evolutionary pathway, modern amphibians have become highly modified from the ancestral tetrapod. The skeletons of frogs are highly specialized for leaping, and salamanders are greatly altered owing in large part to the developmental retardation of certain skeletal growth patterns. Caecilians are strange, wormlike creatures whose fossil record reveals a history marked by reduction and finally total loss of the limbs.

As we have seen, Amniota, the other major lineage of tetrapods, is the lineage with the amnion. Usually the amnion is surrounded by a hard or leathery shell, like the eggs of birds or turtles. However, even human embryos grow within an amnion that develops inside the mother's womb. The evolution of the amniotic egg represented the next stage in the tetrapod transition to land by allowing the egg to be laid on land. The evolutionary paths for birds, dinosaurs, and humans lie within the phylogenetic hierarchy of Amniota.

The exact number of fingers and toes was variable in the earliest tetrapods but the number eventually stabilized at five fingers and five toes in the ances-

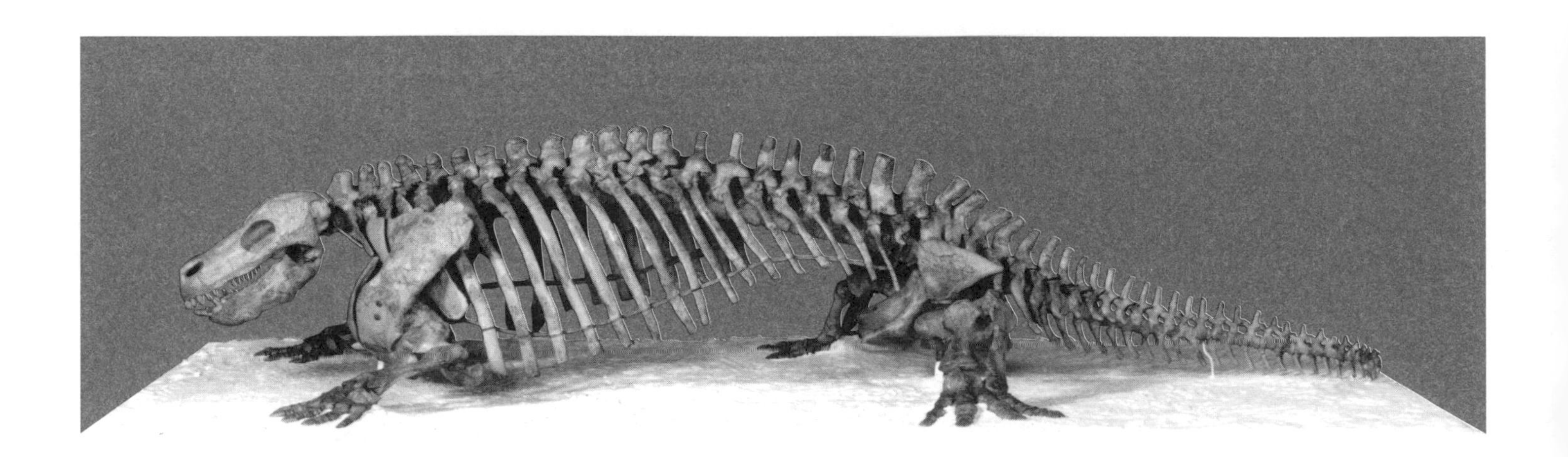

Diadectes phaseolinus **shows the general form of fossils lurking near the ancestry of amniotes.**

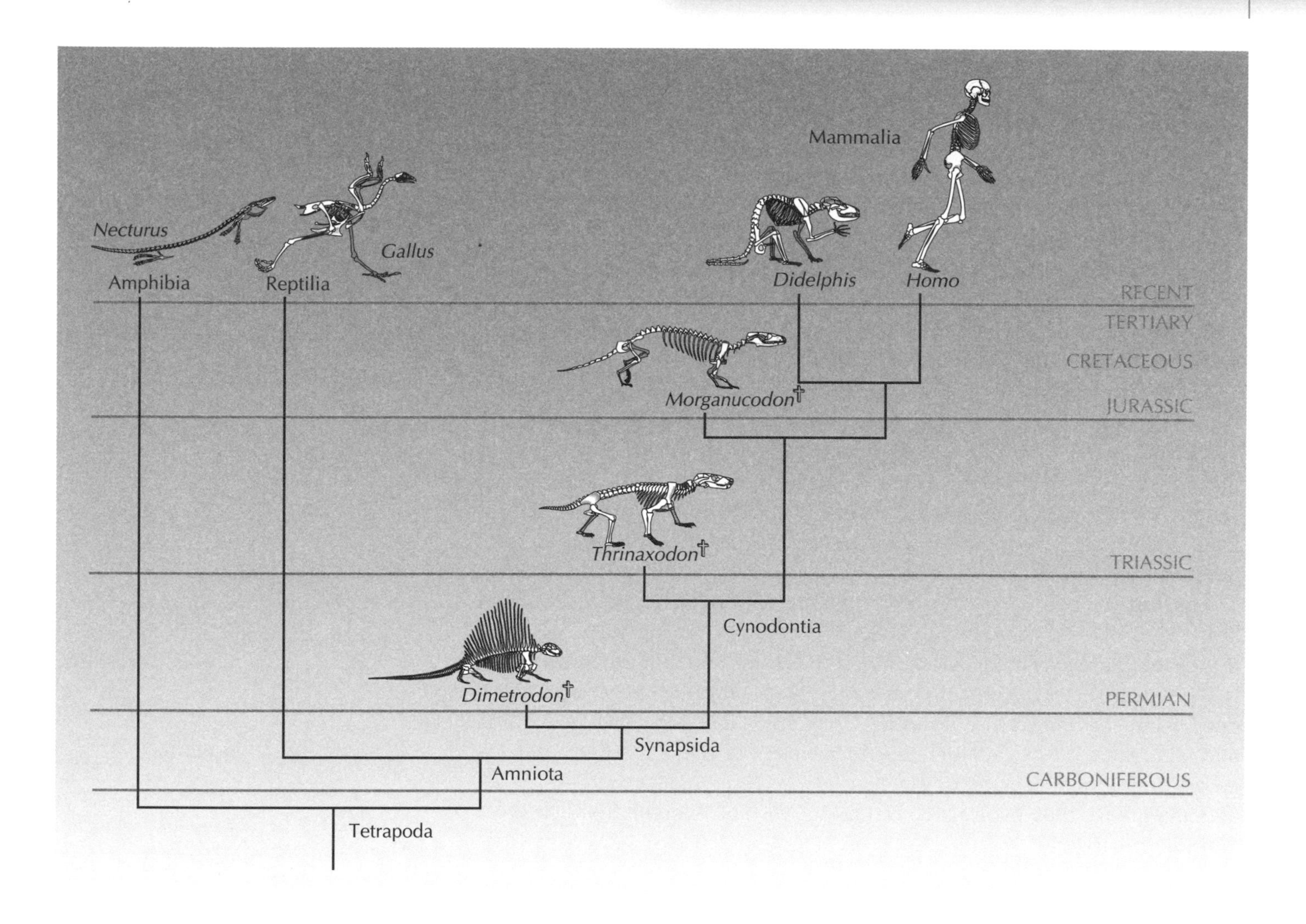

tral amniote, as it evolved greater agility and speed on land than before. The vertebral column was also strengthened to withstand the greater forces generated during locomotion on land. Whereas their fishlike relatives are more or less neutrally buoyant, in effect weightless while in the water, the terrestrial amniotes had to confront the problem of weight and gravity. In addition, the earliest amniotes have the beginnings of a sophisticated joint between the head and neck, enabling the head to bend and twist from side to side, and to take items from the ground. Fishlike vertebrates often suck food items into their mouths with a large gulp of water, but amniotes must be more agile to catch prey crawling or flying by.

The relationships among amniotes have been controversial for decades. In our years at Berkeley, working with Jacques Gauthier, we were among the first to apply cladistics to the problem of amniote phylogeny.[10] Subsequent studies confirm our map that there are two major living lineages diverged from the ancestral amniote, namely, Synapsida and Reptilia, and with this split we see the evolutionary pathways of humans diverge from that of dinosaurs and birds. Synapsida includes humans and all other

About 325 million years ago, the synapsid lineage diverged onto its own trajectory. From that time onward our own evolutionary history was distinct from that of dinosaurs and birds.

mammals, plus a host of extinct species. Up until this time, our own lineage shared more than 3 billion years of common evolutionary history with birds and dinosaurs. But about 325 million years ago, the synapsid lineage diverged onto its own trajectory, and from that time onward our own evolutionary history was distinct from that of dinosaurs and birds. In the context of life's 4 billion years of history, we are not such distant relatives of dinosaurs after all.

Synapsid history is documented in great detail by a dense fossil record and, because it is our own history, it has been intensively explored.[11] Early synapsids were largely predatory, like their ancestors, although there were some early herbivorous experiments. Synapsids eventually evolved sophisticated means of locomotion, some being able to run, gallop, leap, climb, and even fly. In the process, the principle movement of their vertebral column transformed from the primitive side-to-side sigmoidal motion of fishlike vertebrates to a more symmetrical up-and-down movement. With symmetrical spinal movement breathing cycles became coupled to running cycles, increasing the efficiency of high-speed locomotion, a feat carried to is greatest level in modern cheetahs.

Mammals, the living synapsids, are warm-blooded and have huge brains,[12] unlike all other vertebrates except birds. Richard Owen had used this resemblance as a basis for arguing that birds and mammals are closely allied. But more recent phylogenetic maps indicate that mammals are deeply internested in the hierarchy of synapsid genealogy, whereas birds and dinosaurs lie within the reptile division of the amniote hierarchy. Large brains and endothermy are points of resemblance between birds and mammals that evolved convergently within different hierarchies of amniote phylogeny.

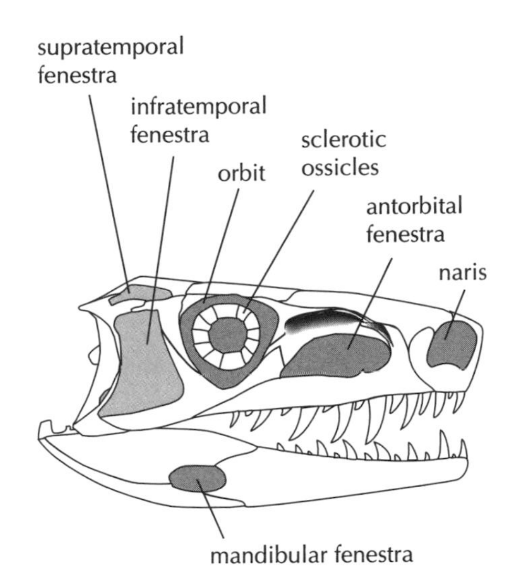

Side view of skull

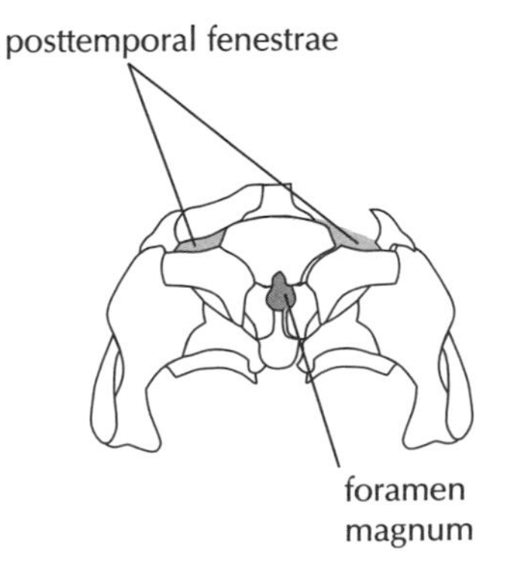

Back view of skull

There are a number of fenestrae, openings in the skull, in reptiles.

REPTILIAN PAST

As we shall see, there is abundant evidence, from both fossils and from all anatomical systems in living species, that birds and dinosaurs belong on evolutionary paths of the reptile lineage. Also sharing common ancestry with the ancestral reptile are living turtles, lizards, and crocodylians, plus a vast diversity of extinct species.[13] By studying living reptiles, we can reconstruct that the ancestral reptile probably had color vision and was strongly diurnal in its habits, because nearly all living reptiles are. Like our own synapsid lineage, the brain increased in relative size during the history of reptiles, although the expansion involved a different region of the brain.

A pervasive theme in reptile history involves elaboration of the jaws and feeding system. The evidence for this lies in a series of new holes in the skull, known as fenestrae. The term *fenestra* (plural, *fenestrae*) means "window." Skull fenestrae are simply openings between bones, and their evolution enabled a great expansion and strengthening of the jaw musculature. The first large fenestra appeared at the back of the skull and is known as the posttemporal fenestra. The first amniote already had a tiny posttemporal fenestra for the passage of a blood vessel. In Reptilia, this hole enlarged as the expanding jaw muscles invaded it.

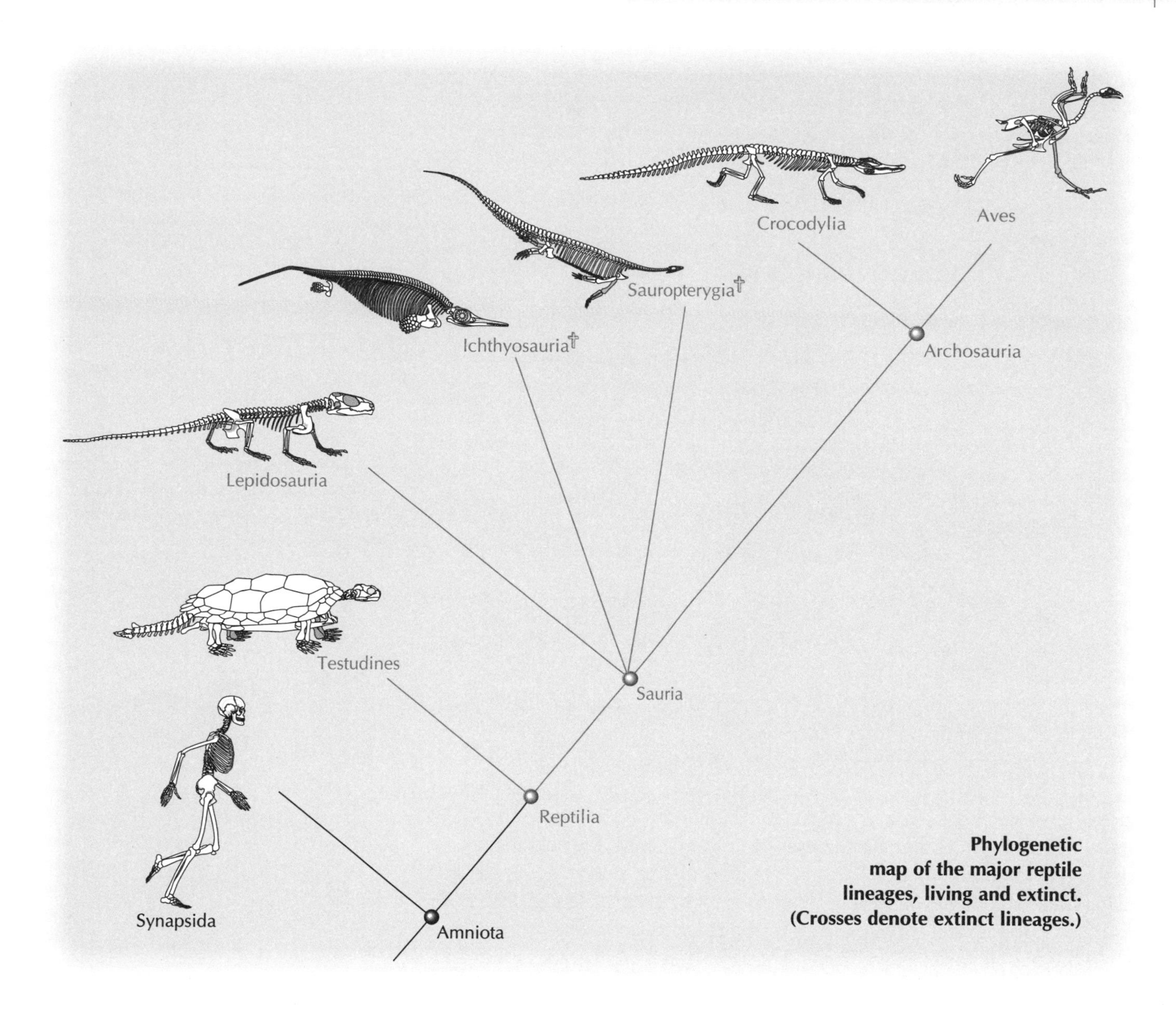

Phylogenetic map of the major reptile lineages, living and extinct. (Crosses denote extinct lineages.)

Two lineages with living members descended from the ancestral reptile. One includes turtles or Testudines. Turtles are highly distinctive in replacing teeth, which they lack entirely, with a horny beak, and in developing a bony shell. Modern birds also have replaced teeth with a beak, and on this basis some early naturalists argued that turtles and birds are allied. But possession of a beak is virtually the only unique resemblance between birds and turtles, aside from the features that both inherited from the ancestral reptile. Both lineages have long fossil records that indicate an extensive hierarchy of separate relationships. Both primitive turtles and primitive birds like *Archaeopteryx* had teeth. When all of the fossils are mapped onto a phylogeny that includes modern species, we

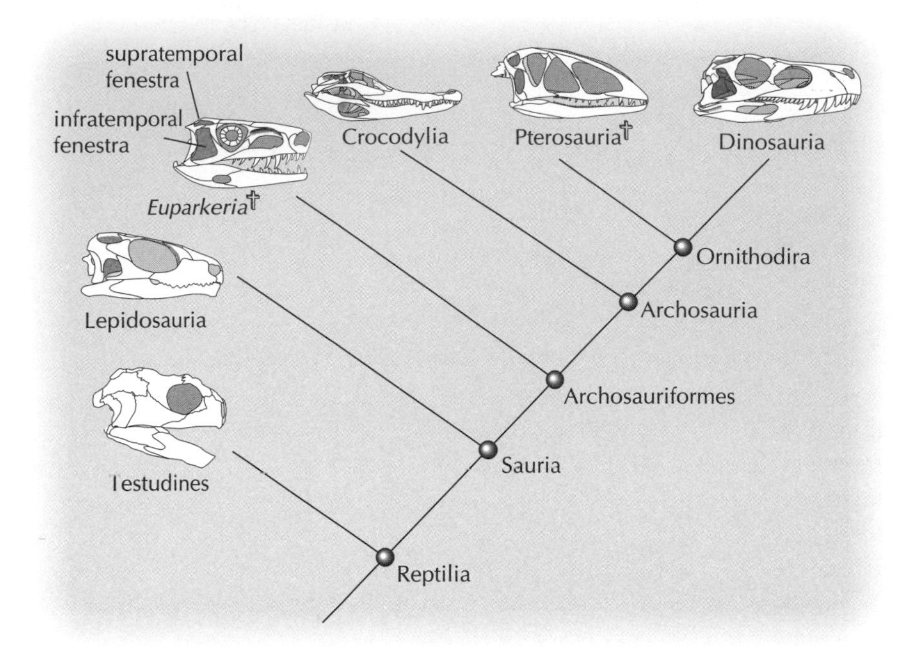

The pattern of fenestrae is one of many features that offer clues about reptile relationships. (Crosses denote extinct lineages.)

can see that the loss of teeth is the result of convergent evolution, rather than descent from a common toothless ancestor.

The sister group of Testudines is Sauria, which today includes lizards, snakes, crocodiles, birds, and a diversity of extinct Mesozoic species. Saurians further increase the size of their jaw muscles, adding additional fenestrae in the skull. On the side of the head behind the eyes are two openings, the supratemporal fenestra and the infratemporal fenestra. Saurians also have long slender limbs, making them more agile and faster than other reptiles. Because they also possess these features, this is the pathway leading toward dinosaurs.

Two saurian lineages have living species. These are Lepidosauria and Archosauria. Lepidosaurs include snakes and lizards,[14] which are distinguished by their extensive covering of overlapping scales. Living lizards can climb, swim, and even parachute from high branches. Most lizards are predatory. In capturing insects and small vertebrates, lizards have evolved a remarkable diversity of equipment. Some, like the chameleon, have protrusable tongues to grab passing insects. Others like the Gila monster have a modified salivary gland that secretes highly toxic venom. In many snakes, the upper and lower jaws come apart at the front of the mouth, allowing them to swallow objects far larger than their head. Genealogically, snakes are simply lizards that have lost their limbs and elongated their bodies via the addition of many vertebrae to their backbones. The most primitive living snakes—pythons and their immediate relatives—still possess remnants of hind legs that afford evidence of their evolutionary descent from running lizards.

As we saw in Chapter 10, early nineteenth-century naturalists confused dinosaurs with lizards, and the confusion sometimes persists thanks to the popular misnomers "thunder lizards" and "terrible lizards." While indeed there are many resemblances, modern phylogenetic maps indicate these features to be inheritances from the common saurian ancestor, rather than evidence placing dinosaurs within the lizard lineage. Modern cladistics only reinforces Owen's original insight that Dinosauria is a distinctive group. The ancestral species of Archosauria probably lived during the latest Permian, around 250 million years ago, a good 25 million years before the oldest known dinosaur.

ARCHOSAURS RULE

Since Robert Broom's work on *Euparkeria*, naturalists have agreed that birds, crocodylians, dinosaurs, pterosaurs, and a host of other extinct reptiles descended from the early archosaurs. Archosaurs had inherited long limbs and a skull with temporal fenestration from their more distant saurian ancestors. From the ancestral reptile they inherited a posttemporal fenestra and color vision. The amniotic egg had been passed down from the ancestral amniote, and the distinctive organization of the hands and feet from the ancestral tetrapod. From the ancestral sarcopterygian, archosaurs inherited a highly stereotyped limb organization, from the ancestral osteichthyan bony ribs and a bony shoulder girdle, from the ancestral gnathostome jaws and paired appendages, a vertebral column from the ancestral vertebrate, and so on and on, back to the first single-celled life form.

Euparkeria is the closest thing we have to the actual ancestor of archosaurs. Modern phylogenetic maps now plot *Euparkeria* as the extinct sister lineage of Archosauria, together comprising the more inclusive lineage Archosauriformes. This lineage was distinguished from other saurians by even more fenestration of the skull. Two new openings were added. The antorbital fenestra lies in front of the eye along a greatly elongated snout, while the mandibular fenestra perforates the side of the jaw below and behind the eye. Both fenestrae may have housed enlarged jaw muscles, but the antorbital fenestra may have also housed a pneumatic expansion—an air sac—expanding from the nasal passage.

The vertebral column and its primitive side-to-side sigmoidal undulation were modified in early archosaurs for a more symmetrical mode of locomotion, known as parasagittal gait. The vertebral column flexed in an up-and-down direction with the limbs

EUPARKERIA CAPENSIS

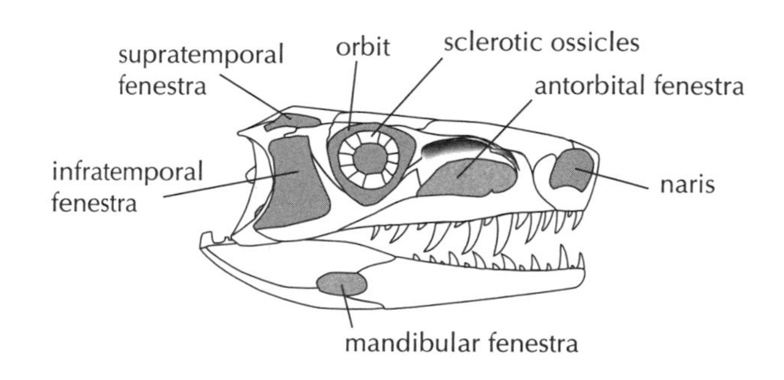

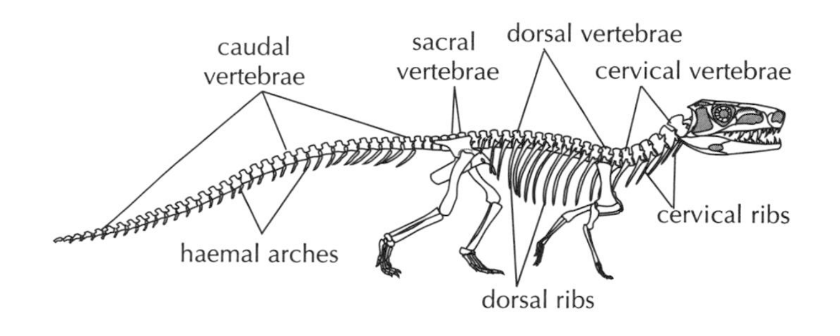

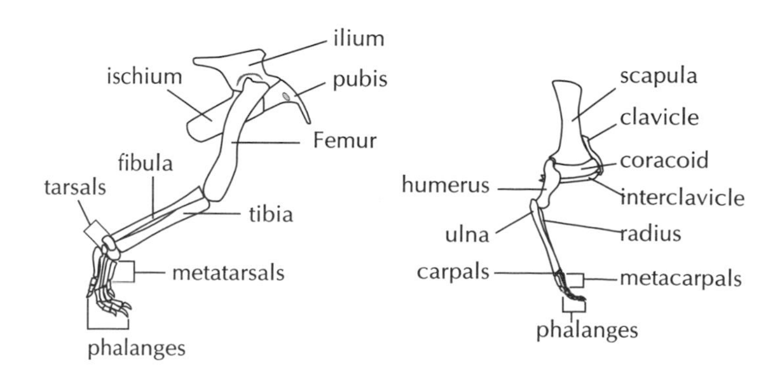

Basic archosaur skeletal anatomy, illustrated by *Euparkeria*.

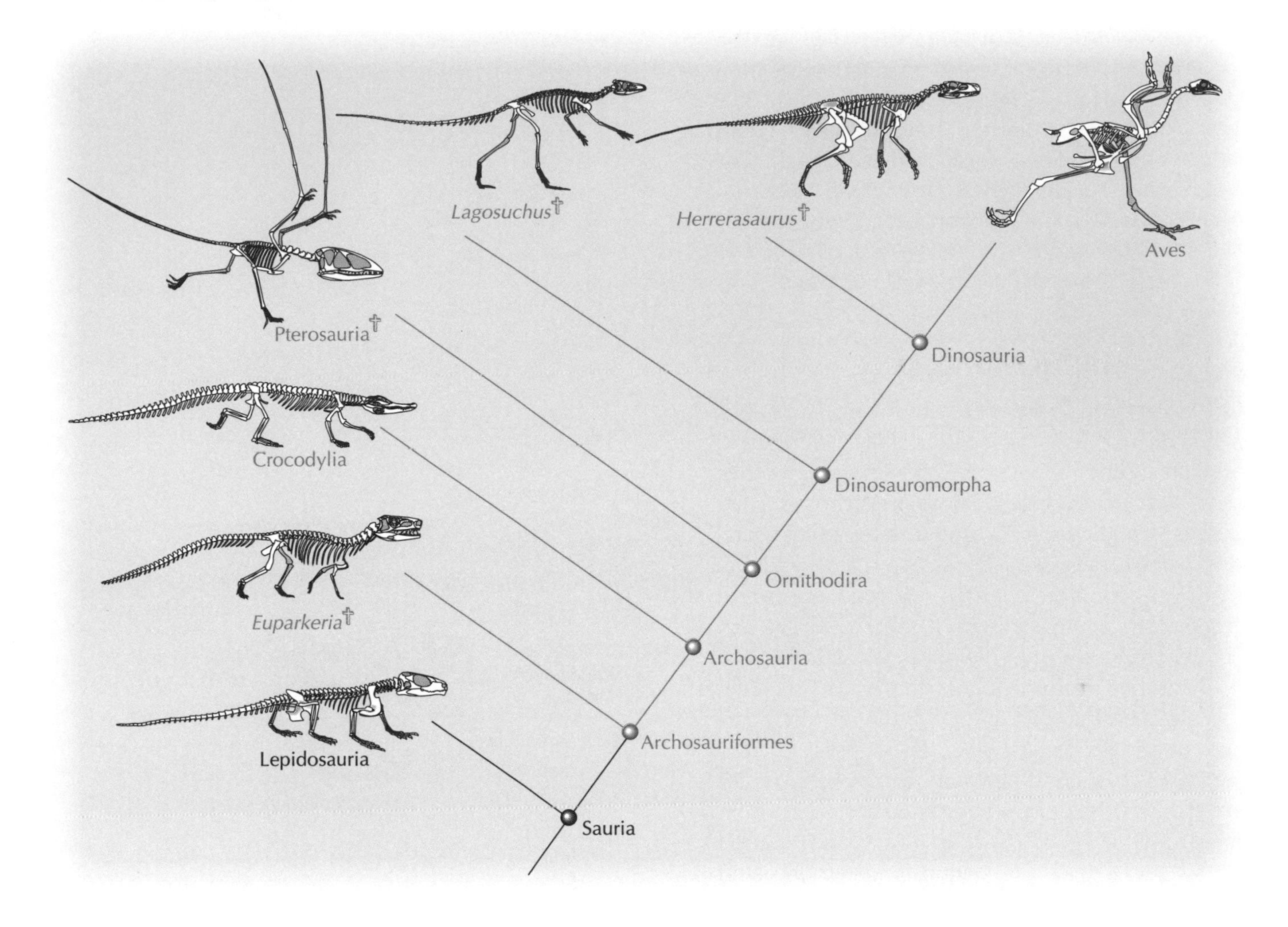

Phylogenetic map of archosauriform relationships. (Crosses denote extinct lineages.)

held more vertically beneath the body. This transformation began in the ancestral archosaur, and its various descendant lineages carried the trend further to varying degrees. Early archosaurs could probably gallop at high speeds. Living crocodylians can do this while they are young, but they lose the ability as they mature. In addition to greater speed, these changes probably enabled early archosaurs to range widely in search of food and mates.

The evidence summarized so far does not indicate which particular pathway of descent from the ancestral archosaur was followed by birds. Is the road to birds via dinosaurs, or via some other archosaurian lineage? If the distinctive avian features like bipedality and flight cannot be arranged into the hierarchy of one of the known lineages, then the popular twentieth century thecodont hypothesis of avian origins could well be true. However, if birds can be mapped onto one of the known archosaur pathways, then this would prove the thecodont hypothesis wrong.

One group of paleontologists, led by Alick Walker[15] of the University of Newcastle-upon-Tyne and Larry Martin[16] of the University of Kansas, has maintained that birds and crocodylians are more closely related to each other than to dinosaurs or pterosaurs. They invoke the theory of homoplasy, arguing that similarities birds share with dinosaurs and pterosaurs evolved convergently. The main lines of evidence they cite are similarities between the teeth in modern crocodylians and in primitive toothed birds like *Archaeopteryx*, and in the shared presence of pneumatic cavities around the braincase.

If the teeth and braincase similarities offer valid evidence that birds evolved from crocodylians, then a complete phylogenetic map should show birds branching from within the hierarchy of crocodylian evolution. But when all the data from modern and extinct archosaurs is taken into account, this is not the case. A series of independent cladistic studies of different segments of the crocodylian lineage, including all modern species plus a long fossil record of extinct crocodylian relatives, was recently conducted by Jim Clark of George Washington University, Michael Benton[17] of the University of Bristol, Mark Norell[18] of the American Museum of Natural History, Chris Brochu[19] of the University of Texas, and several other paleontologists. Molecular data on modern species, plus information on the entire skeleton in modern and fossil species was used to generate a detailed series of hierarchical maps of crocodylian relationships. None of these phylogenetic maps includes the evolutionary pathway for birds. While there are a few points of resemblance, when all available data are studied, the overwhelming conclusion is birds are *not* a part of the crocodylian hierarchy.

Instead of evolving flight, crocodylian history saw the elaboration of cursorial, quadrupedal running styles in some descendant lineages, while others adopted an aquatic habitat to varying degrees. Modern crocodylians spend most of their lives swimming and feeding in the water. With this ecological shift, the antorbital fenestra was closed over as the skull became specialized so that only the eyes and nostrils would protrude above the water. Adult crocodilians have short legs compared to the length of their bodies, and tail-driven swimming has become their dominant mode of locomotion. As hatchlings and juveniles, crocodylians have relatively longer legs and can gallop. But as they age, the body and tail grow faster than the legs, and galloping and bipedality are sacrificed in favor of a powerful trunk and tail for swimming. Birds must lie within the hierarchy of some other archosaur lineage.

Within archosaurs, the sister lineage to crocodylians is Ornithodira. Right from the start, ornithodirans began to experiment with bipedality.[20] This is reflected in the simplification of their ankle, which forms a simple hinge joint. In early archosaurs, the bones of the ankle interlocked and moved in a series of complex rotations as the body rotated past the foot, which sprawled outward. In early ornithodirans, the body moved over the foot, flexing the ankle in a simple hingelike motion. In addition, the hindlimbs are elongated and are now considerably longer than the forelimbs. The neck is also longer and has become slightly S-shaped, holding the head higher than the backbone. It is through Orinthodira that the evolutionary path toward dinosaurs and birds can be found.

From the ancestral ornithodiran we can trace two descendant lineages. One of these is Pterosauria, the flying archosaurs sometimes allied to birds by early naturalists. A tiny, Late Triassic ornithodiran named *Scleromochlus* is probably the most primitive known member of the pterosaur lineage.[21] This fast-running terrestrial animal documents a preflight stage of its history. The oldest flying pterosaur is *Eudimorphodon* from the Late Triassic of Italy. Its forelimbs were transformed into wings, primarily through the elongation of the fourth finger, which supported a flap of skin that attached along the arm and to the side of the body. But the detail structure of the wings in pterosaurs and birds is very different. Pterosaurs went on to diversify throughout the Mesozoic. The giant pterosaur *Quetzalcoatlus* is the largest flying creature known, with a wing span of about 40 feet (12 m). But as with crocodylians, nowhere within the hierarchy of Pterosauria can we place birds. Although points of similarity like wings can be found, the similarities diminish as we compare these structures in detail. Owen, Huxley, and virtually everyone since has agreed that, when all available information is taken into account, flight arose independently in birds and pterosaurs.

The sister group of pterosaurs is known as Dinosauromorpha. As the name suggests, this is the evolutionary trail of dinosaurs.[22] The Middle Triassic *Lagosuchus* bears a detailed resemblance to dinosaurs in the structure of its pelvis, very long, graceful hindlimbs, and shortened forelimbs. The structural disparity between the small hands and huge feet suggests that dinosauromorphs moved almost entirely on their hindlimbs. The massive pelvis and sacrum carried more of the animal's weight than when it was evenly distributed on all fours. The head of the femur is bent inward to fit into the hip socket, and there is a small crest of bone over the top of the hip socket, enabling the knees to be carried close against the body. The tibia, fibula, and metatarsal bones are elongated, and the structure of the feet indicates that the earliest dinosauromorphs stood and walked on the balls of their feet, not flat-footed like their ancestors. Only three toes touched the ground, and three-toed trackways dating back to the Triassic document the narrow gait of dinosauromorphs. The feet struck the ground near the midline of the body. With their knees rotated in close to their flanks, and with the up-and-down flexure of their vertebral column, dinosauromorphs had a swift, bounding mode of bipedal running.

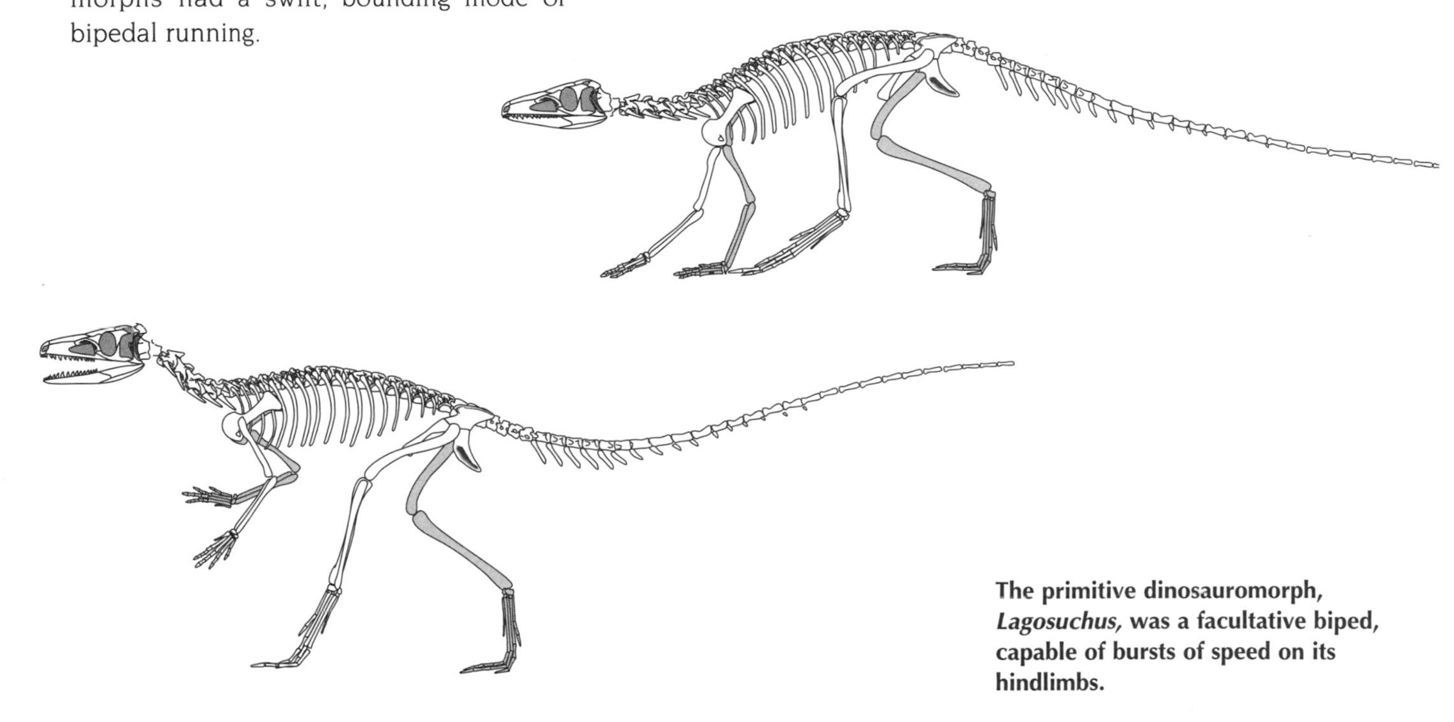

The primitive dinosauromorph, *Lagosuchus,* was a facultative biped, capable of bursts of speed on its hindlimbs.

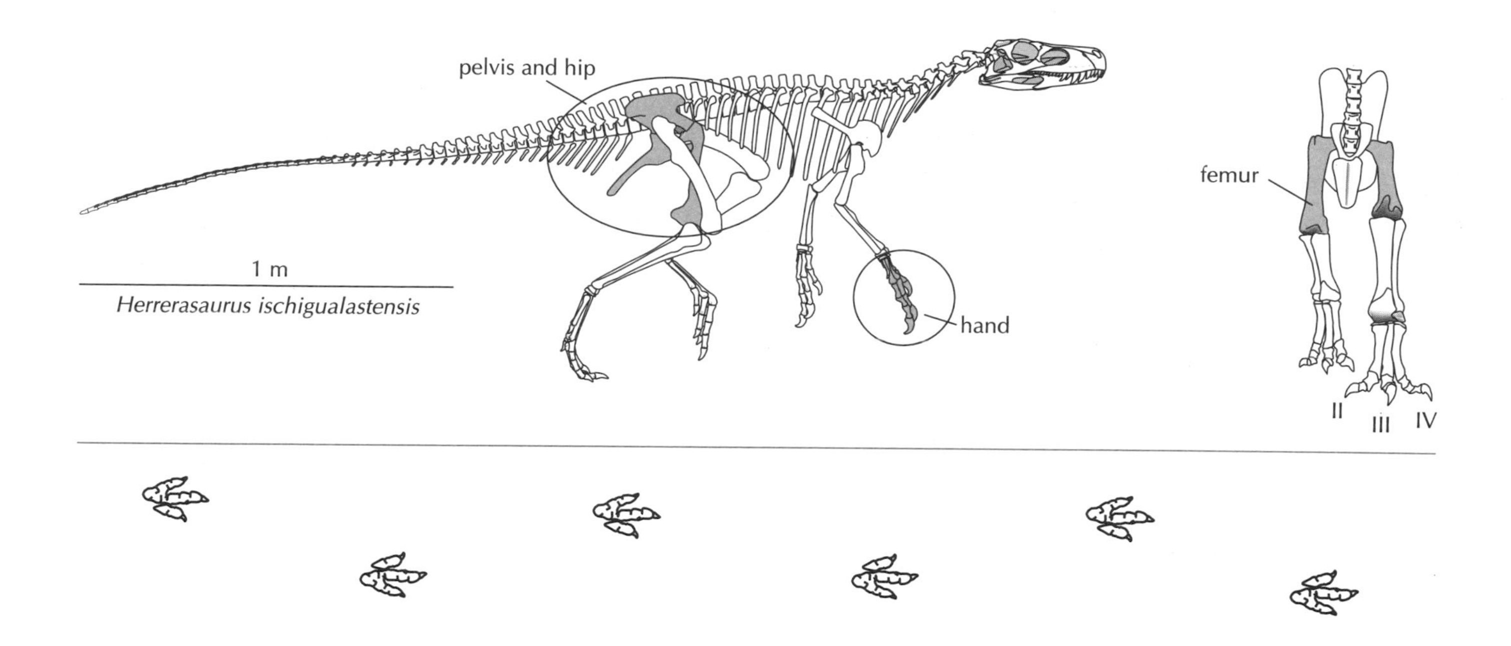

Diagnostic regions of the dinosaur skeleton are the pelvis, hip, and hand, illustrated here by *Herrerasaurus.* The hip modifications locked dinosaurs into a narrow gait. The hands could grasp, much as our own hands can.

DINOSAURIA: PORTRAIT OF THE FOUNDER

Dinosauria presents a further elaboration of several of these trends. The hands of early dinosaurs were largely freed from their primitive role in locomotion and took on other functions, becoming highly distinctive in the process. The thumb could bend inward to oppose the other fingers of the hand. This is similar, but not identical, to the situation in the human hand. In both cases, one end of the first metacarpal bone, a bone in the palm, is offset to permit the thumb to be pressed *against* the other fingers instead of just bending parallel to them. In humans the offset occurs at the base of the first metacarpal bone, where the first metacarpal fits against the wrist. In dinosaurs the offset occurs at the other end of the bone, where the thumb joins the palm. The primitive dinosaur thumb was less mobile than the human thumb, but their hands could nonetheless grasp objects, and among Mesozoic vertebrates this was a revolutionary innovation.

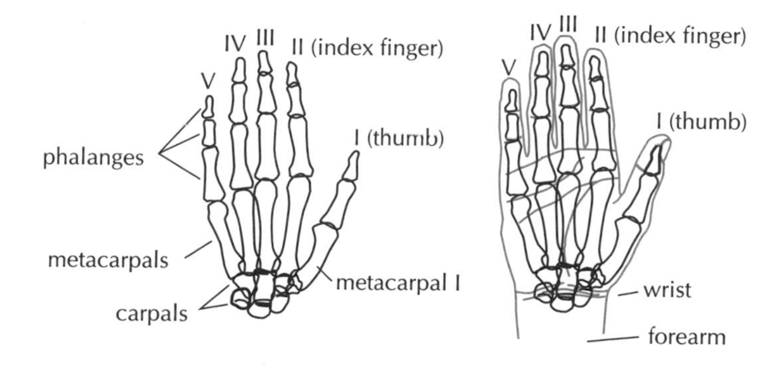

The bones of the human hand, as they relate to the skin of the palm.

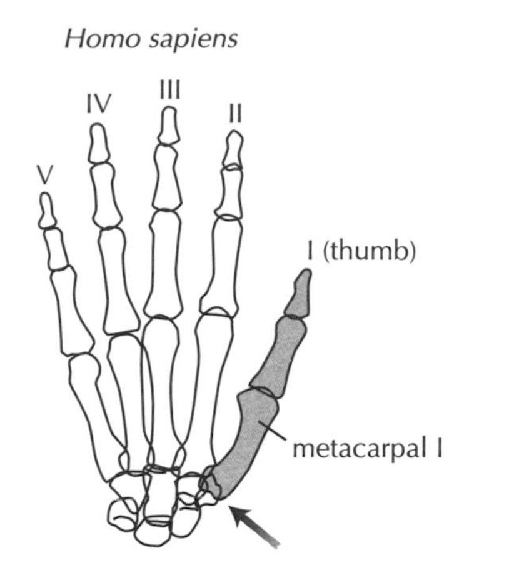

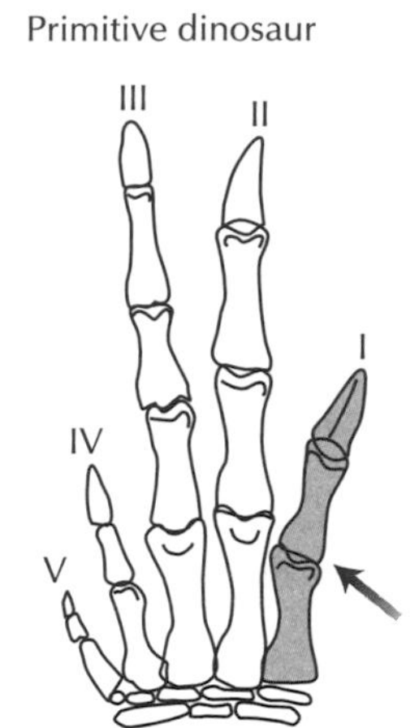

The opposable thumb of humans *(left)* is formed by a joint at the base of metacarpal I, whereas in dinosaurs the offset bending occurs at the first knuckle of the thumb.

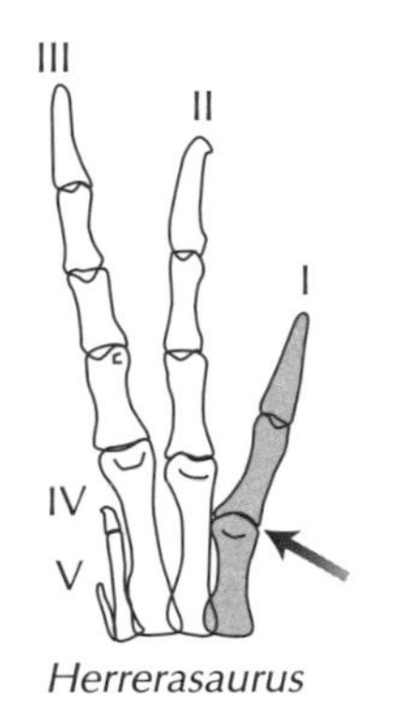

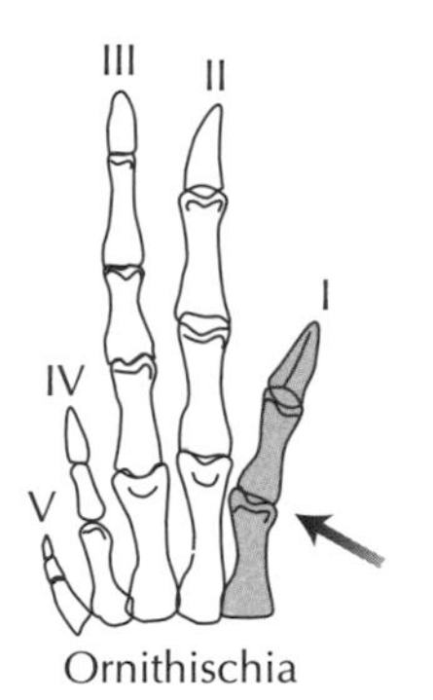

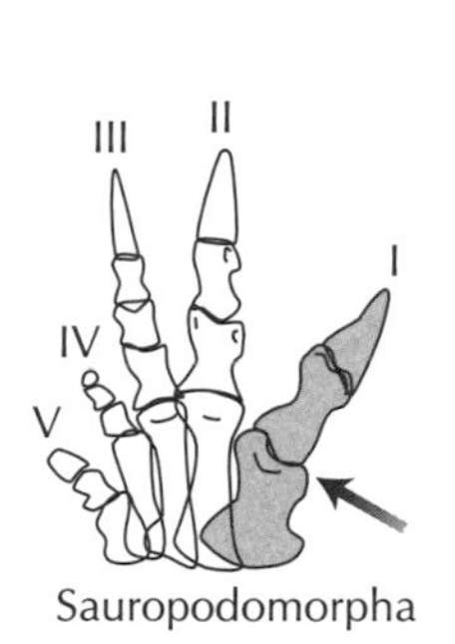

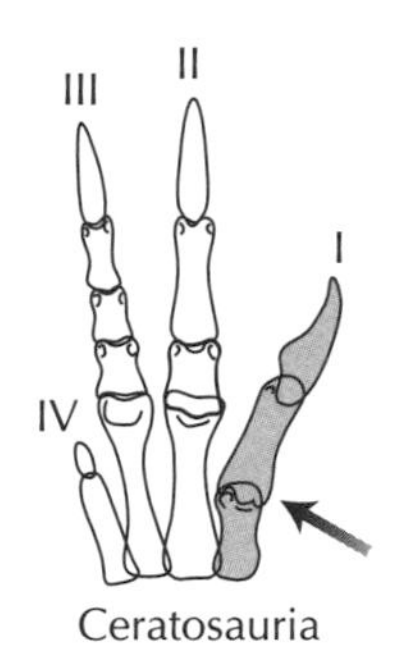

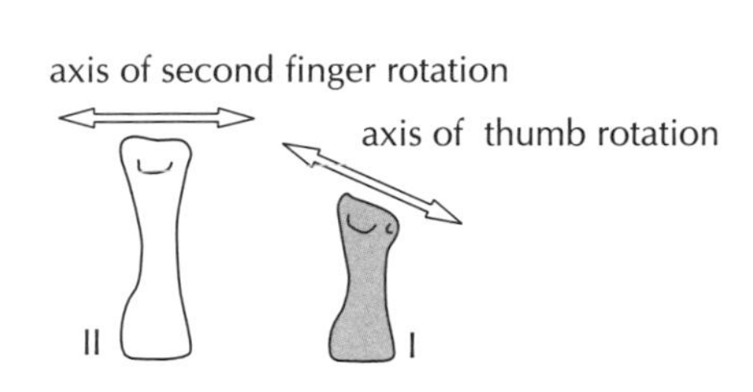

The hands of early dinosaurs *(above)* all show the characteristic offset of the thumb.

This twisting of a single bone might seem like a trifling evolutionary step, but it reflects a more complex underlying genetic change. Over much of vertebrate history, the front and hindlimbs were built on similar structural plans that present almost mirror images of each other. In most vertebrate lineages, evolutionary changes affected both front and hindlimbs together, not one to the exclusion of the other. The Giant Panda—a living mammal species—presents a famous example of this phenomenon. Pandas feed exclusively on bamboo leaves, which they strip from the stalks with a unique bony strut that protrudes from one of the wrist bones near the base of the thumb. Remarkably, there is a similar, nonfunctioning bony protuberance on the corresponding bone of the ankle, near the base of the first toe. It is evident from this and similar examples that a genetic linkage exists between the fore- and hindlimbs. Only the bony strut in the wrist is functional, but the genetic linkage between the two limbs brought about a corresponding structural change in the ankle. In most tetrapods, this link-

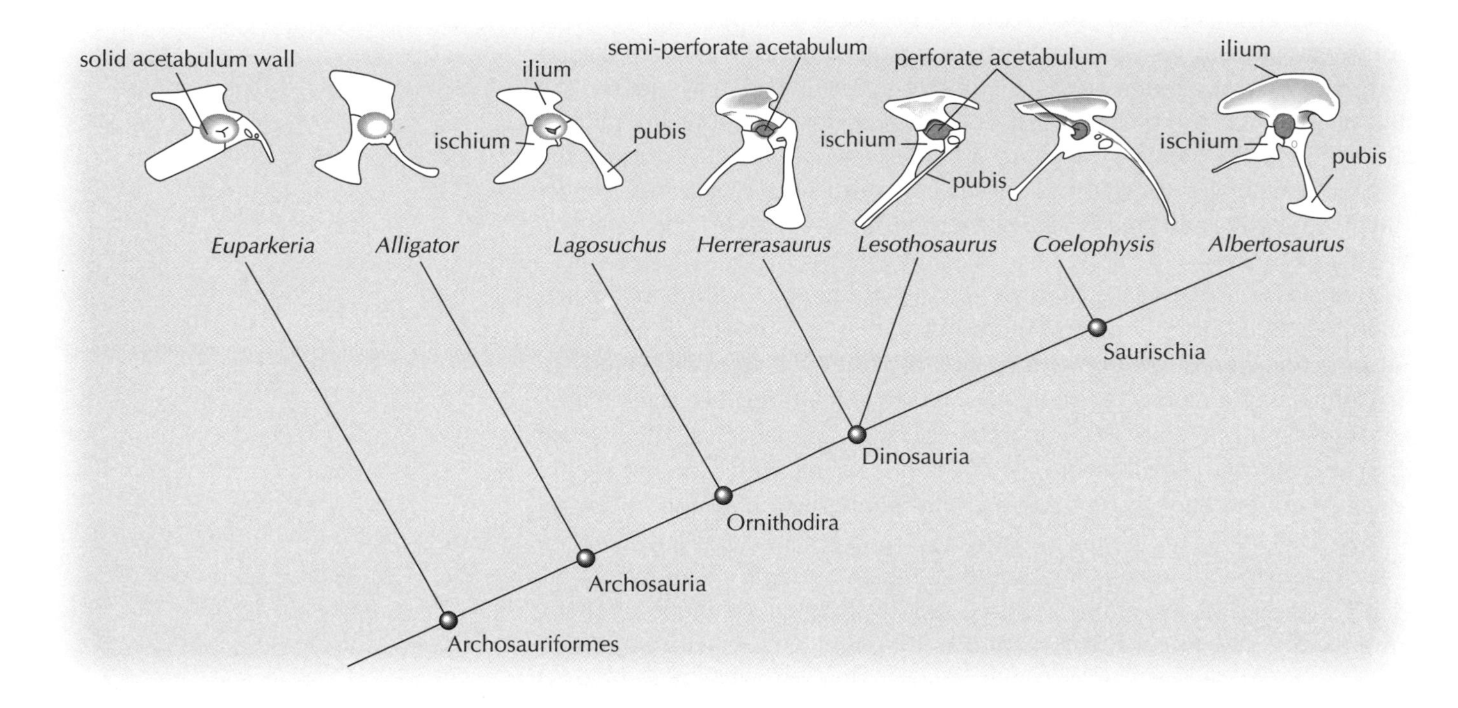

These archosaur pelves show a primitive closed-socket acetabulum, compared with the perforated acetabulum (dark green) of dinosaurs.

age usually caused heritable, evolutionary changes in one limb to be manifested in the other. In the ancestral dinosaur, this linkage was broken. Unlike the foot, the dinosaurian hand is markedly asymmetrical. The outer two fingers each lost one phalanx and the remaining parts of the digits became almost vestigial structures. In contrast, the thumb, index, and middle fingers were robustly constructed, and the strength of the hand appears to have been focused there.

With the hands and forelimbs no longer supporting the body, the hindlimb played a much greater role in locomotion. Evidence of this is found in the pelvis and sacrum of dinosaurs, which Richard Owen had found so distinctive. The ilium, the upper bone of the pelvis, was especially expanded and more massive. The ilium forms both the roof of the hip socket and part of the attachment to the vertebral column. The vertebrae of the sacrum, which lie between the right and left halves of the pelvis, are fused together in adult dinosaurs. In addition, the specialized ribs coming off of the sacral vertebrae are massively expanded to provide a stronger attachment to the ilium. Quite simply, the bigger pelvic frame indicates bigger pelvic musculature for a more powerful "motor" to move the legs.

In addition, the hip socket, or acetabulum, provides evidence of a major reorientation in the posture and movement of the femur. In other tetrapods, the acetabulum forms a closed cup, and the femur presses into the socket which acts as the fulcrum for the hip. In dinosaurs, the hip socket has a hole in the middle, a condition referred to as a perforate acetabulum. Instead of extending out to the side and pressing straight into the socket, the dinosaur femur is oriented more vertically, its head bent inward almost at a right angle and pressed against the top and back of the thickened acetabular rim. A robust supraacetabular buttress and antitrochanter projects over the perforated socket, to withstand the redirected forces of the inturned head of the femur, The shaft of the femur and the leg were held in a plane roughly parallel to the vertebral column, the parasagittal plane, and the knee was held close against the body. Bony bumps, or trochanters, on the femur expanded to provide stronger attachment points and greater leverage for the massive muscles originating on the enlarged pelvis.

Birds share all of these features of the hindlimb. Richard Owen had described the perforated acetabulum in *Archaeopteryx*, as well as the other features. The hands of *Archaeopteryx* are very different from the hands of early dinosaurs, but even so there are marked similarities, like the offset in the thumb. Birds also display many of the features that distinguish Dinosauromorpha, Ornithodira, Archosauria, Sauria, Reptilia, and so on. Hence, there appears to be a hierarchy of similarities between birds and Mesozoic dinosaurs.

The ancestral dinosaur was small compared to the "fearfully great" image conveyed in the name of the group. It probably weighed about the same as an adult human. The head was long and narrow, with a pointed snout. Its mouth extended literally from ear to ear, and its jaws were lined with sharp, serrated teeth that curved toward the back of the mouth. A deft predator, it could have hunted a wide range of prey items, from insects to animals larger than itself. It could potentially swallow items as large as its head, but this is no indication that it actually did so. The rare examples of stomach contents in early dinosaurs are of much smaller animals.

It was habitually bipedal and able to run rapidly for considerable distances in pursuit of agile, fast prey. The ancestral dinosaur had large eyes and large optic lobes in its brain for processing visual information. These provided a sophisticated sense of sight, including color perception, acute long-distance vision, and refined sensitivity to smallest movements. Its hearing was also well developed. Because the head was held high off the ground on a long, flexible neck, these sensory receptors could be very rapidly directed and redirected over wide fields to quickly locate and track potential prey. Sight and hearing were probably the principal sense organs, with a relatively less-developed sense of smell.

The oldest unquestionable dinosaur fossils were collected in South America from Late Triassic rocks (approximately 230 million years old). Several different localities in Argentina and southern Brazil have produced several different species of early dinosaurs. Of these, only *Herrerasaurus*[23] has become reasonably well known and, owing largely to the incompleteness of the others, exact relationships among the basal species are not yet understood. In North America, rocks of nearly the same age, or perhaps only slightly younger age, have also yielded fragmentary bones of early dinosaurs. During the early 1980s we accompanied a group from Berkeley's Museum of Paleontology, who discovered *Chindesaurus*,[24] a close relative of the South American dinosaurs while collecting in the Petrified Forest National Park. With this discovery, dinosaurs seemed to appear abruptly in the fossil record, already diversified into several species, and distributed to both hemispheres.

Each of these early dinosaurs has unique anatomical specializations that preclude it from direct ancestry to other dinosaurs. Because it must have taken some length of time for the specializations to evolve, and because dinosaurs have a global presence at their earliest appearance, it is likely that the species ancestral to all other dinosaurs lived a few million years earlier than the oldest known fossils, during the Middle or Early Triassic (between 235 and 245 million years ago). A few tantalizing bone fragments collected from Early Triassic rocks might represent something even closer to the ancestral dinosaur than the species named above, but more complete fossils are needed to tell whether these are truly dinosaurs. Even if unequivocal evidence is eventually traced as far back as the Early Triassic, Dinosauria, the icon for ancientness in our culture, originated only after 90 percent of the history of life had already passed.

Determining *where* the ancestral dinosaur species lived is even more difficult to narrow down than *when* it lived. Throughout the Triassic Period, today's continents were welded together into a continuous land mass known as Pangaea. Unlike today, back then there was little or no land positioned at either the north or south rotational poles, there were no polar ice caps, and the continental masses that had collided to form Pangaea were positioned closer to the equator. As a result, the Triassic climate was on average probably warmer and seasonal temperature fluctuations less extreme than today. Without either oceanic or climatic barriers, dinosaurs dispersed throughout Pangaea. Owing both to their mobility and to the evidence of an even earlier, undiscovered history for the lineage, the fact that the oldest known dinosaur fossils come from South American doesn't necessarily mean that Dinosauria originated on that continent. The ancestral species might have lived anywhere on Pangaea during the first half of the Triassic.

A SINGLE HIERARCHY

There are many differences between this picture of the ancestral dinosaur and modern birds, but because the two are separated by more than 200 million years we should expect some profound differences. More significantly, there are numerous unique and detailed similarities between birds and early dinosaurs. Moreover, these similarities lie in an internested hierarchy of novel resemblances that link dinosaurs back in time via the ancestral archosaur to the ancestral amniote, to the ancestral tetrapod, the ancestral vertebrate, and ultimately to the ancestral cell. It is also true that points of similarity seem to link birds with other groups. But when data from the entire skeleton are examined, when fossils and modern species are examined collectively, and when the map of vertebrate phylogeny is considered as a whole, there is only one hierarchy into which birds fit. So, genealogy, not homoplasy, appears to offer the most powerful explanation of the similarities between birds and dinosaurs.

Does this necessarily mean that birds are dinosaurian descendants? The scientific world view that we encountered at Berkeley is to obsessively test conclusions. Rather than accept the easy answer, we should test the dinosaur-bird hypothesis further, by mapping the hierarchy of dinosaur evolution. If birds are really the living descendants of Mesozoic dinosaurs, we should be able to map a distinctive trail of hierarchical clues through Mesozoic fossils. In the next chapter, we will see if birds can be plotted onto a specific part of the map of dinosaur phylogeny.

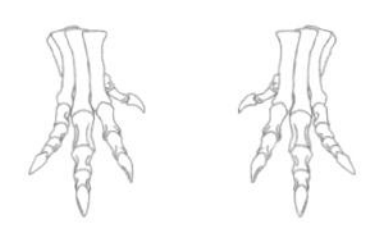

CHAPTER TWELVE

The Evolutionary Map for Dinosaurs

Many different groups of dinosaurs have been implicated in avian origins. In fact, so many different branches on the dinosaur tree have names with the root *ornitho-*, meaning "birdlike," that they are hard to keep straight. But even granting that some dinosaurs show similarities to birds, many others look nothing like them. It is difficult to recognize any special resemblances between *Triceratops* and a bird, for example. Even though we have traced a hierarchy of relationship linking birds and dinosaurs this far, some paleontologists challenge that comparing birds to dinosaurs as a whole presents only the confusing mosaic of resemblances expected of convergent or homoplastic evolution.

Another challenge of a sort comes from Late Triassic rocks in Texas, where a small animal named *Protoavis*—meaning the primordial bird—was unearthed by a researcher in Lubbock. *Protoavis* is said to be a long lost ancestor of birds that links their pathway of evolution to dinosaurs, but not via *Archaeopteryx*, nor through dinosaurs like Huxley's *Compsognathus* or Ostrom's *Deinonychus*. The discovery of *Protoavis* resurrected the theory of homoplasy, but with a different twist. Birds may be descended from dinosaurs after all, but not from any of the usual suspects.

Once again we meet with allegations that can't all be true, so how can we choose between them? If birds have rightfully inherited the "family" name Dinosauria, they will all lie along a single branch in the hierarchy of dinosaur relationships. If not, we will find only conflicting points of similarity, randomly adorning different evolutionary pathways. We can test these alternatives by following an evolutionary map for dinosaurs, starting with the ancestral dinosaur species, and tracing all its descendant lineages to their natural ends. All we need is a map.

At about the time we arrived at Berkeley as new graduate students, Kevin Padian arrived as a new assistant professor of paleontology. Padian had studied pterosaur evolution at Yale under John Ostrom's supervision, during the height of the battle over *Deinonychus*. And Padian brought the excitement of bird origins to Berkeley, where he served as dissertation supervisor for Jacques Gauthier, as well as for one of us. Gauthier's dissertation included a cladistic analysis of the relationships among dinosaurs, using the methods described in an earlier chapter.[1] When he published this work in 1986 it created a new round of controversy on bird origins because it was the first attempt at a strictly hierarchical map of dinosaur genealogy, and it supported a dinosaurian ancestry for birds.

A decade later, Gauthier carried the debate back to Yale University, filling the professorship vacated by John Ostrom's retirement. To read the newspapers, you might think that that decade, which saw dozens of new fossil discoveries and refinements in mapping technique, only stoked the flames of controversy over dinosaur relationships. Some paleontologists challenge Gauthier's work, pointing to the persistence of controversy as an indication that cladistics simply doesn't work, and journalists are quick to publish these allegations. But others argue that today's cladistic map of dinosaur relationships is basically accurate and in need only of minor polishing.

So, while no one today doubts that areas on the map of dinosaur genealogy are controversial and in need of further work, are these just refinements of a fundamentally sound cladistic structure, or should we dump the last decade of work and start over?

MAPPING DINOSAUR HISTORY

By the end of the nineteenth century, there was general agreement that the dinosaurs known to Richard Owen all belonged to two grand sister lineages known as Ornithischia, the bird-hipped dinosaurs, and Saurischia, the lizard-hipped dinosaurs.[2] As we will see, both names are more fanciful than accu-

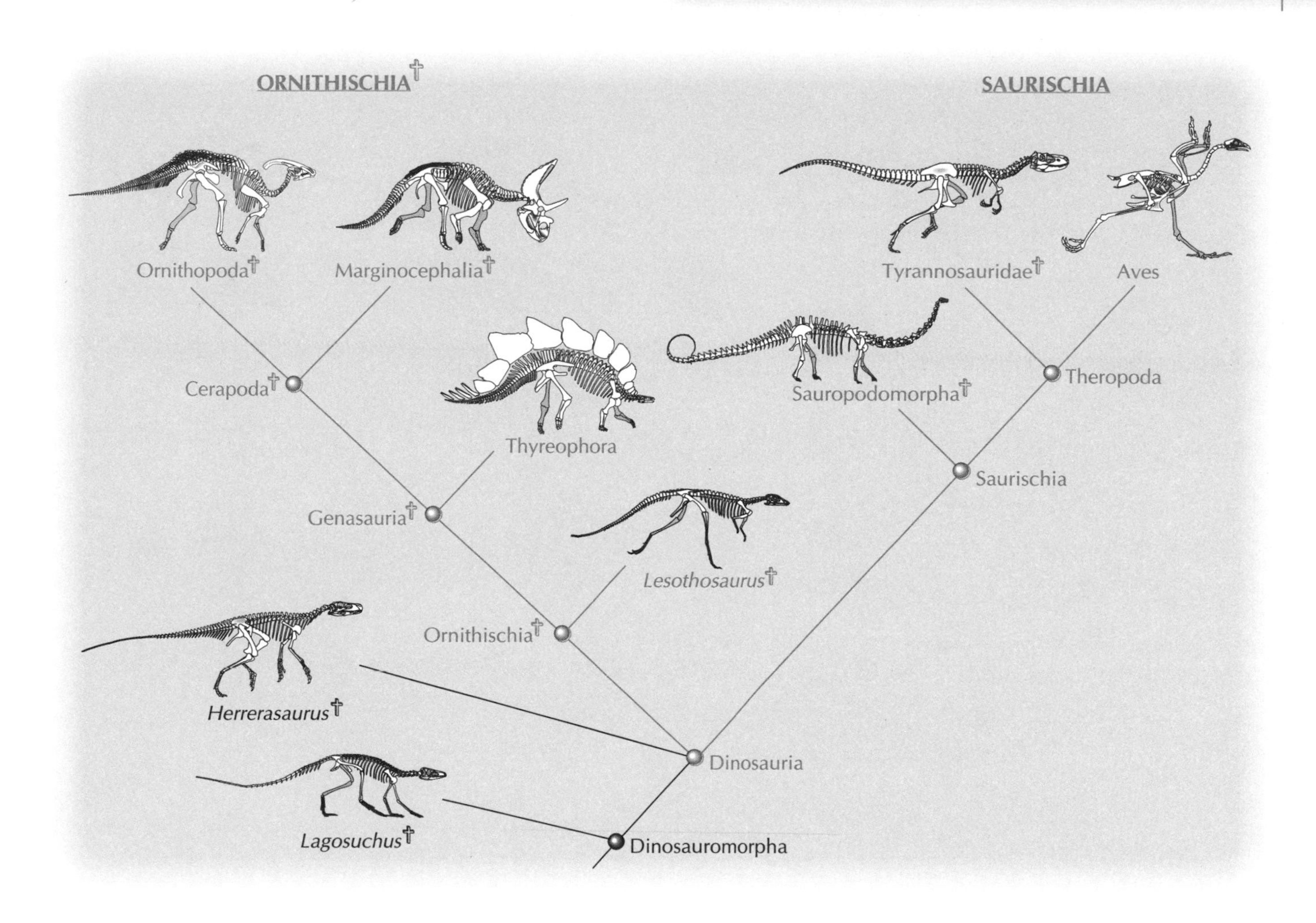

Phylogenetic map or cladogram showing the relationships of the major lineages of dinosaurs (dark green). (Extinct lineages are indicated by crosses.)

rately descriptive of the lineages they represent. Nevertheless, Gauthier's cladistic analyses and all subsequent analyses confirm this basic division of dinosaurs, and even modern critics of the bird-dinosaur hypothesis agree. Ornithischians and saurischians are both recognized as dinosaurs because they have a thumb that can grasp, along with the fully perforated acetabulum and sharply in-turned femoral head indicative of upright, parasagittal gait. All dinosaurs discovered subsequently are members of one or the other of these two distinctive sister lineages.

Despite the unanimity on these two major features of dinosaur history, there are some controversial points near the beginning of the dinosaur map. In particular, the position of *Herrerasaurus*, whom we met in the last chapter, is a little uncertain. Recent analyses disagree on whether *Herrerasaurus* is the first cousin of Dinosauria, a basal member of Saurischia, or a member of one of the lineages within Saurischia. If the former position holds true, *Herrerasaurus* would be a member of the larger group Dinosauromorpha, but it would not be a proper dinosaur, because it did not share the common ancestor unique to ornithischians and saurischians. But the position of this one species has

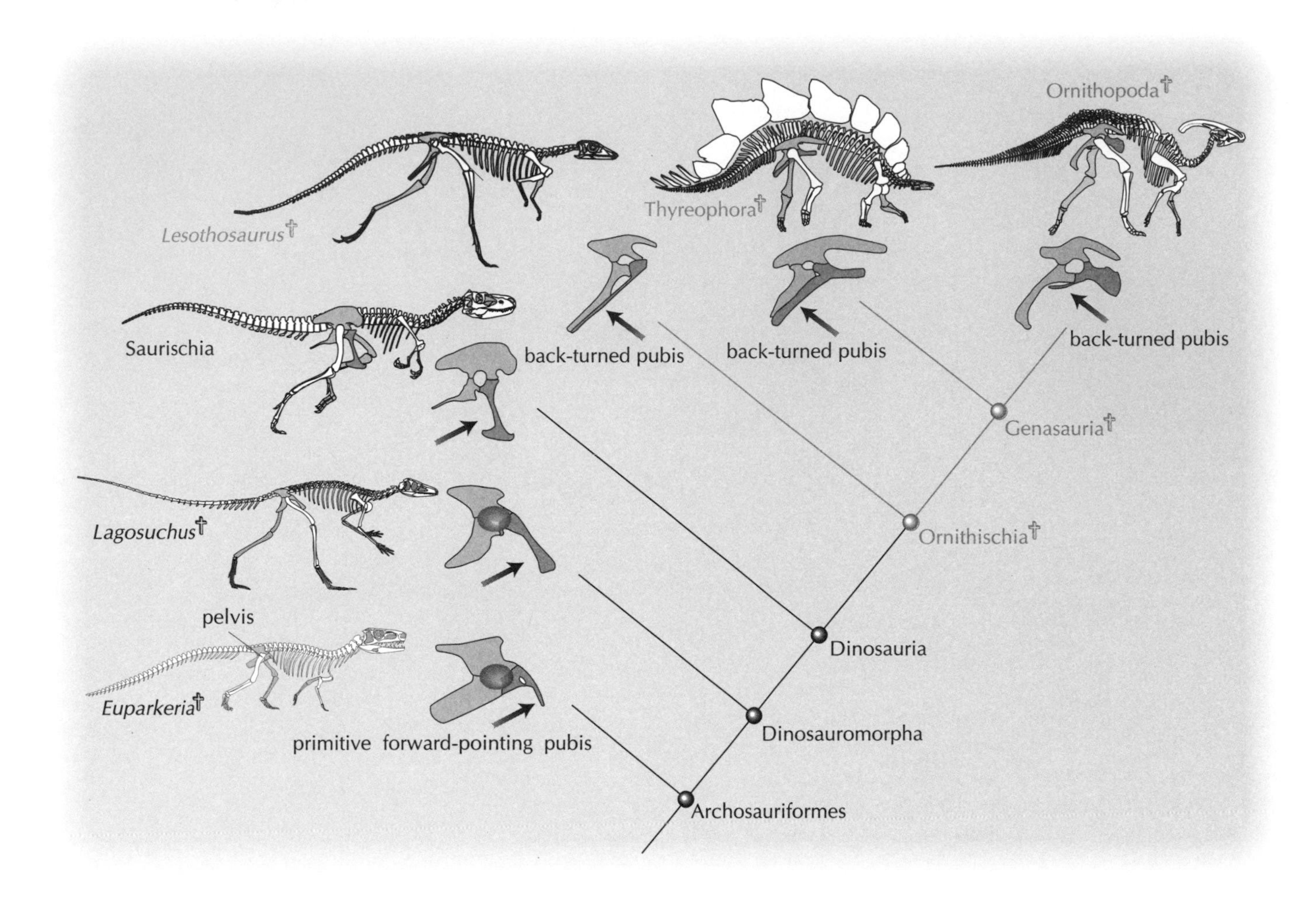

Ornithischian dinosaurs have a distinctive pelvis (green) in which the front bone, known as the pubis (dark green) is turned backward, making room for the long digestive tract that it takes to digest a diet of bulk vegetation. (Extinct lineages are indicated by crosses.)

little bearing on the distinctiveness of Ornithischia and Saurischia, each of which has its own unique features. The question now is whether one of these two great dinosaurian sister lineages gave rise to birds.

With such a suggestive name, Ornithischia might seem the place to start in seraching for avian ancestry. The name "ornithischian" refers to a birdlike ischium—a bone of the pelvis. However, it is a different pelvic bone, the pubis, not the ischium, that is birdlike in being back-turned. Despite the misnomer, we can still ask whether the pubis is a mere point of resemblance or if it marks a longer trail to birds. The oldest ornithischians are from Late Triassic deposits in Argentina, and they are known from slightly younger deposits in North America and South Africa. Early ornithischians[3] were similar to other early dinosaurs in being small bipeds, but several unique features mark their lineage, which mostly reflect a shift in diet. Ornithischian teeth and jaws were designed for tearing and grinding plants, and the ribs and pelvis housed an enlarged digestive tract. A new bone, the predentary, forms the front of the ornithischian lower jaw. In all but the oldest ornithischians, the frontmost teeth are gone and a horny beak rims the front of the mouth. In all but one of the very earliest ornithisichians, the teeth are set in from the margins of the mouth, where they were probably covered by fleshy cheeks that assisted chewing.

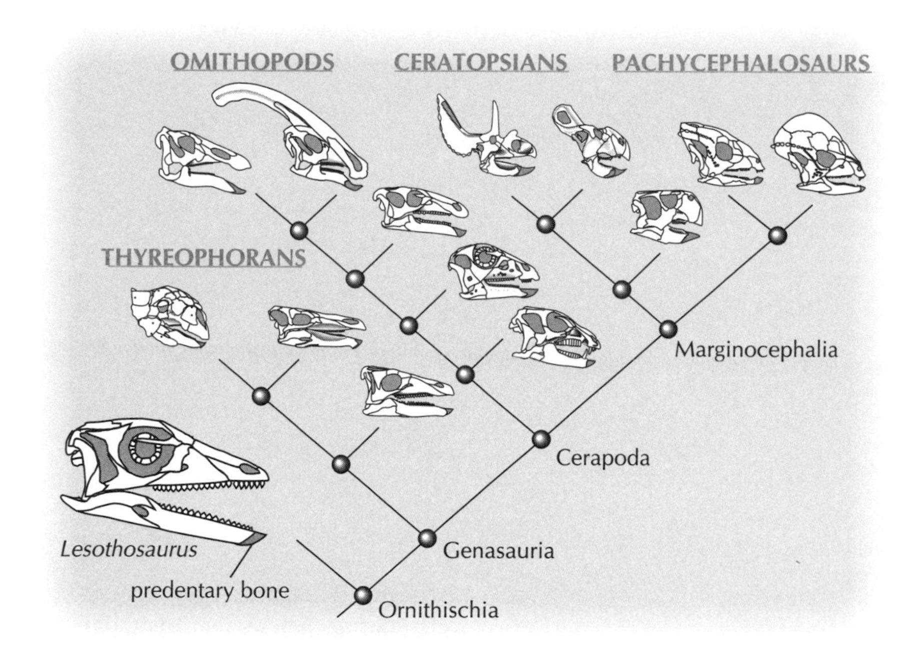

The predentary bone (dark green), a unique feature of ornithischians, contributed to their ability to crop and strip vegetation. The predentary bone is present in all known members of the lineage.

Abdominal expansion is indicated by the characteristic modifications of the pubis. In most other reptiles the pubis points down and forward, but in ornithischians it is rotated backward, making room for a longer intestinal track to digest the relatively insoluble cellulose of plant cells. One last modification is a network of ossified tendons along the backbone. Normally, tendons are strong, flexible, ropelike fibers that attach the fleshy body of a muscle to a bone. In ornithischians, bone formed within some of the tendons along the back, forming a mesh that permitted some up-and-down flexing and extension of the vertebral column, but it prevented any adverse rotation. Perhaps this increased rigidity of the vertebral column offered a sturdier framework from which to suspend their enlarged gut.

Ornithischia is a diverse lineage and within it are several different evolutionary paths. The ornithischian lineage with the longest fossil record is Ornithopoda. This name means birdlike foot, in reference to the three-toed ornithopod foot. The lineage was so christened by Yale's preeminent dinosaur specialist of the nineteenth century, O. C. Marsh, at a time in which only a small number of dinosaur fossils were known. More recent discoveries indicate that virtually all early dinosauromorphs left three-toed tracks, so the namesake feature that caught Marsh's attention is more widely spread than he knew when he coined the name.

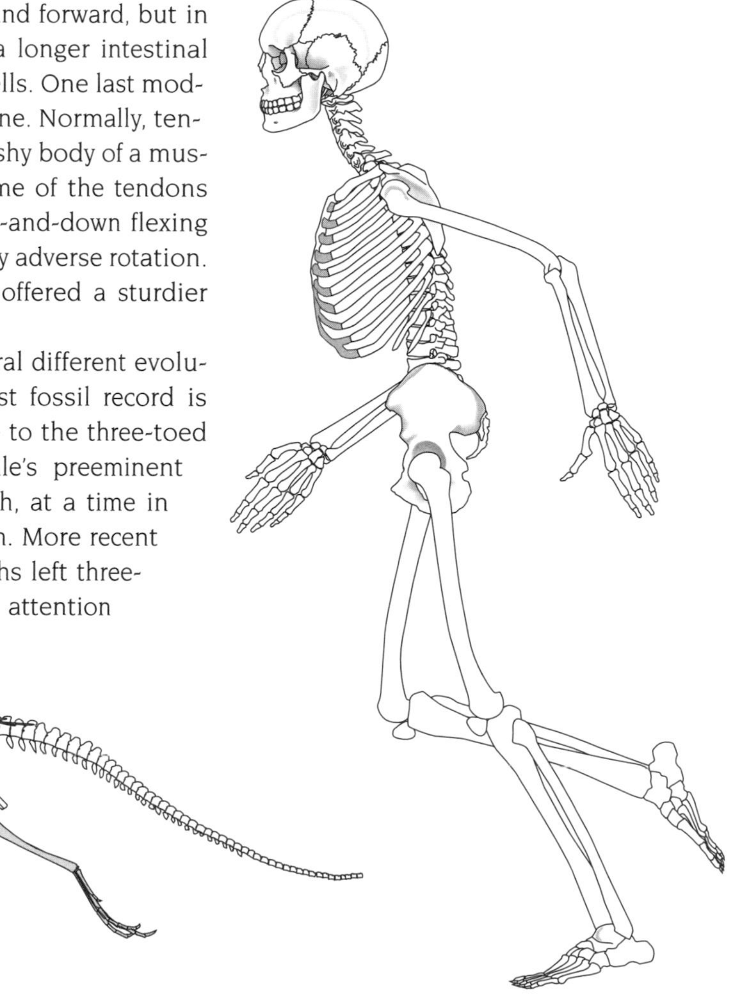

The oldest ornithischians were all relatively small. Shown here is the Early Jurassic *Lesothosaurus*, compared to a modern 6-foot-tall human for scale.

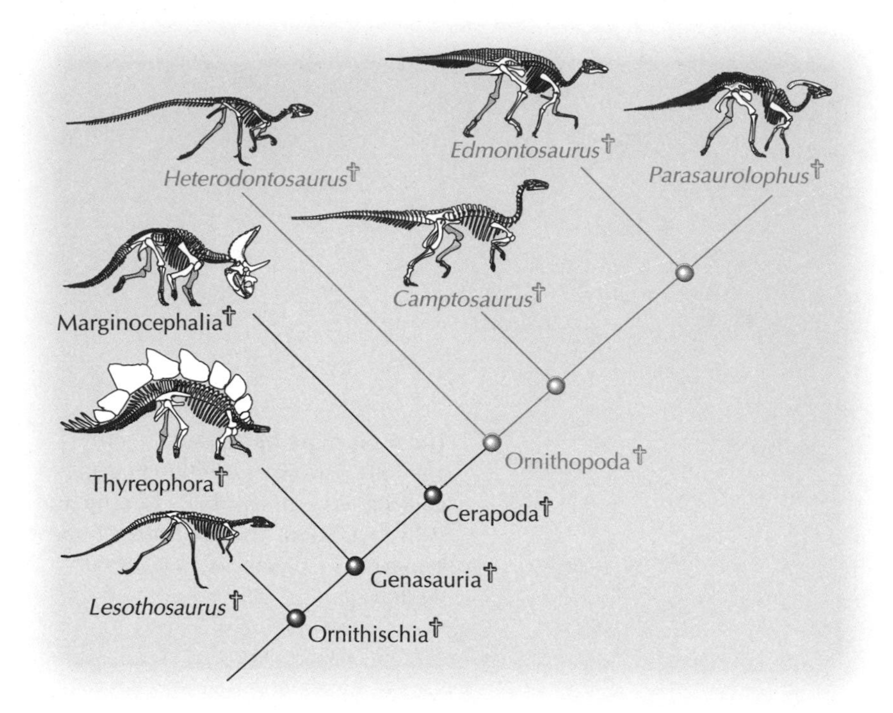

Phylogenetic map showing the relationships among representative ornithopods (dark green). (All the lineages depicted here are extinct.)

Ornithopod history has been mapped in detail by David Weishampel of Johns Hopkins University, David Norman of the University of Cambridge, John Horner of Montana State University, Paul Sereno of the University of Chicago, and their colleagues.[4] Ornithopods probably existed in greater numbers and are more abundantly fossilized than any other dinosaur. By their first appearance in the Early Jurassic, ornithopods had already dispersed around the world. Early members were small, about 3 to 6 feet in length. They were distinguished by unique dental modifications which, in some of the more derived members of the lineage, produced a great increase in the grinding capability of the teeth. This was accomplished by increasing the numbers of teeth and the rate at which they were replaced throughout life. Some later ornithopods had as many as 1,000 teeth in the mouth. There was also an evolutionary increase in body size. Some of the Cretaceous forms reached nearly 40 feet in length, and adults weighed several tons. Richard Owen's *Iguanodon*, which lived during the Early Cretaceous, is one of the more derived, large-bodied ornithopods. The name for Dinosauria might have been very different had Owen first studied the comparatively tiny Early Jurassic ornithopods. Apart from the resemblances found in dinosaurs ancestrally plus the back-turned pubis, birds and ornithopods bear no special similarities. Recent mapping efforts have found only points of resemblance between the two, and the giant ornithopods only become more different from birds with time. Despite their enticing name, ornithopod dinosaurs are not the ancestors of birds, and available evidence indicates that the lineage became extinct in the terminal Cretaceous event.

The sister lineage of Ornithopoda is Marginocephalia, the "margin-headed" ornithischians, who had weird skull modifications that make them highly distinctive.[5] Various segments of their history have been mapped by Paul Sereno,

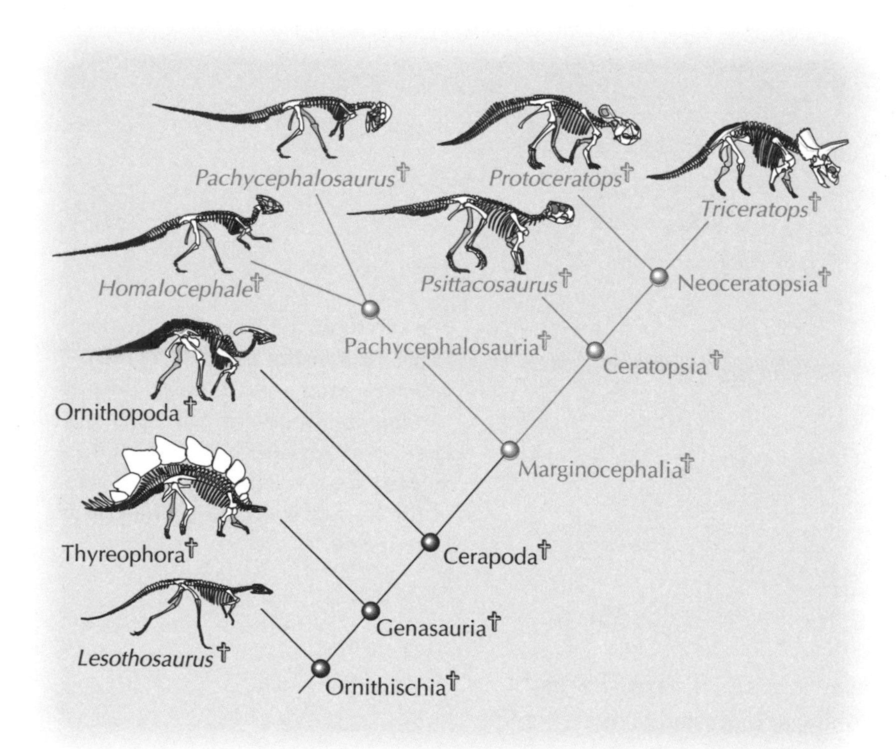

Phylogenetic map showing the relationships among representative marginocephalians (dark green). (All the lineages depicted here are extinct.)

Peter Dodson of the University of Pennsylvania, Catherine Forster of Cornell University, Tom Lehman of Texas Tech University, and a host of associates. The two major lineages of marginocephalians are Ceratopsia and Pachycephalosauria. Some authors derive the latter directly from ornithopods, but this point of disagreement doesn't affect our search for the ancestry of birds. Pachycephalosaurs, the "thick-headed" ornithischians evolved high domes of thickened bones over the top of the brain. The name is not particualry descriptive of early flat-headed forms, but it is apt for later members of the lineage. So extreme is their head thickening that, when pachycephalosaurs were first discovered, paleontologists thought them to be pathological. Others considered them to be dinosaur "knee-caps." As more complete specimens were recovered, it was clear that these bony domes were parts of the skull, and mechanical analyses suggest that pachycephalosaurs used their heads as battering rams. Comparable head-butting and flank-butting behavior occurs in modern musk ox, mountain sheep, and goats, usually in battles with members of their own species over territory and mates.

The other marginocephalian lineage is Ceratopsia—the horned ornithischians.[6] When first discovered in the nineteenth century, the only known skulls possessed horns. Since then, there have been numerous ceratopsian species discovered that lack horns and are primitive in other respects, so this is another name that is misleading. One distinctive feature of the lineage is the rostral bone, a unique structure that forms the upper part of the beak above the predentary bone. Even the primitive hornless ceratopsians have a rostral bone, indicating that the acquisition of a powerful beak preceded the evolution of

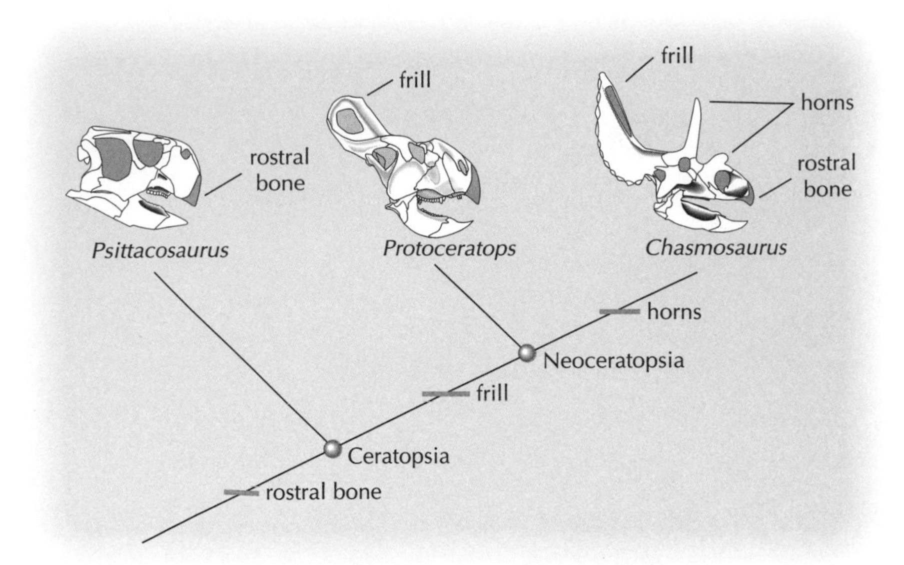

The ceratopsian lineage is distinctive in having a rostral bone (dark green), which created a sharp beak. Later in ceratopsian history the frill arose, and still later horns appeared. All three of these animals are ceratopsians, but the name of the lineage is accurately descriptive of only one.

frills and horns. Early ceratopsians were relatively small, like the early ornithopods and pachycephalosaurs. When running at high speeds they were bipedal, but at lower speeds they probably moved on all fours. Later ceratopsians were rhino-sized animals that reverted habitually to quadrupedal locomotion. With increased size, the shelf at the back of the head also expanded into a fan-shaped sheet of bone protruding up and backwards from the head, reaching more than 4 feet long in some species. Variable patterns of horns and projections from the cheek region appeared, and the edges of the frill became elaborately ornamented in some species as well. Marginocephalian history led to a terrific diversity of form, but most of this was in a

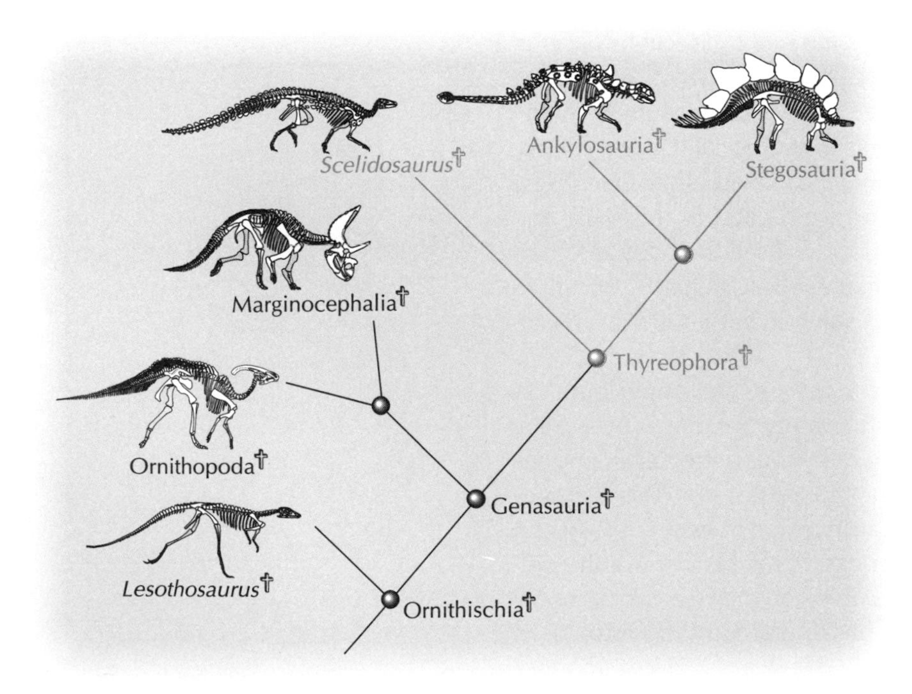

Phylogenetic map showing the relationships among representative thyreophorans (dark green). (All the lineages depicted here are extinct.)

different direction from the evolutionary pathway taken by birds. Marginocephalians survived until the very end of the Cretaceous, but all available evidence indicates that they were extinct when the Tertiary dawned.

The third major lineage of ornithischian dinosaurs is Thyreophora, whose name means "shield-bearers" in reference to their body armor. There are several thyreophoran lineages,[7] the most distinctive of which are the ankylosaurs and stegosaurs, whose histories have been studied by Peter Galton of the University of Bridgeport and Walter Coombs of Western New England College, Teresa Maryanska of the Polish Academy of Sciences, and others. The thyreophoran fossil record extends back to the Early Jurassic, where its early members were small, like other early dinosaurs. They possessed an armor shield of bony scutes which floated in the skin along the back and sides of the body. Descendant lineages like the ankylosaurs became fully armored and were completely covered with a patchwork of bony scutes. Ankylosaurs were most common in the Late Jurassic and Early Cretaceous, and they were squat, lumbering quadrupeds that approached 2 tons in weight. Stegosaurs, the plated dinosaurs, have a distinctive paired row of platelike scutes along either side of the backbone, from the head to the tip of the tail. In some cases, an additional row or two of smaller scutes lies on either side of these, and tail scutes may form spikes of varying lengths. Despite the diversity, nowhere in the thyreophoran lineage is there evidence of a close relationship to birds. The last thyreophorans died out in the terminal Cretaceous event along with the other surviving members of the ornithischian lineage.

So, although ornithischians inherited from the ancestral dinosaur a number of unique similarities with birds, beyond these we find only points of resemblance, like the back-turned pubis, that might link them to birds. Moreover, there is no evidence to suggest that birds ever possessed a predentary bone or ossified tendons, nor do any known features place birds within the hierarchies of Ornithopoda, Thyreophora, or Marginocephalia. If birds descended from the ancestral dinosaur, it was not via the ornithischian branch of the family tree. Ornithischia became extinct in the terminal Cretaceous event.

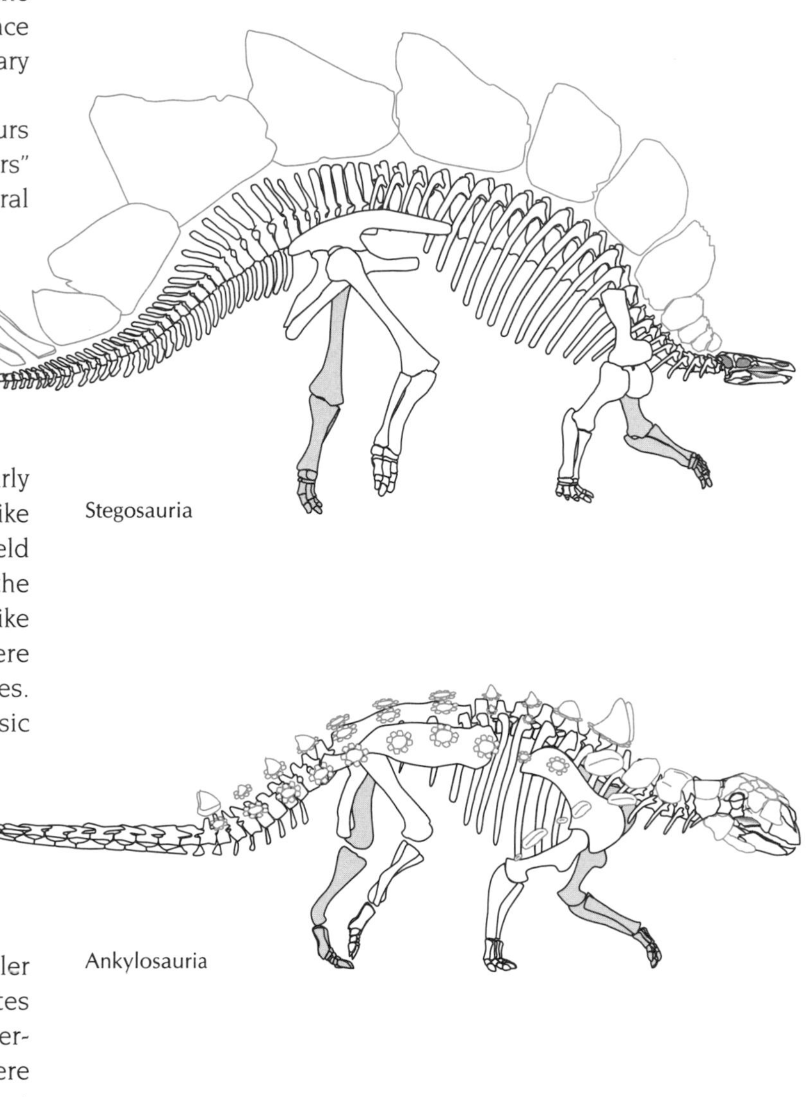

All thyreophorans had body armor in the form of scutes (dark green) that floated in the skin. Each of several descendant lineages had its own distinctive scute pattern.

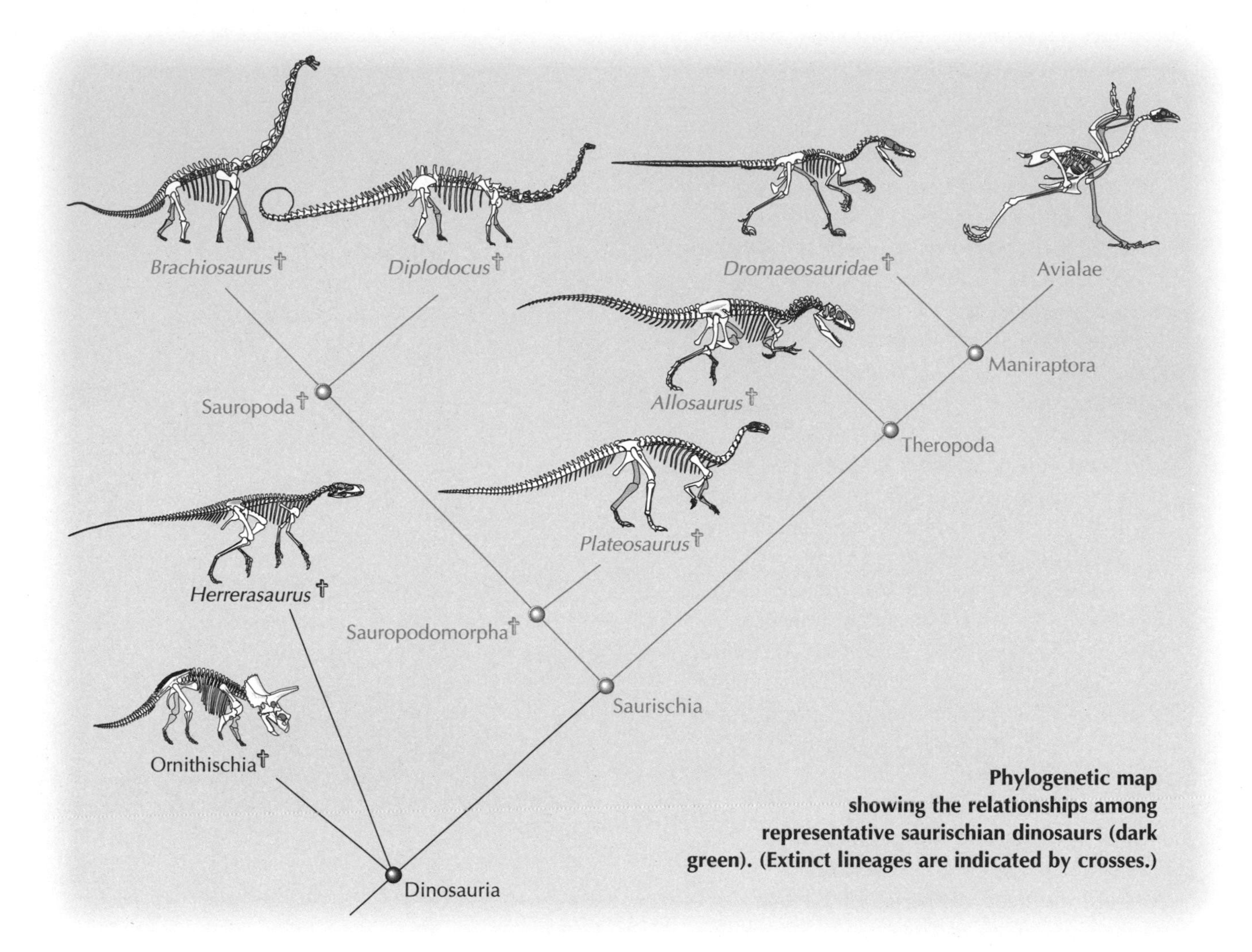

Phylogenetic map showing the relationships among representative saurischian dinosaurs (dark green). (Extinct lineages are indicated by crosses.)

SAURISCHIANS

If Thomas Huxley was right that the evolutionary road to birds runs through dinosaurs, then it must lead into saurischian, rather than ornithischian dinosaurs. This became one of the central questions that Jacques Gauthier asked in his cladistic dinosaur-mapping expedition.[8] Does Saurischia, as commonly constituted by twentieth-century paleontologists, include all descendants of the ancestral saurischian? In other words, are birds the descendants of saurischian dinosaurs and if so, which saurischians are the closest relatives of birds?

One distinctive feature shared by all members of the saurischian lineage is a very long neck in which each vertebra is elongated, and in some cases additional vertebrae are added to the neck from the rib cage. The ribs on the neck vertebrae are also lengthened, each extending backward along several vertebrae. This combination of long vertebrae and long overlapping cervical ribs enabled the neck to move smoothly and function as an integrated unit. Saurischians also have a distinctive hand. The second or index finger is the longest, instead of the third finger as was the case in

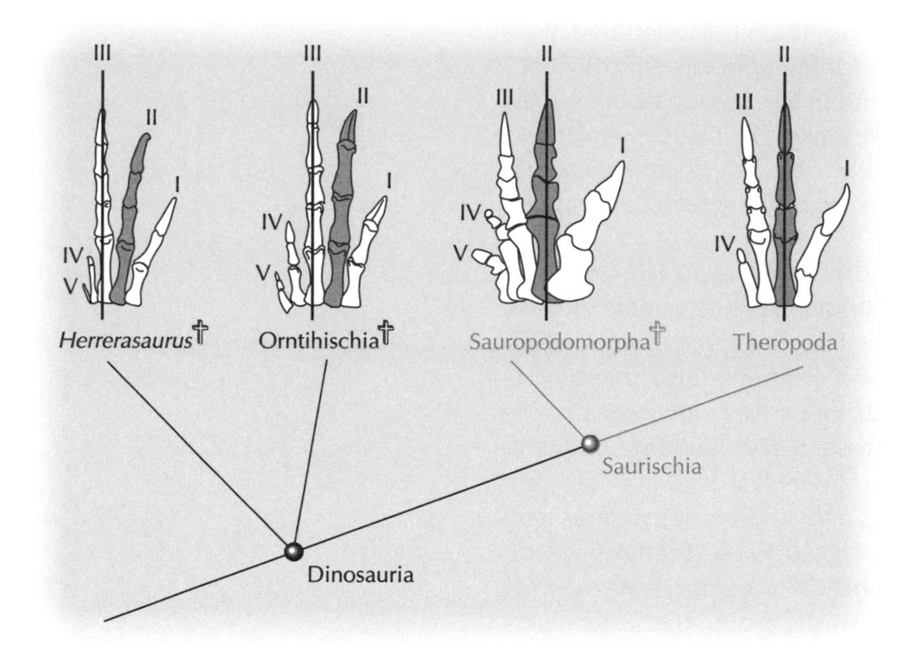

Compared to other dinosaurs, saurischians have a distinctive hand, in which digit II is longest. The longest finger defines the axis of symmetry of the hand, on either side of which are shorter digits. (Extinct lineages are indicated by crosses.)

ancestral dinosaurs, and the thumb was equipped with a large, recurved claw. While the long neck is obviously birdlike, a powerful grasping hand seems unexpected in the ancestor of a delicate bird. But the hand in the early saurischians resembles birds in its axis of symmetry—the second finger is the longest, with shorter digits arrayed on either side. In most other reptiles the axis of hand symmetry runs through either the third or fourth finger.

The earliest saurischian fossils come from Late Triassic deposits of Africa, South America, Europe, and North America. The largest Triassic saurischians were about 15 feet in length and weighed perhaps 200 pounds, but most were

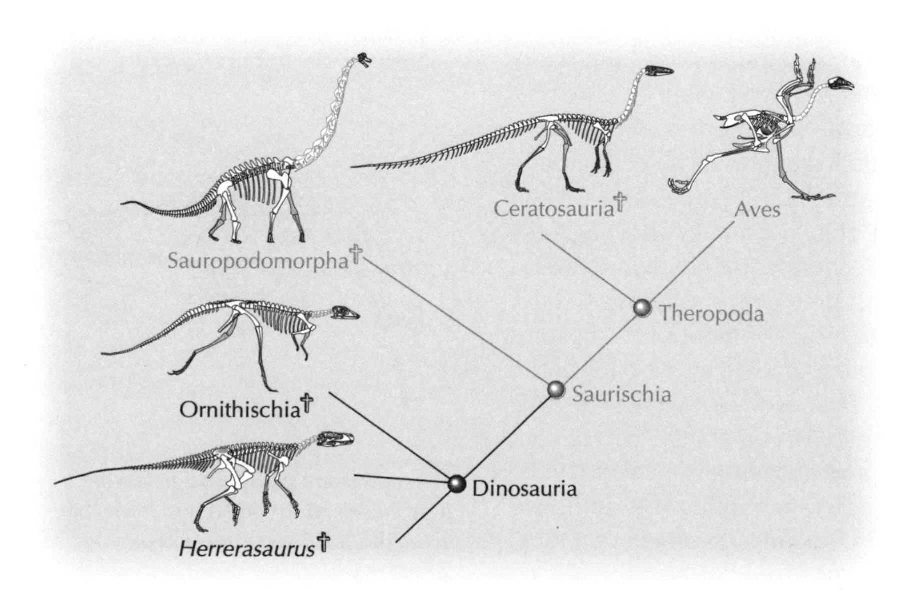

Saurischians are distinguished by a long neck, compared to other members of Dinosauria. (Extinct lineages are indicated by crosses.)

much smaller. Like other early dinosaurs, all Triassic saurischians were bipedal. Two different saurischian lineages, Sauropodomorpha and Theropoda, have been mapped and virtually all saurischians belong to one or the other. These too have unfortunate names if we look at their strict translations. The sauropodomorphs would have "lizard-like" feet and theropods would have "mammal-like" feet, if their names were accurately descriptive, but neither name is.

Over most of their history, sauropodomorphs were herbivorous.[9] Their teeth were blunt and spatula-shaped, few in number, and unable to chew vegetation. Instead, they cropped and stripped foliage from stems, which was swallowed for processing in a muscular gizzard whose presence is indicated by clusters of large smooth stones found in the ribcages of well-preserved specimens. An overwhelming theme in sauropodomorph history is size. During the Jurassic, they became the largest land animals ever, reverting to quadrupedality, and evolving unbelievably long necks and tails in the process. One of the most distinctive features of all sauropodomorphs is that they have tiny heads in comparison to other dinosaurs. In later, more derived members of the group, the head seems ridiculously small compared to the immense body. Giants like *Supersaurus* may have reached 130 feet (40 m) in length. A single neck vertebra of *Ultrasaurus* is over 3 feet long,[10] and there were perhaps as many as 17 individual vertebrae in the neck, though not all vertebrae were equally long. The limbs were columnar and elephantine in proportion, to support a bulk estimated in the very largest species to approach 100 tons. This is nearly 10 times the weight of an adult male African elephant. Despite the name of this lineage, in most of its members the bones of the toes and feet were reduced to stubs, and it must have seemed that their bodies were set upon four great posts instead of arms and legs. What could be less birdlike? Several species survived in the southernmost parts of North America and in South America until the end of the Cretaceous, but none crossed the K-T boundary alive.

THEROPODS

This leaves us with one last dinosaur lineage—Theropoda. Most of the unique features of basal theropods are associated with a predatory lifestyle, and they further enhanced an inherited body plan already well equipped for this task. Added to the ancestral armament of sharp teeth and claws, theropods had a kinetic or flexible lower jaw, in which a mobile joint between the bones of the lower jaw enabled it to bend downward and outward. Some paleontologists argue that this was for swallowing prey items larger than their own heads, while others contend that it was a buffer of flexibility in dealing with struggling prey, preventing the slender jaw bones from snapping. Also distinctive is the attachment of the head to the neck, in which a large, ball-shaped occipital condyle at the back of the skull fits into a deep socket formed by the first two neck vertebrae. This bony joint supported wide motion of the head, with a stable, strong connection to the neck. Placing the eyes, ears, and nose in a skull that can be rapidly directed from right to left, up or down, amplifies the ability to extract precise spatial information from light, sound, and smell. The long,

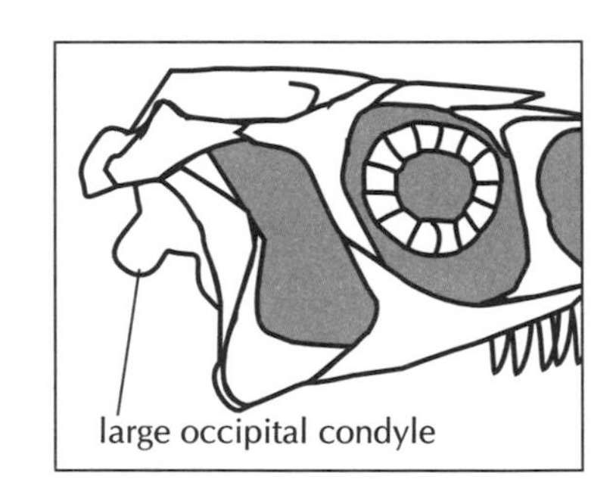

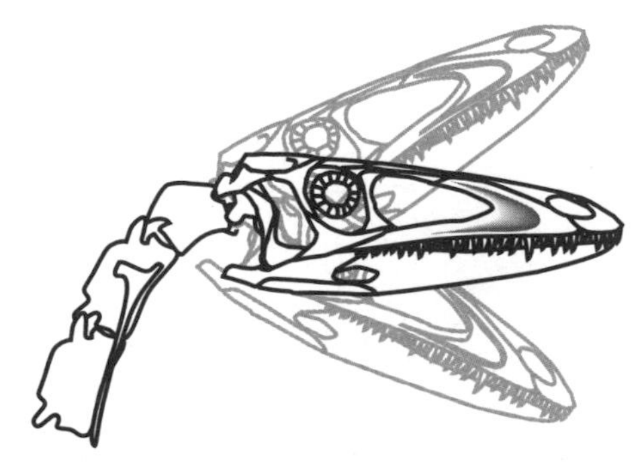

Theropods are distinctive in having a joint between the head and neck that gives the head great mobility.

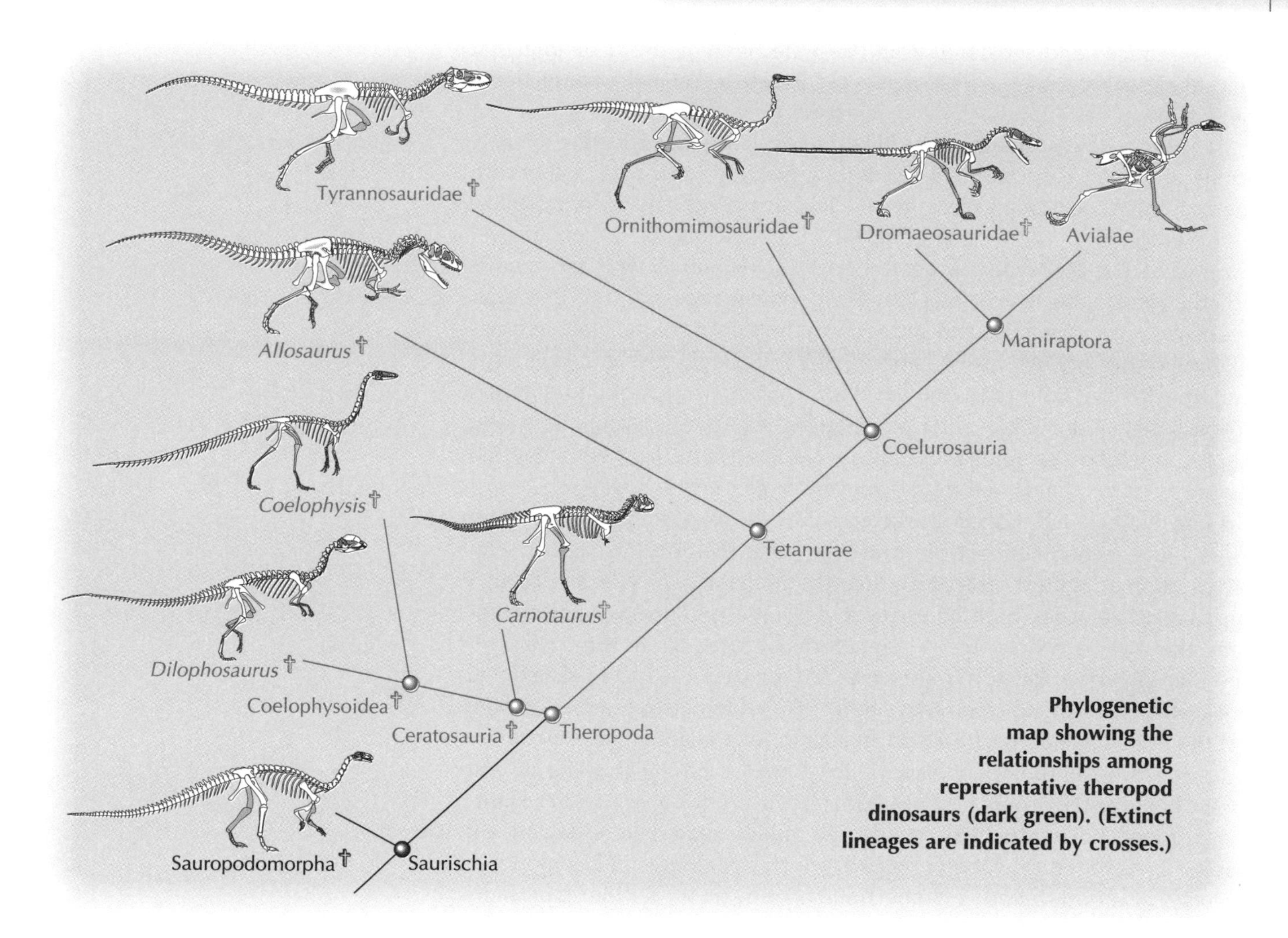

Phylogenetic map showing the relationships among representative theropod dinosaurs (dark green). (Extinct lineages are indicated by crosses.)

mobile neck inherited from saurischian ancestors amplified head mobility even further.

Early theropods also have hands designed for snagging and raking flesh.[11] The second and third fingers were elongated, and the mobile thumb worked with them in concert. Each of the three grasping fingers was tipped by a strongly curved raptorial claw equipped at its base with a large tubercle that increased the leverage of the muscles in the forearm as they closed the fingers around an object. The fourth finger was reduced and the fifth finger was lost altogether in all adult theropods but *Eoraptor*.[12] Throughout their history, theropods have been obligate bipeds, with a pelvis and hindlimb modified to withstand the entire burden of swift, forceful running. Although the ancestral dinosaur was probably an habitual biped, theropods carry this trend to a far greater degree than other dinosaurs. Additional vertebrae are incorporated into the sacrum for a stout attachment between the backbone and pelvis, and the pelvis itself was enlarged to support greater thigh muscles. This also happens in some of the giant sauropodomorphs and ornithischians, but even small theropods have a stronger pelvis. In the foot, the first or "big" toe became reduced and separated from the bones forming the ankle joint. On the outside of the foot, the fifth digit is reduced to a vestigial, nonfunctional splint of bone.

Many of these characters are birdlike. For example, all birds have a highly mobile joint between the head and a long neck, and many have jaws with kinetic joints. Even small birds have a massive pelvis and sacrum. Unlike early theropods, in the foot of adult birds, the fifth toe is entirely absent. However, all five digits are present in bird embryos, and all five lie in contact with the developing ankle joint. As development proceeds, the first digit breaks away from the ankle and slides down the side of the foot, before twisting around the back. As this happens, digit V is gradually lost and adult birds wind up with only four toes. The development or ontogeny of bird embryos recapitulates an ancient evolutionary pattern. As we will describe in a later chapter, nowhere else among reptiles do these two patterns of ontogeny and phylogeny in the bones of the foot coincide in precisely this way.

A last distinctive theropod character is the one discovered by John Hunter in his brutal experiment on the arm of a bird. Theropod bones are thin-walled, tubular, hollow structures. Like the frame of a bicycle, the tubular construction provides both lightness and a high bending strength to withstand the high levels of force generated in fast locomotion. Hunter discovered that the avian skeleton is hollow and Owen cited the hollow skeleton of *Archaeopteryx* as evidence that it is a bird.[13] Paleontologists commonly claim that the hollow skeleton evolved as an adaptation to lighten the skeleton for flight. However, all theropods have hollow bones, even *Tyrannosaurus rex*, but no one believes they could fly. The hollow skeleton obviously serves the function of flight, but it would be nearly 100 million years before descendant theropods recruited the tubular skeleton for this new mode of locomotion.

Critics of a bird-theropod connection correctly point out that pterosaurs and some small mammals also have hollow skeletons, and that a tubular skeleton must have evolved convergently several times.[14] They argue that convergent features offer no insight into the relationship between birds and theropods. As we have already seen, this similarity between birds, pterosaurs, and mammals constitutes a mere point of resemblance rather than a mappable hierarchy that includes birds. However, in the case of theropods, there are many additional special similarities to birds. By following the trail of anatomical clues, we can test whether hollow bones offer evidence of common ancestry for birds and theropods, as we trace the map of theropod genealogy to its natural conclusion.

The fossil record of theropods is not very good, probably because their hollow skeletons rarely withstood the dynamics of sedimentation and burial. Most known theropod skeletons are incompletely preserved, and gaps spanning tens of millions of years still punctuate our knowledge of theropod history. This is the same general problem that faced Darwin and Huxley, although in their time the gaps were more on the scale of 100 million year intervals. So, while many problems in our map of theropod history remain, today's resolution is far better than it was a century ago.

By the Late Triassic, two theropod lineages had arisen, namely Ceratosauria and Tetanurae. Ceratosaurs, which were first mapped and named by Gauthier,[15] are the best-known Triassic and Early Jurassic theropods, having a global distribution at their earliest appearance in the fossil record.[16] Possibly the richest dinosaur locality ever discovered is the *Coelophysis* bone bed at Ghost Ranch, New Mexico. There, dozens, perhaps even hundreds, of *Coelophysis* individuals, including juveniles and adults, were buried together en mass in a Triassic grave.[17] The Late Jurassic *Ceratosaurus nasicornis* is the last-known member of the lineage in North America, but more recent discoveries in South America, Madagascar, and India may indicate that Ceratosauria had a Gondwanan distribution in the Cretaceous.

Currently, there are only about a dozen or so species of ceratosaurs, but they document a 170 million year history, from the Triassic into the Late Cretaceous, so our record of this lineage is highly incomplete. In several skeletal features, a few of the smaller ceratosaurs show additional similarities to birds, including fusions between bones in the feet and further strengthening of the pelvis and sacrum. But there are other theropods that share even greater degrees of resemblance to birds, and the available evidence indicates that ceratosaurs disappeared from the northern hemisphere before the Cretaceous began, and from the southern hemisphere at its end.

The *Coelophysis* quarry at Ghost Ranch, New Mexico, when it was reopened in 1981, forty years after it had been worked by Edwin Colbert of the American Museum of Natural History. The large white plaster-jacketed block at the left end of the quarry contained dozens of skeletons.

STIFF-TAILED TETANURINES

The trail of clues to avian ancestry leads into the tetanurine, or stiff-tailed theropods, that make up the sister lineage to ceratosaurs.[18] There is much yet to be discovered about their early history, which probably extended into the Triassic, judging from the antiquity of their sister lineage. But the oldest informative tetanurine fossils currently known are from the Late Jurassic, almost 100 million years after the lineage most likely originated. Tetanurines are more birdlike than other theropods in virtually all parts of their skeleton. Breathtaking new specimens from China may even offer evidence linking the origin of feathers, or at least "protofeathers," to an early evolutionary stage in tetanurine history.

The tetanurine snout was more delicately built than in other theropods. The teeth, if present at all, lie entirely in front of the eye. Over much of their history, there was a gradual loss from the back of the jaws to the front, and several lineages evidently lost their teeth independently. The tetanurine forelimb took on a striking of resemblance to birds. The wishbone or furcula

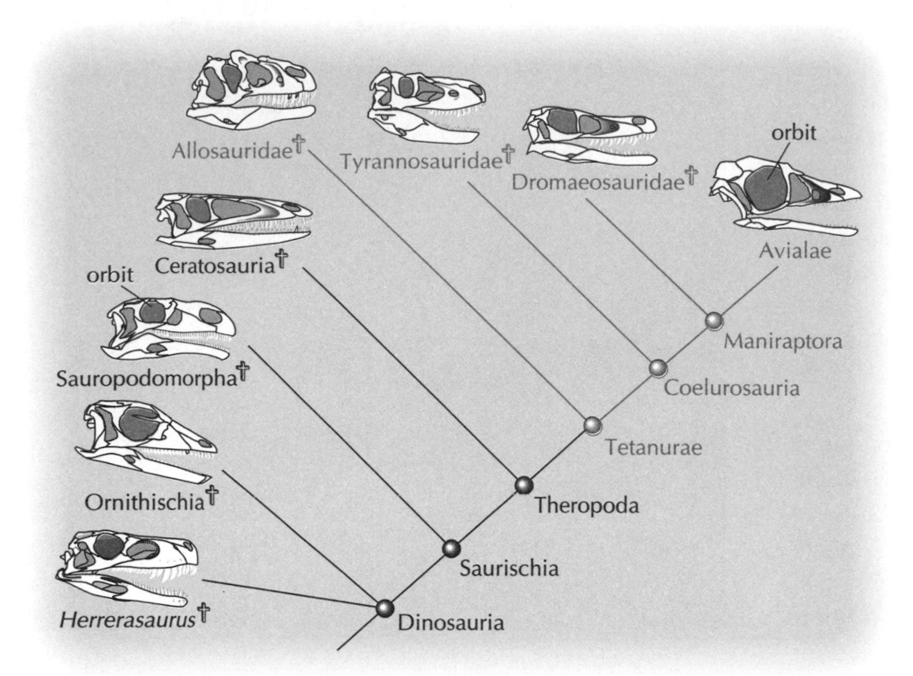

In tetanurine dinosaurs, the teeth lie entirely in front of the orbit (dark green), which held the eyeball. This began a trend in which the tooth row was successively shortened, from back to front, as the teeth were reduced in size.

appeared early in the history of this lineage, along with more powerful arms and hands. The wishbone occurs today only in birds, prompting some biologists to argue that it is an adaptation for flight. But many nonflying Mesozoic tetanurine dinosaurs have a wishbone, so it would appear that the furcula was only secondarily co-opted into taking a role in flight. The wishbone extends between the two shoulder joints and the breastbone or sternum, and when the wishbone appeared, the sternum became a rigid, bony structure to anchor large pectoral muscles. The arms and hands were also longer, the hand now consisting of only three adult fingers. Together, these changes enormously enhanced the reach and power of the forelimb.

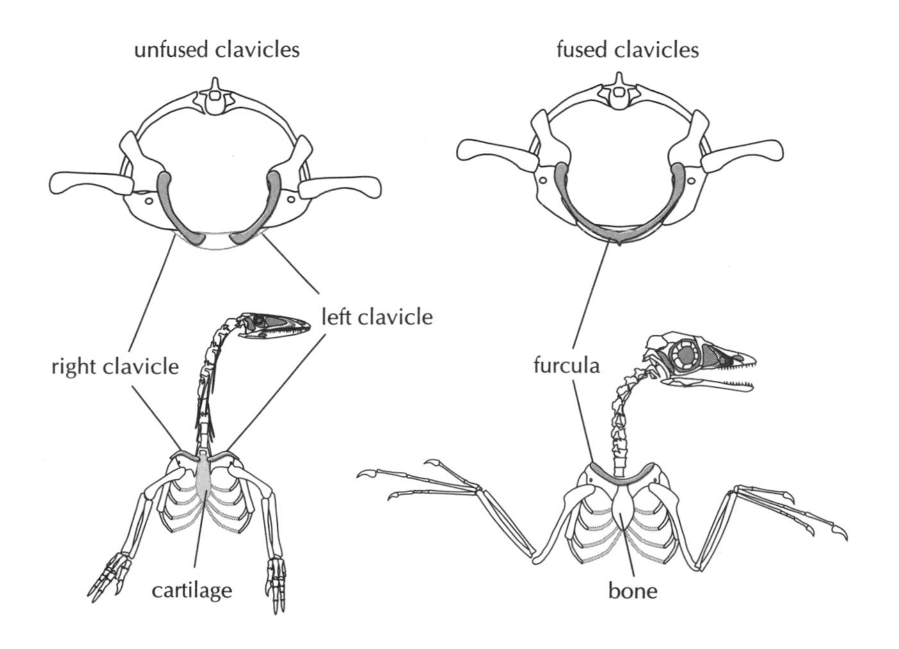

The collar bones, or clavicles, became fused together in tetanurines, forming the wishbone, or furcula. Although often thought to be linked to flight, the furcula is present in dinosaurs like *Allosaurus* and other extinct tetanurines, which no one believes could fly.

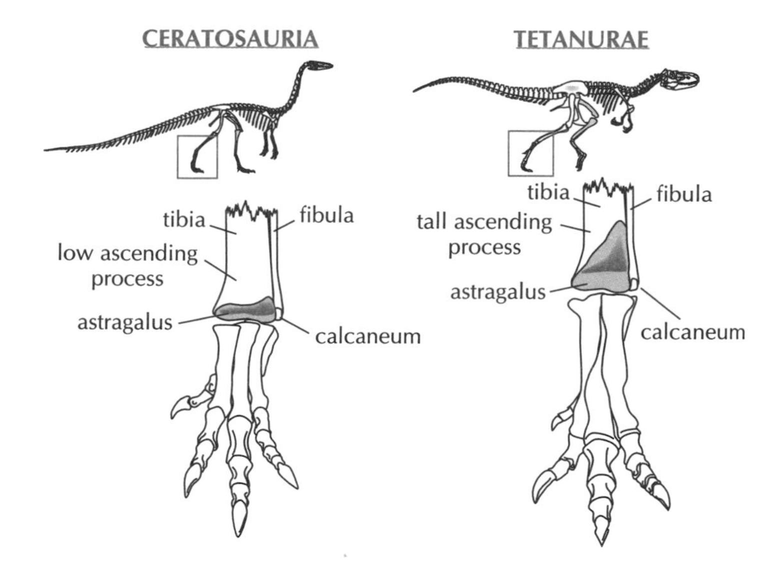

Tetanurine dinosaurs are distinctive in having a sheet of bone extending from the ankle joint upward over the shin bone. This sheet is known as the ascending process, and it is still visible in the drumstick of young birds as a sheet of cartilage.

More subtle resemblances to birds are found in this lineage as well. For example, in the ankle, a tall sheet of bone laps up onto the front of the tibia or shin bone. This ascending process is an easy feature to spot on the drumstick of a young bird, and you can generally find it on the legs of chickens or turkeys that you eat, right about where you hold the drumstick when taking a bite out of it. In more mature birds, the ascending process usually becomes fused indistinguishably to the tibia, and this may function to solidify the bones above the ankle joint. Only birds and extinct tetanurine dinosaurs possess this ascending process, whatever its function.

Lastly, the entire rear half of the tetanurine tail has become stiffened, in a highly distinctive fashion that earned the line its name. Tails are not often completely preserved in Mesozoic theropods. But when they are, the rear is stiffened, forming a ramrod-straight structure, with a mobile base. Even in postmortem rigor mortis, where the neck becomes arched backwards by the stiffening muscles, the end of the tail is straight. The death postures of the exquisite Solnhofen limestone specimens of *Compsognathus* and *Archaeopteryx* show that the tail was rigid. The stiff tetanurine tail probably served to enhance dynamic stabilization during fast locomotion.

Owen's *Megalosaurus* and Huxley's *Compsognathus* are among the basal members of Tetanurae, and for many years not a whole lot more was known of the early history of the line. But spectacular recent discoveries are rapidly filling out our knowledge of this history and they indicate an unsuspected diversification of tetanurines in the Jurassic and Cretaceous. From 1986 to 1990, Philip Curry of the Tyrell Museum, Drumheller directed a joint Sino-Canadian expedition into the People's Republic of China whose specific focus was the dinosaur faunas of central Asia. These were extremely successful expeditions and they recovered several new theropods from Jurassic and Cretaceous rocks. Among these were *Monolophosaurus*, a large crested theropod from the Middle Jurassic of Xinjiang, China, and *Sinraptor*, which are both probably allied to *Allosaurus* or to Owen's *Megalosaurus*.[19,20]

In 1995, Rudolfo Coria and Leonardo Saigado announced the discovery of a gigantic theropod exceeding even the size of *Tyrannosaurus rex*, exceeding 40 feet in length and approaching 8 tons. *Giganotosaurus* was found in Late Cretaceous sediments, but its skeletal structure indicates that it was part of a lineage that had diverged from the others very early in tetanurine history.[21] In 1996, Paul Sereno of the University of Chicago and a group of his associates announced the discovery of excellent new specimens of *Afrovenator* and *Charcharodontosaurus*.[22] *Charcharodontosaurus* also rivaled *Tyrannosaurus* in size. This third giant theropod was mapped along with *Allosaurus*, Curry's *Sinraptor*, and *Giganotosaurus*, into a lineage christened Allosauroidea. These are exciting discoveries of an unsuspected diversity of Mesozoic theropods, but everyone agrees that the new discoveries represent extinct side branches of tetanurines, not the ancestors of birds.

A breaking discovery at the time of this writing, from Early Cretaceous rocks in the Liaonang province of northeastern China, may provide evidence of the first feathers in what is unequivocally a nonflying, basal tetanurine dinosaur. Like the Solnhofen deposits, the beds that yielded this specimen have turned up a bonanza of fossils with some soft tissues preserved. One specimen being displayed in the media has a fringe of featherlike structures along its back bone. Could these be protofeathers? The bird-dinosaur hypothesis predicts that structures intermediate between scales and feathers might be found in a nonflying theropod. As with the discovery of *Archaeopteryx*, the pace of publicity has outstripped the pace of scientific evaluation and publication.

Among tetanurines, we can follow the trail of anatomical clues into Coelurosauria,[23] a lineage distinguished by some rather subtle characteristics. For example, in the foot, the middle bone becomes pinched, as all three of the metatarsal bones are more tightly packed to produce a stronger foot.

In coelurosaurian theropods, the middle bone of the foot, metatarsal III (dark green), is pinched between the others, as the foot bones were more tightly packed together to make a stronger foot.

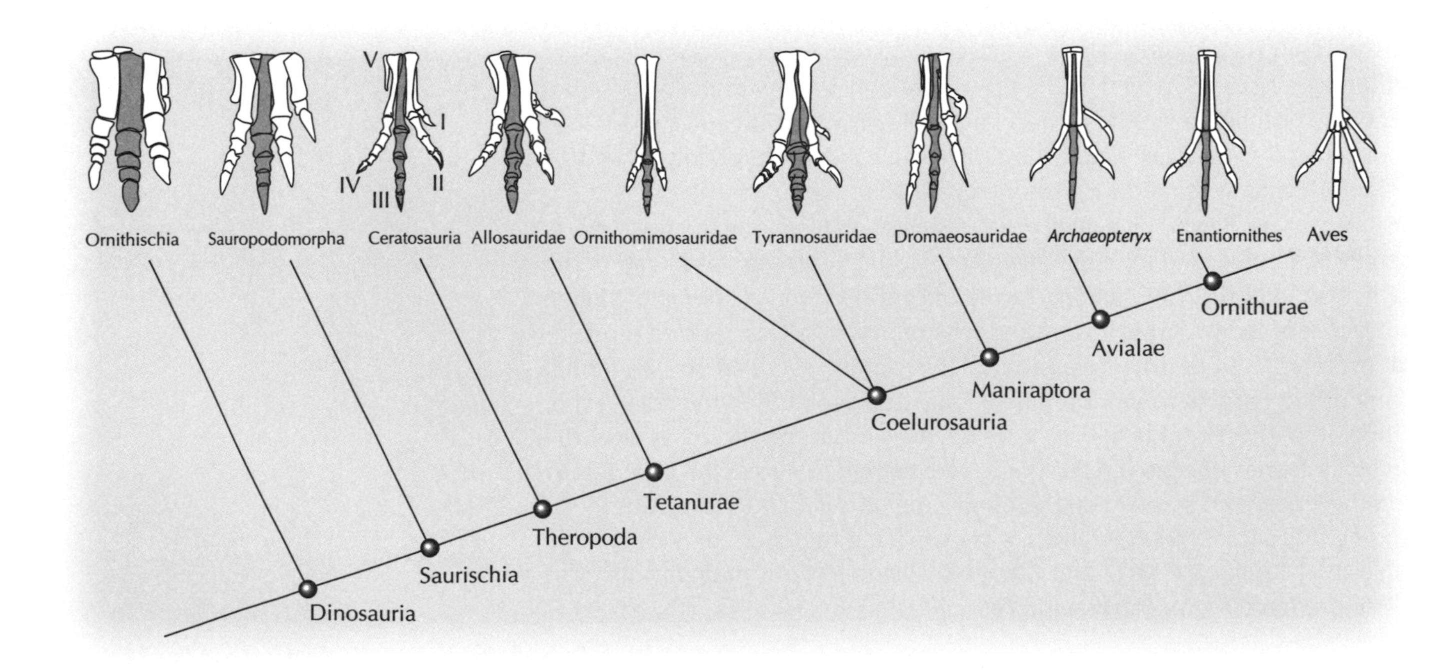

Descendants of the ancestral coelurosaur form a cluster of evolutionary paths that lead to tyrannosaurids, ornithomimosaurs, oviraptorids, maniraptorans, therizinosauroids, and several others. Here again the exact relationships are a little unclear, largely because we have not had time to digest and map abundant new information. But there has been time for researchers to appreciate that each line has features that distinguish them from other tetanurines like *Allosaurus*.

The first lineage to consider is the great Tyrannosauridae—*Tyrannosaurus rex* and its closest relatives—which probably also approached 8 tons. Tyrannosaurids are paradoxical in their behavior, as well as in their exact placement on the map of theropod relationships. Although they evolved to huge size, their forelimbs became dwarfed to the point that we can only speculate as to their function. Because tyrannosaurids have long, slender, birdlike feet, along with other resemblances in the pelvis and neck, they may be closer to birds than any of the lineages describes so far. This is a case where gaps in the fossil record make the precise relationships of these huge modified tetanurines uncertain. But this point of uncertainty is unimportant from the standpoint of bird origins. No one doubts that the tyrannosaurid lineage died out at the end of the Cretaceous.

The modern ostrich came to mind when O. C. Marsh described the first ornithomimosaur—the bird-mimic saurian. These tetanurine theropods were lightly built, with long cursorial hindlimbs and very long arms with slender hands. Most distinctive is the loss of teeth; the mouth is instead bordered by a birdlike beak. However, the oldest and most primitive ornithomimosaur, *Harpymimus*, retained some teeth, documenting a transitional stage in this evolutionary loss that was independent of that in birds.[24] The ornithomimosaur brain is very large and birdlike. In fact, it very nearly reaches the relative size of the brain in modern flightless birds, such as the ostrich and emu. The eyes of ornithomimosaurs were also huge, suggesting a further improvement in visual acuity. Primitive members of the lineage were medium-sized theropods, about 10 feet in length, but some later species range up to 20 feet long and were 6 feet tall at the hip. The largest is *Deinocheirus mirificus*, known only from its arms, but these are 7 feet in length! Ornithomimosaurs are indeed very birdlike, but other theropods show even more unique similarities, and current evidences indicates that the lineage died out at the end of the Cretaceous.

The *Oviraptor* lineage would surely win the contest for the dinosaur with the weirdest head.[25] A dozen and a half species are known from the Late Cretaceous of the Northern Hemisphere. They are small, sleek cursorial bipeds, that ranged up to perhaps 15 feet in length. They were also toothless, like ornithomimosaurs, but their shortened jaws and beak were much more powerful, similar to the beaks of modern parrots. The skull is shortened and very deep. Bizarre crests and pneumatized outgrowths of the head are highly distinctive of the lineage. Owing to their highly unusual skulls and jaws, we can only speculate upon their exact diet. None has been preserved with gastroliths, and it is hard to picture them digesting foliage. But if oviraptorids were herbivorous, as has been claimed, they must have eaten seeds and fruits more than leaves. Their hands and feet, which have the usual theropod armament of recurved claws, suggest that oviraptorids were predatory.

The name *Oviraptor* means "egg stealer." But a recent discovery of associated oviraptorid embryos and adults revealed that this was the most misleading name for any dinosaur lineage. During the Central Asiatic Expeditions in the 1920s by the American Museum of Natural History, an adult *Oviraptor* was found preserved near nests that contained dinosaur eggs presumed to belong to *Protoceratops*. The discovery suggested that *Oviraptor* was preying upon the *Protoceratops* eggs, hence the name egg stealer. In 1993, American Museum of Natural History paleontologists discovered nests containing the eggs and embryos of *Oviraptor*.[26] As it turns out, most of the eggs once thought to belong to *Protoceratops* are actually eggs of *Oviraptor*.

At one of these sites, a skeleton of *Oviraptor* was preserved actually sitting on a nest of eggs in a brooding posture exactly like that of many modern birds.[27] There is no evidence of postmortem transport, making it improbable that some other factor could account for the lifelike association of the adult on the nest. The eggs are arranged neatly and systematically, implying that the eggs were manipulated and positioned by the parents into a specific configuration as is typical of most modern birds. This specimen provides the first direct evidence of the history of parental brooding so characteristic of modern birds. Paleontologists are now asking whether this discovery bears on another question. Modern birds sit on their eggs to keep the nest warm. What was *Oviraptor* doing on its nest if it couldn't generate its own body heat? No doubt this specimen will be closely studied by everyone evaluating evidence for warm-bloodedness in extinct dinosaurs.

So, ornithomimosaurs and oviraptorids and possibly several other coelurosaur lineages bear special resemblances to birds, and the trail to birds is growing much warmer. And there is still one more lineage to consider that is even more birdlike.

MANIRAPTORANS

The somewhat different evolutionary paths traced by John Ostrom and Jacques Gauthier met in maniraptoran theropod dinosaurs. Gauthier's expedition to map saurischian phylogeny uncovered abundant new anatomical evidence supporting Ostrom's claim that *Deinonychus* and *Archaeopteryx* are indeed closely related. Owing to the weight of this evidence, Gauthier coined the name Maniraptora—the raptorial-handed dinosaurs—for the evolutionary path that includes Dromaeosauridae and its sister lineage Avialae.[28] The most famous members of the dromaeosaur lineage are *Deinonychus* and *Velociraptor*. Avialae, named by Gauthier for the winged theropods, is the lineage that includes *Archaeopteryx* plus modern birds and everything else descended from their last common ancestor. On the cladistic map, *Deinonychus* and birds are members of sister lineages.

The maniraptoran lineage is distinguished from other tetanurines by its skull, arms, and tail. The skull was simplified through the loss of a bone called the prefrontal, which was previously situated above and in front of the eyeball, along the rim of the eye socket. This seemingly obscure anatomical detail is significant because it may signal the presence of a large salt gland or nasal gland above the eyeball. This structure is unique to birds among living species. It helps maintain a balanced concentration of blood salts, preventing dehydration and death from a prolonged or extreme imbalance. In tetrapods, the kidneys

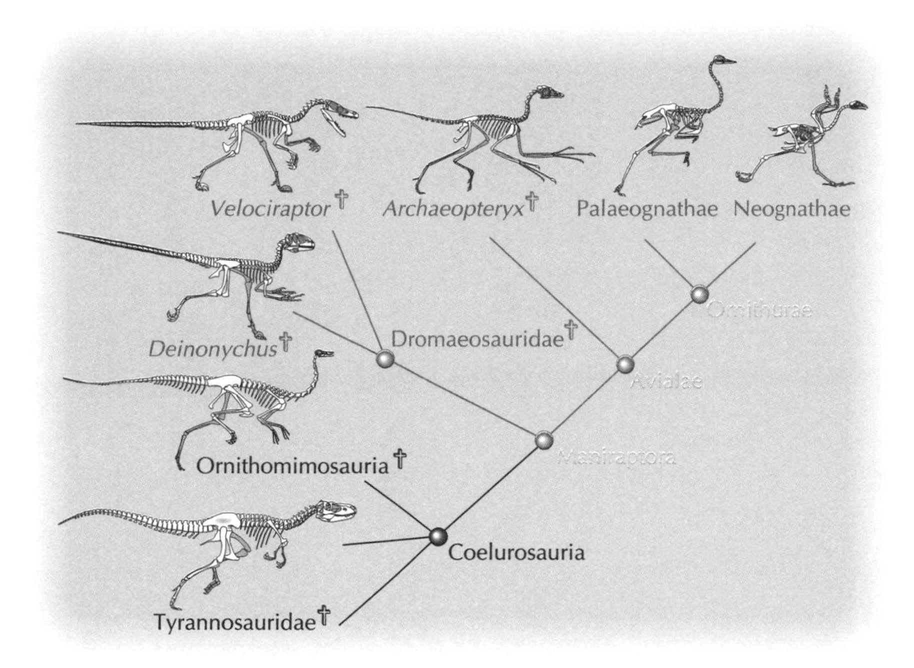

Phylogenetic map, or cladogram, showing the relationships among representative maniraptoran dinosaurs (dark green). The palaeognath and neognath lineages comprise the two major sister lineages of living birds. (Extinct lineages are indicated by crosses.)

also play a big role in this. The kidneys of warm-blooded mammals and birds are enhanced, probably to compensate for rapid salt buildups driven by their higher body temperature and faster moisture loss. In mammals, sweat glands help the kidneys excrete excess salts. But in birds, there are no sweat glands. Instead, the kidneys are augmented by a salt gland, which removes and excretes salt from the blood in very high concentrations. The salt gland permits many birds to live in the arid deserts of the American Southwest and Mexico, where they survive dehydration without developing poisonous levels of blood salts. Marine birds survive by drinking sea water because they are able to excrete excess salt through the nasal gland. It is this gland that enables birds to exploit many of the diverse habitats that they occupy today. The subtle loss of the prefrontal bone in maniraptorans may signal an important evolutionary step in avian physiology.

No one would suggest that maniraptors like *Velociraptor* and *Deinonychus* could fly. Yet these dinosaurs had most of the skeletal modifications that later proved essential for flight in modern birds.[29] Their arms are even more elongated, to roughly three-fourths the length of the hindlimb. The hands are extremely long and slender. The short thumb reflects an extensive grasping capacity, and the tips of all three fingers are armed with trenchant, recurved claws—an elaboration of the pattern found in the ancestral theropods. The muscles operating the maniraptoran forelimbs originated from a powerfully built shoulder girdle. Although a bony sternum and wishbone arose in the ancestral tetanurine, they are noticeably stronger whenever they have been preserved in basal maniraptorans. This strong armature enhances power both when the hands reach out to grab prey and when they pull the quarry back toward the mouth.

More subtle clues to the relationship between dromaeosaurs and birds are visible in the forelimbs. The ulna and third metacarpal bones are bowed, instead of being straight. More importantly, there is a distinctive bone in the

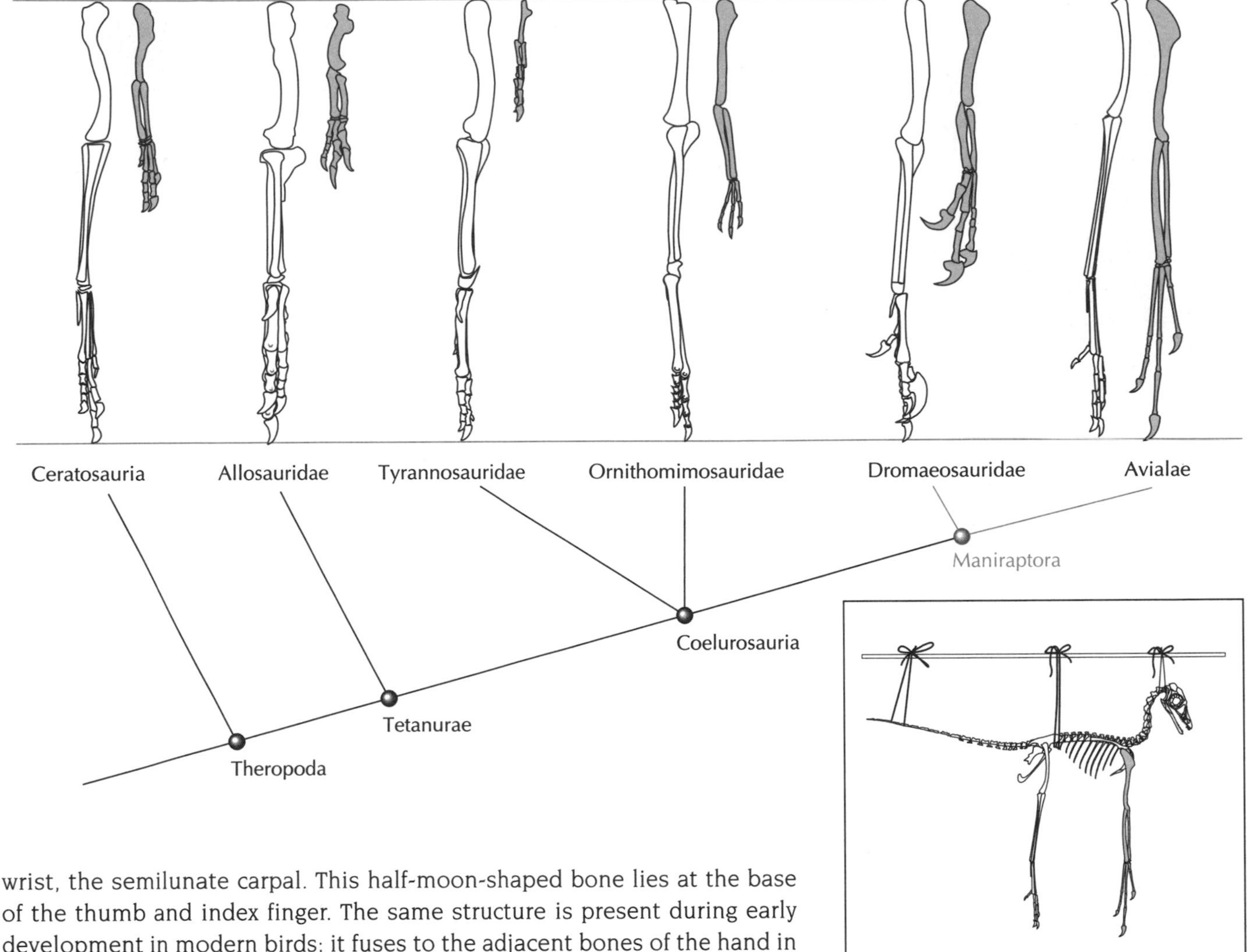

Maniraptoran dinosaurs are distinguished by longer arms than in other theropods. Their arms are three-fourths the length of the hindlimbs or greater. The skeleton of *Archaeopteryx* is inset to illustrate limb proportions as they relate to the entire skeleton.

wrist, the semilunate carpal. This half-moon-shaped bone lies at the base of the thumb and index finger. The same structure is present during early development in modern birds; it fuses to the adjacent bones of the hand in adults. In living birds, the semilunate carpal is important in directing movement of the hand in a fan-shaped motion during the wing's flight stroke. In basal maniraptorans it may have helped snap the hands forward to quickly grab fleeing prey. The joint surfaces and muscle scars on the bones suggest that basal maniraptorans may have been able to fold their hands against the body much like modern birds while resting. For John Ostrom, this bone was one of the keys to allying *Deinonychus* and *Archaeopteryx*. For Gauthier, it was another element in the hierarchy of features linking birds and all theropods.

The pelvis and tail also exhibit subtle evidence for close relationship. The pubis has rotated backwards, superficially like its orientation in ornithischian dinosaurs. In the tail, the transition point between mobile vertebrae at its base and stiff vertebrae at its end lies near the pelvis. Almost the entire tail was stiff, and much of the musculature that once attached to the tail has shifted to the pelvis. Taken together, these features indicate a more forward center of gravity in maniraptorans. Birds carried this trend further by suspending the center of gravity between the wings during flight.

In Avialae, we reach the point on the map representing the last common ancestor shared by *Archaeopteryx* and living birds.[30] As we have seen, the

Solnhofen specimens of *Archaeopteryx* preserve long flight feathers, extending backwards from the hand and arm to produce the airfoil wing that enabled flight. The snout is slender and pointed, and the teeth are reduced in size and number, foreshadowing the origin of the avian beak. There is also an enlarged brain—one of the most characteristic features of birds among modern tetrapods. The arms and hands are nearly half again as long as they are in dromaeosaurs, to support of their new mode of locomotion. Corresponding modifications in the hindlimb produce a more solid structure with fewer separate elements that can withstand the forces generated in landing, as the bones of the ankle and foot begin to fuse into a solid structure. In addition, the big toe or hallux has moves onto the back of the foot, affording a degree

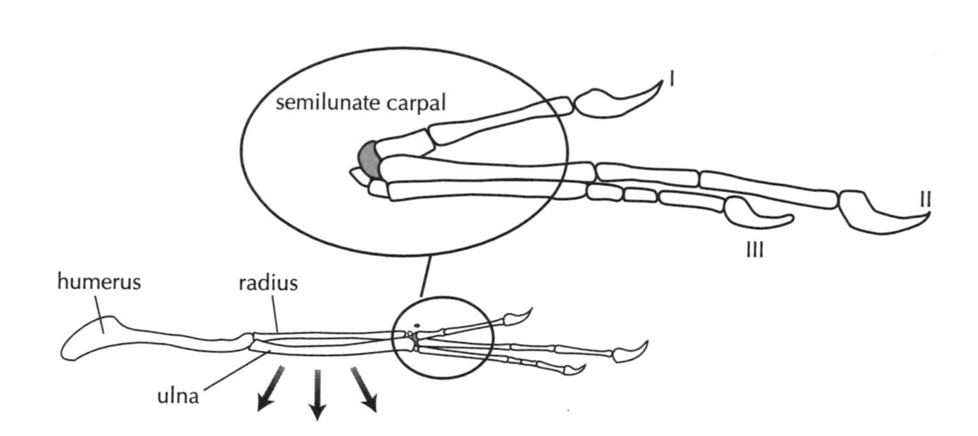

(At left) **The forearm in maniraptorans is distinctive in having a bowed ulna and a wrist with the semilunate carpal. Together with the wishbone and hollow skeleton inherited from more primitive theropods, maniraptorans had most of the major components used for flight in modern birds.**

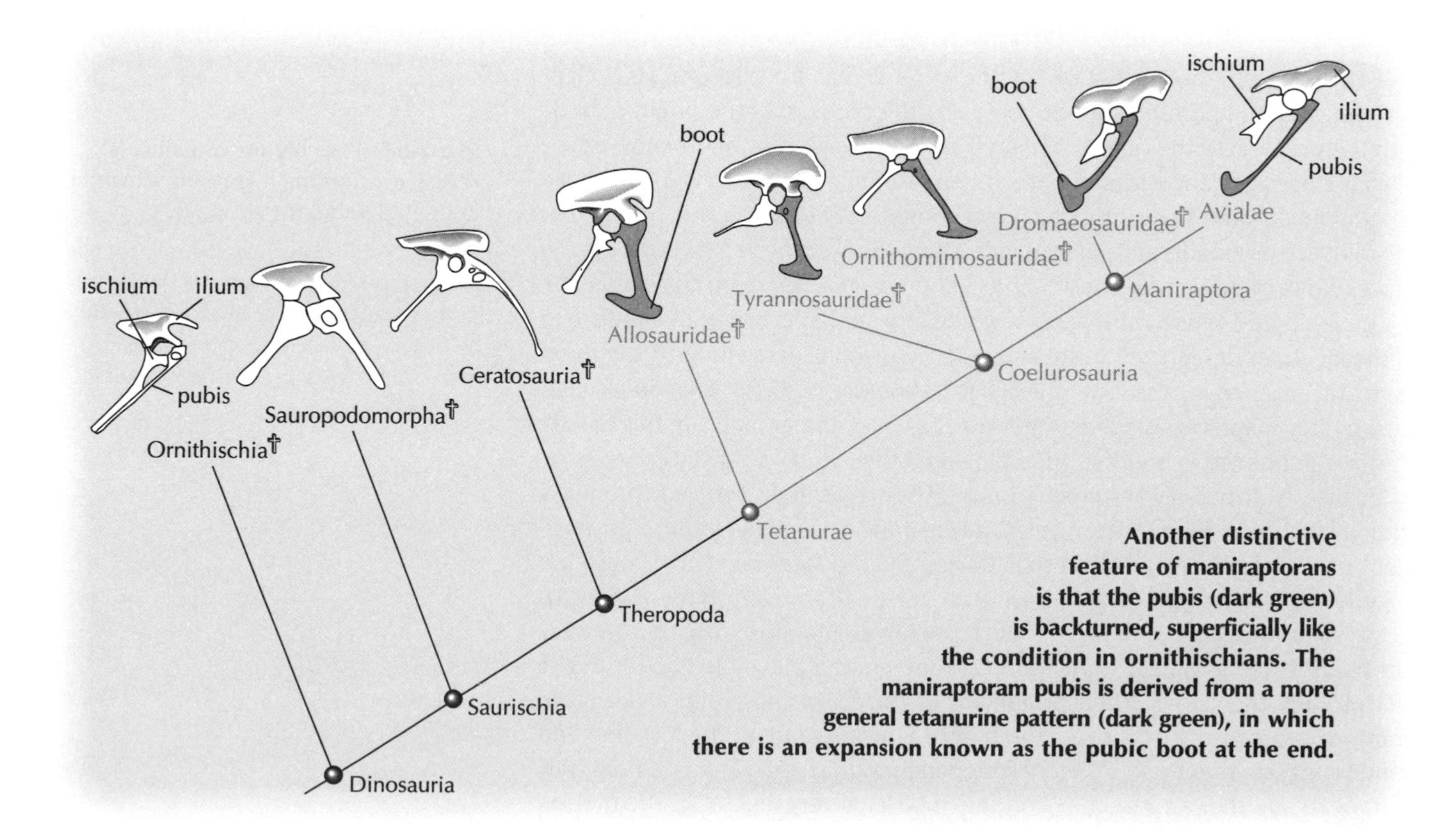

Another distinctive feature of maniraptorans is that the pubis (dark green) is backturned, superficially like the condition in ornithischians. The maniraptoram pubis is derived from a more general tetanurine pattern (dark green), in which there is an expansion known as the pubic boot at the end.

of grasping capability. The tail is also shortened to no more than twenty-three separate vertebrae. Thanks to these and other characters, for the last century almost everyone has agreed with Richard Owen that *Archaeopteryx* is a bird.

PROTOAVIS—TRIASSIC BIRD?

A challenge to the idea that birds are avialian, maniraptoran, tetanurine theropod dinosaurs came from Dr. Sankar Chatterjee of Texas Tech University, with the discovery of a fossil for which he coined the named *Protoavis texensis*.[31] It was collected from Late Triassic rocks, predating *Archaeopteryx* by 75 million years, and pushing the origin of birds to the earliest stages of dinosaur evolution. Chatterjee maintains that *Protoavis* is on the direct path from dinosaurs to birds, and that *Archaeopteryx* and the other theropods described above represent an unrelated side path on the map of avian evolution. The scientific community has been skeptical of Chatterjee's proposal because many of the bones from the skull are flattened and difficult or impossible to interpret. Moreover, there is reason to suspect that *Protoavis* represent a death assemblage of different animals instead of associated parts of a single kind of animal. The hindlimb has a primitive unfused ankle and foot, lacking a tall ascending process, and it may belong to a ceratosaur or something closer to *Herrerasaurus*. It is doubtful that the elements comprising the hand are from the same creature, if in fact they are hand bones. Only Chatterjee and Larry Martin of the University of Kansas have defended this view; everyone else wants to see more complete specimens and more data before they consider abandoning the cladistic map that summarizes so much actual data.

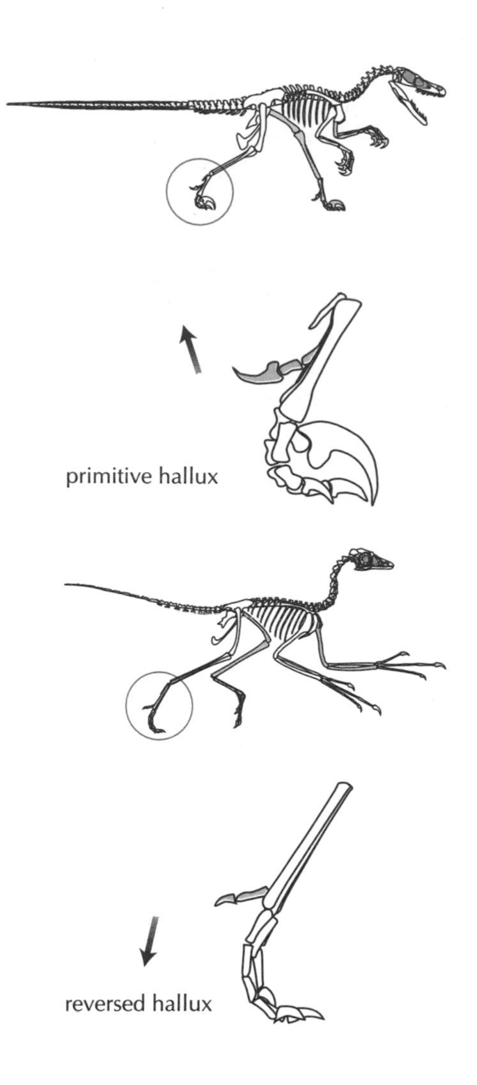

In avialians, the big toe or hallux is reversed, affording a grasping capability not found in other theropods.

THE ORIGIN OF FLIGHT

One of the points made clear by Gauthier and Padian in collaborative work at Berkeley is the importance of not confusing the problem of the origin of birds with the problem of the origin of flight. Even though we may never know exactly what stages the ancestors of birds passed through as they evolved flight, we can still read from the evolutionary map compiled by Gauthier and his successors that the preponderance of evidence favors the hypothesis that birds are the descendants of extinct Mesozoic theropod dinosaurs. So, given that birds are, in a genealogical sense, dinosaurs, it would appear that powered flight arose in a lineage that descended from a bipedal, cursorial, fast running theropod dinosaur, smaller in size but much like *Velociraptor* or *Deinonychus* in general design. Our map suggests that flight evolved from the ground up, but exactly how this happened is another question altogether.

Intuitively, it has always seemed more likely that flight evolved through a gliding stage, and John Ostrom faced severe criticism with his ground-up argument on that basis. A question that Ostrom had to face was, "How could the beginnings of the flight stroke arise in a cursorial animal?" Birds appear to move their wings in a unique way to achieve lift—how did this evolve? Although it was a difficult question, Ostrom found a plausible answer in the diet of early theropods. Their armament of trenchant claws and sharp teeth points to a predatory existence. The powerful architecture of the forelimb and hands suggests that the arms were snapped forward to grab at a potential prey item or to smother an insect.[32] And this motion is very much like the motion of the flight stroke.

There is now a growing body of experimental information that clarifies how birds use their wings during flight.[33] To generate lift, the forelimbs are thrown forward and somewhat downward. Adding to this argument is independent evidence based on the physics of aerodynamics. Several researchers recently demonstrated that even small increases in the surface area of the hands can generate significant levels of lift, provided that the surface is shaped like an airfoil. The long hands of early maniraptorans and avialians could provide lift, if the hands were being accelerated forward. This is not to say that early maniraptorans were able to fly, but that they may have used the aerodynamic properties of their hands for balance and maneuverability while chasing prey or escaping over broken ground. Many living birds can run very fast and are highly maneuverable even when running across uneven surfaces. Running birds can use their wings to help maintain balance, and young birds even flap their wings to gain greater speeds while running. Even though the forelimb is committed to flight in most adult birds, it can still aid terrestrial locomotion at points in their lifetimes. Differences between the situation in birds and Mesozoic dinosaurs have also come to light, and it is clear that the flight stroke is considerably modified over the likely movements of nonflying dinosaurs. But, this at least shows the plausibility of the argument that flight evolved from the ground up.

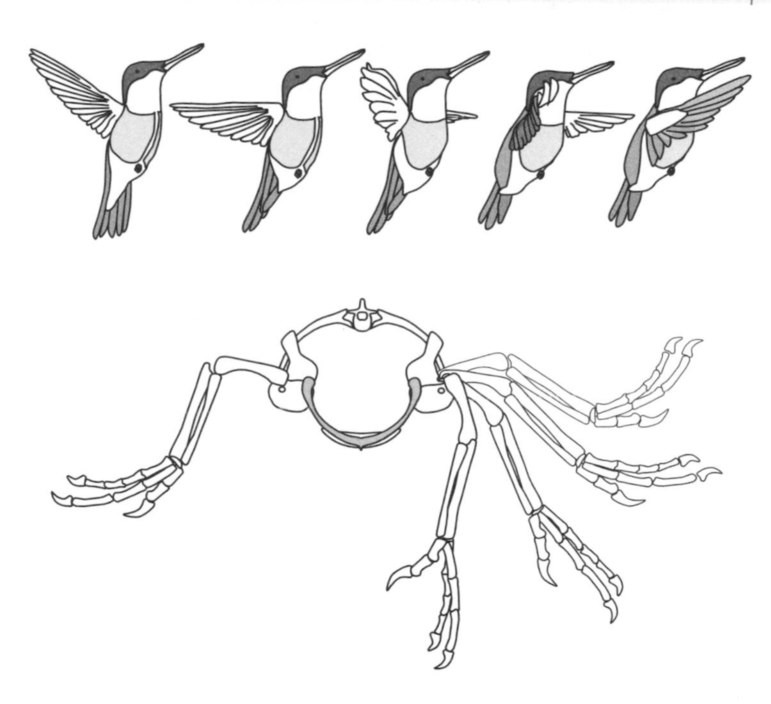

The flight stroke of birds, looked at from the side in a hummingbird, is very much like the raptorial, prey-grabbing motion of the arms in more primitive maniraptorans, viewed here from above.

With a more rigorous, hierarchical map of dinosaur genealogy, Gauthier and Padian were able to highlight the sequence of changes documented in theropod history to show that the ground-up theory was in fact plausible.[34] As we have just seen, over theropod history there was a series of changes in which the arms became successively longer and more powerful, and the shoulder girdle was heavily reinforced. All of these changes are consistent with both a predatory existence and the earliest stages of the evolution of flight. Once there was reason to suspect that the true historic pathway of avian descent was from the ground up, the ecology and physics behind such a transition were not so hard to understand. After all, human invention has taken the creation of flying machines from the ground up as well.

Not surprisingly, another argument about the origin of flight is emerging that represents a consensus of the trees-down and ground-up hypotheses. This argument recognizes the long temporal gap that separates the Late Jurassic *Archaeopteryx* and the various Cretaceous dromaeosaurs. If our map of theropod history is correct, then the histories of both lineages must have extended to a common ancestor that lived in or before the mid-Jurassic. It is possible that from a cursorial dinosaurian ancestor, the predecessors of avialians became arboreal—small tree-climbing maniraptoran dinosaurs—and that the transition to powered flight actually did take place in the trees. While there is no direct evidence of such intermediate tree-climbing maniraptorans, some paleontologists argue that the claws of *Archaeopteryx* were designed for climbing, implying a nonpreserved tree-climbing phase in their history. Obviously, more needs to be learned in order to understand the whole process that occurred as flight evolved in extinct theropods.

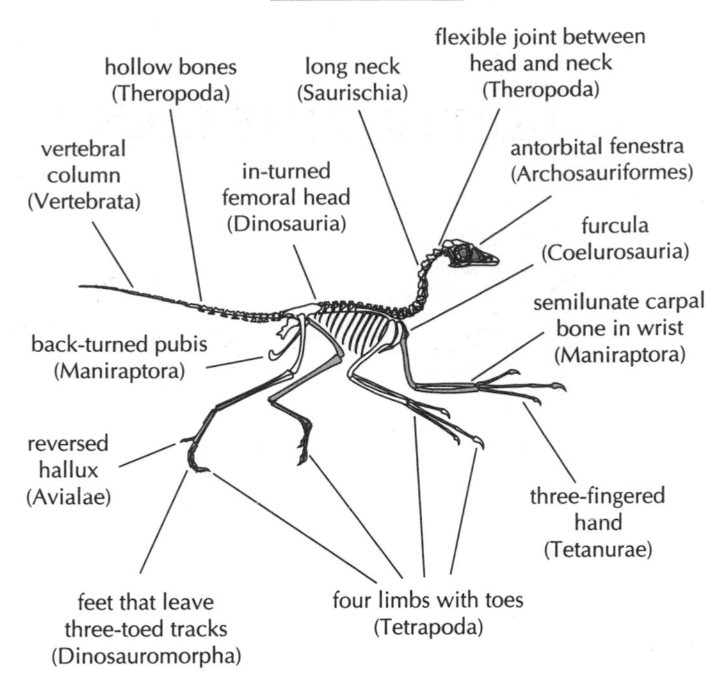

The various features of birds, illustrated here by *Archaeopteryx*, can be traced to historic times of origin. For example, from the ancestral vertebrate, birds inherited a vertebral column; from the ancestral tetrapod, four limbs with wrists/ankles and fingers/toes; feet that leave three-toed trackways from the ancestral dinosauromorph; an in-turned femoral head from the ancestral dinosaur; a long neck from the ancestral saurischian; a flexible joint between the head and neck from the ancestral theropod; and so on.

DOWN THE ROAD

Whereas points of resemblance might be found between birds and pterosaurs, or birds and ornithischians, or birds and ceratosaurs, we have now mapped the position of birds within a single hierarchy of relationships. Although it really begins at the origin of life, in the last two chapters we have traced the hierarchy of avian relationships from the origin of Vertebrata, through the emergence of vertebrates onto the land, and onto the dinosaurian pathway of reptilian evolution. Cladistic techniques enable mapping of the most characteristic features of birds backward in time, matching each feature back to the particular ancestor in which it arose. Ironically, Gauthier mapped many of the characters that Richard Owen had used to verify the avian affinities of *Archaeopteryx*, like a wishbone and a hollow skeleton, to positions in the hierarchy of evolutionary relationships among extinct theropod dinosaurs.

Paleontologists following Gauthier's trail have unearthed additional evidence that birds are avialian, maniraptoran, tetanurine, theropod, saurischian dinosaurs. From the ancestral dinosaur, birds inherited an in-turned femoral head and a perforate acetabulum, and from the ancestral saurischian a long neck and a hand in which the second finger is the longest. To the ancestral theropod, birds owe their hollow bones and a flexible joint between the neck and head. The wishbone, a three-fingered adult hand, and the ascending process of the ankle can be traced back to the ancestor of tetanurines. The ancestral maniraptoran added still longer arms with a semilumate carpal in the wrist. And the ancestral avialian added still longer arms, flight feathers, and the ability to fly. In summary, Huxley was right—the evolutionary road to birds passes through dinosaurs.

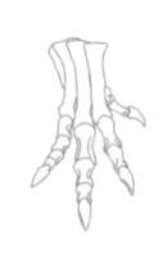 

CHAPTER THIRTEEN

Death by Decree

In the same paper in which Richard Owen coined their name, he also claimed that dinosaurs are extinct in an effort to disprove the idea of evolution.[1,2] But with so much evidence linking birds and dinosaurs, why have so many evolutionists, from Huxley to Ostrom, agreed that dinosaurs are extinct?

The revolution catalyzed by Darwin's theory of evolution shed a bright new light on dinosaurs. Dinosauria represented a lineage that could evolve. Using *Compsognathus* and *Deinonychus*, Huxley and Ostrom showed that dinosaurs not only *could* evolve, but that they *did* evolve, both birdlike size and birdlike features. So, if birds *did* evolve from Mesozoic dinosaurs, as the preponderance of evidence suggests, how can we say that dinosaurs are extinct?

The belief that dinosaurs are extinct is one of the great ironies of paleontology. Richard Owen is sometimes reviled for fighting throughout his life against evolution. Yet, even though modern science recognizes Darwin as the victor in this battle, the world did not go on to adopt a Darwinian view of dinosaurs. If Owen is looking down from the Hereafter, he must be gratified despite the bad press. Most people still accept his anti-evolutionary view, that dinosaurs are extinct.

LINNAEUS AND THE LINNAEAN SYSTEM OF NATURE

This paradox arose as scientists tried to map the evolutionary history of various lineages using the preevolutionary Linnaean system of classification and nomenclature, which for centuries was about the only system available. Up until this point in the book, we have followed a system that is strictly hierarchical and that attempts to plot all available information onto a single map of relationships, using cladistic mapping techniques. As we saw earlier, shared evolutionary novelties are the basis for phylogenetic mapping, and the map of dinosaur history that we've followed to this point is our best current approximation of their relationships. But this system was not always in use, and in its place was the Linnaean system of classification.[3]

As beginning graduate students, the prospect of studying the "science" of classification loomed before us like a barren desert of endless boredom. There were countless unpronounceable Latin names to learn and regurgitate, along with the many ranks—genus, family, order, phylum, etc.—that were assigned to each clam, leaf, or bone fragment that we came across in an exam or on a field trip. But it was a desert that we had to cross in order to reach the professional world. The classification of organisms is the basic language that scientists use to communicate about dinosaurs and all other organisms, and we couldn't participate in that world without mastery of its lingo. Like any other language, the system of animal classification has complex rules, countless exceptions to the rules, and a vast vocabulary. Moreover, classification systems inevitably change over time and for many technical terms an intricate maze of implied meanings has evolved over the years. So we had to understand not only modern classification, but also the history of classifications. The only way to make it through the exams, and to get a foothold in the professional world, was to muscle your way through—memorize the glut of arcane terms and rules that had accumulated over the centuries and that could not easily be categorized and dealt with more efficiently.

The system of classification used by Owen, with which he both founded Dinosauria and proclaimed it extinct, was developed in the previous century by the great naturalist Carolus Linnaeus (1707–1778). Linnaeus was a botanist, and he became as famous as Newton and Galileo for the resounding endurance of his influence. Even during his lifetime Linnaeus was enormously famous and influential. He may be the only scientist whose death and service to his country were recorded by a European sovereign in a speech from the throne. In 1778 King Gustavus III of Sweden eulogized Linnaeus at his funeral in Uppsala, saying "I have lost a man, whose renown filled the world, and whom his country will ever be proud to reckon among her children. Long will Upsal remember the celebrity which it acquired by the name of Linnaeus."[4] A medal was struck in honor of Linnaeus, and his picture still appears on Swedish currency.

As his system of plant classification developed, Linnaeus extended his interest to practically all organisms known at the time.[5] Linnaeus called his classification *Systema Naturae*, the Natural System, although just what he meant by "natural" was never clear.[6] Before Linnaeus and for many years after, natural historians argued over what criterion should be used to classify organisms. For animals, some argued that fur or feather color was best. Others maintained that the number of fingers and toes should be used. Still others proposed that habitat or way of life were best. The problem is that different criteria produced different classifications. Late in his life, Linnaeus admitted that he had spent decades trying to articulate criteria and principles for classifying organisms, but that he had failed.[7] In the absence of clear guidelines, intuition had been his guide.

The Linnaean system grouped organisms that basically looked alike and, given some key character, it established a naming system to help naturalists discuss nature in a precise and efficient fashion. Referring to groups based on their names, instead of listing all their various characters, created a shorthand for scientific communication. For example, Linnaeus coined the name "Mammalia" for a group of organisms whose members possess an extensive and unique suite of characteristics. In defining the name, Linnaeus enumerated what he considered to be the essential characteristics: "Mammals have a heart with two auricles and two ventricles, with hot red blood; that the lungs breathe rhythmically; that the jaws are slung as in other vertebrates, but "covered," i.e., with flesh, as opposed to the "naked" jaws of birds; that the penis is intromittent; that the females are viviparous, and secrete and give milk; that the means of perception are the tongue, nose, eyes, ears, and the sense of touch; that the integument is provided with hairs, which are sparse in tropical and still fewer in aquatic mammals; that the body is supported on four feet, save in the aquatic forms in which the hind limbs are said to be coalesced into the tail."[8] It is obviously easier to use the word "mammal" than to list all these features every time you want to refer to the group. Of course, this only works if everyone in the conversation shares a common understanding about what the name means.[9]

Carolus Linnaeus (1707–1778), the founder of the Linnaean system of classification, published his first great classification 101 years before Darwin's *On the Origin of Species* appeared in print.

In the Linnaean system, named groups are also given ranks based on their distinctiveness. The categories genus, family, order, class, phylum, and kingdom form a successively more inclusive hierarchy of ranks. A cluster of similar species would be grouped together in the same genus, whereas species that are sufficiently different would be placed in a separate genus. Similar genera would be ranked together in the same family, similar families grouped under a single order, and so on. If a species

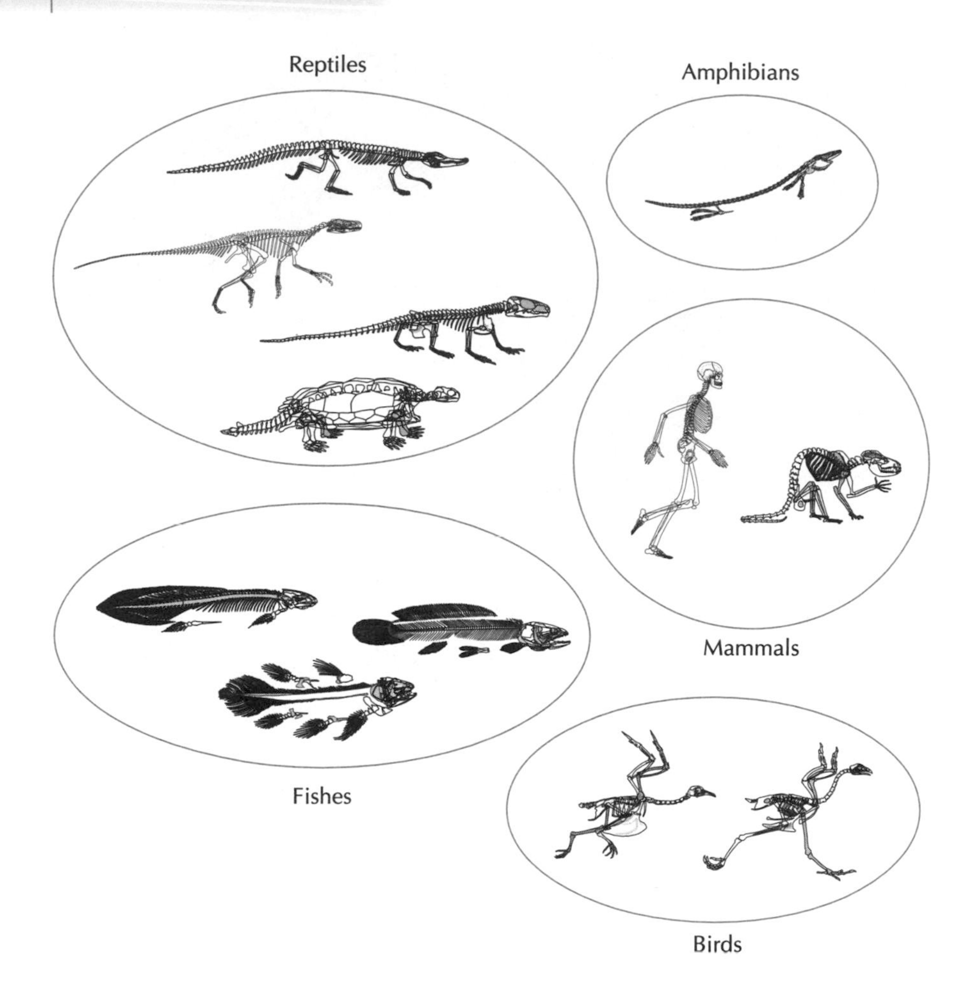

The Linnaean classification divided vertebrates into five nonoverlapping classes. While this does pigeonhole all the vertebrates into a convenient filing system, it is a poor reflection of their relationships to one another.

proved especially distinctive, it might also be placed in its own genus and family, or perhaps even order, to highlight this distinctiveness. The more distinctive the group, the higher the rank, and the more subjective the process became. Birds and mammals were each assigned the rank of class to emphasize how distinctive and different these two groups are. So long as groups are assigned ranks, they need not be arrayed in a strictly hierarchical scheme based on shared inherited features. For example, the kingdoms Plantae and Animalia were regarded as fundamental divisions that are entirely separate but equal in rank. Before scientists understood that species are linked genealogically, there was no reason to unite all Life into a single hierarchy of relationships.

The Linnaean ranking scheme depends on the idea of fundamentally distinct types of organisms—groups that are separate but equal in rank. Among vertebrates, for example, Linnaeus recognized the class Osteichthyes, class Amphibia, class Reptilia, class Mammalia, and class Aves, to be nonoverlapping categories representing separate but equal classes.[10] Membership was defined by distinctive features, such as fur and mammary glands for mammals, scales and cold blood for reptiles, and feathers for birds. Linnaeus strove to discover the characteristics of essential importance to the group. But without clear principles for guidance, biologists fought bitterly over what a natural classification really represented and which criteria should be regarded as essential in building a classification. And, without some objective measure of difference, they also fought over what ranking should be assigned to any given group. The result has been continual turmoil and revision in our system of classification. Consequently, ever since Linnaeus's day, many scientists have claimed that classification is mere pigeonholing, and that the arguments over how to classify any particular organism generate only heat, not enlightenment. Systematics and classification are for stamp collectors, not scientists.

The Linnaean system is what was available to Darwin as a student. Even after the Darwinian revolution was underway, the Linnaean system remained enormously successful because it was at least partly hierarchical. It provided a convenient means of conveying nature's diversity. After Linnaeus's death, naturalists expanded the Linnaean classification to include newly discovered living species as well as fossils. The classification became all-inclusive and rapidly grew into one of the most general tools in the naturalist's repertoire. For more than two centuries, the Linnaean system of classification has provided a basic language for communication about nature's diversity. And, this preevolutionary system was still in general use when we entered graduate school.

DARWIN AND NATURAL CLASSIFICATION: THE ROOTS OF CONFLICT

Darwin noted that, even in the most ancient written records, humans recognized that organisms resemble one another to varying degrees.[11] They classified organisms into smaller groups contained within larger groups. Primates are placed within the larger group Mammalia, which in turn is contained within Amniota, Tetrapoda, Vertebrata, and so on. But unlike the constellations of stars, this arrangement of groups is not entirely arbitrary. Species that look most alike are grouped together, and those that are different are grouped separately.

To pre-Darwinian naturalists, the classification of species was simply a scheme for arranging living objects that looked most similar, a convenient tool to "sort out" organisms. To Darwin, much more was implied by the shared resemblances of organisms. A shared history of descent, the one known cause of close similarity in organic beings, is what the general system of classification revealed. The bond among members of a group is relationship, "propinquity of descent," though it can be hidden in various degrees by the modifications which make the different groups so distinctive. To Linnaeus and Owen, organisms were grouped together simply because they looked alike. But to Darwin and his followers, organisms are grouped together because they are descendants of a common ancestor.

The Darwinian view cast a very different light on what classified groups represent and on how to build classifications.[12] The groupings were generally seen to represent the branches of the evolutionary family tree. The "naturalness" that Linnaeus groped for but failed to identify is genealogy. Ever since Darwin, scientists have worked to see that each group, whether

Unlike a Linnean classification, this map of vertebrate phylogeny shows the relationships among all its members.

it be a genus, family, or higher group, contains only related forms. But as this work has progressed, it has become clear that the Linnaean system of classification can never provide a completely accurate representation of relationship, because it is not completely hierarchical. It was never intended to reflect evolutionary relationship. Thanks to the newly developed maps of vertebrate phylogeny, we have realized that many of the groupings established through Linnaean methods fail to depict genealogy. Instead, they reflect ecology, geography, or some other criterion.

The Linnaean classification would work perfectly well if the gaps between groups were always distinct. For example, if only one group were designed for life in the water, one for life on land, one as a predator, one as a flyer, and so forth, classification would be a simple process. But the variability among organisms inevitably seems to cross these convenient boundaries. Some tetrapods still live mostly in the water, while other tetrapods never go near it. Lungfish can live buried in their burrows at the bottm of dried ponds and breathe air for years, while other fish will quickly suffocate outside water. Rarely can a group be defined by a single character shared among all its members and no other species. Even when groups seem highly distinctive and sharply separated from each other, as living birds differ from lizards and crocodylians, the distinction often

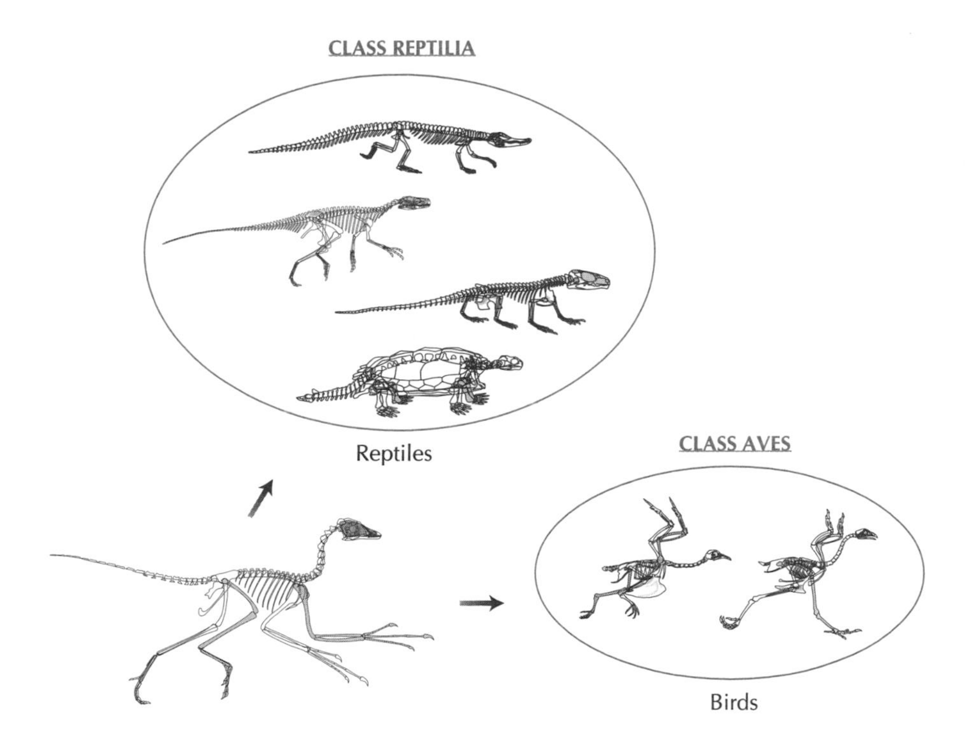

In the Linnaean system of separate but not overlapping ranks, *Archaeopteryx* could be either a reptile or a bird, but it couldn't be a member of both classes despite its genealogical tie to both.

becomes blurred when fossils are considered. *Archaeopteryx* is the classic example. Nineteenth-century scientists asked, "Is it a feathered reptile or a reptilelike bird?" Fossils blurred the seemingly sharp and objective boundary. Under the Linnaean system *Archaeopteryx* can not be *both* a reptile and a bird, even if birds have reptilian ancestors.

Naturalists had long noted that gaps exist between groups of equal rank, and as we saw earlier the existence of these gaps represented a basic challenge to the theory of evolution. But when fossils narrowed the gaps and offered evidence in support of Darwin's theory, it posed a real dilemma for Linnaean classification. Whether *Archaeopteryx* was segregated into its own class or lumped into either Reptilia or Aves, the solution was an uncomfortable one, because either approach arbitrarily broke the genealogical bond. Some scientists advocated splitting, some lumping, and the two camps fought bitterly over how to handle any particular case.

This argument is important because, if classifications are to represent genealogies, splitting versus lumping taxa poses a problem that directly affects our understanding of history. Classifying *Archaeopteryx* as a bird in the class Aves breaks its connection to reptiles. Classifying it within the class Reptilia severs its connection to birds. Placing it in a class by itself would tear apart both connections. Paleontologists sometimes comment that they are fortunate that so many distinct gaps still exist between different groups, for without them classification would be impossible. Reading between the lines, what they are also admitting is that Linnaean classification is stronger when based on less information. When used as an evolutionary tool, Linnaean classification has a difficult time dealing with new discoveries like *Archaeopteryx*.

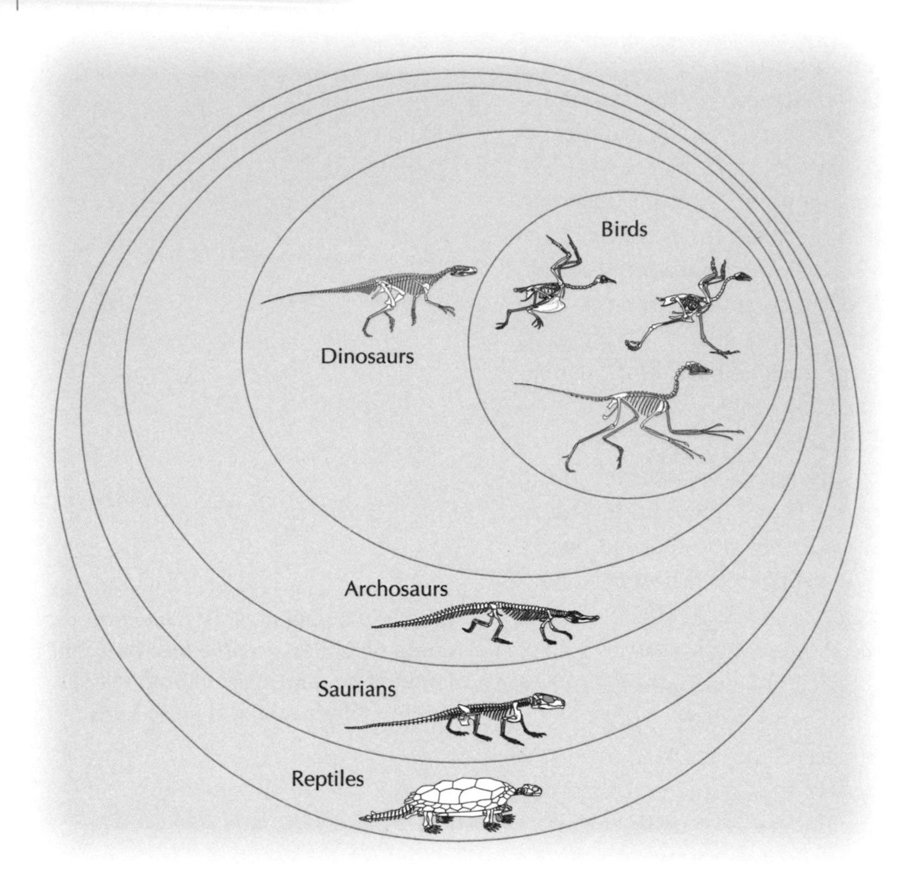

In contrast to the Linnaean system, in the phylogenetic system *Archaeopteryx* is a bird, a dinosaur, an archosaur, a saurian, a reptile, and so on.

IMPLICATIONS OF EVOLUTION

Darwin's theory has become the most central principle of biology, and today virtually all biologists interpret classifications to reflect evolutionary history. Since Darwin's time, scientists have discovered the mechanism of inheritance. With computers they can decipher the structural features of double helix DNA molecules and use DNA itself as evidence for classifying organisms. DNA evidence is even used routinely in the British and American legal systems. Where would modern medicine be if DNA had never been discovered? Genetic engineering and cloning are now possible. Their future potentials are so vast that we can't foresee where biotechnology, which transplants genetic information from one species into the genetic mechanism of another, will have taken us a century from now. So transformed is humankind by the theory of evolution, that it is difficult to imagine what our lives might be like today in the absence of the cascading discoveries it has spawned.

All the same, a number of influential twentieth-century biologists have commented that the Darwinian revolution did not lead to a similar revolution in the way organisms are classified, despite acceptance that classification should reflect genealogy. We had learned of this paradox as undergraduates and we had read about some attempted solutions. But the solutions seemed only to introduce new problems, and at first we shared the general scientific response to the entire issue, which was to ask, "So what?" And then we met a Berkeley graduate student named Kevin de Queiroz. Now a curator of herpetology at the Smithsonian Institution, de Queiroz had a resounding influence across our community as he explored the paradox in the way evolutionary biology was studied. Most evolutionary biologists have continued to use Linnaean methods, lumping organisms together based on overall similarity, and splitting groups into different ranks to reflect their differences. Even though they may express their ideas in evolutionary terms, their methods for detecting the underlying pattern of relationship were devised long before Darwin's theory emerged. Other scientists had argued that a classification designed optimally to reflect evolutionary relationships would be a far better tool than traditional Linnaean classifications, but de Queiroz showed us how powerful such a tool could be.

Before Darwin's theory of evolution, which stipulated that species could transform, there was no reason to develop a system that depicted the dynamic properties of lineages. To better reflect what we have learned about evolutionary history, the Linnaean system has been tinkered with, modified, revised, and overhauled. New rules for classification have been added, and a Linnaean Commission has published a *Code of Taxonomy* for more than a century. Since Darwin's *On the Origin of Species* was published, the Code has evolved into a governing system for classification that rivals the American tax code in its mind-boggling complexity. But despite countless alterations, it remains painfully evident that the Linnaean system was designed to classify static, unchanging objects. To Linnaeus, organisms were separately created and "permanent." He had no idea that species could become extinct, much less that they could transform as part of evolutionary lineages. Darwin's *Origin* was still a century in the future when Linnaeus published the basic structure of his classification system. Kevin de Queiroz argued that it was time to developed a system designed for studying evolution, using the idea of descent with modification as axiomatic, and deriving from that axiom the best evolutionary tools possible.[13]

THE PHYLOGENETIC SYSTEM

Berkeley became a hotbed for overthrowing the Linnaean system while we were graduate students there. The hierarchical map of genealogy that we have been using in this book, known as the phylogenetic system, is what was proposed in its place. The German naturalist Willi Hennig had founded the field of phylogenetic systematics in works on insect relationships that he published in the 1950s. In 1966, his major book on phylogenetic systematics was translated into English, and over the next decade his methods were refined, mostly by a small group of ichthyologists led by Colin Patterson and associates at the British Museum (Natural History), and by Donn Rosen, Gareth Nelson, and their associates at the American Museum of Natural History.[14]

At first the movement was no more than a small network of a few dozen scientists scattered across the United States and Europe. Kevin de Queiroz and Jacques Gauthier were among its first supporters at Berkeley, often to the dismay of some of the faculty, whose careers were deeply rooted in the Linnaean system. But by applying phylogenetic methods to some long-standing problems in reptile evolution, they were able to offer compelling demonstrations

Willie Hennig, who founded phylogenetic systematics. It took about 30 years for Hennig's view to catch on, but it is now the most widely used method to reconstruct genealogy.

of the difference between the two systems. De Queiroz worked on mapping the relationships among modern lizards,[15] while Gauthier focused on the phylogeny of dinosaurs and the origin of birds, and the two collaborated in a great deal of this research. They argued that by merely superimposing a secondary evolutionary interpretation on top of a Linnaean classification, biologists were risking many mistakes, and Dinosauria was a classic example. As Gauthier put it, explaining why Dinosauria became extinct is like explaining why Napoleon crossed the Mississippi. The Linnaean system had misled scientists into seeking an explanation for something that had never happened. Gauthier and de Queiroz argued that it was time to break with the past and construct a phylogenetic system to reflect Darwin's concept of evolution.[16]

Ancestry, rather than overall similarity, formed the fundamental basis of the phylogenetic system of classification. Many Linnaean names, like Dinosauria, Saurischia, and Theropoda, are preserved to provide a linkage to historical Linnaean schemes. However, the meaning behind those names shifted from a static concept based on physical characteristics to a dynamic one based on ancestry, and the practice of ranking lineages was abandoned entirely. In the phylogenetic system, groups must include the last common ancestor of a lineage plus all its descendants, no matter what form the descendants might eventually assume through evolution. Whereas the Linnaean system was only partly hierarchical, the phylogenetic system is exclusively hierarchical. In the phylogenetic system, anything born to a vertebrate is a vertebrate, anything born to a tetrapod is a tetrapod, and anything born to a dinosaur inherits that name, plus all the others.

A system based on ancestry is at least potentially stable, because organisms can't escape their history. One's ancestry can never be altered, and the phylogenetic system remains loyal to Darwin's fundamental evolutionary concept—all species share common ancestry. And by linking particular names to particular ancestors, the precise meanings of the names is potentially stable. Discovering ancestors and historic relationships—the process of phylogeny reconstruction—is a different question, and it is not always a simple task. The phylogenetic map of organisms is still under construction, as we will see in the chapters ahead. But despite the difficulties that face phylogeny reconstruction, the basic idea that ancestry provides a stable criterion for an evolutionary system of classification is now being put into practice on a global scale.

This was a radical shift in perspective and one that was deeply upsetting to many scientists when we were graduate students. It would mean, for example, that dinosaurs are not extinct! And similar revelations faced researchers studying many other lineages. At about the same time as the war over an asteroid impact at the K-T boundary was underway, a debate over the phylogenetic system stormed across the community, although it obviously didn't gather nearly the same level of media coverage. At Berkeley the debate was so strong that it led to several formal seminars that involved students and faculty from many different departments. One of the seminars was led by Kevin Padian, who carried an historic perspective that brought Richard Owen into the spotlight of our discussions. As the group discussed the phylogenetic system, Gauthier discovered the striking similarity of the modern debate to the debates that had raged in England a century before. Not only was the relationship between birds and Mesozoic dinosaurs once again being challenged, but the very role that the theory of evolution should play in science was again at stake.

A ROSE BY ANY OTHER NAME?

Owing to the fundamentally nonevolutionary design of the Linnaean system, even evolutionists like Thomas Huxley and John Ostrom, were trapped into arguing that dinosaurs are extinct. But instead of dying out, dinosaurs were merely defined out of existence. In the Linnaean system, with its foundation of defining characteristics, only birds could have feathers, and birds belonged to a class entirely separate from reptilian dinosaurs. The name Dinosauria, as originally defined by Richard Owen, referred only to giant extinct Mesozoic species, and Owen refused to believe that they could transform into something with feathers. But, under the phylogenetic system this doesn't necessarily mean that the dinosaurian lineage is extinct. It may be true that living descendants are not so "fearfully great" as *Megalosaurus* or *Iguanodon*. But beneath their feathers, they retain many attributes that were inherited from their Mesozoic ancestors. Consequently, birds have legitimately inherited the evolutionary titles of their ancestors. We now tell our students that birds are card-carrying avialian, maniraptoran, coelurosaurian, tetanurine, theropod, saurischian dinosaurs, and don't you forget it! Because in doing so, you would be denying them their rightful claims to a proud and distinguished ancestry.

So, from today's cladistic perspective, not all dinosaurs became extinct at the end of the Cretaceous. The avian dinosaurs flew over whatever it was that affected their huge cousins at the K-T boundary. Subsequently, dinosaurs evolved into the most specious lineage of land-living vertebrates ever to appear. Today, living dinosaur species outnumber those of all the other major branches of the tetrapod family tree. Once an icon for obsolescence, Dinosauria now appears as one of Mother Nature's greatest success stories.

This isn't simply a question about what names to apply to which organisms. Once the relationships of a lineage have been phylogenetically mapped out, the next step is to reevaluate interpretations of its history that were based on Linnaean classifications. Dinosauria is a marvelous example of how Mother Nature can turn science on its head. Mapping the phylogenetic relationships of dinosaurs indicated that Owen's original conception of dinosaurs as huge, lumbering, extinct reptiles is only partly correct. Some dinosaurs fit that bill, but in fact, the majority do not.

To explore the implications of this new interpretation of dinosaurs, we now return to the map of dinosaur phylogeny and follow it to the present. The evolutionary evidence represented by anatomical signposts on the map will show that a diversity of dinosaurs probably crossed the K-T boundary unscathed, and that only recently have they been threatened with mass extinction.

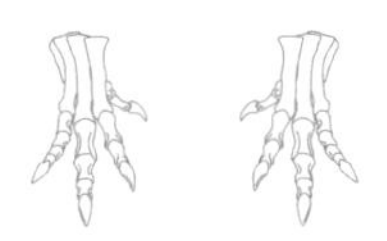

CHAPTER FOURTEEN

The Road to Jurassic Park

When we were graduate students, it seemed that *Archaeopteryx* was the sole source of information about early bird evolution. Our textbooks said little about other Mesozoic birds simply because not much was known about them. Avian history was dark and murky for a span of 85 million years following *Archaeopteryx*, and it was not until the beginning of the Tertiary that the trail of fossils leading toward today's birds resumed.

Recently, an explosion of new discoveries illuminated the darkness. In the last fifteen years, new Mesozoic birds have been collected in many parts of the world. South

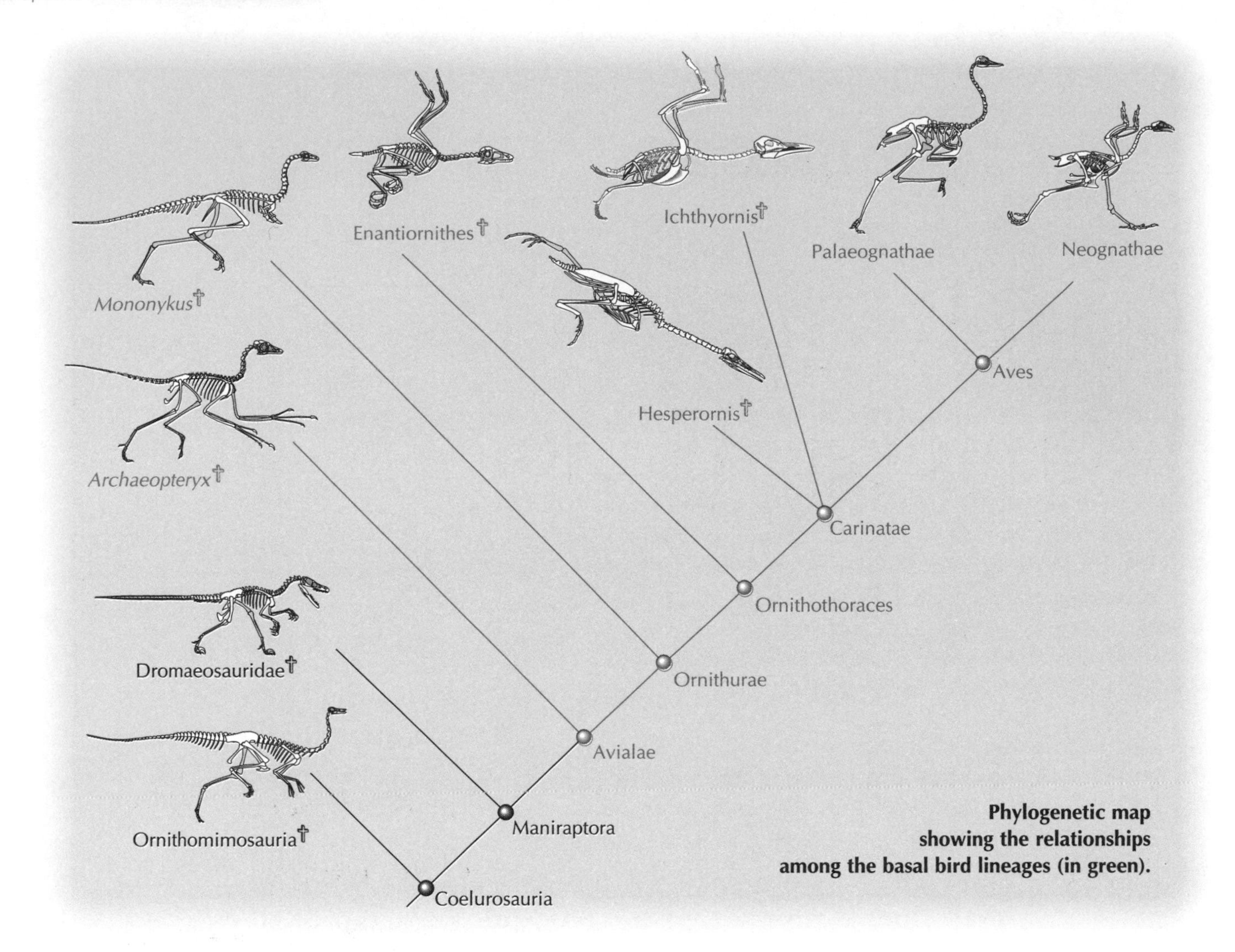

Phylogenetic map showing the relationships among the basal bird lineages (in green).

America, Spain, and Asia are producing a wealth of new Cretaceous fossils, including complete skeletons of mature birds, entire nests of eggs, embryonic skeletons, and, as we saw in Chapter 12, there are new fossils preserving possible feathers. With every professional paleontological meeting, it seems, someone announces a new Mesozoic bird. Others bring pictures or specimens representing new species, but show them only to a privileged few. Thanks to the press, the new discoveries are feeding rumors and excitement on a global scale, as the growing league of bird-watchers race to add one of the rarest species to their "life list"—a Mesozoic bird.

Through the blinding flash of new discoveries, the Mesozoic roots of bird evolution are dimly coming in to focus, and what was terra incognita in our student days is now crossed by several major highways. But some highly contentious issues remain, and different paleontologists have very different visions of early avian history. The new phylogenetic maps are like a *Consumer Reports* study on the paleobiology of birds and other Mesozoic dinosaurs. By matching the timing and sequence of evolutionary events, these maps can test evolutionary cause and effect. By plotting living birds on the same map with their extinct relatives, new insights into the genetics, embryology, physiology, and behavior of both living and extinct species are emerging. For example, the phylogenetic maps offer insights into whether Mesozoic dinosaurs were warm-blooded and whether or not DNA fragments recovered from Cretaceous fossils belonged to dinosaurs. Could Michael Crichton's *Jurassic Park* be more than a fantasy? As we will see, some popular ideas about dinosaurs are endorsed by phylogenetic mapping, while others earn much lower *Consumer* ratings.

EVOLUTION AND DEVELOPMENT

For more than a century, domestic chicks, quail, and a few other birds have been faithful biological laboratory animals. A great deal is now known about their genetic makeup, including how some of their genes direct embryological development and growth.[1] We even know how to experimentally manipulate and alter the growth patterns of embryos in the laboratory. But without a phylogenetic map, most scientists who have done this work did not realize that they were working with dinosaurs, or that their research on modern avian biology might unlock ancient mysteries.

Knowledge of modern genetics and embryology increased as we entered the information age. The genome is a little like a computer's hard drive. Not all of its programs are used at once, and some may lie dormant for long periods. There may even be old files and outdated software that are no longer used. But with a mistaken command, older programs can still be activated, often harmlessly but occasionally with unfortunate consequences. Genomes also preserve remnants of ancient developmental programs. In a few cases, embryologists have discovered ways to turn them back on. Unlike the science-fiction fantasy *Jurassic Park*, we will probably never be able to resurrect an extinct species in the laboratory. But by using embryological techniques to activate a bird's "genetic memory," scientists have induced modern species to regrow some ancient structures.

In a number of regions of the skeleton, rearrangements of bones occur during development that recapitulate similar evolutionary rearrangements in early theropod history. If birds are not descendants of Mesozoic theropods, why does their development history reflect so much of early theropod evolutionary history?

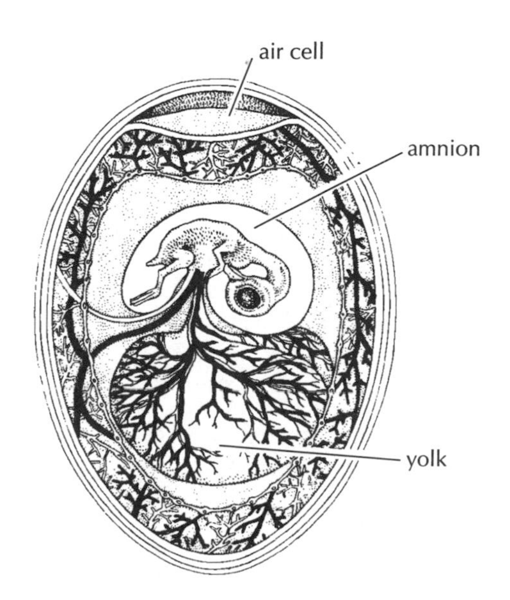

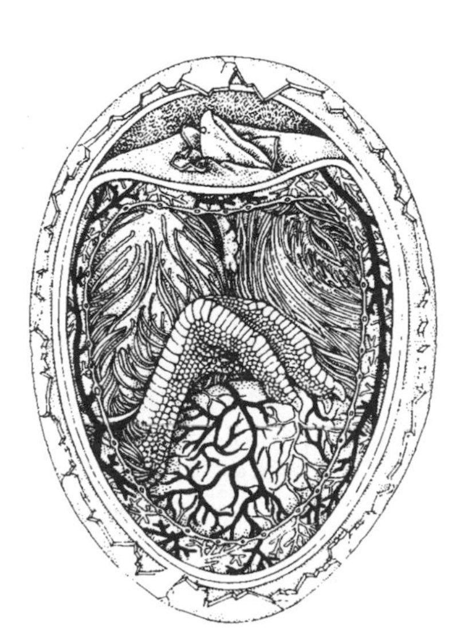

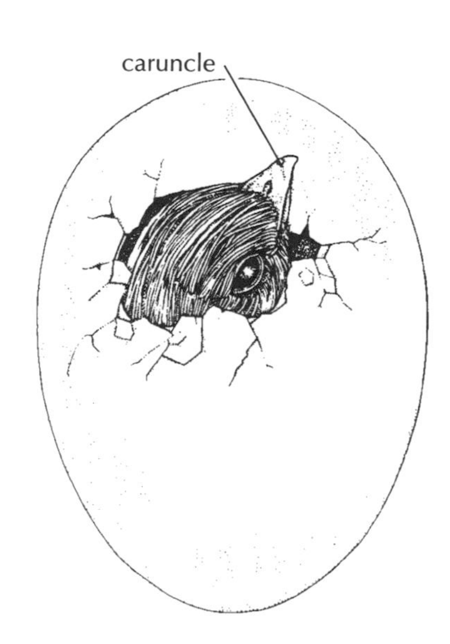

As old programs are triggered, either in nature or the lab, we sometimes get a glimpse of the past, when "throwbacks" to more primitive structures, known as *atavisms*, develop. Atavistic structures are surprising because they otherwise appear only in more distantly related, primitive species. For example, young ducks, geese, swans, and ostrich have a claw on one of the fingers. These usually disappear as the bird's mature plumage develops, but occasionally claws on the wings persist into adulthood. To preevolutionary biologists, birds with claws were difficult to explain. To evolutionists, however, these structures record history by reflecting the reptilian ancestry of birds.

The most significant embryological ties to the past are the parallel transformation sequences that can be observed in both the developmental and evolutionary histories of an organism. Evolutionary transformations documented in the fossil record are often mirrored or recapitulated in the embryonic tissues of a developing embryo.

Consider the distinctive avian foot. As we saw in Chapter 12, a lot of subtle evolutionary changes occurred as the distinctive feet and legs of living birds evolved from more primitive dinosaurs. The ancestral dinosaur had a foot with five toes. Each toe was supported by its own metatarsal bone (the metatarsals are the main girders of the foot), which was connected to the tarsal bones that form the hingelike ankle joint. In the theropod lineage, metatarsal I (above our "big toe") was shortened. The first toe remained func-

During theropod evolution, the first toe, the fifth toe, the ascending process, and the fibula were all rearranged to produce the foot structure characteristic of living birds.

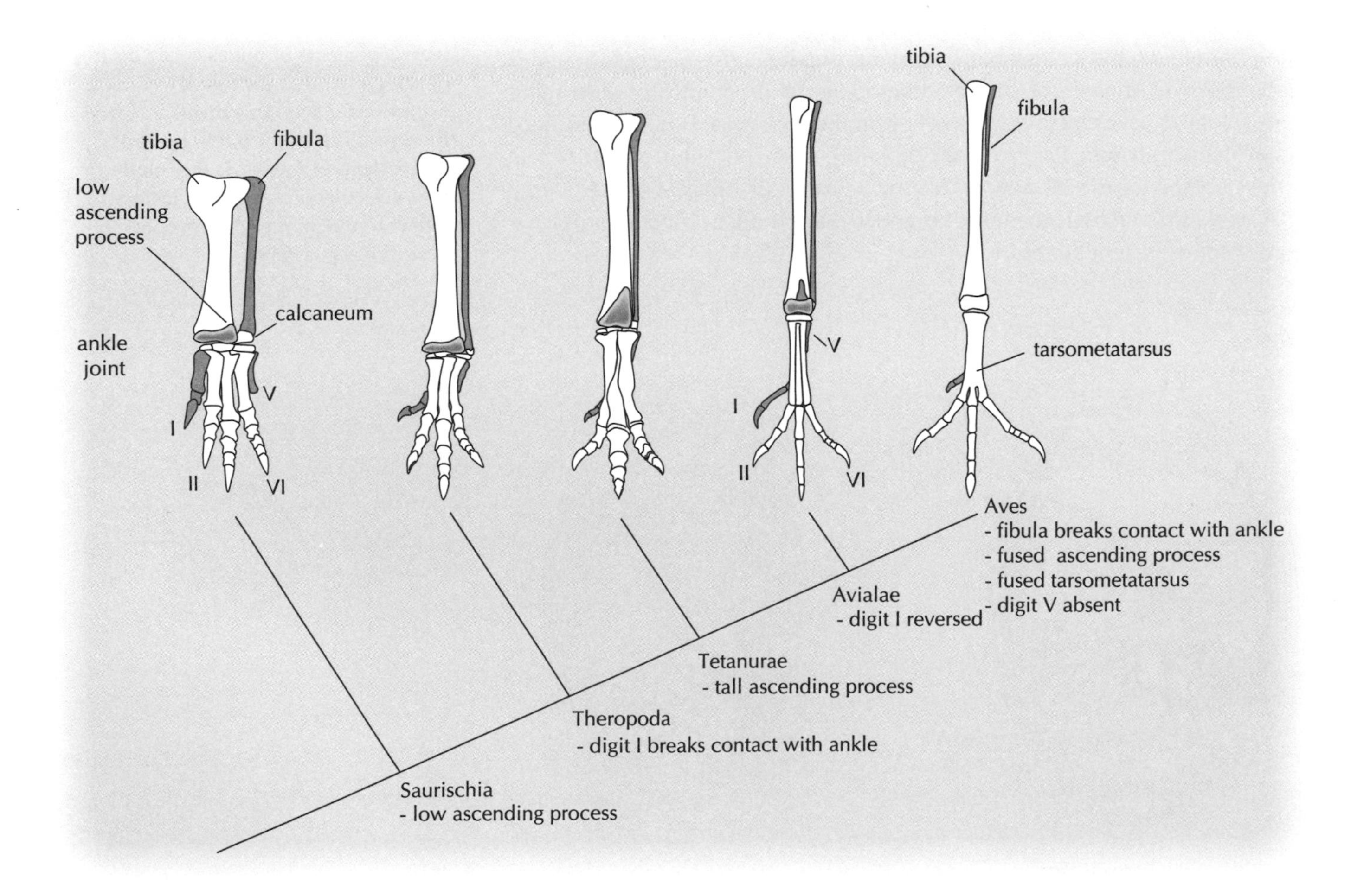

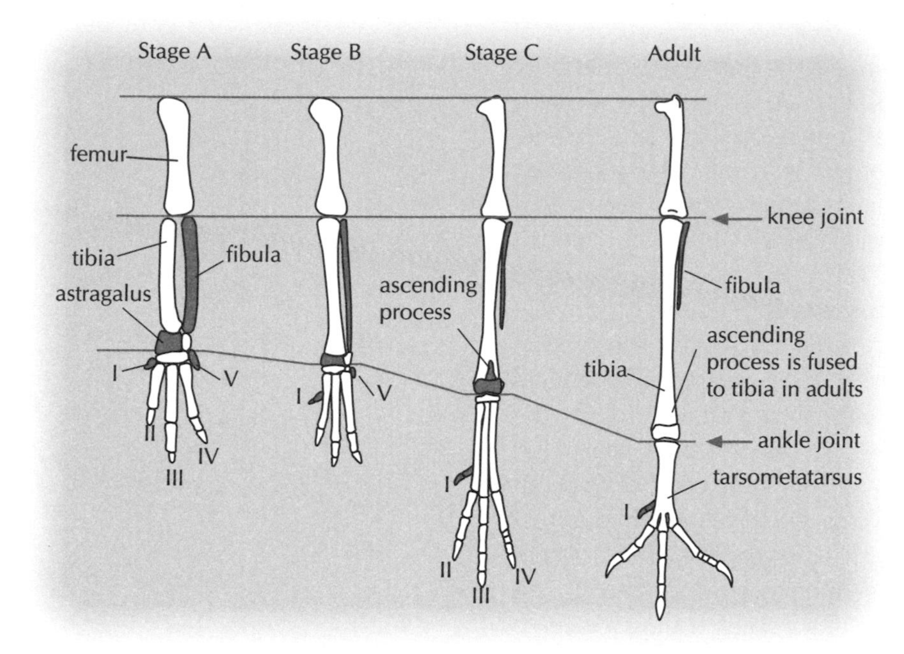

During bird development, the bones of the foot undergo rearrangements that are similar to those occurring in theropod evolution.

tional, but it was no longer connected directly to the ankle bones. At the same time, digit V (our "little toe") became reduced, losing all its phalanges so that only a thin metatarsal splint remained attached to the ankle. In avialians, digit I rotated around to the back of the foot, affording a crude grasping capability. The remaining large metatarsal bones (II, III, IV) began to fuse to each other and to some of the ankle bones, forming a single compound bone called the tarsometatarsus. The tarsometatarsus is made up of several bones that historically were separate and independent. Still later, digit V disappeared altogether, leaving three toes directed forward (digits II, III, and IV) and one directed backwards (digit I) as in modern birds.

Comparable changes occur early during the development or ontogeny in most birds. These changes generally mirror the same sequence documented in evolutionary history, although the transformations take place in embryonic tissues (like cartilage) rather than bone. While still inside the egg, evidence of five toes is visible as the hindlimb starts to grow. The cartilaginous beginnings of five metatarsal bones arise, all in contact with the developing ankle cartilages, in the same configuration found in dinosaurs ancestrally. But soon digit I separates from the ankle and slides down the side of metatarsal II, later rotating around back to afford a grasping capacity. The growing cartilages for the remaining three metatarsals (II, III, IV) eventually coalesce as they turn to bone, to form the compound tarsometatarsus. The tiny remnant of metatarsal V eventually disappears so that no trace of a fifth digit is seen in adults. In all of these changes, the developing embryo repeats or recapitulates the same changes that occurred during the evolutionary history of its ancestors. In the avian foot, ontogeny recapitulates phylogeny.

If adult birds have only four toes, why do their embryos begin development with five? To preevolutionists, this was difficult to explain. But to evolutionists, these recapitulated sequences in bird development are historic relics that express ancient genetic programs.[2] Birds have five toes at the start of their lives because their ancestors did. Darwin was the first to make sense of recapitulations and atavisms.[3] He explained in *On the Origin of Species* that a genealogical relationship is the fundamental source of biological similarity among different organisms, regardless of whether that similarity is expressed between adults or between an adult and an embryo. Atavisms and recapitulations are to embryology what fossils are to paleontology.

The discovery of a recapitulation is what led to the discovery of a close genealogical tie between birds and extinct dinosaurs. Carl Gegenbaur—a great nineteenth-century evolutionary morphologist and embryologist—became an ardent evolutionist as soon as he read Darwin's *Origin*. Gegenbaur was the first to study both the early development of the ankle in a modern bird and the ankle bones of the Jurassic theropod *Compsognathus*.[4] During growth, the "solid" foot of adult birds begins with all the parts that remained separate throughout life in *Compsognathus*. Among reptiles, only *Compsognathus* exhibited this pattern; hence dinosaurs were the closest reptilian cousins of birds. As we saw in Chapter 10, Thomas Huxley extended Gegenbaur's embryological observations to the entire hindlimb, and became the most vocal advocate of the nineteenth century for the bird-dinosaur connection. So, with the importance of comparing patterns of ontogeny and phylogeny in mind, we now return to our tour of dinosaur phylogeny, rejoining our evolutionary map at the beginnings of birds.

EVER SINCE THE JURASSIC

More than a century after its discovery, our most direct evidence for the origin of birds still comes from *Archaeopteryx*, the oldest flying bird from the Late Jurassic Solnhofen limestones. A consensus has emerged that *Archaeopteryx* was probably a good runner and an adequate flapping flyer.[5] But it probably could not flap over long distances, nor could it glide especially well. The hand and wrist bones were unfused, and the shoulder bones retained a number of primitive features suggesting only limited flight capability. These differences also imply that the arms of *Archaeopteryx* may not have been used exclusively for flapping flight. Just how the astounding flight capabilities of modern birds evolved from such a primitive ancestor was once as murky as the fossil record of birds. But the new Mesozoic discoveries preserve transitional stages, so we can now map the evolution of modern flight capabilities, step by step, through the Mesozoic.

Only one other extinct bird has been reported from Jurassic rocks, a toothless bird named *Confuciusornis*, recently described by Larry Martin of the University of Kansas and colleagues Zhongue Zhou and Lian-Hai Hou of the Chinese Academy of Sciences.[6] They took the presence of a Jurassic toothless bird as evidence that the bird-dinosaur hypothesis was wrong. They argued that its occurrence indicated that bird origins must predate *Archaeopteryx* by millions of years. Still more anomalous is that the alleged primitive dinosaurian cousins of birds—the dromaeosaurs like *Deinonychus*—are known only from younger rocks. If birds descended from dinosaurs, they argued, the sequence of fossils was wrong. But more recent workers studying the age of the Chinese fossils report that the rocks are of Early Cretaceous age, some 20 million years younger than previously thought,[7] so the toothless bird is younger than its more primitive toothed cousin. In addition, a fossil from the Late Jurassic of

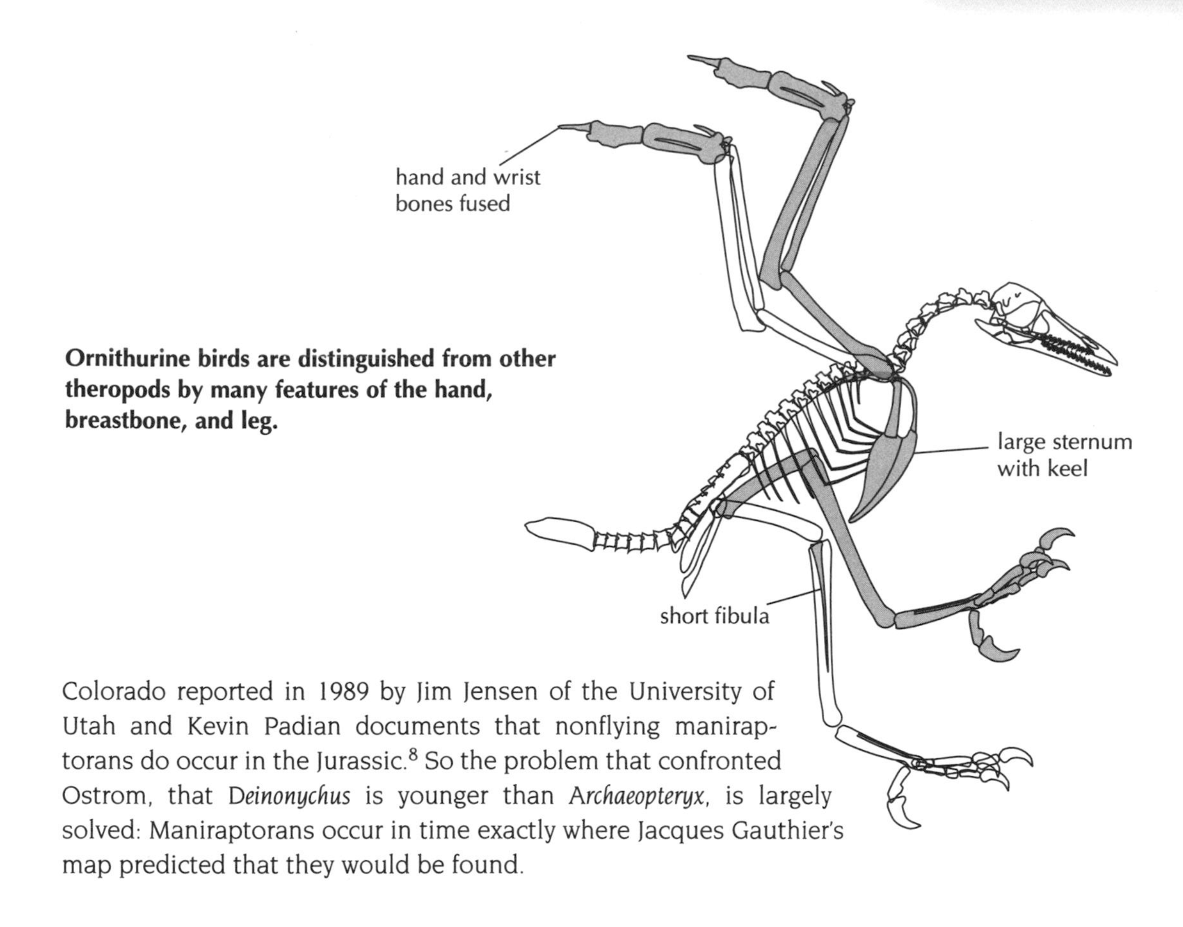

Ornithurine birds are distinguished from other theropods by many features of the hand, breastbone, and leg.

Colorado reported in 1989 by Jim Jensen of the University of Utah and Kevin Padian documents that nonflying maniraptorans do occur in the Jurassic.[8] So the problem that confronted Ostrom, that *Deinonychus* is younger than *Archaeopteryx*, is largely solved: Maniraptorans occur in time exactly where Jacques Gauthier's map predicted that they would be found.

ORNITHURAE: POWERED FLIGHT

Just as the discovery of small, birdlike dinosaurs closed the gap between *Archaeopteryx* and more typical Mesozoic theropods, new discoveries of Cretaceous bird fossils are beginning to close the gap between *Archaeopteryx* and modern birds. The lineage christened Ornithurae by Jacques Gauthier[9]—in reference to the birdlike tail that marks the lineage—includes all birds more closely related to living species than to *Archaeopteryx*. In the last decade, three species of Early Cretaceous ornithurine birds have been discovered, including *Ambiortus* from Mongolia, and *Chaoyangia* and *Gansus* from China.[10] Dr. Luis Chiappe of the American Museum of Natural History is leading the effort to map these new discoveries on the phylogeny of dinosaurs.

The skeleton of ornithurine birds implies distinctly more powerful and possibly more sustained flight than in *Archaeopteryx*. The breastbone or sternum is a large, shield-like bone at the front of the chest. The sternum in other tetrapods is rather inconspicuous, but it is the largest bone in the skeleton in most ornithurines. A high central keel is also present. If you have ever carved a turkey or chicken for dinner, you have encountered its central ridge of bone between the wings, where the white meat of the powerful breast muscles attach. The keel adds broad, strong attachment surface for the massive flight muscles, which generate the power stroke that keeps the bird airborne.

A complementary change occurs in some bones of the wrist and hand, which fuse into a rigid skeleton for supporting the primary flight feathers of the wing. The ornithurine hand is a solid structure built from historically

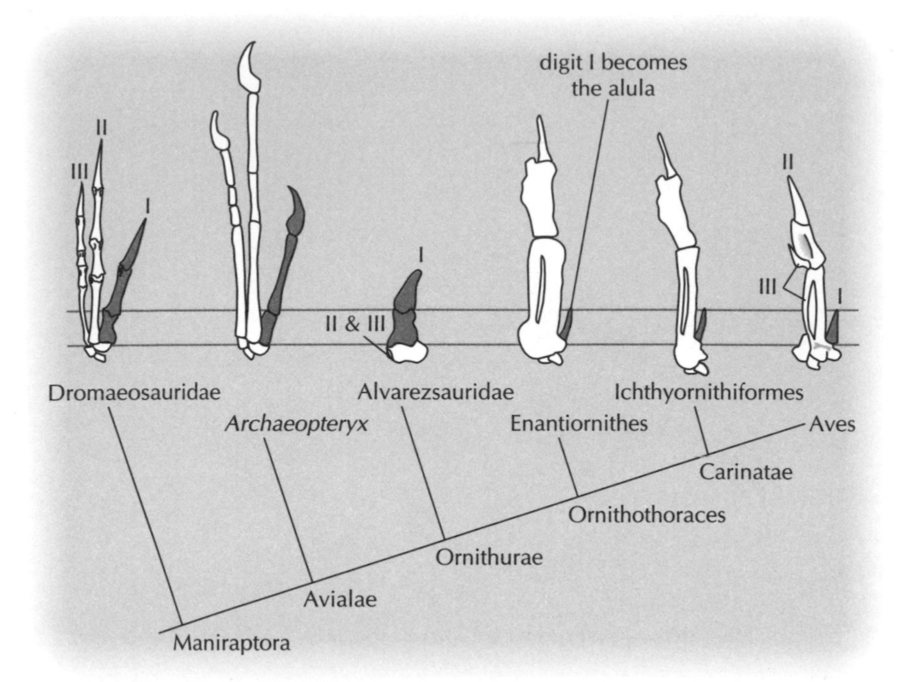

In ornithurine birds, many of the bones of the hand became fused together to form a solid structure.

separate elements. In bird embryos, they form as separate elements that fuse soon after hatching—another recapitulation.

In mutant strains of the domestic chicken, these elements of the hand and wrist never fuse, suspending development at a stage resembling the ancestral structure. These mutations are usually tied to recessive genes,[11] which cause lethal complications soon after hatching. But in normal ornithurines, most of the separate bones of the wrist and hand fuse together, forming a strong hand for flapping. These natural mutants suggest that a simple genetic mechanism may have triggered the complex evolutionary change that occurred in the ancestral ornithurine.

A third distinctive ornithurine feature is found in the arrangement of shin bones. In most tetrapods, the tibia and fibula extend from the knee to the ankle. This is also the condition seen in young ornithurine embryos, like the chicken. Subsequently, however, the tibia grows at an accelerated rate, consuming nearly all the nutrients available for generating that part of the limb skeleton. The fibula becomes a mere thin splint of bone below the knee in adults, leaving only the massive tibia to support the body between the knee and ankle. The "handle" of the drumstick is formed by the tibia, while the fibula is buried in the meat of the leg.

In a famous laboratory experiment, the fibula of a chick was induced to grow all the way down to the ankle, producing a leg with the primitive configuration of bones found in the distant ancestors of birds.[12] To accomplish this, a thin plate of mica was placed between the developing tibia and fibula at a very early stage, equally partitioning the available nutrients. Both grew to the same length, reaching the ankle. Some of the ankle and foot bones even remained separate, similar to the ancestral condition. The leg did not look exactly like that of *Compsognathus* or *Velociraptor*, but the resemblance in the distribution of parts was still striking.[13] This experimental simulation of a more primitive structure may reflect the operation of a simple evolutionary mechanism.

MONONYKUS SHAKES THE TREE

A bizarre new Mesozoic path on the ornithurine map was recently discovered in the Gobi desert by Perle Altangerel from the Mongolian Academy of Sciences, Mark Norell and Luis Chiappe from American Museum of Natural History and James Clark from George Washington University.[14] Several specimens of a turkey-sized bird named *Mononykus olecranus* provided the first evidence that a highly aberrant, flightless lineage of birds evolved during the Cretaceous. *Mononykus* was startling because its forelimbs are profoundly shortened, and there is only one massive finger in its hand. The arms of *Mononykus* seem more suited to digging than flight. Whatever it was doing with its arms, *Mononykus* could certainly not fly.

So weird is this little creature that its identification as a bird prompted a storm of criticism from a host of paleontologists.[15] While acknowledging that the hip and shortened fibula are birdlike, Zhang Zhou argued that these are convergent similarities, reflecting its bipedal way of life rather than its ancestry.[16] If this sounds like déjà vu, it is because Harry Seeley launched the same criticism at Thomas Huxley a century ago, to attack the hypothesis that birds are closely related to dinosaurs. But like Seeley, the critics have no alternative genealogical hypothesis. If *Mononykus* is not a bird, then where does it fit? Why does it have a keeled sternum, like other ornithurines? Why does *Mononykus* have fused wrist bones, a bony sternum, and a pelvis with a back-turned pubis, like other mainraptors? Why does *Mononykus* have a shortened tooth row, a stiff tail, and a tall ascending process in the ankle, like other tetanurines? Why is its skeleton hollow and its foot equipped with a first toe set far below the ankle joint, like other theropods?

The weird, dwarfed forelimbs of *Mononykus* are the source of controversy, because they seem so unbirdlike. However, other theropod lineages have evolved dwarfed forelimbs. Tyrannosaurids are the most famous extinct theropods with dwarfed forelimbs, and *Carnotaurus*, a recently discovered basal theropod from

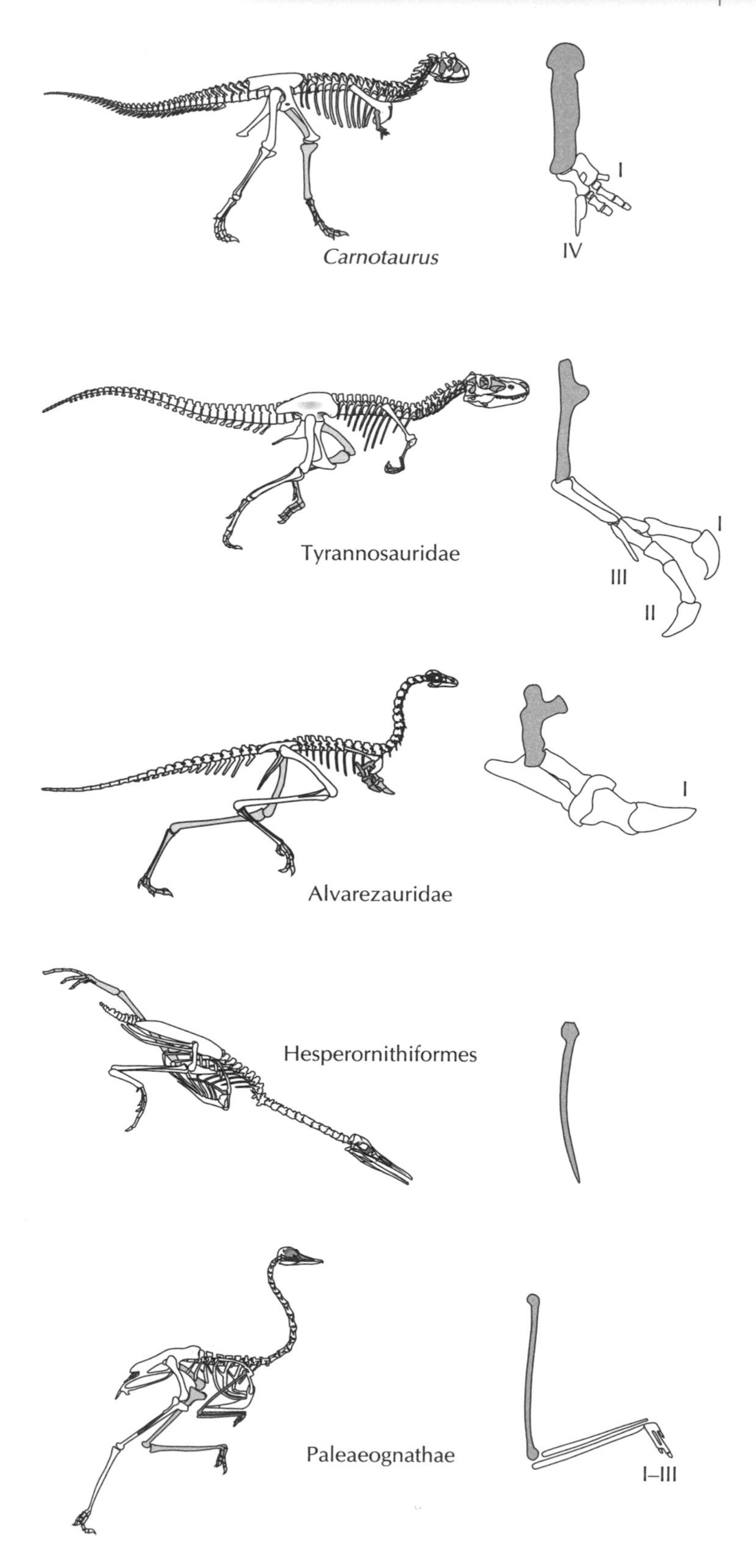

Many theropods besides *Mononykus* have independently evolved dwarfed forelimbs. A very simple genetic mechanism may have controlled these seemingly complex evolutionary changes.

Argentina, also had dwarfed forelimbs. And dwarfed forelimbs characterize a diversity of living birds, including some rails, the penguins, kiwis, ostrich, emu, and rhea, as well as the extinct Cretaceous foot-propelled diver named *Hesperornis*, a more recently extinguished lineage of flightless ducks from Hawaii called moa-nalos, and other flightless fossil birds of the Cenozoic. Moreover, the arms, wrists, and hands in flightless birds differ in the shape and function of the bones. Convergent evolution produces structures that function similarly but differ radically in anatomical details. So everyone agrees that forelimb dwarfism must have evolved several times within birds and their extinct theropod relatives. The alternative would be to argue that tyrannosaurids, penguins, and flightless ducks are each other's closest relatives, and no one has ever suggested this.

Embryologists discovered long ago that "dwarfing" of the forelimb is not a terribly complicated phenomenon. A diversity of natural mutations produce limb dwarfing, and other mutations have been experimentally induced in the laboratory. Several drugs induce limb dwarfing, including thalidomide, which was inadvertently and tragically discovered to dwarf the limbs in human babies whose mothers took it during pregnancy. Invariably, these drugs must be applied at a specific time during development to have this effect. Natural forelimb dwarfing and complete winglessness have also been studied closely.[17] Naturally occurring genetic mutants have been bred in the lab to create pure strains whose members carry the mutant gene. One such strain of laboratory chickens displays the lethal wingless syndrome, which is tied to a simple recessive gene. The shoulder bones develop normally, but the wings are either reduced to small nubs or completely absent.

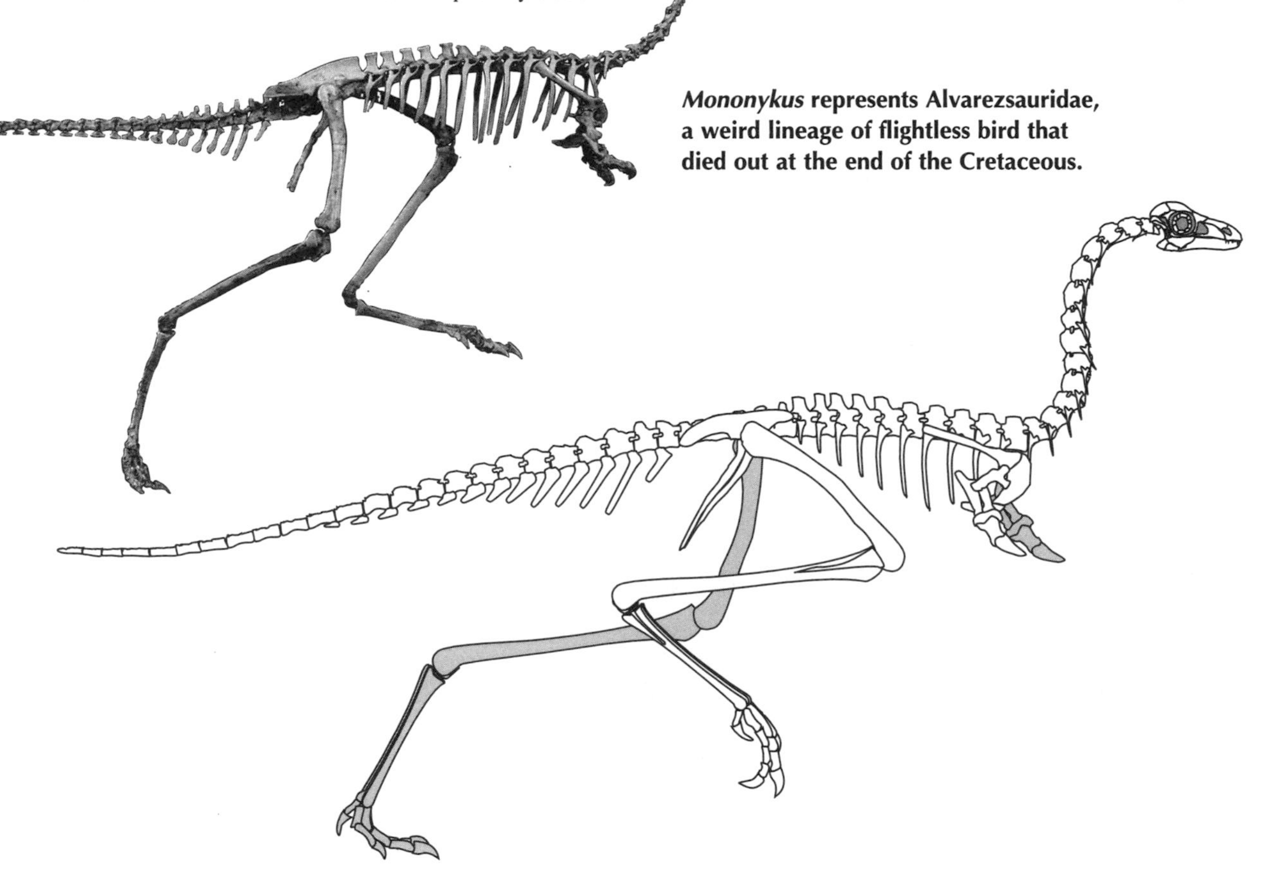

Mononykus **represents Alvarezsauridae, a weird lineage of flightless bird that died out at the end of the Cretaceous.**

The legs are less severely affected, with only minor malformations in the foot. There are other complications of this syndrome, and the chicks die soon after hatching. Thus, one specific gene can't be generally responsible for forelimb dwarfing in theropods. Nonetheless, it shows how a simple genetic mechanism might have induced forelimb dwarfing several times over the course of theropod evolution.

Two flightless birds similar to *Mononykus* were recently discovered in Late Cretaceous deposits of Argentina. The first, named *Alvarezsaurus*, was originally interpreted as a nonavialian theropod. But further study of its skeleton and the discovery of *Mononykus* have established that *Alvarezsaurus* and *Mononykus* are close relatives.[18] The second was recently announced by Fernando Novas of the Argentine Museum of Natural Sciences, who named it *Patagonykus*.[19] These three taxa mark a lineage—Alvarezsauridae—that was widely distributed by the end of the Mesozoic.

ORNITHOTHORACES: INTO THE TREES

The lineage known as Ornithothoraces includes all birds that are closer to modern birds than to Alvarezsauridae. Here, we see an important step toward the evolution of flight capabilities of modern flapping birds. Early members were small. They are distinguished from their more primitive relatives by

A series of changes affected the sternum and coracoid bone in ornithothoracan and carinate birds, in association with more maneuverable and powerful flight.

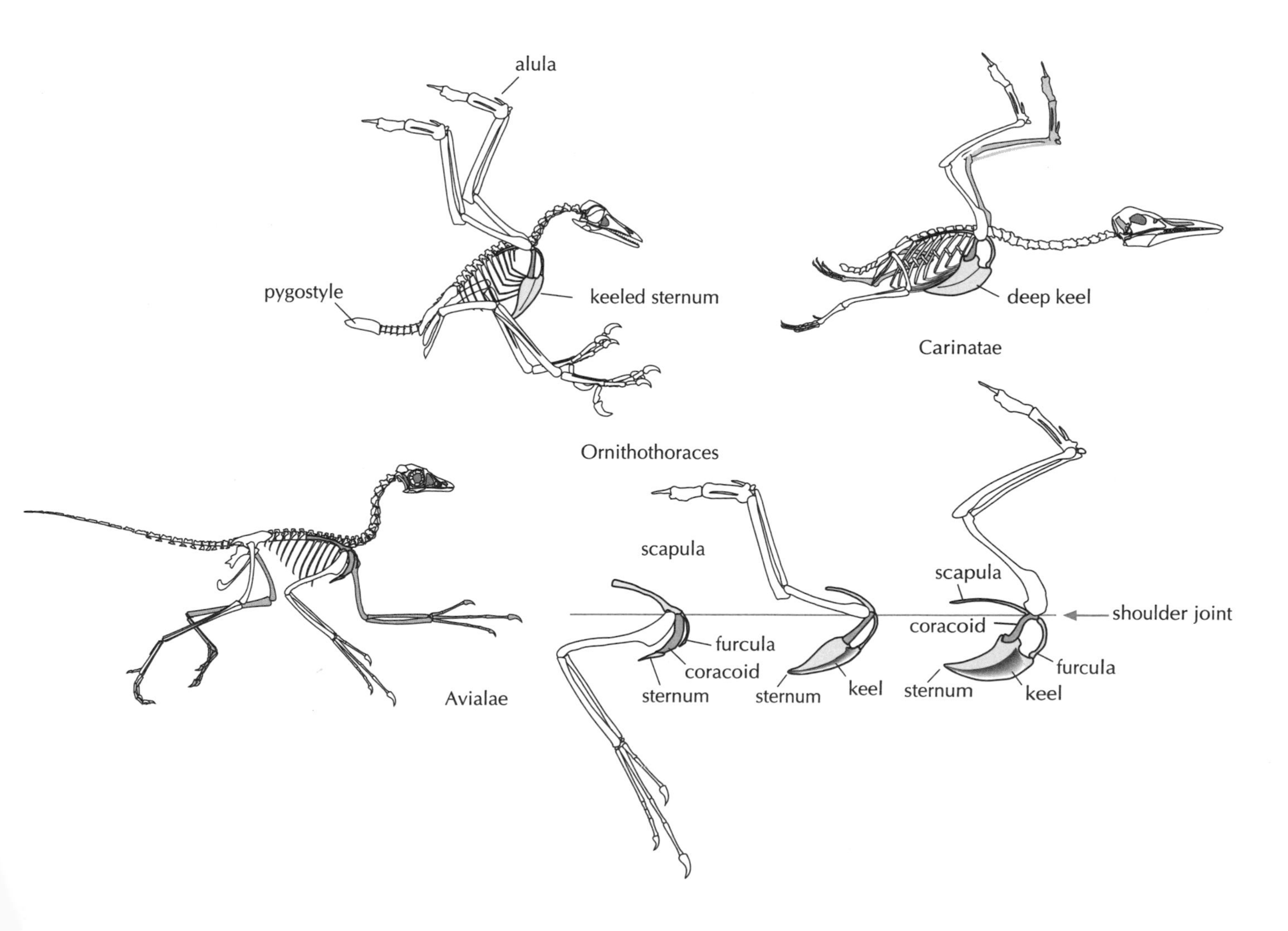

skeletal modifications that suggest enhanced flight maneuverability.[20] Specifically, they had the ability to fly slowly without stalling. This is critical for navigating through complex three-dimensional spaces, such as flying through branches to land and perch.

One shoulder bone, called the coracoid, transformed into a robust strut that displaced the shoulder joint upward, away from the sternum. This increased the leverage of the flight muscles generating the power stroke, principally the pectoralis muscle. It also created a pulley system for a muscle anchored to the sternum that raises and repositions the wing in flight—the supracoracoideus muscle. These enhancements probably locked the arm into functioning exclusively for flight.

A recent discovery in Early Cretaceous rocks of Spain by José Sanz and his colleagues shows that the wings of ornithothoracan birds are equipped with the bastard wing or alula.[21] The bastard wing is formed by a feather that grows from the tip of the tiny thumb, and it creates a wing-slot along the leading edge. By extending the thumb, the slot can be opened to help prevent turbulent flow that causes stalling. The alula and its wing-slot were critical features in the evolution of low-speed, maneuverable flight.

Finally, the vertebrae in the rear half of the tail fused during early development to form the pygostyle, the solid bony structure that supports the tail feathers. As noted earlier, Richard Owen first observed that the pygostyle begins development as a series of separate elements, which correspond to the individual tail vertebrae in *Archaeopteryx*. The separate elements consolidate early in development to produce the pygostyle, recapitulating the evolutionary sequence from separate to fused tail vertebrae. Consequently, the bony tail of adult ornithothoracan birds is much shorter and has fewer parts than in *Archaeopteryx*.

The pygostyle enabled the tail feathers to move rapidly over broad arcs. The tail feathers could now be fanned and rapidly reoriented both vertically and horizontally during flight to control lift and direction precisely. Overall, early ornithothoracans were more powerful and maneuverable flappers than *Archaeopteryx*, more capable of the complex navigation required to fly among the branches and land in the trees.

The pygostyle also marked the beginning of a subtle shift in the way birds walk. In *Archaeopteryx* and more primitive bipedal dinosaurs, the long bony tail anchored massive muscles that pulled the thigh backward. This retraction of the limb at the hip provided

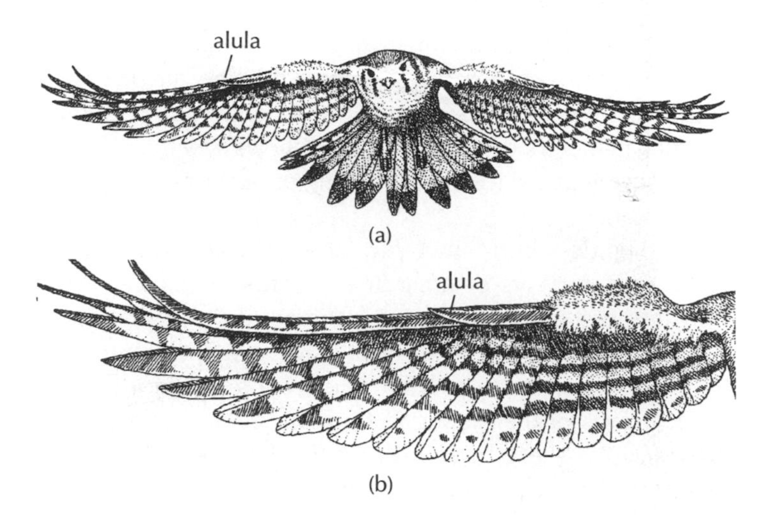

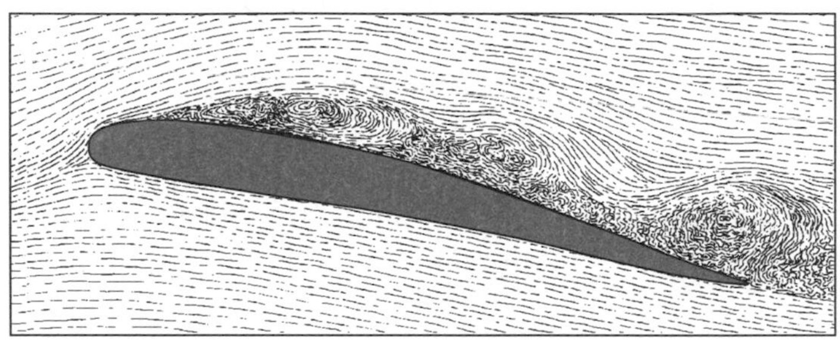

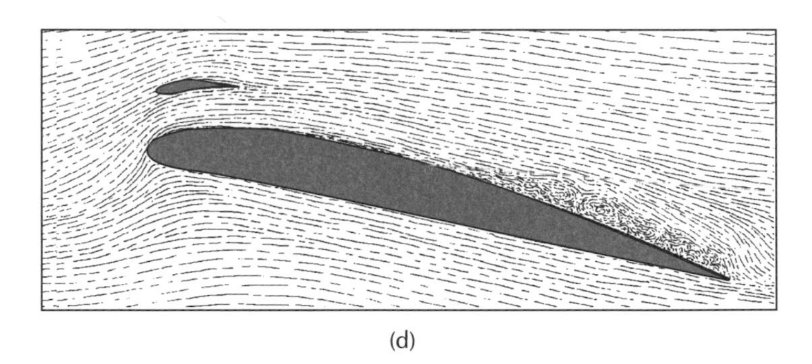

The alula is a feather extending from digit I in ornithothoracan birds, like this falcon (a, b). Cross sections through two wings in slow flight demonstrate turbulence above the wing that can cause stalling (c), and the effects of the alula which enhances slow-speed flight (d).

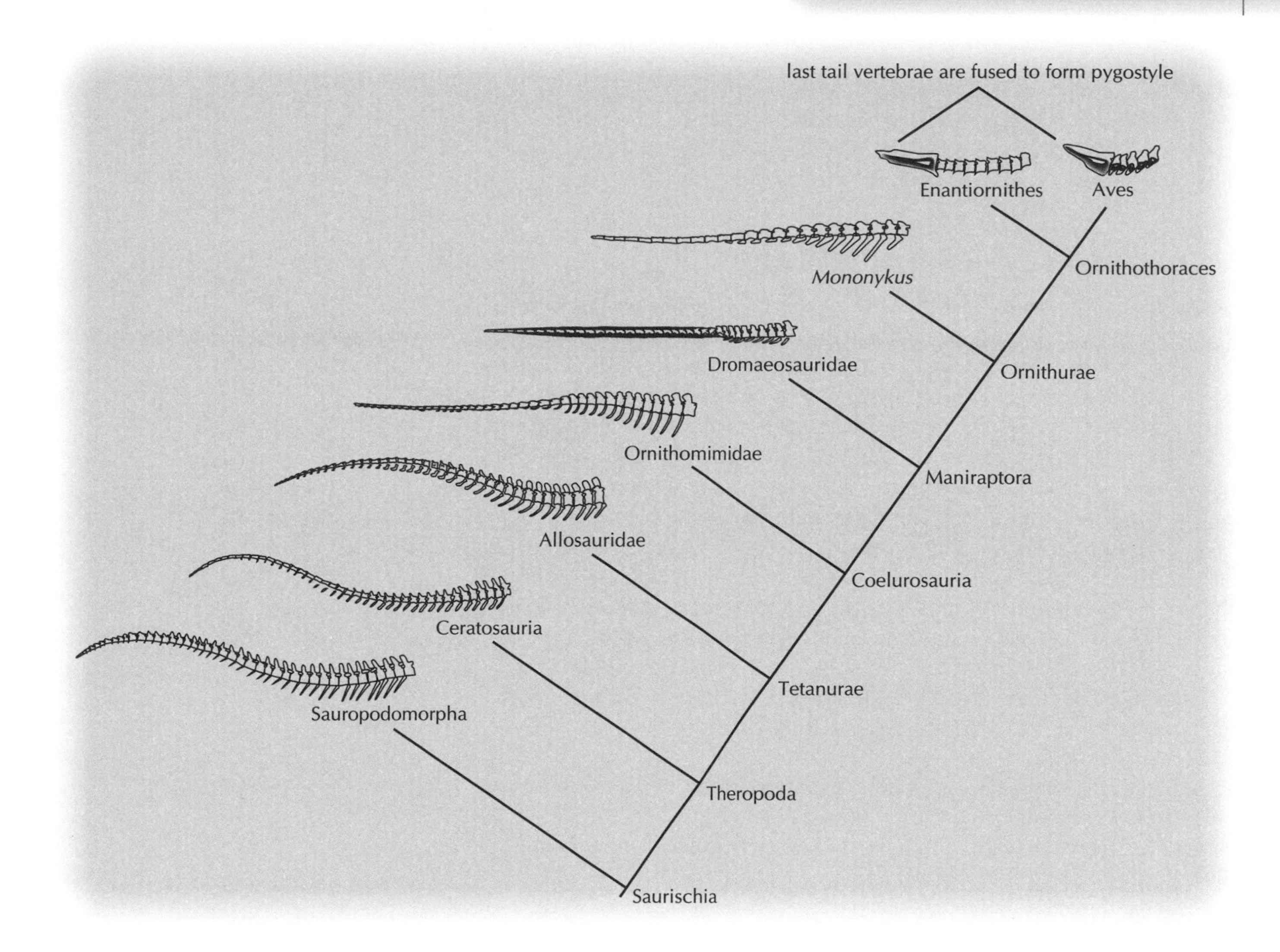

The vertebrae forming the end of the tail in ornithothoracan birds became fused together to form the pygostyle.

the power stroke for walking and running, much as it does in living crocodiles and lizards. But with the reduced ornithothoracan tail, the muscles that retract the thigh were also reduced. These muscles are still present in modern birds, but are tiny in comparison to those of crocodiles. Therefore, the range of movement in the bird hip is limited, and the thigh rotates across a somewhat smaller arc than in more primitive theropods. To compensate, greater movement takes place at the knee, which in ornithothoracans is flexed and extended over a greater arc.

Ornithothoracan birds also have a simplified foot. In their ancestor, the fifth metatarsal disappeared and with it went all vestiges of the fifth toe. As noted earlier, this transformation is recapitulated in bird development, where a thin split is briefly present, only to disappear before hatching.

The oldest ornithothoracan is probably *Noguerornis*, which was discovered in the 1980s in Early Cretaceous deposits of northern Spain that are slightly younger than the Solnhofen limestone.[22] *Noguerornis* was a finch-sized bird with large wing surfaces that possibly enabled the powerful bounding type of flight typical of sparrows and finches. A slightly younger but more complete ornithothoracan from Spain is *Iberomesornis*.[23] About the same size as *Noguerornis*, it had slightly longer arms and a perching foot that may signal the origin of arboreality in birds. Several other basal ornithothoracans, known from Early Cretaceous rocks of China and Mongolia, offer evidence that a global diversification of flapping birds was under way by that time.[24]

ENANTIORNITHES

Entirely unknown twenty years ago, the enantiornithine lineage is now known to have been diverse and widespread in the Cretaceous.[25] Representative fossils were actually collected over a century ago, but they were misidentified as nonavian dinosaurs or as members of modern bird lineages. Enantiornithine birds have been recovered primarily from terrestrial deposits, suggesting that the lineage radiated widely across the continents, but they never evolved into shorebirds or seabirds. Enantiornithines are found in Early Cretaceous deposits in China and Spain, as well as Late Cretaceous deposits in North America, South America, and Australia. Over the course of its 70 million year history, the lineage diversified into species with many different sizes and shapes. Early enantiornithines were small, sparrow-sized birds that were strong flyers. By the end of the Cretaceous there were turkey-sized forms with wingspans of more than 3 feet, stilt-legged waders, and powerful runners.

Enantiornithines are sometimes referred to as "opposite birds" because their feet grew in an opposite pattern from other birds. Alan Feduccia of the University of North Carolina and several colleagues argue that Enantiornithes, plus *Archaeopteryx* and *Confuciusornis*, comprise a lineage named "Sauriurae," which is the sister lineage to all other birds.[26] They claim that fusion among the foot bones begins near the ankle and proceeds downward. In other birds, fusion begins in the middle of the foot and grows in both directions. But these are points of resemblance instead of evidence for close relationship, because, as we have seen, apart from being members of the Avialae lineage, *Archaeopteryx* and enantiornithines lie within different hierarchies. Based on its position on the phyloge-

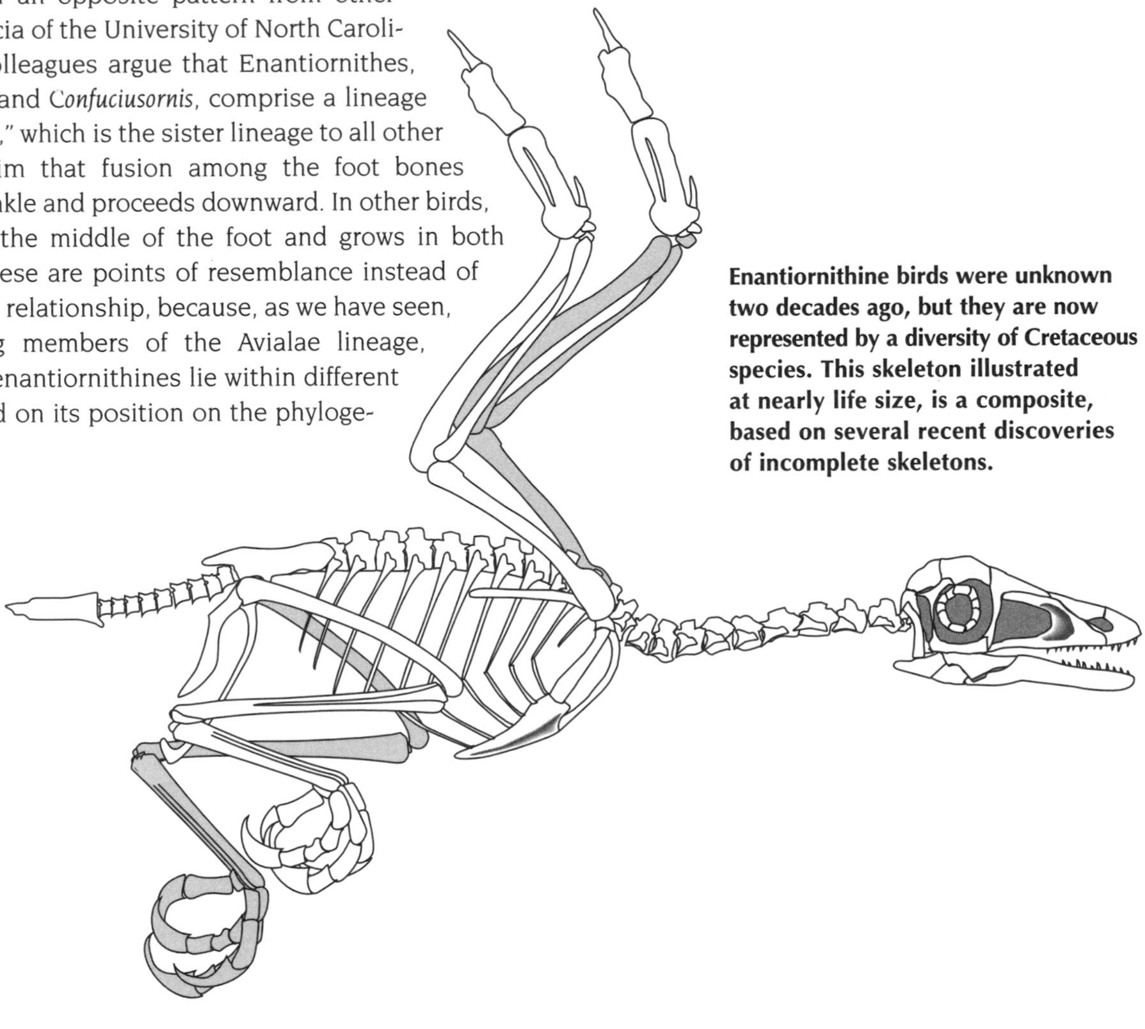

Enantiornithine birds were unknown two decades ago, but they are now represented by a diversity of Cretaceous species. This skeleton illustrated at nearly life size, is a composite, based on several recent discoveries of incomplete skeletons.

netic map, fusion in the foot bones characterizes all ornithothoracan birds, and variations on the pattern of fusion such as that in enantiornithines evolved within the group. There is no evidence to suggest that "Sauriurae" is a natural group, so it cannot be plotted on the evolutionary map of early birds.

A flightless lineage of extinct ornithothoracan birds that can be mapped is *Patagopteryx*, from Patagonia. This chicken-sized bird lived during the Late Cretaceous and was probably a good runner, like its more primitive dinosaurian relatives. The pelvis lost the pubic boot so distinctive of its more primitive tetanurine relatives. Therefore, *Patagopteryx* may be a step closer to modern birds than the enantiornithines.[27]

In contrast to what we learned as grad students, a rich diversity of birds shared the landscape with their more famous Cretaceous relatives such as *Deinonychus*, *Velociraptor*, and *Tyrannosaurus*. But like them, alvarezsaurids, enantiornithines, and *Patagopteryx* all became extinct at or near the end of the Cretaceous.

CARINATES: AIR AND WATER

Carinate birds are advanced in the structure of their skulls and flight apparatus. The respiratory passages through the snout, in front of the eyes, may have housed large structures known as turbinates. We will discuss these interesting structures later, when we investigate the issue of warm-bloodedness in dinosaurs. The sternum of carinates has a greatly deepened keel, indicating another step in the evolution of powerful flight muscles. The trunk is also short and stout compared with more primitive dinosaurs, with less than twelve vertebrae between the base of the neck and the pelvis. This short stout trunk enhanced powered flight by providing a rigid armature for attaching larger flight muscles and absorbing greater forces during landing.

Several carinate lineages are known from fossils found predominantly in marine rocks.

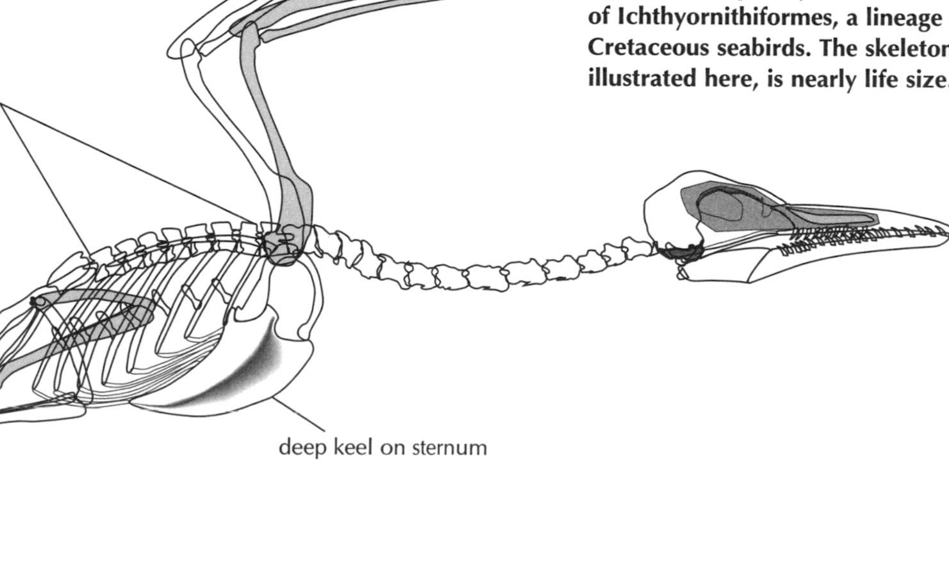

Several of the most characteristic features of carinate birds are the shortened trunk and a deep keel on the breastbone. Pictured here is *Ichthyornis*, the most completely known member of Ichthyornithiformes, a lineage of Cretaceous seabirds. The skeleton, as illustrated here, is nearly life size.

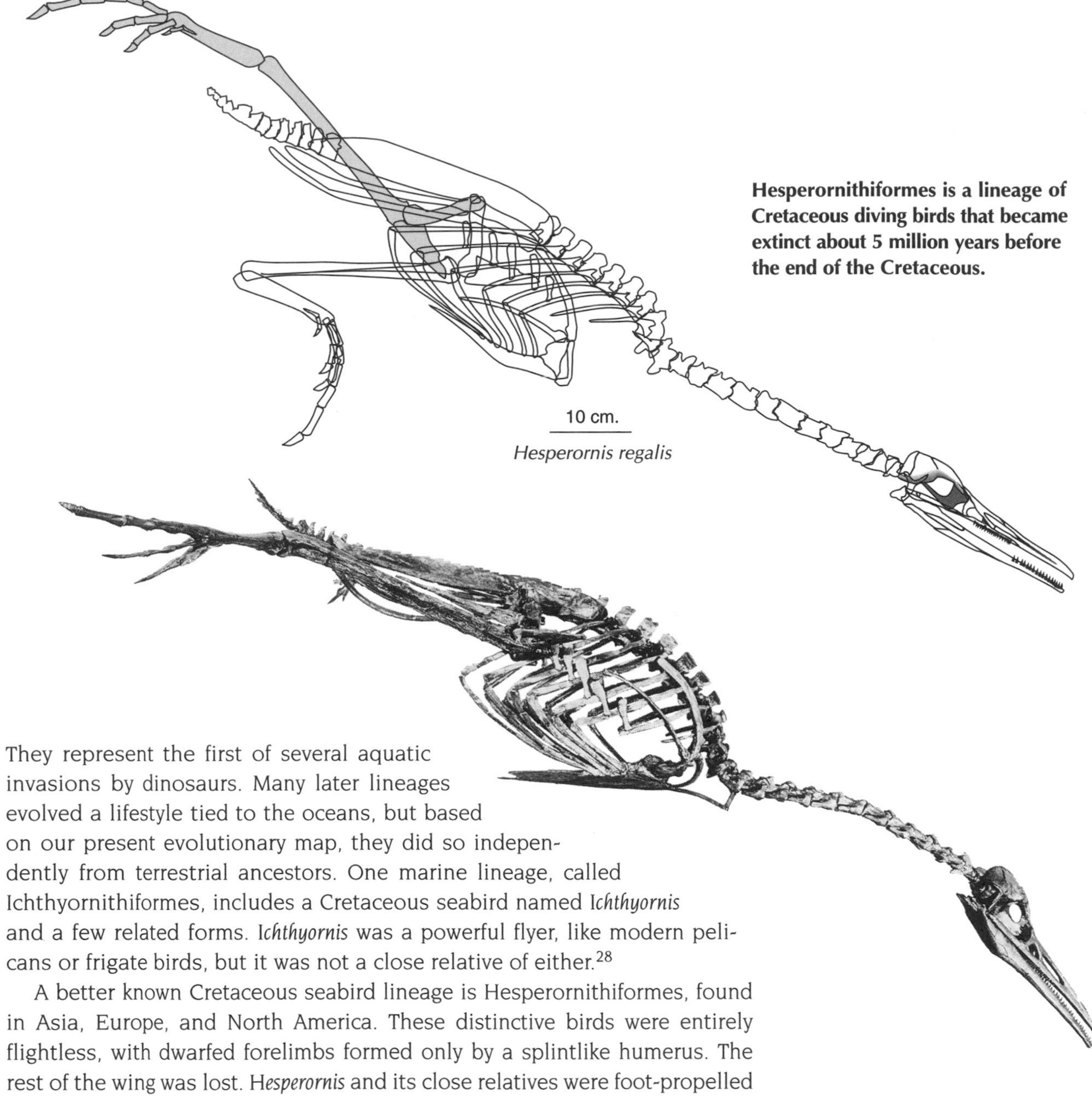

Hesperornis regalis

Hesperornithiformes is a lineage of Cretaceous diving birds that became extinct about 5 million years before the end of the Cretaceous.

They represent the first of several aquatic invasions by dinosaurs. Many later lineages evolved a lifestyle tied to the oceans, but based on our present evolutionary map, they did so independently from terrestrial ancestors. One marine lineage, called Ichthyornithiformes, includes a Cretaceous seabird named *Ichthyornis* and a few related forms. *Ichthyornis* was a powerful flyer, like modern pelicans or frigate birds, but it was not a close relative of either.[28]

A better known Cretaceous seabird lineage is Hesperornithiformes, found in Asia, Europe, and North America. These distinctive birds were entirely flightless, with dwarfed forelimbs formed only by a splintlike humerus. The rest of the wing was lost. *Hesperornis* and its close relatives were foot-propelled diving birds.[29] The pelvis was elongated and the feet were located at the back of the body, like modern penguins and loons. They probably used their feet for propulsion under water or while swimming across the surface. These birds foraged for food along the coastlines of the shallow seaways that crossed North American and Asian continents during much of the Cretaceous, living on fish and other marine organisms. They must have been very ungainly on land and, apart from nesting, spent most of their lives in the water.

The oldest carinate is probably *Ambiortus* from the Early Cretaceous of Mongolia.[30] Known only from fragmentary fossils, *Ambiortus* was a terrestrial bird, like most of the lineage. The fossil records of both Ichthyornithiformes and Hesperornithiformes begin in the Early Cretaceous and disappear about 5 million years before the end of the Cretaceous.

SCARCE AS HEN'S TEETH

Aves is the lineage that includes the common ancestor of all living bird species, and all its descendants.[31] As we will see at the height of its diversity, a few thousand years ago, Aves included more than 12,000 and perhaps as many as 20,000 species of birds. A distinctive characteristic of Aves is the complete loss of teeth. Modern birds even lose the egg tooth, which evolved at the same time as the amniotic egg, to help the embryo break out of its shell. Although many people refer to an "egg tooth" in modern birds, the structure in question is really the caruncle, a horny spike that grows at the tip of the beak. The true egg tooth is a real tooth, with layers of dentine and enamel, like other teeth. Even this tooth is absent in birds, and no natural atavisms—birds with teeth—have ever been observed.

Replacing the teeth is a horny beak or bill. As the hands and arms of birds became increasingly modified and committed to flight, they lost the grasping ability of earlier theropods for capturing food and manipulating objects. This is where the strong toothless beak comes in. The upper bill became movably hinged to the skull in front of the eyes, and a series of levers coupled its movement to that of the lower jaw. As the lower jaw dropped, the upper beak raised. This provided unparalleled versatility, and a vast diversity of specializations evolved from this basic mechanism through subsequent modification in the beaks and the lever system between them. Modern birds can tear flesh, build nests, drink nectar, and perform many other functions with their beaks.

In birds, teeth may be gone but they are not forgotten. In a laboratory experiment, embryologists induced embryonic tissues around the mouth of an unhatched chick to differentiate into tooth buds—the first stage of tooth development.[32] Thus, the genetic program for growing teeth has been conserved in birds, but a regulatory gene has switched off the program or blocked it in some other way. As the genetic control for development is mapped in greater detail, we may see embryologists reengineer other ancient structures in some modern birds.

In addition, there is a greatly enlarged brain in Aves compared with other theropods. This potentially fascinating distinction is as yet largely unstudied, but the structures responsible for integrating sensory and motor information are clearly enlarged. More volume means more neurons, and more neurons provide greater computing power for gathering, filtering, and coordinating sensory information. Greater computing power also enhances the integration of muscular actions and responses to environmental stimuli.

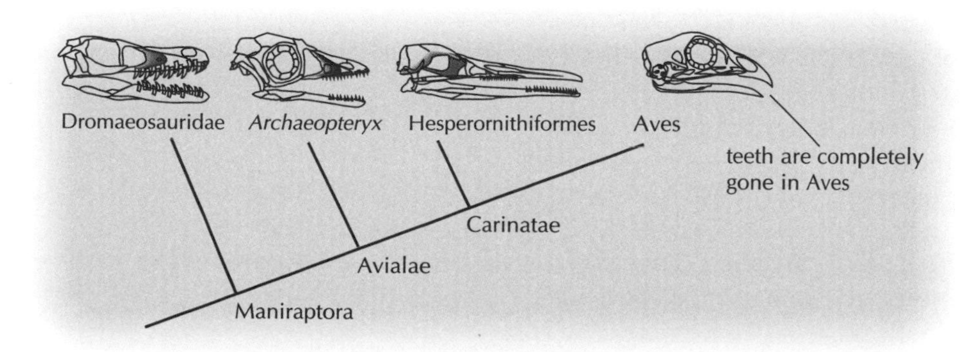

One of the most characteristic features of Aves is the complete loss of teeth.

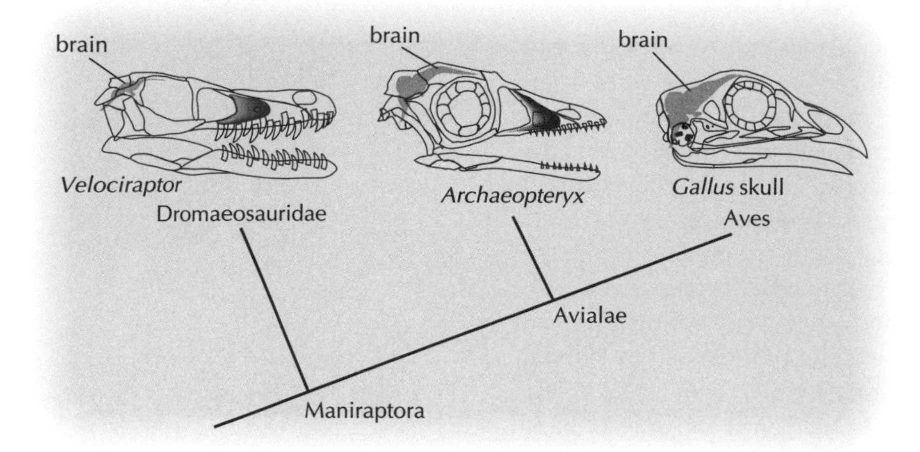

Living birds have a larger brain than their relatives among extinct theropods.

WARMER BLOOD

A war erupted in the early 1970s over whether dinosaurs were warm-blooded. By the time we reached graduate school, the media and scientific press had escalated the issue to global proportions. After a quarter century of debate, where does the scientific community stand today?

First, we should take a closer look at the problem. Today, only mammals and birds are warm blooded. This means that they maintain a constant body temperature that is comparatively high, and that they do this through the generation of heat via the metabolic activities of the cells. Warm-blooded animals are technically known as endotherms. They have an increased capacity for longer periods of high activity and a greater independence from ambient environmental temperatures. This has helped endotherms to colonize all parts of the globe. The alternative physiological condition, from which mammal and bird physiology evolved, is known as ectothermy. In ectotherms, body temperature varies more directly with the ambient environmental conditions. It may be higher or lower than in a given endotherm, depending largely on environmental conditions. Bursts of high activity cannot be sustained, and environmental conditions tend to be more limiting for ectotherms.

For all its seeming advantages, endothermy comes at a high cost. Even with layers of insulation, like feathers in birds and hair in mammals, heat is continually being lost to the environment. To keep their bodies at a constant temperature, the metabolic "engines" of endotherms burn about five to ten times the amount of oxygen as do ectotherms of comparable size. So, endotherms usually require constant supplies of food, water, and oxygen to sustain themselves. The question of our generation was: Were dinosaurs endothermic like mammals and birds, or ectothermic like other reptiles?

In one sense, the question of whether dinosaurs evolved warm bloodedness can be decisively answered from our map of dinosaur phylogeny. Birds, living dinosaurs, are unquestionably warm-blooded. So, some dinosaurs—in fact most dinosaurs—are endothermic. But such an answer still begs the question of *when* in dinosaurian history endothermy arose. Were all dinosaurs endothermic, or only birds, or some combination in between? Did endothermy evolve in gradual steps or suddenly in one step?

The charge in this paleontological debate was led by Bob Bakker, who provocatively claimed that *all* dinosaurs are endothermic.[33] Like Richard Owen, Bakker was influenced by anatomical clues for an upright posture, such as the in-turned femoral head. But Bakker went beyond Owen's conclusions and argued that upright posture *demanded* endothermic physiology in *all* dinosaurs. Bakker argued that advanced metabolic machinery was required because he thought that muscular activity is what held the limbs against the body. But by measuring the metabolic costs of locomotion in living reptiles, mammals, and birds, experimental physiologists found that simply having an upright posture does not require an elevated metabolic rate. The cost of standing, in terms of the amount of oxygen consumed, was hardly different in a mammal or a lizard.[34] Moreover, modern crocodylians are capable of spectacular bursts of activity by pulling their hindlimbs into an erect position when doing the "high walk"—a fast walk that is almost a run. They just can't do this for long periods because their muscles quickly fatigue from the buildup of lactic acid. So, an upright posture doesn't necessarily indicate endothermy.

Bakker also argued that relative abundances of fossil carnivores and herbivores can be used to infer physiology, but this has been a limb that few other paleontologists are willing to crawl out on. Few are willing to accept that the fossil record, so dependent on the vagaries of preservation and containing such huge gaps, preserves the actual abundance of different kinds of dinosaurs in living communities, or that census numbers of fossils necessarily bear on the physiology of the population. As we saw in Chapters 7 and 8, the completeness of the stratigraphic record is quite variable and in many locations probably preserves little of an ancient community's structure. Our sample size of fossils and extinct communities is so small that few scientists accept the relative abundance of fossils as a very sound datum.

Additional arguments have been advanced by different researchers in support of endothermy in all dinosaurs.[35] Speeds inferred from Mesozoic trackways are alleged to indicate that dinosaurs moved at higher than average velocities, which in turn was alleged to signal higher metabolic levels. But most trackways are of dinosaurs walking at slow speeds and only a few short trackways preserve any evidence of higher speeds—for whichever dinosaur it was that left them. The geochemistry of fossil bones was also presented in support of warm-bloodedness.[36] However, so little is known about the geochemical effects of being buried, and so few dinosaur fossils have been studied in this way, that few scientists trust the results. For a time, scientists argued that the biogeography of Mesozoic dinosaurs and their global distribution suggests that nonavian dinosaurs were ectotherms. As we saw earlier, the discovery of a rich fauna of dinosaurs from Alaska's north slope, which was inside the Arctic Circle even in the Cretaceous, indicated that the biogeographic pattern is more complicated than was previously thought.

More recently, two independent lines of skeletal evidence have been presented as evidence of endothermy in at least some Mesozoic dinosaurs. Both suggest that the ancestral carinate was physiologically closer to modern birds in the way it breathed and the way it grew, implying elevated metabolic levels. This connection was discovered by John Ruben of Oregon State University, who argued that to sustain high levels of activity for prolonged periods, endotherms needed a constant high supply of oxygen.[37] Birds and mammals have a supercharged heart and lung system that pumps more blood and more air than their cold-blooded relatives. Birds are even aided by the blind air sacs and pneumatic cavities that branch throughout their bodies and even invade their bones. Modern birds can deliver high quantities of oxygen to the muscles even during extreme exercise, by breathing faster to pass more air across their respiratory membranes. Birds deliver about fifteen times more oxygen to their muscles than ectothermic reptiles, whereas mammals deliver six to ten times that amount. The increased oxygen supply enables the cells to metabolize aerobically, in contrast to the anaerobic metabolism of cold-blooded vertebrates. This cleaner metabolic engine permits long, sustained periods of high activity without the debilitating build up of lactic acid. Modern birds could not sustain long flights without having this aerobic metabolism sustained by their powerful heart-lung system.

Ruben suggests that a series of scrolled bones or cartilages called respiratory turbinates, which lie inside the nasal chamber along the respiratory passage in birds and mammals, is correlated to endothermy. He argues that small, warm, fast-breathing birds or mammals can potentially lose a lot of water vapor very quickly. As cool air enters the lungs, it becomes warmed and saturated with condensing water vapor. Without a mechanism to retain body moisture while exhaling, dehydration would severely restrict the scope of activity unless an unlimited supply of water were available. Birds and mammals circumvent this limit through the respiratory membrane that covers the elaborate turbinates. Living carinates all have respiratory turbinates. *Hesperornis* may also have had turbinates, judging from the shape of the nasal chamber in front

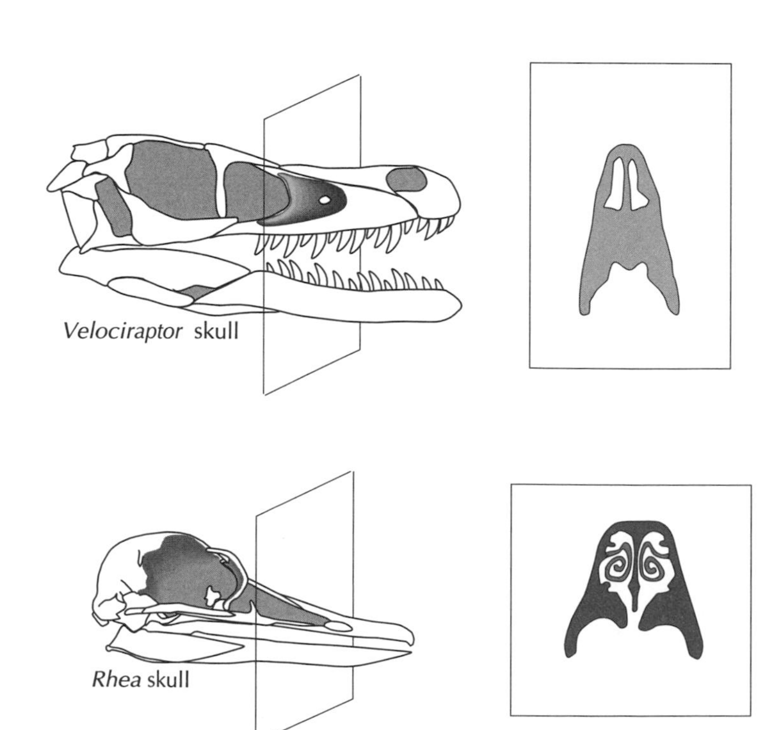

In noncarinate dinosaurs, such as *Velociraptor*, the respiratory passageways are simple tubes. In carinates, the passageway is filled with scroll-like turbinate bones, which capture water from exhaled air, to help prevent dessication during breathing.

of its eyes. However, turbinates are easily lost during decay and burial, and so far, despite claims to the contrary, they have not been verified in any Mesozoic dinosaur. On the mammal side of the amniote tree, moreover, there is evidence from CAT scanning of Mesozoic fossils that the turbinates arose in several steps, so the picture may not be so clear as Ruben's argument suggests.

More bony clues pointing to endothermy have been suggested by the microscopic patterns of bone growth, which may indicate whether a bone underwent continuous, uninterrupted growth, or if there were periods in which growth stopped. This line of evidence was first championed in the 1970s by Armand de Ricqlès of the University of Paris, and Anusuya Chinsamy of the South African Museum recently carried the technique further.[38] Both have studied modern birds along with Mesozoic dinosaur specimens. Bones of enantiornithines and more distant relatives grew discontinuously, with long dormant periods that produced growth lines visible in microscopic cross sections of the bones.[39] Similar growth lines are found in modern ectothermic reptiles—crocodylians, lizards, and turtles—in which there is seasonal growth. But in carinates, no growth lines are reported. Chinsamy and colleagues argue that an elevated metabolic rate and warmer body may have insulated carinates from seasonal temperature fluctuations that can affect the physiology and growth in ectotherms. So far, not many species of either living or extinct dinosaurs have been studied, so these conclusions await further testing.

ANCIENT DNA

In Michael Crichton's fantasy *Jurassic Park*, DNA was recovered from the stomachs of Cretaceous mosquitoes that were trapped in amber shortly after sucking dinosaur blood for their last supper. A recent claim by Scott Woodward of Brigham Young University and colleagues that fragments of DNA molecules had been extracted directly from Cretaceous dinosaur bones[40] seemed to put us on the scientific road to *Jurassic Park*. A modern technique called polymerase chain reaction (PCR) enables biologists to amplify extremely tiny amounts of DNA into samples large enough that researchers can measure the characteristic sequences of nucleic acids. With such a genetic blueprint for an extinct dinosaur, could we implant this DNA into a living egg and clone a Mesozoic beast?

Using PCR, Woodward and colleagues measured several short DNA sequences. But when they were compared to those of modern reptiles, the sequences showed no special similarities to any. Blair Hedges of the University of Pennsylvania and Mary Schweitzer of the University of Montana later demonstrated that the DNA sequences are probably human—an artifact of human contamination of the original sample.[41] More importantly, Hendrik Poinar of the University of Munich and his colleagues reported evidence that DNA quickly decays, in a process known as racemization.[42] Over a scale of hundreds to thousands of years, DNA is relatively stable. But on a scale of millions of years, racemization leads to severe deterioration of original structure. Apart from environments like amber, which embalm and preserve the soft tissues, it looks unlikely that we will ever obtain DNA from Mesozoic dinosaurs.

Even if we could somehow recover Mesozoic dinosaur DNA from amber, we would need not just DNA fragments but a complete, intact genome.

Building a living sauropod or ceratopsian from fossil DNA fragments would be like trying to build and launch the space shuttle using something as fragmentary as the Dead Sea scrolls for an instruction manual. A second problem is that complex feedback mechanisms exist between the DNA in different nearby genes, as well as with various parts of the host cells and surrounding tissues. Only through this feedback can the right switches be thrown at the right time to produce a functioning, viable living organism. Without the proper and specific feedback from the egg containing the cloned DNA, there is little hope that development will proceed very far before something goes wrong and the embryo dies.

So, even though we are a long way from cloning extinct dinosaurs using Mesozoic DNA, a door to the genetic past may stand slightly ajar thanks to embryology of their modern descendants. But regulating the growth of a bird embryo is an enormously complex process that presents science with a vast, unexplored terrain. Much of the ancient genetic history of birds may be erased, and the scientific road to *Jurassic Park* now appears infinitely long.

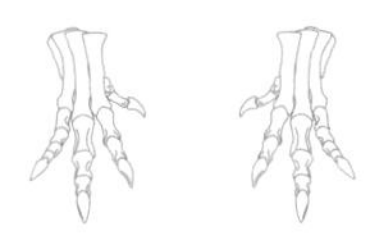

CHAPTER FIFTEEN

Crossing the Boundary

The early history of birds is hotly debated because it offers both a new perspective on the Cretaceous-Tertiary extinction and on one of the greatest adaptive radiations in vertebrate history. Adaptive radiation is evolutionary diversification on a grand scale, often occurring in a relatively short geologic time span. Lineages split again and again, rapidly filling environmental niches with a new dynasty of novel species.

Tropical forests and islands have hosted some of the greatest rapid adaptive radiations in history. Organisms that somehow invade an island at the right time or find an unoccupied level of forest canopy may exploit a wide new landscape of opportunities for themselves and their offspring. The famous adaptive radiation of finches and tortoises on the Galápagos Islands profoundly shaped Darwin's ideas on evolution and natural selection. Archipelagos, lakes, caves, and rift valleys may all serve as spawning grounds for adaptive radiations by providing isolation from parent populations and new ecological opportunities for the proliferating species. The impact of an adaptive radiation upon the local economy of nature can be far reaching.

Nonavian dinosaurs underwent an adaptive radiation during the Mesozoic that generated the hundreds of species and enormous diversity of shapes described earlier.[1] But that was just the beginning. The adaptive radiation of avian dinosaurs was even greater, yielding more than 9,000 living species and thousands more that arose and became extinct over the course of their Tertiary history. Because birds occupy most of the paths on the dinosaurian map, paleontologists have begun to ask how the diversity of the *entire* dinosaurian lineage was affected at the end of the Cretaceous? When did the great adaptive radiation of birds begin, and what environmental factors may have triggered it?

Alan Feduccia of the University of North Carolina and his associates contend that birds underwent a Mesozoic proliferation before being nearly wiped out at the end of the Cretaceous.[2] Only a single lineage crossed the K-T boundary, but this was followed by an explosive adaptive radiation of birds in the 5 to 10 million years following the great extinction. Like the Phoenix, modern birds rose from the ashes, reborn from a lone Tertiary survivor. According to this hypothesis, that post-extinction explosive radiation over the last 65 million years generated the 9,000 species alive today. Feduccia argues that birds and mammals both experienced parallel, contemporaneous episodes of Mesozoic diversification and extinction before radiating explosively in the Tertiary. An extraterrestrial event, like an asteroid impact, probably shaped the common course of evolution for both. These biologists reject the idea that birds are the descendants of Mesozoic dinosaurs. Instead, they argue that the hierarchy of similarities outlined in the preceding chapters represents convergent evolution, and that the roots of avian evolution remain shrouded in mystery. Once again, we cross paths

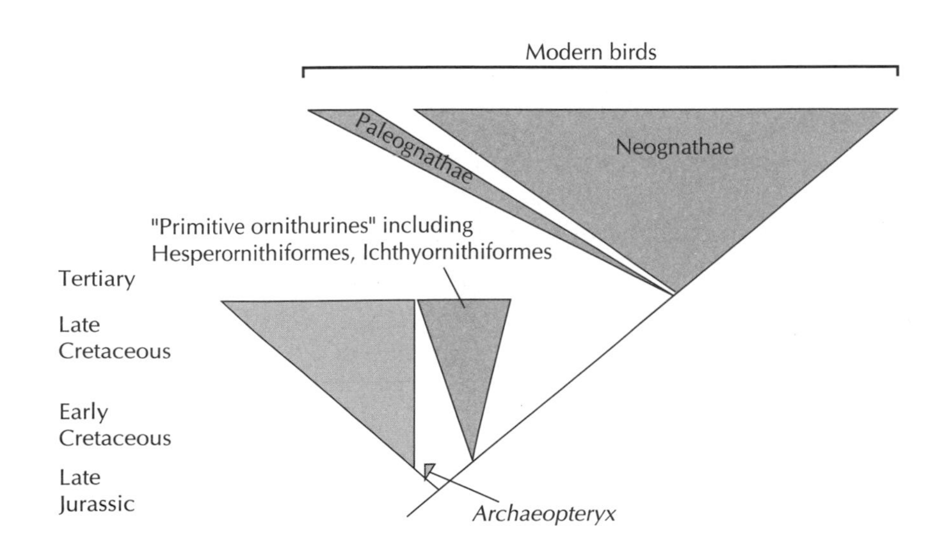

Alan Feduccia's controversial hypothesis about the history of avian diversity depicts an episode of Mesozoic diversification that was nearly chopped off at the K-T boundary. The catastrophic extinction was followed by an "explosive" Tertiary radiation of modern birds from a small group of survivors.

with scientists using the theory of homoplasy to explain the resemblance between birds and dinosaurs.

The opposing view actually represents a much older idea recast in a modern phylogenetic mold. It points to the beginning for modern avian diversification on the map of theropod dinosaurs in the Early or mid-Cretaceous. Although there were losses at the end of the Cretaceous, birds flew across the K-T boundary relatively unscathed. The modern architects of this view are Joel Cracraft and Luis Chiappe of the American Museum of Natural History. Cracraft was a pioneer in using cladistic methods for phylogenetic analysis, which he applied to mapping the early history of birds.[3] Chiappe, who has handled nearly all of the known Mesozoic bird specimens, is extending Cracraft's map to include new discoveries of fossil birds.[4] By plotting Mesozoic birds on the evolutionary map with the help of computers, these and other scientists point to evidence for an ancient diversification of modern birds.[5] Although birds suffered some Late Cretaceous losses, the common ancestor of living species is twice as old as suggested by Feduccia's Phoenix hypothesis. If this hypothesis is true, avian history was shaped by much slower, terrestrial processes.

Did birds rebound from a catastrophic near miss with extinction? Did the history of birds parallel that of mammals in an explosion of Tertiary diversification? Both of these hypotheses are predictive and can be tested to some extent. To see how this is done, it is helpful to first examine how biologists map diversity through time and deal with the incompleteness of the fossil record.

DISCOVERING GHOSTS

One of the grandest pre-Darwinian portrayals of the history of life's diversity is the division of the geologic timescale into three successive ages. First came the Age of Fishes (Paleozoic), followed by the Age of Reptiles (Mesozoic), and then the Age of Mammals (Cenozoic = Tertiary + Quaternary). When these names were coined, there were no tetrapod fossils known from Paleozoic rocks, so the Age of Fishes seemed a fitting title. Reptiles first appeared in younger Mesozoic rocks. Mesozoic bird and mammal fossils were unknown, so the Age of Reptiles, which became The Age of Dinosaurs in Richard Owen's hands, was another appropriate title. The Age of Mammals was named because giant reptiles were absent and instead mammals were abundantly preserved throughout the Cenozoic fossil record.

The discovery of Mesozoic mammals, which did not belong in the Age of Reptiles, was a shock to the nineteenth-century scientific community. A few years later, the discovery of *Archaeopteryx* produced the same shock all over again. As the fossil record grew, more fossils violated the boundaries between ages. The scientists most surprised by these discoveries were those who accepted the sequence of fossils as representing successive episodes of divine creation and biblical floods. But evolutionists had predicted the discovery of "missing links" between all the paths on the map of life. They argued that the known fossil record for any lineage is only a small sample of its complete history. Hidden paths of relationship extend deeper into the past to connect all fossil and modern species on a single map of genealogy. Since the discovery of *Archaeopteryx*, this inference has been tested and confirmed by countless fossil discoveries.

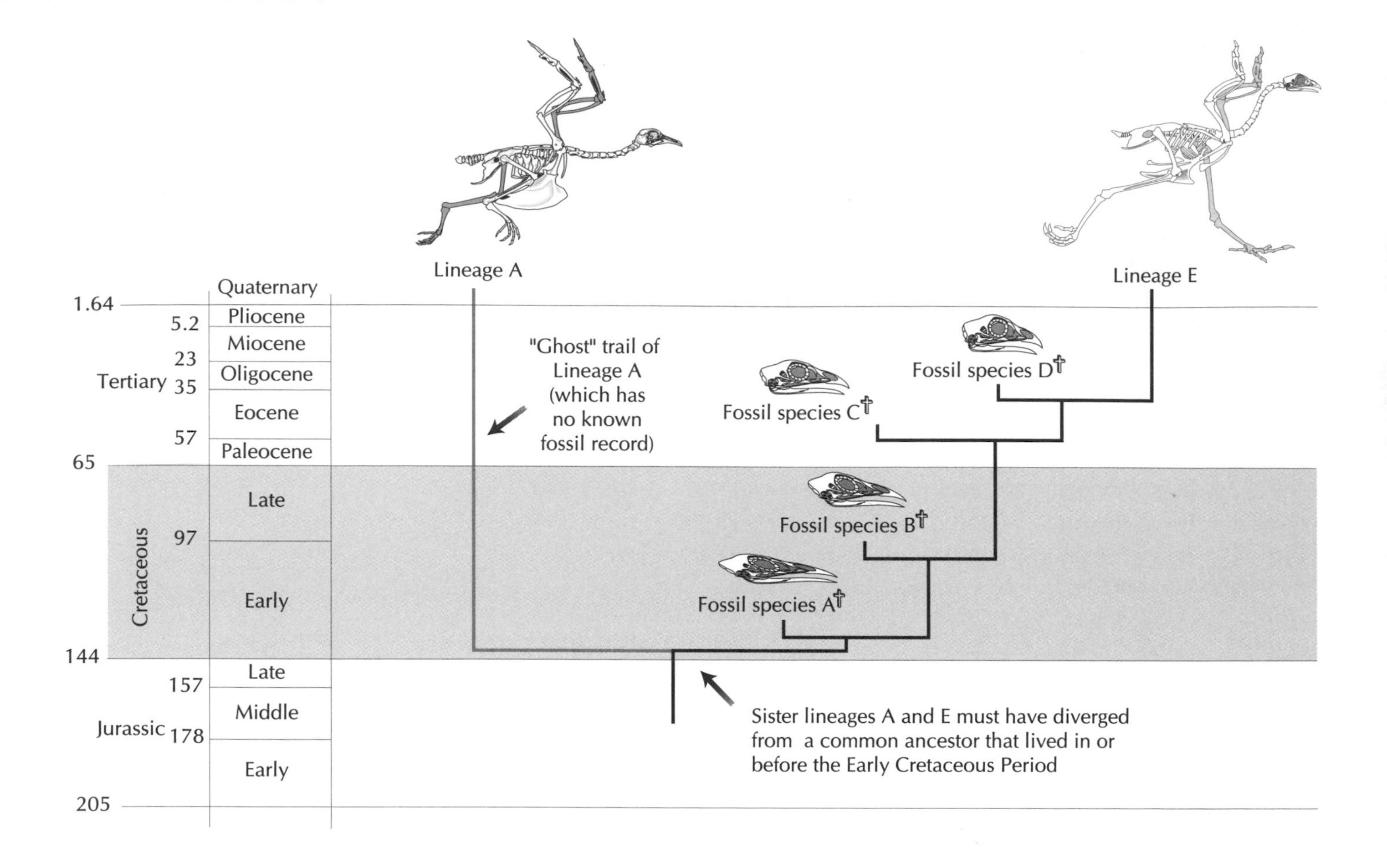

If living lineages A and E are sisters, then both must have been present in the Cretaceous, and both must have survived across the K-T boundary. Although we have found no fossils for lineage A and have no idea of just how diversified it was in the past, we can connect its history in time to lineage E, which did leave a fossil record. The ages at left denote millions of years before present. (Extinct species are denoted by crosses.)

Emerging from modern phylogenetic mapping is an even more vivid appreciation for these "hidden" lineages than the nineteenth-century evolutionists had. One fossil, accurately plotted on the phylogenetic map, can potentially provide new evidence on the time of origin for several lineages. Imagine two sister lineages, originating from an ancestral species that lived during the Cretaceous, and both represented by living species. One lineage leaves fossils from the Cretaceous onward, while the sister lineage leaves no fossils at all. Because they are sister lineages (a hypothesis subject to independent phylogenetic testing), then both must have been present together from the Cretaceous onward. Although only one line is documented directly by Mesozoic fossils, the map indicates that both lineages successfully crossed the K-T boundary.

The sister lineage that left no Mesozoic fossils is sometimes referred to as a ghost lineage, based on the work of Mark Norell at the American Museum of Natural History.[6] Although ghost lineages cannot be seen directly in the fossil record, they exist on the phylogenetic map. One of the great strengths of any mapping enterprise is the capacity it generates to extrapolate between known species. Once plotted on the map, the ghostlike lines of inferred relationship can be identified and counted, revealing the diversity within tetrapods that was not captured by the preservation or discovery of fossils. Ghost lineages predict what the fossil record should eventually yield, so they can be tested through both field work and further phylogenetic analysis.

The two competing hypotheses of avian diversification can be tested using the phylogenetic map for Cretaceous and living birds. Likewise, the claim that birds and mammals shared parallel histories can be examined by comparing phylogenies for both lineages. So, with ghost lineages in mind, let's now examine the histories of mammals and birds.

MESOZOIC MAMMALS

Richard Owen played a major role in validating the first discoveries of Mesozoic mammals.[7] Two tiny jaws of Jurassic mammals were discovered in the Stonesfield Slate of Britain, in 1812. Georges Cuvier identified them as Mesozoic mammals, linking them to living marsupials like the opossum. But many other contemporary naturalists rejected as impossible the idea that mammals lived in the Age of Reptiles. Yet, several more specimens were unearthed over the next few years. These fossils were surprising to Owen, but after personally inspecting them and comparing them to modern mammals, Owen convinced the scientific community that mammals were indeed present in England during the Age of Reptiles. Since then, many more Mesozoic mammals have been discovered. Scientists now universally accept that mammals were not only present in the Mesozoic, but that they had achieved a global distribution long before the era ended.[8]

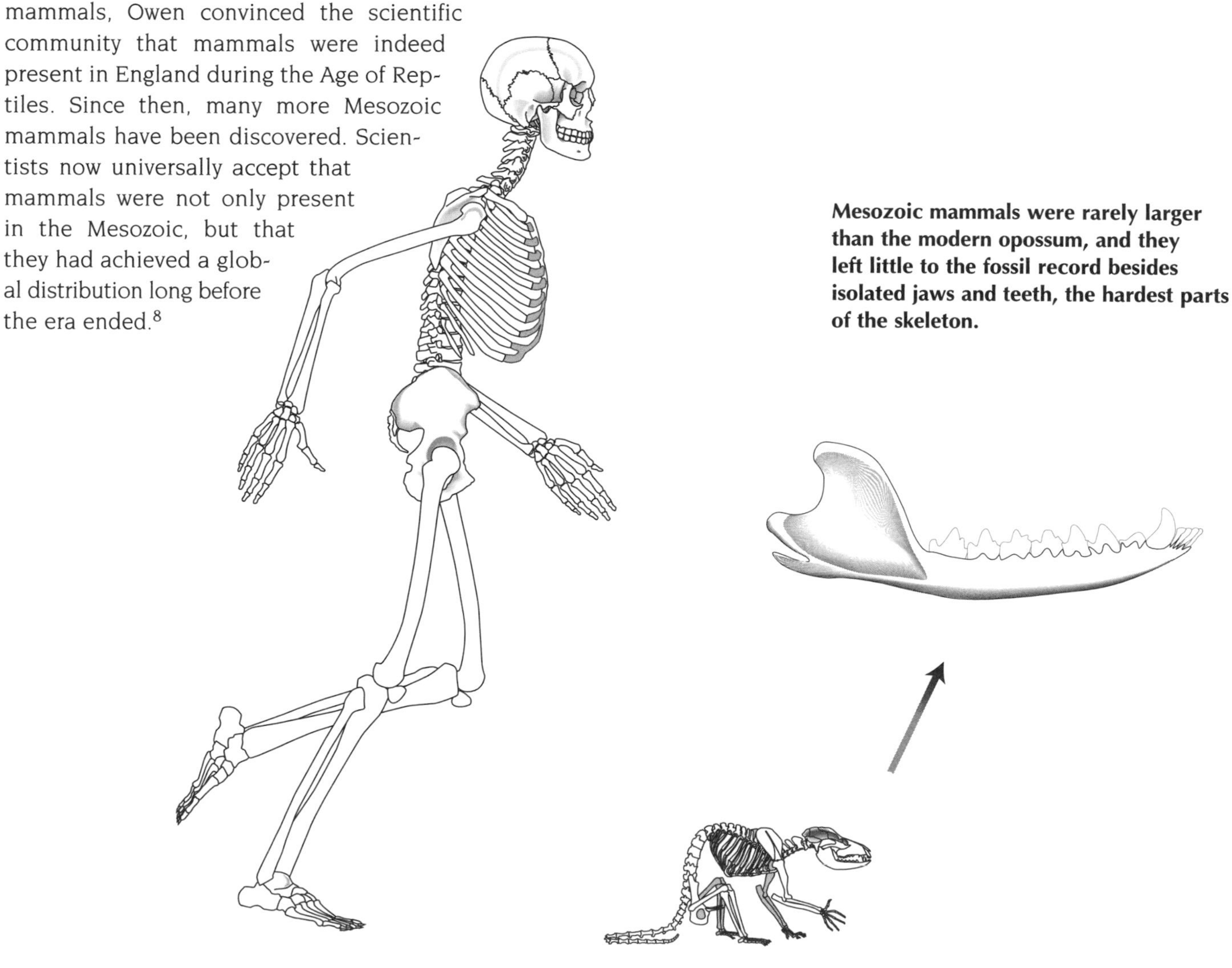

Mesozoic mammals were rarely larger than the modern opossum, and they left little to the fossil record besides isolated jaws and teeth, the hardest parts of the skeleton.

Mesozoic mammals were uniformly small. Most were the size of modern mice and shrews, so their bones were poorly constructed for preservation in the fossil record. But their tiny teeth are much more durable. The mammalian dentition evolved to unparalleled levels of complexity and diversity during the Mesozoic. Teeth are very hard and last long after the rest of the skeleton is destroyed by natural processes. They can even pass through the intestinal tract of a predator and remain identifiable. So mammalian teeth are often fossilized, but their very small size makes them difficult to find. About 30 years ago, a technique known as screen washing was developed to promote the recovery of tiny fossil teeth. Large volumes of sediments are sieved, producing a concentrate rich in tooth-sized particles. The concentrate is then sorted under a microscope to pick out the teeth and other fossils. Although tedious and labor-intensive, screen washing has enabled paleontologists to successfully recover teeth from dozens of Mesozoic mammal species that have now been named.

Some unbelievably rich fossil sites in several parts of the world provided complete jaws, skulls, and skeletons of Cretaceous mammals. These have been especially important for phylogenetic mapping. When only teeth were known, it was difficult to precisely map the relationships among Mesozoic mammals. A recent discovery from Greenland by Farish Jenkins of Harvard University and associates highlighted the problem.[9] Individual teeth of an extinct lineage named Haramiyidae had long been known from Late Triassic and Early Jurassic deposits in many parts of the world. After many years of speculating on what haramiyids looked like and who they were related to, Jenkins discovered a complete skull. The upper and lower teeth of the animal had been mistakenly identified as separate species, based on isolated teeth collected by earlier paleontologists. One species was named for the upper teeth while the other was named for the lower. The skull showed that the two "species" were parts of the same animal. Complete specimens provide a much stronger basis for phylogenetic analysis than teeth alone.

It has long been clear that the Age of Mammals started well before the Age of Dinosaurs ended. But the true diversity of Mesozoic mammals is only now becoming clear as the Mesozoic fossils are plotted on phylogenetic maps. One important assemblage of fossil mammals, dating back about 85 million years, was collected over the last two decades by the late Russian paleontologist Lev Nessov.[10] Without the assistance of a field vehicle, Nessov hitchhiked and walked vast distances while prospecting for fossils in remote regions of the Kyzylkum Desert south of the Aral Sea. He usually collected only what he could carry out with him. As a result, Nessov is one of the few paleontologists who praised Mesozoic mammals for their small size! Although he couldn't collect the dinosaur skeletons he found, he brought a great diversity of beautifully preserved Cretaceous mammals to Saint Petersburg.

Nessov's American colleague David Archibald, who we met in Part I, was the first to plot these fossils on an evolutionary map of mammals.[11] Some fossils preserve characteristics found only within the ungulate lineage. Today, ungulates are highly diversified and include horses, rhinos, tapirs, pigs, deer, elephants, hyraxes, whales, sea cows, and aardvarks. Archibald's phylogenetic analysis placed some of the Kyzylkum fossils at the base of the ungulate line. Before Nessov's discoveries, ungulates were known only from Cenozoic rocks of North America. Most paleontologists thought that ungulates originated in

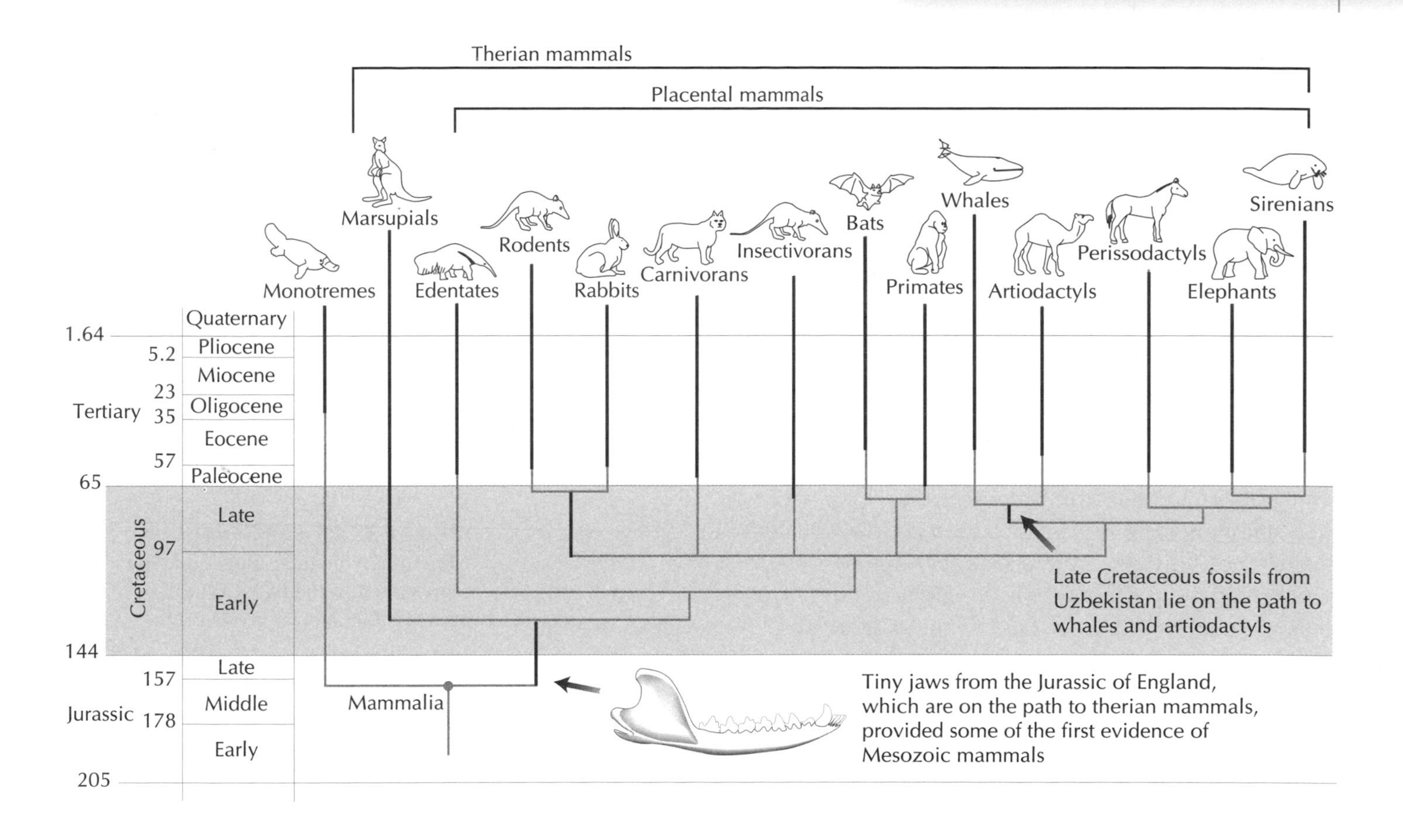

The solid black lines indicate the range of each lineage, based on known fossils. The green lines are ghost lineages that represent the approximate duration in time for each lineage, based on what we know about their relationships.

North America and later spread to Asia. But now our map indicates that the origin and early diversification of ungulates began in Asia during the Cretaceous, where they crossed the K-T boundary. Ungulates later spread to North America via land bridges at the Bering Strait, and by island hopping across the North Atlantic, which was much narrower than it is today.

In mapping Nessov's fossils as Cretaceous ungulates, Archibald drew ghost lineages for many other mammalian groups into the Cretaceous, extending their suspected ranges 20 million years further into the past. Based on current maps, the lineages including modern sloths, rabbits, rodents, tree shrews, bats, and carnivorans had all originated and began to diversify by the end of the Cretaceous. It now looks as though even our own primate lineage originated in the Cretaceous and survived the K-T boundary. We haven't found their Cretaceous fossils yet, but the map predicts that with time and sweat paleontologists may eventually do so. Mapping both fossils and ghost lineages against time contradicts Feduccia's argument—modern mammalian diversity is rooted deeply in the Age of Dinosaurs.

MAPPING CRETACEOUS BIRD DIVERSITY

The Mesozoic fossil record of birds reflects the same problems, only worse. Globally, there are only a few dozen named species of Mesozoic birds. It is difficult to establish the precise estimates due to the fragmentary nature and uncertain identifications of a large percentage of specimens. Shorebirds and waterbirds have the best fossil records, because they live close to potential burial sites. Terrestrial birds must withstand more abrasive transport by streams and rivers. The rarity of terrestrial bird fossils is a global problem that can be best appreciated on a local scale. For example, over the last three decades about 150,000 vertebrate fossils have been collected from rocks spanning the K-T boundary in Alberta and Montana by Berkeley's Museum of Paleontology. Only about 30 are birds, representing five different species.[12] Even more than mammals, our knowledge of Mesozoic birds has been hard-won despite decades of searching. Only now are we beginning to see the acceleration of Mesozoic bird discoveries that began with mammals three decades ago.

Another problem is that important regions of bird phylogeny remain unmapped. This is ironic because birds have been a favorite subject of naturalists for centuries. A vast scientific literature of observations exists about breeding behavior, songs, nesting, migratory pathways, physiology, diet, embryology, and genetics. Many of the smaller roads on the map of avian phylogeny are now charted in some detail, but the connections between some of the major avian highways are highly controversial. So it is hard to determine what unique and diagnostic features were present in the ancestors of many large modern groups. What would the ancestral chicken or duck look like if you discovered it? Today, no one has trouble distinguishing between ducks and chickens, but we have yet to sort out exactly what the basal members of these groups looked like or to unequivocally recognize their oldest fossil relatives.

With these caveats in mind, what can we say about how birds fared during the K-T transition? The best-known Cretaceous bird fossils belong to Hesperornithiformes, Ichthyornithiformes, and Enantiornithes, which we met in Chapter 14. These and several others became extinct at or before the end of the Cretaceous. However, other Cretaceous fossils

(Below left)* The anseriform lineage, which today includes ducks may be represented in the Late Cretaceous by a fossil known as *Presbyornis.

***(Below right)* If the relationships of fossils from central Asian and the Antarctic peninsula are correctly mapped, then the lineage that included today's loons (like this Great Northern Diver) had achieved a global distribution by the end of the Cretaceous.**

found around the world do not belong to any of these lineages, and they document an unsuspected diversity of Late Cretaceous birds. Some have been placed on the evolutionary pathways to living bird lineages, and they offer direct clues about avian survivorship across the K-T boundary.

Expert opinions on precise placements of Cretaceous birds on the avian map vary and are largely untested by phylogenetic analysis. Nevertheless, the consensus view is that four living lineages were present. The first is known as Anseriformes, which today includes about 151 species of ducks, geese, and screamers. Most anseriforms live and feed along the margins of bodies of water where they hunt for small animals and aquatic plants. With thick, waterproof plumage, anseriforms have navigated the world's oceans and waterways and are found over much of the world. Many are migratory, flying thousands of miles each year. Others have dwarfed forelimbs and are flightless. Many oceanic islands have endemic species, which live there and nowhere else. Fossil anseriforms have been found in marine and freshwater deposits throughout the Tertiary. From Late Cretaceous rocks of Vega Island along the Antarctic peninsula, a fragmentary skeleton has been identified as *Presbyornis*,[13] which is otherwise known from well-preserved Tertiary fossils. *Presbyornis* is thought to be either a relative of the ducks or a more primitive anseriform.[14] Another Cretaceous *Presbyornis* specimen is reported from Mongolia.[15] Only the fused metatarsal and ankle bones (the tibiotarsus) of one foot were found.

A second modern lineage whose path begins in the Cretaceous includes the four living species of modern loons—Gaviiformes.[16] Throughout their known history, loons have been waterbirds. Modern loons are foot-propelled divers, whose hindlimbs resemble *Hesperornis* in mechanical design. Loons have pointed bills and torpedo-shaped bodies for efficient swimming. They chiefly eat fish, but also dine on other marine life. Unlike *Hesperornis*, loons retain powerful wings and can fly great distances. With dense, compact

(Below left) This fulmar represents the modern procellariiform lineage—the tube noses. Notice the tube on the upper surface of the bill.

(Below right) This European avocet represents the charadriiform lineage, which unquestionably extends into the early Tertiary and may be represented by Cretaceous fossils.

plumage they winter mostly at sea, living and feeding on the water for months at a time. The oldest fossil loon is *Neogaeornis wetzeli*, from the Late Cretaceous of Chile. This specimen was long mistaken for *Hesperornis*, until the U.S. National Museum's preeminent paleoornithologist Storrs Olson recognized it as a loon. Additional *Neogaeornis* material from Seymour Island, Antarctica, strengthens this identification. A third Cretaceous loon was recently discovered in Uzbekistan, from rocks about 20 million years older than *Neogaeornis*. If this identification is correct, then loons had dispersed widely across the globe long before the end of the Cretaceous.

The third Cretaceous lineage is Charadriiformes, which today includes a vast diversity of birds that spend most of their lives around water. Most modern charadriiforms migrate; most are strong flyers; and many can dive for food. They eat crabs, mussels, insects, fish, snails, lizards, seeds, and sometimes vegetation. The approximately 366 living charadriiform species represent many distinctive lineages and include auks, avocets, coursers, curlews, gulls, murres, oystercatchers, plovers, puffins, sandpipers, skimmers, stilts, terns, and woodcock among others. Not all were present in the Cretaceous, but nearly a dozen species have been named for Cretaceous fossils found in North America and Asia.[17] Owing to their incompleteness, and to revisions of age estimates for the rocks from which many of the fossils have come,[18] these identifications are possibly the most problematic and controversial among all the Cretaceous birds. However, charadriiforms are unequivocally represented by fossils from the first 10 million years of the Tertiary.

The fourth modern lineage represented by Cretaceous fossils is known as Procellariiformes—the tube noses—which includes the 92 living species of albatrosses, shearwaters, and petrels. All of its modern members have hooked, deeply grooved bills with nostrils enclosed in a narrow tube that conveys excess salt secreted by the salt gland. With long, narrow, pointed wings, they are excellent flyers. Some have very wide wings for soaring great distances over water. The wingspan of the albatross reaches 10 feet, the greatest of any living bird. The Cretaceous fossil record of procellariforms is

All living birds belong either the neognath lineage, represented here by the domestic chicken *Gallus gallus,* or to the palaeognath lineage, represented here by the rhea, *Rhea americana.*

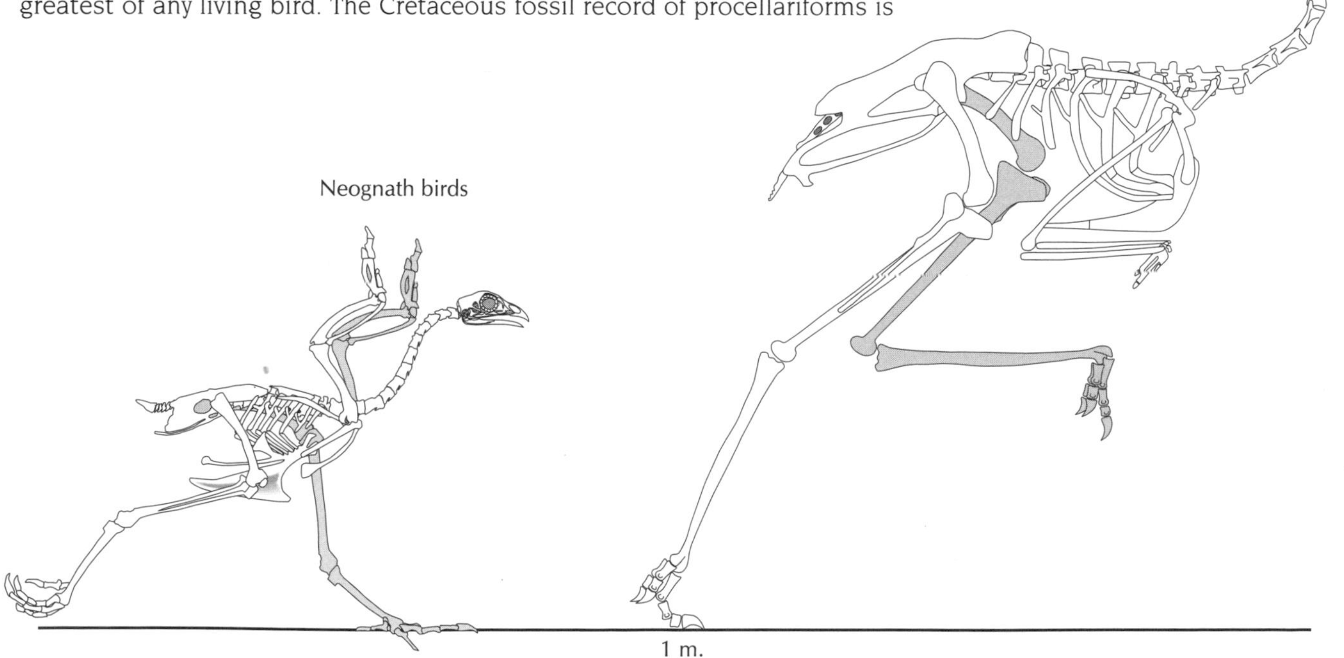

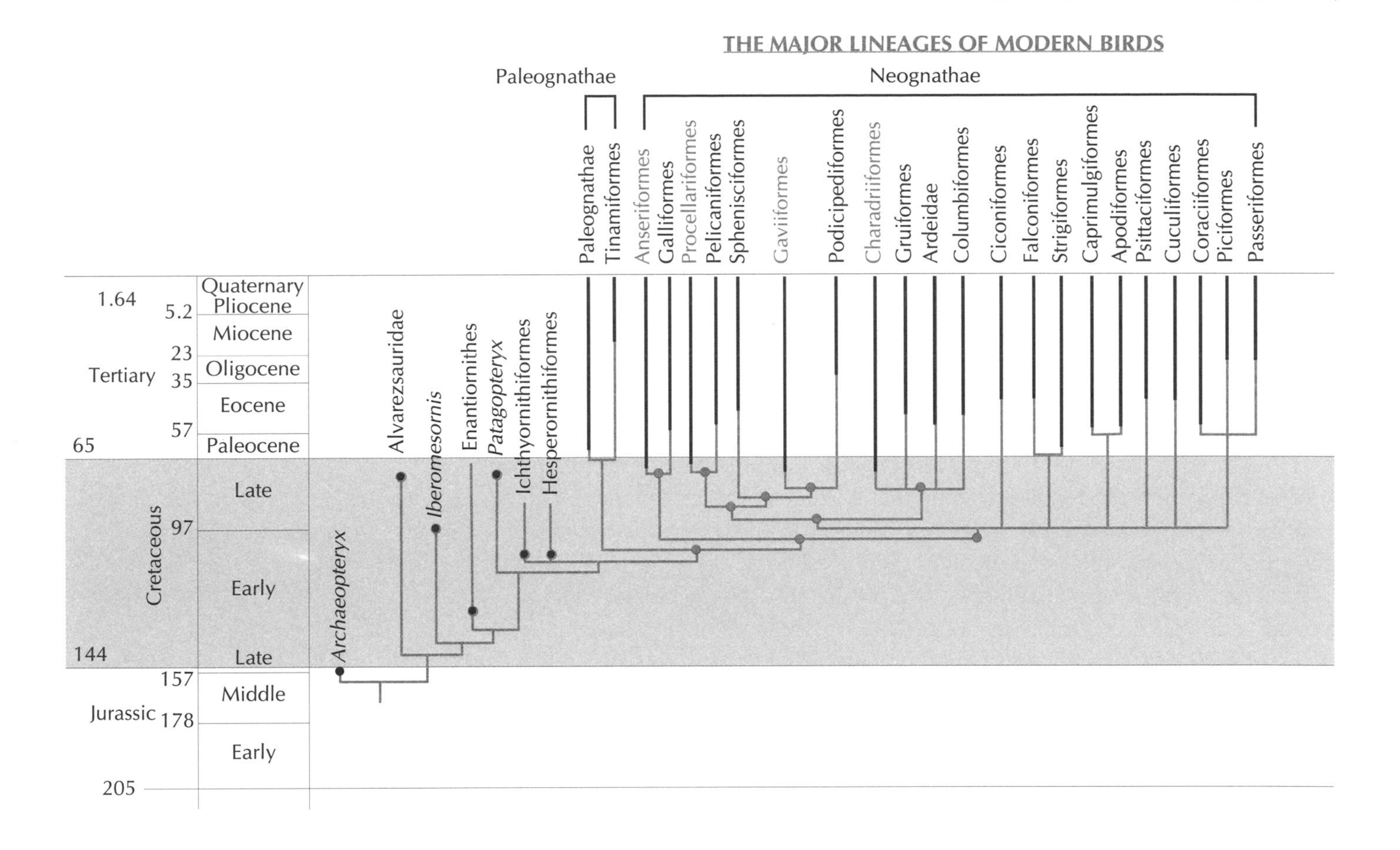

widely acknowledged in the literature. But it seems to rest largely upon fragmentary wishbones from New Zealand and Mongolia.[19]

If this map of relationships is correct, then nearly 20 major living bird lineages may have crossed the K-T boundary, even though very few are yet known from Cretaceous fossils.

Additional modern lineages identified from Cretaceous fossils include cormorants, pelicans, tropicbirds, flamingos, and relatives of chickens and turkeys, as well as others. But, while possible, these identifications are based on even less secure evidence than the four lineages just mentioned.

Adding ghost lineages that connect the known fossil species provides a more accurate estimate of the diversity of dinosaurs that crossed the K-T boundary than counting fossil species alone. Conservatively, if we accept as correct any one of the fossils now identified as anseriform, procellariform, gaviiform, or charadriiform, then at least three major evolutionary paths of living birds were present in the Late Cretaceous. One of these is called Palaeognathae, which includes the modern ostriches and its relatives. The other is its sister lineage, Neognathae, which includes all other birds. Furthermore, the neognath lineage must have split into two distinct lines during the Cretaceous. The first includes the anseriforms plus the galliform birds (chickens, turkeys, and relatives). Its sister lineage includes all other neognathous birds.

If all four of the alleged Cretaceous groups are correctly identified, then an additional dozen or more of today's major living lineages must have originated in the Cretaceous. One lineage includes many of today's large fish-eating birds—the greebs, loons, pelicans, frigate birds, gannets, cormorants, albatrosses, and petrels. From what we know of their relationships, at least three of these must have split off before the loons. Another Cretaceous ghost lineage leads to today's pigeons, doves, cranes, herons, rails, gulls, auks, and other shorebirds. At least two major evolutionary trails had split for these birds during the Cretaceous. Another ghost lineage extending into the Cretaceous leads to the living raptorial birds the hawks, falcons, and owls. Still another leads to living swifts and hummingbirds, and still another leads to living kingfishers, woodpeckers, toucans, and songbirds. It is hard to say how diverse any of these lineages had become in the Cretaceous—only that the stem lineages for a great diversity of living birds were present.

Alan Feduccia contends that the Cretaceous specimens are so incomplete that "regretfully, many must simply be ignored."[20] But despite being incomplete, phylogenetic systematists argue that their positions on the map of life can be generally plotted. After all, they can be identified as birds based on their hollow bones, the fused elements in the foot and ankle, and details of the shapes of other bone fragments. In some cases, they can be placed fairly precisely within Aves, fulfilling the aim of phylogenetic analysis. A major weakness of Feduccia's Phoenix hypothesis is that it fails to account for much of the Cretaceous fossil bird record.

Even if we were to accept Feduccia's claim that most of the Cretaceous birds are too fragmentary to identify accurately, still other new mapping techniques trace the ghosts of many living bird lines into the Mesozoic. Alan Cooper of Victoria University, New Zealand and David Penny of Massey University, New Zealand examined the question of bird survivorship across the K-T boundary, by comparing DNA sequences from pairs of modern birds whose relationships were uncontroversial, using the differences in sequence as a proxy for the rates of evolutionary divergence.[21] They calibrated the speed of evolutionary change using a well-established early Tertiary fossil record for the lineages. The only Cretaceous identification Cooper and Penny accepted was that of a loon, and even this identification was not critical to their results.

From these calibrated speed estimates, Cooper and Penny calculated the minimum divergence time of the two members for each pair. By combining distantly related pairs of birds into quartets, they estimated the minimum divergence times between the pairs, and further refined age estimates for the original pair-wise comparisons. These were conservative estimates because if errors were made regarding relationships—if the pairs were more closely related than they thought—it would result in an artificially young divergence date. In addition to the lineages listed above, Cooper and Penny estimated that the parrot lineage and the passeriform lineage, which include several thousand living species of perching birds and songbirds, were both present in the Cretaceous and survived to the present.

AT THE BOUNDARY

When all the available evidence from ghost lineages is taken into account, it looks like both birds and mammals had diversified and become globally distributed by the end of the Cretaceous. But what happened at the K-T boundary? Was there a catastrophic loss of diversity?

Let's look first at the record for mammals. As we saw in Chapter 7, only a few localities preserve vertebrate-bearing sedimentary rocks just above and below the K-T boundary, and all are in western North America. The most complete sequences of terrestrial rocks leading up to the boundary are in Montana and southern Alberta, where we can track species that were fossilized in rocks of the Judith River Formation into the overlying Hell Creek Formation. These two formations intermittently span the last 10 million years of the Cretaceous.

Dave Archibald and William Clemens have shown that in this region, the marsupials, which today include the kangaroos, opossums, and other pouched mammals, were hardest hit. Only one of eleven marsupial species present in the Cretaceous survived into the Tertiary. But Archibald also points out that marsupial fossils are rare, so sampling artifacts may significantly inflate the severity of loss.[22] Marsupials disappeared altogether from North America very early in the Tertiary, but elsewhere marsupials evolved into about 259 living species, plus numerous extinct Tertiary species that have been found in many parts of the world. Given this success elsewhere, it is doubtful that the K-T loss of marsupial diversity in North America was representative of the rest of world, but without more evidence we can only speculate.

Rodentlike multituberculates were present in the Late Jurassic and Cretaceous, and had a global distribution. In North America, only 50 percent of the Cretaceous multituberculate species survived into the Tertiary. Locally at least, this group also took a big hit, but elsewhere in the world, the Tertiary record of multituberculates is not well known. So the loss at the boundary in Montana may or may not represent global diversity trends. Multituberculates survived in the Northern Hemisphere for 20 million years after the K-T extinctions before finally disappearing in the mid-Tertiary.

The placentals comprise about 93 percent of modern mammalian diversity. By Archibald's estimates they were not affected. We noted earlier that all six placental species found in the Cretaceous survived into the Tertiary. Based on what is known of placental relationships in general, there must have been a still greater diversity of placentals present elsewhere in the world.

Overall, in western North America as many as 55 percent of the mammalian species may have become extinct, but it is impossible to say that this number is representative of the world. It is clear that mortality was highly selective and varied from lineage to lineage. Available data indicate that placentals, which spawned the vast majority of post-Cretaceous mammalian species, were largely unaffected at the K-T boundary.

How did Dinosauria fare at the K-T boundary? Most paleontologists agree that the nonavian dinosaurs all became extinct at or before the end of the Cretaceous, despite occasional claims to the contrary.[23] The evidence for their post-Cretaceous survival consists almost entirely of isolated teeth from rocks as much as 10 million years younger than the boundary. But all the physical evidence points to these rare teeth being reworked from older Cretaceous deposits. In every case, the teeth consist of isolated, worn, and broken specimens with abrasions that record an extensive history of transport by water. Reworking is a common geological process, and dinosaur teeth are the densest and most durable parts of the skeleton. As the younger Paleocene streams cut down into underlying Cretaceous sediments, they occasionally exhumed older remains, breaking them up and abrading them by tumbling the pieces further downstream before their second burial. The alleged Paleocene teeth are always more worn than the teeth collected directly from Cretaceous rocks, and no articulated skeletal parts have been found in Paleocene rocks. Without stronger evidence, it looks like all the nonavian dinosaurs were extinct by or before the end of the Cretaceous.

Was their extinction rapid or gradual? Many paleontologists have now spent numerous field seasons trying to answer questions about changes in dinosaur diversity in North America. The pace of discovering new species is now slowing, and the current picture is more complete than for any other region. Despite the completeness, there are conflicting interpretations.

As noted earlier, David Fastovsky of the University of Rhode Island and Peter Sheehan of the Milwaukee Public Museum interpreted their observations of the latest Cretaceous Hell Creek Formation to be consistent with a geologically instantaneous extinction of all the nonavian dinosaurs, along with 88 percent of the terrestrial tetrapods found in the Cretaceous.[24] However, Archibald and Laurie Bryant (now at the Bureau of Land Management) measured dinosaur diversity by surveying a broader time span, combined with a more fine-scaled stratigraphic sequence, and came to a different answer.[25] They tabulated a minimum of 32 species of non-avian dinosaurs in the Judith River Formation, but in the overlying Hell Creek Formation there are only 19 species—a 40 percent decline in the diversity of nonavian species during the 10 million years leading up to the K-T boundary. The proportion of this decline was greatest among ornithischians, with roughly 66 percent of these species disappearing before the end of the Cretaceous. But only about 25 percent of the nonavian saurischians disappeared during this 10 million year interval. There is evidence to suggest that the 19 nonavian dinosaur species from the Hell Creek died out in a very short time near the boundary. So, in western North America, the end for some species of nonavian dinosaurs may have been relatively quick, but their numbers had apparently been waning for several million years before the end of the Cretaceous. Whether this pattern holds for the rest of the world is uncertain. Globally, there are approximately 120 named species of nonavian dinosaurs that can be accounted for during the last 10 million or so years of the Cretaceous,[26] but we do not know exactly when within this interval most of them died out.

Among avian dinosaurs, there simply were not enough specimens represented in the Berkeley collections for Archibald and Bryant to meaningfully estimate changes at the boundary. On a global scale, the Hesperornithiformes and Ichthyornithiformes were evidently gone 5 million years before the end of the Cretaceous. Feduccia's picture of a catastrophic extinction at the end of the Cretaceous is distorted because he failed to account for significant losses of avian diversity leading up to the boundary. Alvarezsaurids, Enantiornithines, and a few others made it to the end of the Cretaceous, or nearly so, but no further. But despite the losses leading up to and at the boundary, a dozen or so separate lineages of living birds trace their origins back across the K-T boundary, and must have survived whatever happened at the end of the Cretaceous. The available data suggest that the adaptive radiation of birds was well under way before the Mesozoic ended.

THE BOTTOM LINE

From a modern phylogenetic perspective, dinosaurs not only crossed the K-T boundary, they survive today in great abundance. But even though this aspect of dinosaur history has radically changed, some other things have remained the same. As in the nineteenth-century, many implications of the theory of evolution have been overlooked by the proponents of the theory of homoplasy. Those scientists still reject that birds are related to dinosaurs, arguing that the similarities in both ontogeny and phylogeny reflect convergent evolution. However, they offer no well-founded alternative hypothesis of relationship. If birds are not dinosaurs, then where do they fit on the tree of life? Although claiming to be evolutionists, these scientists use methods that were championed a century ago by anti-Darwinians seeking to discredit the theory of evolution. Without a body of supporting anatomical or genetic evidence, saying that birds evolved independently is little different from saying that they were separately created.

There are similarities in the patterns of diversification for birds and mammals. But the patterns measured with phylogenetic maps suggest that both lineages originated long before the Mesozoic ended and steadily increased their diversity far into the Tertiary. The jury is still out in deciding exactly what happened at the K-T boundary. More meaningful estimates of changes in diversity across the K-T boundary will only come with path-by-path analyses, once more fossils are known, and as the individual evolutionary trails and ghost lineages are mapped in greater detail.

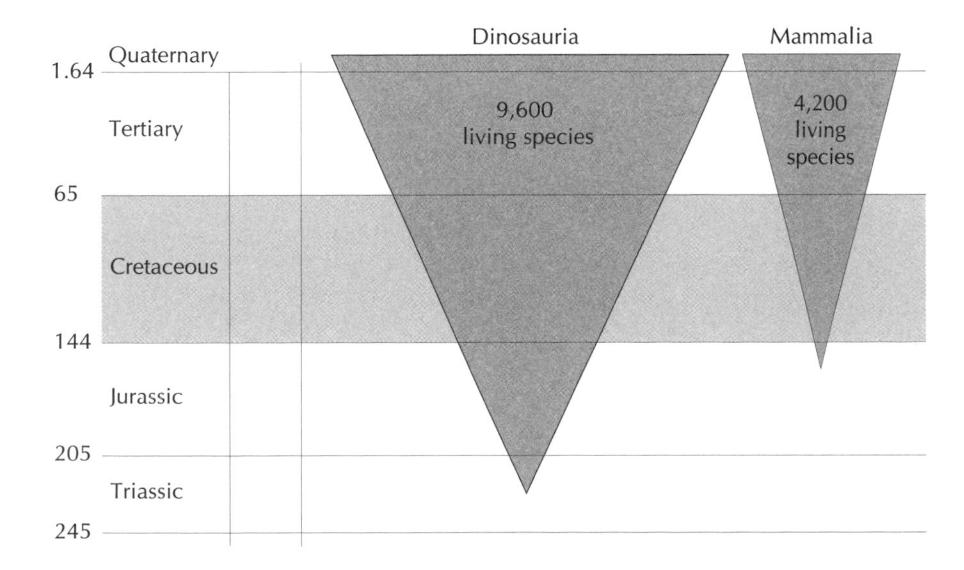

From the general patterns of their histories of diversification, it is evident that the Age of Mammals played out its entire history in the shadow of the dinosaurs.

Alan Feduccia wrote an influential book whose title, *The Age of Birds*, conveys the important point that birds have outnumbered mammals throughout the Cenozoic. But he overlooks the fact that, from the Triassic to the present, dinosaurs have *always* been represented by more species than mammals. The Age of Dinosaurs is not over yet.

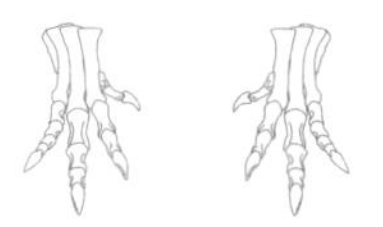

CHAPTER SIXTEEN

Diversification and Decline

The period from the mid-Cretaceous to the end of the Tertiary witnessed an unprecedented diversification among vertebrates, as the world became populated by thousands of species of birds. One recent encyclopedic treatment by Charles Sibley and Burt Monroe of Yale University lists 9,702 living species—about ten times the number of known Mesozoic dinosaurs.[1] However, archeological research is revealing that today's avian diversity is well below its all time peak, and that several thousand more species were alive just a few thousand years ago. But before examining the portentous recent losses, we will survey the historic diversification of birds and investigate the circumstances behind the greatest exploitation of the land and sky by any vertebrate lineage. An appreciation for the history of diversification of living birds will help to clarify a series of mass extinctions possibly more severe than the one at the K-T boundary for the dinosaur lineage.

ISOLATION AND DIVERSITY

Multitudinous organisms populate the world today, most of whom originated through different modes of speciation. Speciation is the great producer of diversity. In the case of birds and most other tetrapods, speciation involved geographic isolation of small populations followed by reproductive isolation. All birds reproduce sexually, like most other vertebrates, and this helps drive diversification. Apart from identical twins, every individual is different, with its own complement of DNA, and every individual has its own unique history in time and space. But as long as the members of a population or different populations can potentially interbreed, gene flow among them has a homogenizing effect from generation to generation. Over time, the boundaries of variability may grow, shrink, or drift, yet we are still left with a single species.

But if two populations become geographically isolated from each another in a way that prevents gene flow, each population has a smaller gene pool than before. Over time, the populations become increasingly different. Through natural selection and genetic drift, descendants are modified in response to their hereditary and environmental circumstances, as genes are mixed from generation to generation. With sufficient time, the populations diverge to the point that, even if they were to again share overlapping territory, they would probably not successfully interbreed. Where before there was only one species, there are now two. Each has its own distinct geographic range, and these ranges do

The islands and archipelagos of the western Pacific Basin were the most prolific sites for the adaptive radiation of birds over the last 100 million years.

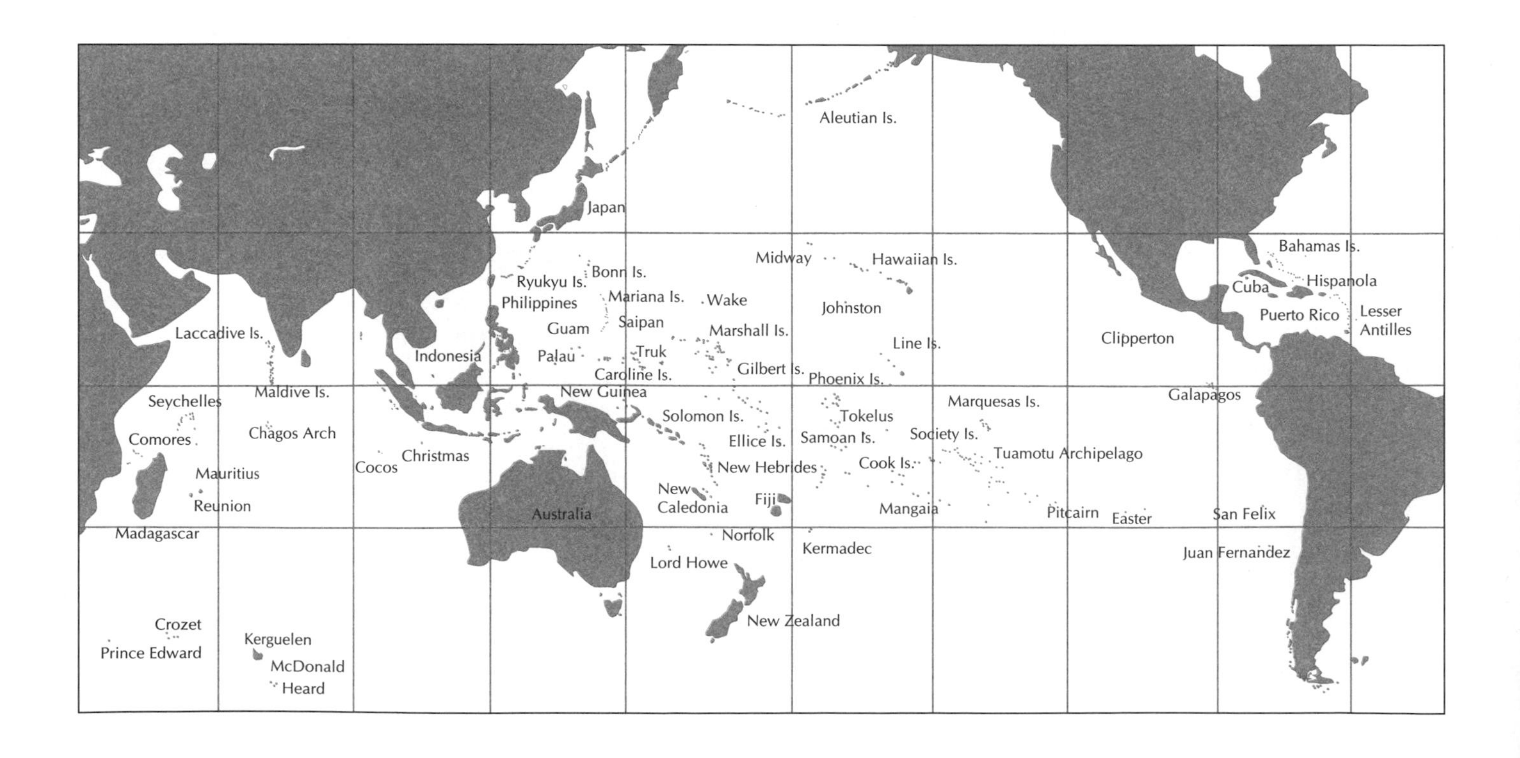

not usually overlap. The diversification of birds is dominated by patterns indicating a history of geographic isolation and speciation. In turn, this history is intricately tied to the dynamic history of the Earth's crust and climate.

The most fertile regions for the adaptive radiation of new bird species have been the islands of the western Pacific and Indian Oceans, as well as the forests of the Andes and great Amazon Basin. Other regions around the Pacific rim, like western North America and southeastern Asia, were also great centers of diversification. Many of these regions have endemic bird species—native forms confined to a small region. But they also host migratory and wide-ranging species. Each island or patch of forest can host its own unique assemblages of fauna and flora.

In the first part of the book, Cretaceous volcanism was implicated as a possible cause for K-T extinctions. This same volcanism may also have been responsible for much of the avian diversity that evolved on both sides of the K-T boundary. Recall that heat emanating from the Earth's core generates convective plumes of magma that rise through the mantle and disrupt the overlying crust through tectonic movement. Between 120 and 125 million years ago, a superplume of hot magma boiled away from near the Earth's molten outer core, rising through mantle below what is now the western Pacific. As the superheated upwelling approached the surface, the most volatile components erupted through the ocean floor, creating submerged volcanoes and mountain ranges that spread over much of the Pacific basin.[2] The tips of some of these peaks are the seamounts and volcanic islands of Polynesia, Melanesia, Micronesia, the Malay Archipelago, and others of the region. Many of these islands have their own unique species of birds.

Oceanic islands typically emerge above the waves rather barren and lifeless. But with time, they provide unique opportunities for colonization. Lichen and fern spores reach their shores quickly with the wind. Their roots churn organic material into the volcanic rubble, and soils eventually accumulate. Many islands develop a rich substrate that sustains great floral diversity. And with the base of the food chain established, the way is paved for the influx of herbivores and carnivores.[3] From the mid-Cretaceous onward, islands began to emerge from the waves, like new condominiums awaiting their first residents.

There are numerous ways for tetrapods to cross saltwater barriers and reach emerging islands, but the happenstance process of island colonization means that each island may be unique. A few species can withstand the trip through salt water and float or swim to the islands that they colonize. Tortoises probably floated to the island chains, like the Galápagos and Aldabra islands, where they evolved to giant size and each island became home to its own distinctive population. Spiders diversified enormously on islands. As tiny juveniles, they can "balloon" over vast distances on air currents with a strand of spun silk. Spiders have been collected at altitudes above 30,000 feet in the aerial plankton by planes towing netlike windsocks. Insects hitch rides with birds and other travelers. Bats also fly onto islands, carrying parasites and seeds of the fruits they eat. Seals and sea lions swim to islands. Rafts of branches, logs, or vegetation mats wash out to sea with storm tides, complete with microorganisms, insects, and small vertebrates. But few mammals or lizards, and virtually no amphibians have successfully crossed wide saltwater barriers, without human help. Birds have been far more successful, flying or swimming out to fish in enriched coastal waters or island lakes.

A continental island is one that separated from its parent continent by tectonic forces or rising sea levels, isolating a sample of the biota found on the mainland. Madagascar is a continental island, whose isolation began nearly 70 million years ago, producing fantastic variations on African species present at the time of separation. Whole continents have become islands, following the break-up of Pangaea during the Mesozoic. South America was an island for much of the Tertiary, until it became tied to North America at the Isthmus of Panama. India was also an island for tens of millions of years, before colliding with Asia and creating the Himalayas. Australia still is an island. During isolation, each continent evolved a unique assemblage of bird species.

Tectonic activity on the eastern side of the Pacific basin catalyzed biotic effects via the formation of the Andes of South America and the Sierra Nevada of North America. Increased elevations along the western edges of these continents disrupted more uniform habitats and generated greater habitat diversity. The shallow Cretaceous seas that had covered North America and Asia receded, opening new lands to pioneering species. Novel terrestrial communities arose as the seabed emerged. In the forests and mountains, species could become isolated as a fire or a river shifting its course separated previously continuous populations of organisms in two. Many tropical forest birds will not cross open water even though they have the strength to do so. Vegetation and rainfall shifts due to elevation changes also partition the environment into different islandlike zones and foster speciation among birds. From the mid-Cretaceous onward, the crust in and around the Pacific basin presented a huge, rapidly changing landscape that became the world's greatest center of bird diversification over the next 100 million years.

DIVERSITY RISING

By far the greatest adaptive radiation was among the small terrestrial birds in forests and woodlands on continents and islands. But waterfowl, shore birds, and wading birds also became highly diversified, leaving the best Tertiary fossil record. Many shorebirds nest in great colonies, and some spectacular fossil localities have produced tens of thousands of their bones. But overall, the Tertiary pace of evolution among the seabirds has been slower than that of their woodland cousins, possibly because water birds are generally strong and wide-ranging flyers. The phylogenetic map for modern birds is still under construction, and there is a big knot of unresolved relationships among several of its major pathways. Some major lineages have proper names, which we will introduce. For the rest, we will simply number them, and briefly describe their diversity and history.

PALEOGNATH BIRDS

Two major evolutionary pathways trace back to the ancestral species of Aves.[4] These are Palaeognathae and Neognathae, which together contain all living birds.[5] Palaeognathae includes the ratite (Ratitae) lineage—the Ostrich, which is the most massive and tallest of living birds, and ten related living species,

***(On the facing page)* Phylogeny of modern birds. All living species belong to either the neognath or the paleognath lineages. Among neognaths there are twelve distinct pathways, but the relationships among these are not well established.**

Aves

Paleognathae

Neognathae

TECHNICAL NAME	COMMON NAMES OF MEMBERS
Ratitae	Cassowary, Emu, Kiwi, Ostrich, Rhea
Tinamiformes	Tinamou
Anseriformes	Ducks, Geese, Screamers, Swans
Galliformes	Chachalacas, Chickens, Curassows, Grouse, Guans, Guineafowl, Megapodes, Pheasants, Quail, Turkeys
Procellariformes	Albatrosses, Diving-petrels, Shearwaters, Storm-petrels
Pelicaniformes	Cormorants, Frigate birds, Gannets, Pelicans
Sphenisciformes	Penguins
Gaviiformes	Loons
Podicipediformes	Grebes
Ciconiformes	Adjutants, Flamingos, Hammerheads, Ibises, Jabiru, New World Vultures or Condors, Openbills, Storks, Shoebills
Gruiformes	Bustards, Coots, Cranes, Gallinules, Kagu, Mesites, Limpkins, Monias, New World Sungrebe, Rails, Roatelos, Seriemas, Sunbitterns, Trumpeters
Ardeidae	Herons, Egrets, Bitterns
Columbiformes	Pigeons, Doves
Charadriiformes	Auks, Avocets, Coursers, Crab plovers, Curlews, Godwits, Gulls, Jacanas, Jaegers, Lapwings, Lilly Trotters, Murres, Oystercatchers, Painted-snipe, Plains-wanderers, Phalaropes, Plovers, Pratincoles, Puffins, Sandgrouse, Sandpipers, Seedsnipe, Sheathbills, Skimmers, Skulas, Snipe, Stilts, Stone-curlews, Terns, Thick-knees, Woodcock
Falconiformes	Caracaras, Eagles, Falcons, Old World Vultures, Osprey, Secretary bird
Strigiformes	Owls
Coliidae	Mousebirds
Tournicidae	Buttonquail
Musophagidae	Turacos, Plantain-eaters
Caprimulgiformes	Goatsuckers, Nighthawks, Nightjars, Oilbirds, Potoos, Whip-poor-wills
Apodiformes	Humming birds, Swifts
Psittaciformes	Lorys, Lorikeets, Cockatoos, Black-cockatoos, Fig-parrots, Racquet-tails, Rosellas, Parakeets, Lovebirds, Macaws, Parrotlets, Parrots
Cuculiformes	Old World Cuckoos, Coucals, American Cuckoos, Hoatzin, Anis, Guira Cuckoo
Coraciiformes	Bee-eaters, Hoopoes, Hornbills, Kingfishers, Motmots, Rollers, Scimitar-bills, Todies, Trogons
Piciformes	Barbets, Jacamers, Piculets, Puffbirds, Toucans, Woodpeckers, Wrynecks
Passeriformes	Anis, Antbirds, Babblers, Birds of paradise, Blackbirds, Bowerbirds, Bristlehead, Broadbills, Bulbuls, Buntings, Cardinals, Chowchillas, Cotingas, Coucals, Creepers, Crows, Cuckoos, Cuckooshrikes, Currawongs, Dippers, Drongos, Fairy-bluebirds, Finches, Flowerpeckers, Flycatchers, Sharpbill, Gnateaters, Grasswrens, Hawaiian honeycreepers, Hoatzin, Honeyeaters, Hornbills, Humming birds, Jays, Larks, Lyrebirds, Magpies, Manakins, Mockingbirds, Nuthatches, Orioles, Ovenbirds, Parrots, Pipits, Pittas, Plantcutters, Roadrunners, Scrubbirds, Sittellas, Sparrows, Spiderhunters, Starlings, Swallows, Tanagers, Tapaculos, Thickheads, Thrushes, Thrashers, Titmice, Treecreepers, Sunbirds, Vireos, Wagtails, Shrikes, Vangas, Warblers, Wattlebirds, Waxbills, Waxwings, Weaverbirds, White-eyes, Woodcreepers, Woodswallows, Wrens

plus a host of extinct relatives. All ratites are flightless, terrestrial birds, with reduced forelimbs, huge egg-shaped bodies, powerful legs, and small heads perched atop long necks. The islands of New Zealand, Madagascar and the island continents of Australia and South America have each fostered endemic ratite species. Ornithologists long debated whether the ancestor of ratites could fly. Judging from their position on the evolutionary map of dinosaurs, we can see that all of the closest living relatives of ratites can fly, as did a host of their closest extinct relatives. Ratites have the fused wrist and hand and short tail of other ornithothoracine birds, the back-turned pubis of maniraptorans, the long neck of saurischians, and so on. This hierarchy of evolutionary features suggests that the ancestral ratite could fly, and that its descendants became flightless secondarily.

The tinamous (Tinamiformes) form a second pathway on the paleognath lineage, and their history reflects an adaptive radiation throughout varied grass and woodland habitats. The 43 living tinamou species are confined to the region from southern Mexico to Patagonia.[6] Given that ratites can't fly and that the flying tinamous are restricted to South America, it is difficult to explain the geographic distribution of the living palaeognath species as a result of dispersal. We should expect to see fossils in North America and Asia, documenting their passage between Africa, South America, and the Australian region. Despite several false alarms, no such fossils have been found. The geographic pattern of ratite distribution resembles that of many insects and plants—a Gondwanan distribution—where formerly continuous populations became fragmented and separated by the Mesozoic breakup of the southern supercontinent Gondwana. Geography provides additional evidence of the great antiquity of the palaeognath lineage.

The palaeognathous birds include the ratites, like this Ostrich, plus the tinamou lineage, which is not shown.

NEOGNATHS

All other birds are members of the neognath lineage. Our current phylogenetic map depicts twelve major living lineages that trace back to the ancestral neognath. The relationships among these major pathways are not well understood, however. In many cases the individual lineages are highly distinctive. Parrots, for example, are rarely confused for some other type of bird, because a large gap separates parrots from all the others. The oldest parrot fossils are from the Eocene, about 50 million years ago, but they are already distinctly parrotlike.[7] This points to an even earlier time of origin, but no known fossils bridge that gap. Older, more primitive fossils of parrots, as well as the other eleven neognath pathways, are needed to fully resolve the map for this part of avian history.

Pathway 1 includes two sister lineages. We met Anseriformes, the ducks and geese, in the last chapter. Its sister lineage, Galliformes, represents a radiation into the forests, brushlands, and grasslands, of the world. There are 282 living species,[8] who mostly spend their lives on the ground where they hunt seeds, fruits, berries, insects, worms, and small vertebrates. Their flight is powerful but not prolonged. The oldest galliform fossils are from the Early Eocene of Wyoming.[9] However, their sister lineage, Anseriformes, includes the Cretaceous species *Presbyornis*. If *Presbyornis* is correctly mapped as a

This nearly extinct Prairie Chicken is a representative of Galliformes, a terrestrial bird lineage with nearly 300 living species and a global distribution.

The White Pelican *(at right)*—a member of Pelicaniformes—which includes some of the largest living birds.

The King Penguin, *(below right)* a member of Sphenisciformes. Nearly the entire history of the penguin lineage has been played out in the Southern Hemisphere.

member of Anseriformes, then the entire galliform line must extend back into the Cretaceous as well.

Pathway 2 leads to four major branches of birds that live in and adjacent to water. The pelicans (Pelicaniformes) are mostly wide-ranging birds that inhabit temperate and tropical zones. Also belonging to this lineage are frigate birds, cormorants, and gannets. Pelicaniformes are generally distinguished by long bills that support a pouch of skin that hangs below the jaw and neck for carrying food to their young. Nearly every species in the lineage is a strong flier with large wings. Pelicans are among the largest living birds, with bodies nearly 6 feet in length and a wingspan close to 10 feet. There is one flightless member of the lineage, the Galápagos Islands Cormorant, which has dwarfed wings and lives along the island coasts. A fossil cormorant was reported from Late Cretaceous sediments of southern Mongolia, but the identification of this specimen has yet to be verified.[10] Fossils are abundant from the Eocene onward.

The 17 species of living penguins (Sphenisciformes) represent another trail on pathway 2.[11] Penguins hunt the southern oceans for fish, squid, and shrimp, "flying" with powerful short wings through the water with spectacular dexterity. The earliest fossils are from the Eocene of Seymour Island, Antarctica, and the history of the penguin lineage appears to have played out mostly in the Southern Hemisphere. Two other trails branching from pathway 2 are the loons (Gaviiformes), who we met in the last chapter, and their cousins the grebes (Podicipediformes). Twenty-one grebe species can be found today worldwide, including several island endemics. The Colombian Grebe is flightless and its geographic range is, or was, confined to a few ponds and lakes in the Andes. This species is probably now extinct.

Pathway 3 (Ciconiformes) includes the storks, ibises, shoebills, adjutants, jabiru, condors, and several others. The lineage comprises approximately 130 species worldwide, and most of its members dine voraciously on small vertebrates, mollusks, insects, crustaceans, and carrion. Flamingos, whose kinship to the rest has been questioned, use specialized bills to filter blue-green algae and diatoms from the water.

Pathway 4 includes a seemingly eclectic assemblage of pigeons, herons, cranes, and shorebirds, and the interrelationships among these are uncertain. We have already met the shorebirds (Charadriiformes)—the gulls, auks, and their relatives. The crane lineage (Gruiformes) includes 197 species of predators that stalk shores, shallows, forest floors, and grasslands of the world. Larger cranes can soar and undertake long migrations, but several flightless endemic species have evolved on Pacific islands. From Paleocene rocks in South America, we can trace a lineage of predatory birds that ruled the top of the Tertiary food chain. Some could fly, but other giant forms were flightless with reduced wings. One of the largest was *Titanis*, a 10-foot-tall form with a huge skull, a long massive beak, and short wings. It originated in South America and migrated overland to North America about 3 million years ago, as the two continents drifted together and the Isthmus of Panama formed. A second lineage of giant flightless birds—Diatrymidae—is found in early Tertiary rocks of the northern continents. *Diatryma* is its best known member, standing over 6 feet tall and weighing close to 300 pounds. It is generally considered to have been predaceous, but it has recently been depicted in controversial reconstructions as a scavenger or browsing herbivore.[12] Also found along this part of pathway 4 are the rails. Ecologists joke about the "rail rule": seemingly

The Crested Grebe, a member of Podicipediformes.

A Bustard, a member of Gruiformes.

The Wood Ibis, a member of Ciconiformes.

***(Above left)* The Little White Egret is a member of Ardeidae.**

***(Above right)* The Crowned Pigeon—a columbiform bird—is part of a global radiation that has left more than 300 living species.**

every island in the South Pacific has its own distinct species of rail. Early rails flew from island to island, eventually colonizing nearly all of them. Being naturally weak and reluctant flyers, the colonists on many islands became flightless. Food and nesting space around the island's shores were in adequate supply, and predatory mammals and reptiles from the mainland were generally lacking. Each colony became isolated from the others and gene flow among them ceased. Archaeological excavations on islands across Pacifica are turning up new species of extinct rails at an astonishing pace.[13]

Two more adaptive radiations of woodland birds can be traced along pathway 4. One includes the herons, egrets, and bitterns (Ardeidae), with 65 living species that hunt the woodlands and waterways of the world. Among these are several endemic island species. The other woodland radiation includes the 313 living species of pigeons and doves (Columbiformes). Many endemic islands species are found among these as well, including flightless forms like the extinct Dodo. The relationships among all these diverse parts of pathway 4 are unresolved. Most ornithologists agree that the cranes and herons are close cousins, and some believe that storks are related to them as well. Whether the shorebirds are part of this lineage is controversial.

Pathway 5 includes the raptorial owls (Strigiformes) and the eagles, falcons, and osprey (Falconiformes). They are among today's most consummate hunters, sustaining a theropod tradition of predatory excellence. Both lineages have a worldwide distribution that includes endemic

The skeleton of *Diatryma gigantea,* a giant extinct gruiform bird from the Eocene of South America.

Steller's Sea Eagle *(top right)* is a member of Falconiformes, which includes some of today's greatest hunters.

The Plantain-eater *(below)* is a member of Musophagidae, a modest radiation of African woodland birds.

The mousebirds (Coliidae) *(above)* represent another modest radiation of African woodland birds.

The Coquette Hummingbird *(at left)* is a member of Apodiformes.

island species like the Bare-legged Owl of Cuba and the New Guinea Eagle. All have short, hooked beaks and strongly hooked talons, with which they prey on other birds, and other small vertebrates. Some hawks are excellent runners. Spotting their prey from the air, they land and give chase through the undergrowth and on foot before leaping and grabbing the prey with their talons. Falcons are the fastest dinosaurs, diving at speeds up to 175 miles per hour, typically raking their prey by diving and tumbling it out of the sky. The fossil record for pathway 5 extends back into the early Paleocene.

Pathways 6, 7, and 8 include still more woodland bird lineages, who made modest radiations into the woodlands and brushlands of the Old World.[14] These are the turacos and plantain-eaters (Muscophagidae), who live along dense forest edges and in wooded areas near water in sub-Saharan Africa. Their fossil record extends back to the Eocene. There are six species of mousebirds (Coliidae), all living around brushlands and forest edges south of the Sahara. Mousebirds are good flyers but only over short distances. When foraging for fruit, seeds, and insects, they climb through trees and brush using their feet and bill. Their fossil record also extends back to the Eocene. The last radiation includes seventeen species of buttonquail (Turnicidae), which inhabit grasslands and brushlands around the western Mediterranean, much of Africa, southern Asia, Malaysia, Australia, and adjacent islands. Their meager fossil record extends only to the Pliocene, although it must be a far more ancient line.

On pathway 9 are the swifts and hummingbirds (Apodiformes), and the nighthawk lineage (Caprimulgiformes). Both are marvelously diverse lineages of woodland and open range birds. There are 103 species of swifts worldwide. Once again we find some species that are wide-ranging, and others that are endemic to South Pacific islands. Swifts have long pointed wings and perform with great aerial dexterity, capturing all their food—chiefly insects—on the wing. Hummingbirds form a dazzling array of 322 species, most of whom are concentrated in northern South America,[15] where they drink nectar and catch insects in the flowers of the forest and brushlands. The Puerto Rican Emerald hummingbird and the Jamaican Streamertail are among several island endemics. The Cuban Bee Hummingbird is the smallest known dinosaur, weighing about 6 grams. Nightjars, nighthawks, frogmouths, owlet-frogmouths, and potoos comprise the 105 species of night birds.[16] They are often allied with owls, but their affinities are highly controversial. Nightjars and nighthawks include both wide-ranging species that inhabit forests and brushlands, and endemic island species. Frogmouths and owlet-frogmouths live in brushlands and forests around Australia and the Indian Ocean. The potoos and Oilbird live in central Mexico and South America, in forested or semiforested regions. The fossil record for this pathway extends back to the Eocene.[17]

These macaws are members of Psittaciformes, which radiated extensively through the tropical and subtropical forests of the world.

Pathway 10 includes the parrots (Psittaciformes), which today comprise a spectacular radiation of woodland birds. Today there are 360 species of lorys, lorikeets, cockatoos, racquet-tails, rosellas, parakeets, lovebirds, parrots, and macaws inhabiting the forests, savannas, brushlands, and urban areas in tropical and temperate regions around the world.[18] In historic times, parrots ventured up to about 40°N latitude in North America and 35°N in Asia.[19] Numerous endemic species are found today on islands

in the eastern Pacific and Indian Ocean. Most are strictly arboreal and strong flyers, yet most are nonmigratory. They are most diverse in the southern hemisphere, but their fossil record is poor. Island archaeological sites are the richest sources for fossil parrots, where a surprising diversity is now being unearthed.

Pathway 11, the cuckoo lineage (Cuculiformes) comprises several more adaptive radiations that exploited forests, woodlands, brushlands, and deserts. Most of the 143 living species are predatory, and some are parasitic, laying their eggs in the nests of other birds.[20] Young cuckoos toss the host hatchlings out of the nest and are reared by the host parents, whose own progeny have been killed.

This brings us to pathway 12, the last and largest avian lineage. This huge evolutionary highway connects 6,000 living species inhabiting the woodlands and savannas of the world. Most are fearsome predators, although none is large. There are three major trails branching from this path and a host of smaller ones. The first (Coraciiformes) includes the kingfishers and relatives.[21] There are 94 living species that populate wooded streams, lakes, marshes, and swamps. Their large heads support a massive bill, and they hunt while perching near or hovering over water to spot their prey, which can be nearly their own body size and still be swallowed. Also included on this trail are 56 living species of hornbills, that range throughout the woodlands of Africa, tropical Asia, Malaysia, and the Philippines.

Cuculiformes, *(above left)* include woodland birds like this cuckoo, and ground birds like the American roadrunners.

Pound for pound, coraciiformes *(above right)* birds like this kingfisher are among the most fearsome living predators.

Another coraciiform radiation is represented today by the 36 species of trogons, living in humid forests, open woodlands, and mountains of tropical and subtropical regions around the world. Smaller coraciiform trails include the bee-eaters, rollers, hoopoes, woodhoopoes, and motmots. Pound for pound, the coraciiform lineage includes some of the most fearsome of living theropods.

The Ivory-billed Woodpecker, a representative of Piciformes. This species has not been sighted in the wild for more than a decade, and is feared extinct.

Next on pathway 12 are the 215 species of woodpeckers and wrynecks (Piciformes), who inhabit woodlands around the world. The toucan lineage includes fifty-six species that range from Vera Cruz, Mexico to northern Argentina. They are usually brightly colored omnivores with huge bills that they use to catch small birds, reptiles, mammals, large insects, and to eat fruit. The oldest fossils from along this pathway are Miocene in age, but the lineage is probably much older.

Last, and more diverse than all other birds combined, is Passeriformes. Approximately 5,739 species—60 percent of all living birds—belong to the passeriform lineage.[22] This line exploited to an unprecedented degree the woodlands, deserts, forests, brushlands, savanna, swamps and similar habitats, from the sea to the highest mountains. Anywhere that you can find both open ground and vegetation rising above it, you will find a passeriform bird. Some are strong flyers, and some are migratory. They eat insects, spiders, snails, crustaceans, small vertebrates, seeds, fruits, nuts, and other vegetation. The fossil record of Passeriformes extends back only to the Miocene.[23] However, their relationship to other birds projects a ghost lineage at least back into the Eocene, and molecular data projects their lineage as far back as the Cretaceous.[24] Passeriform birds represent by far the most rapid and most diverse adaptive radiation in the entire 230 million year history of the dinosaur lineage. The remaining lineages are far too numerous to describe in any detail, but simply reading their names (see the accompanying list) conveys a vivid idea of their vast diversity.

Nearly 6,000 species of living birds are members of Passeriformes, the small perching birds.

MAJOR GROUPS OF PASSERIFORMES

Ordered by number of species

Tyrannidae	(537 species)	Flycatchers, Tityras, Becards, Cotingas, Plantcutters, Sharpbill, Manakins
Funariidae	(280 species)	Tanagers, Neotropical honeycreepers, Plushcap, Seedeaters, Flower-piercers
Thraupini	(413 species)	Ovenbirds, Woodcreepers
Timaliini	(233 species)	Babblers
Acrocephalinae	(221 species)	Leaf-warblers
Thamnophilidae	(188 species)	Typical antbirds
Emberizini	(156 species)	Buntings, Longspurs, Towhees
Saxicolini	(155 species)	Chats
Strildini	(140 species)	Estrildine finches
Pycnonotidae	(137 species)	Bulbuls
Carduelini	(136 species)	Goldfinches, Crossbills
Nectariidae	(123 species)	Sunbirds, Spiderhunters
Cisticolidae	(119 species)	African warblers
Corvini	(117 species)	Crows, Jays, Magpies, Nutcrackers
Ploceinae	(117 species)	Weavers
Parulini	(115 species)	Wood warblers
Muscicapini	(115 species)	Old World flycatchers
Sturnini	(114 species)	Starlings, Mynas
Oriolini	(111 species)	Orioles, Cuckooshrikes
Malaconotinae	(106 species)	Bush-shrikes, Helmet-shrikes, Vangas
Monarchini	(98 species)	Monarchs, Magpie-larks
Certhiidae	(97 species)	Creepers
Icterini	(97 species)	Troupials, Meadowlarks, New World blackbirds
Zosteropidae	(96 species)	White-eyes
Alaudidae	(91 species)	Larks
Hirundininae	(89 species)	Swallows, River-martins
Troglodytinae	(75 species)	Wrens
Pardalotidae	(68 species)	Thornbills, Scrubwrens, Bristlebirds, Pardalotes
Motacillinae	(65 species)	Wagtails, Pipits
Pachycephalinae	(59 species)	Whistlers, Shrike-thrushes, Shrike-tits, Sittellas
Formicariidae	(56 species)	Ground antbirds
Garrulacinae	(54 species)	Laughing thrushes
Parinae	(53 species)	Titmice, Chickadees
Vireonidae	(51 species)	Vireos, Peppershrikes
Eopsaltriidae	(46 species)	Australo-Papuan robins
Paradisaeini	(45 species)	Birds-of-paradise
Dicaeini	(44 species)	Flowerpeckers
Rhipidurini	(42 species)	Fantails
Cardinalini	(42 species)	Cardinals

(continued)

MAJOR GROUPS OF PASSERIFORMES *(continued)*

Ordered by number of species

Passerinae	(36 species)	Sparrows
Mimini	(34 species)	Mockingbirds, Catbirds, Thrashers
Pittidae	(31 species)	Pittas
Drepanidini	(30 species)	Hawaiian Honeycreepers
Laniidae	(30 species)	True Shrikes
Rhinocryptidae	(28 species)	Tapaculos
Maluridae	(26 species)	Fairywrens, Grasswrens, Emuwrens
Sittidae	(25 species)	Nuthatches, Wallcreepers
Artamini	(24 species)	Currawongs, Woodswallows
Dicrurini	(24 species)	Drongos
Sylviini	(22 species)	Sylvia
Megalurinae	(21 species)	Grass-warblers
Ptilonorhynchidae	(20 species)	Bowerbirds
Polioptilinae	(15 species)	Gnatcatchers, Gnatwrens
Cinclosomatinae	(15 species)	Quail-thrushes, Whipbirds
Viduini	(15 species)	Whydahs
Eurylaimidae	(14 species)	Broadbills
Prunellinae	(13 species)	Accentors, Dunnock
Remizinae	(12 species)	Penduline-tits
Irenidae	(10 species)	Fairy-bluebirds, Leafbirds
Melanocharitidae	(10 species)	Berrypickers, Longbills
Conopophagidae	(8 species)	Gnateaters
Aegithalidae	(8 species)	Long-tailed tits, Bushtits
Climacteridae	(7 species)	Australo-Papuan treecreepers
Regulidae	(6 species)	Kinglets
Pomatostomidae	(5 species)	Australo-Papuan babblers
Cinclidae	(5 species)	Dippers
Philepittidae	(4 species)	Asities
Aegithininae	(4 species)	Ioras
Acanthisittidae	(4 species)	New Zealand wrens
Ptilogonatini	(4 species)	Silky-flycatchers
Fringillini	(3 species)	Chaffinches, Brambling
Callaeatidae	(3 species)	New Zealand wattlebirds
Bombycillini	(3 species)	Waxwings
Corcoracinae	(2 species)	Apostlebird, Australian chough
Orthonychidae	(2 species)	Logrunners, Chowchillas
Menurineae	(2 species)	Lyrebirds
Atrichornithinae	(2 species)	Scrub-birds
Promeropinae	(2 species)	Sugarbirds
Chamaeini	(1 species)	Wrentit
Dulini	(1 species)	Palmchat

SOURCE: C. G. Sibley and B. L. Monroe, Jr., 1990. *Distribution and Taxonomy of Birds of the World*, Yale University Press, New Haven.

DIVERSITY IN DECLINE

Having traced the great post-Cretaceous diversification of dinosaurs, we now return to the issue of dinosaur extinction. First, we need to introduce a geological time period that has never before figured into discussions of dinosaur extinction—the Quaternary. It began about 1.6 million years ago, and extends to the present day. The Quaternary is divided into two epochs. The Pleistocene, which extends from 1.6 MA to about 10,000 years ago, includes the great Ice Ages that our "caveman" ancestors had to contend with. The Pleistocene was followed by the Holocene, which extends from 10,000 years ago until the present. We live in the Holocene. The great adaptive radiation of birds continued virtually unabated throughout the Tertiary. Although many species arose and died out, the overall trend reflected a global increase in the numbers of species. But from Quaternary sediments, evidence has slowly begun to emerge that this trend has reversed.

An episode of mass extinction on a worldwide scale took place during the Quaternary. Like the K-T extinction, dinosaurs were not the only organisms affected. Like the K-T extinction debate, several killing mechanisms have been implicated, and the relative influence of each is hotly debated. Thanks to a finely calibrated time scale for the Quaternary, we can tease apart the proposed causes and effects on a finer scale than is possible for the events of 65 million years ago. And the ability to tell time precisely provides evidence of at least three separate waves of extinction that affected birds during the Quaternary. Some paleontologists argue that there was a single mechanism behind each one, while others argue that the three waves had more than one cause.

The first wave occurred during the Pleistocene, and it produced a great loss of global diversity. Both climatic change and human activity have been implicated, with some scientists attributing all the losses to one or the other cause, and some to a combination of both. There is little doubt that the Pleistocene was a period of climatic fluctuation, as a succession of Ice Age glacial advances and retreats crossed vast regions.[25] From the North Pole southward, the Pleistocene ice sheets advanced and retreated in five major pulses, blanketing large regions of North America and Europe. Ice also advanced down from many of the world's major mountain chains, and global cooling affected the Southern Hemisphere almost as severely as the Northern.

Since the Late Pleistocene, the overall climatic trend has been one of increased warming, although minor pulses of cooling also occurred as our modern climate emerged. At the peak of glacial coverage, the mean annual temperature was 5 to 7°C colder than today. About 14,500 years ago the glaciers began to retreat as a global climatic warming trend took effect. There was one last pulse of cold between 10,000 and 11,000 years ago, which marked the end of the Pleistocene Epoch and the beginning of the Holocene. Since then, the Earth's climate has become generally warmer. By about 4,000 years ago, it came to resemble the climate over much of the world today.[26]

By the end of the Pleistocene, a momentous loss of diversity had decimated the large vertebrates.[27] In North America nearly three-quarters of the larger mammals became extinct, including mammoths, mastodons, saber-toothed cats, horses, tapirs, camels, ground sloths, and the bizarre glyptodonts (giant cousins of the modern armadillo). In South America, a similar proportion of large species became extinct. In Australia, 55 mammalian species are now known to have been lost. Severe losses in Europe claimed the woolly rhino, mammoths, and the giant deer, although many lost European species survived elsewhere in the world, such as the hippo, horse, musk ox, and hyena.

Some of the largest birds were also affected, including the giant flightless cranes like *Titanis*. They had been the dominant terrestrial predators in South America over most of the Tertiary. Although *Titanis* reached North America in the Pleistocene, it perished before the Holocene began. A Pleistocene condor named *Teratornis*, larger than the California and Andean condors, also disappeared. Its bones are among the most commonly preserved in the older deposits of the famous Rancho La Brea tar pits, but it evidently died out by the Holocene. A lineage known as the Gallinuloididae, a cousin of modern turkeys, had also died out by the end of the Pleistocene. Between 20 and 40 species of birds become extinct at the end of the Pleistocene in North America alone.[28]

One group of paleontologists, led by Paul Martin of the University of Arizona, alleges that human overkill was behind the surge of Pleistocene extinctions.[29] Archeological sites have produced the bones of extinct animals in association with human artifacts, along with evidence that people butchered the animals for food. There is little doubt that humans preyed on the extinct species. But Martin maintains that the extinction of the mammalian "megafauna" was a rapid "blitzkrieg" in which the human invaders into North America rapidly fanned out across the continent, from Alaska south and east to the Gulf Coastal plain, and decimated the large mammals in their path. Owing to economies of scale, early human hunters preferentially hunted the large animals first. If climate had been responsible, they argue, then small mammals and birds should have suffered extinction as well.

Opponents to the human overkill hypothesis have marshaled several arguments exonerating our Pleistocene relatives in the worst of the losses: They point instead to the climatic fluctuations of the Ice Ages. First, they argue that small vertebrates *were* affected during the Pleistocene, along with the megafauna. Although the losses weren't as great, the geographic ranges of many small vertebrates became more restricted and entire vertebrate communities were reorganized as a result. To better understand the effects of climatic change on late Pleistocene and Holocene mammals, a massive computer database was recently compiled by Ernest Lundelius of the University of Texas at Austin, and his former graduate students Russell Graham of the Denver Museum of Natural History, and Rickard Toomey and Eric Schroeder of the Illinois Natural History Museum, together with more than a dozen colleagues from around the country. The database, known as FAUNMAP, contains records for nearly 3,000 Quaternary fossil localities in the continental United States and Alaska.[30] Each locality represents a collection of fossil and subfossil mammals that had been identified and curated in North American natural history museum collections. Each locality's age was determined by radiocarbon methods, whose usefulness is confined to the last 50,000 years. FAUNMAP uses a computer to map the distributions of past faunal and environmental characteristics as they changed through time. The database includes a wide range of environmental information, such as the extent of the ice sheet at different times and the precise location of each site.

The FAUNMAP analysis revealed a more heterogeneous environment during the early Pleistocene than in later times, in which a higher diversity of mammals existed than at present. During the Holocene warming episode, some mammals dispersed northward, tracking a northward shift in their preferred environment. But different species responded at independent rates and times, moving in different directions. The retreating glaciers left a landscape that was much more uniform in its topography and faunal composition. As Holocene warming progressed, the loss in environmental heterogeneity corresponded with a loss in diversity. The large animals, with slower generation times, were the most severely affected by the loss of environmental heterogeneity. Their decline may have been helped along by people, but climate was probably the major factor.

Other scientists support this conclusion with the argument that global human populations were very small during the Pleistocene, a point that we will return to in a later chapter. The low state of human technology during the Pleistocene is another factor to consider; humans probably did not have sufficient killing power to have wiped out all the species that went extinct in the Pleistocene. Still other scientists have argued that the timing of events recorded in the archeological record—at least in North America—is more complex than claimed by the overkill proponents. The blitzkrieg should have progressed from northwest to southeast, but instead the opposite pattern of loss occurred.[31]

In Australia, the situation is also complex, and the solution on this continent may or may not have been the same as in North America. Humans had arrived on that continent by 60,000 years ago. Between 60,000 and 18,000 years ago, a major wave of megafaunal extinction had swept over the continent. Human impacts may have been strongest by increasing the frequency of fires. Increased concentrations of charcoal are found in sediments dating back about 30,000 years.[32] Fires altered the vegetation and erosion increased, which further degraded the Australian habitat and accelerated the pace of extinction. But there is also evidence that the Australian climate became more seasonable, with dry intervals, starting about 25,000 years ago[33]. This may have led to a higher frequency of natural fires, and the climatic stress may have had a role in the extinction.

So, even though humans lived during the Pleistocene, their exact role in the Pleistocene losses may have been different in different parts of the world, and climate remains a major suspect throughout the world.

Evidence being unearthed from islands of the Pacific basin, one of the greatest sources of avian diversity, indicates that there was a second wave of extinction that followed the great losses of the Pleistocene. Zooarcheological research indicates that the second pulse postdated the Pleistocene wave of extinction by several thousand years. At the beginning of the Quaternary, for example, there were at least 33 species of ostrichlike ratites. Archeological records indicate that most Pleistocene ratites survived far into the Holocene, some until only a few hundred years ago. But only eleven species survive today.[34] The same pattern holds for a growing list of other birds and mammals that is growing, as collaborating zoologists and archaeologists survey Holocene history in detail. Although the climate approached its current patterns during that time, the loss of species didn't stop and in some cases it accelerated. Owing to its timing—several thousand years after the last glacial retreat—this second wave of extinction is difficult to attribute to climate alone.

CODA

Looking at the history of the entire dinosaur lineage, it is a myth that theropods surrendered their role as top predator at the end of the Cretaceous, at least in some parts of the world. In South America and elsewhere, giant predatory theropods dominated terrestrial faunas for millions of years after the K-T boundary. Today, feathered raptors remain at the top of the food chain in many parts of the world, having inherited the role from some of their Mesozoic relatives. In many ways, the post-Cretaceous history of avian dinosaurs is a continuation of what took place in the Mesozoic.

To observe the communities and social interactions among living birds is to see their intelligence and complexity. To encounter birds in nature as they soar, dive, swim, sing, to see them hunt on the ground, in the trees, or under the ice, is to witness more than we can accomplish with our most elaborate machine. How different from its traditional connotation is the phylogenetic image of Dinosauria, with powerful brains, sophisticated "bio-technology," and a vast diversity of living species. E. O. Wilson describes that "Great biological diversity takes long stretches of geological time and the accumulation of large reservoirs of unique genes." The evidence described above suggests that the great radiation of birds took about 120 million years—nearly twice the duration of avian history accorded by the Phoenix hypothesis. But like all else in life, it would not be sustained forever. After a debilitating setback in the Pleistocene, the decline has continued into the Holocene as dinosaur history has become inextricably entangled with our own.

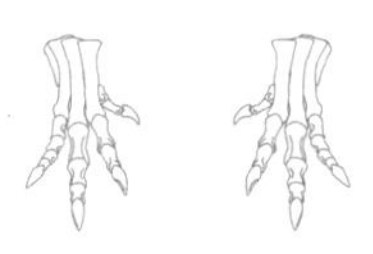

CHAPTER SEVENTEEN

The Real Great Dinosaur Extinction

When we applied to graduate school two decades ago, every applicant was required to take the Graduate Record Exam. Our GRE scores would, it seemed, be the most important numbers in our life. Images of our own academic extinction loomed in our imaginations as we studied. Even now the memory remains vivid and searing. But our spirits were buoyed when we encountered a couple of simple questions about vertebrate evolution. One asked, "Which is the oldest bird? (a) *Australopithecus afarensis*, (b) *Archaeopteryx lithographica*, (c) *Escherichia coli*, (d) *Drosophila melanogaster*, or (e) *Penicillium notatum*?" Paraphrasing the second, "If a human generation averaged 20 years long, how many generations back would you have to go for humans to overlap in time with dinosaurs?"

Both questions can be traced back more than a century, to Richard Owen's description of the oldest bird and his naming of the first-known extinct dinosaurs. Even with a modern evolutionary perspective, the answer to the first question hasn't changed, although the meaning has been altered a bit.

But the answer to the second question has changed. When we took the GRE, the correct answer was about 3,250,000 human generations. The answer was qualitatively the same to university aspirants in Richard Owen's day. Even though the age of the Earth was not then known, it was clear that the Cretaceous ended long ago, and that thousands or millions of generations separated humans and dinosaurs. Today, however, our map of vertebrate phylogeny indicates that the correct answer is 0. Possibly the most meaningful lesson gleaned from mapping dinosaur phylogeny is that our own history has overlapped with dinosaurs for about 20,000 generations, depending on what you choose to call a human.

In previous chapters we traced the rise of dinosaurian diversity, but we now reach a profound reversal of that trend, as we investigate one of the greatest episodes of mass extinction in tetrapod history. This change in fortune for birds and many other species occurred at basically the same time that the world became inhabited by people. Unlike the great K-T extinction, humans witnessed this one.

HUMAN EVOLUTION

Humans are members of the primate lineage, whose fossil record now traces back to the very beginning of the Cenozoic. As we saw in Chapter 15, Lev Nessov's discovery of fossil zhelestid mammals in Cretaceous rocks provided a benchmark for calibrating the age of many of the living mammal lineages, and David Archibald mapped the primate ghost lineage back into the Cretaceous.[1] Although we have yet to find a Cretaceous fossil that is unequivocally a primate, our phylogenetic map indicates that the ancestral primate species probably lived during the Middle or Late Cretaceous, and that one or more of its descendants survived the K-T extinction. Without a better fossil record, we can't say much about early primate diversification, or how they were affected at the K-T boundary, except that the lineage survived. Like dinosaurs, primates underwent a long history of post-Cretaceous diversification, and today there are about 200 living primate species. But over the whole of the Cenozoic, the pace of bird diversification exceeded that of primates 50-fold, to produce today's approximately 9,702 species.

Modern primates include tarsiers, lorises, lemurs, galagos, monkeys, and our own group—Hominoidea. The hominoids today include our own species, *Homo sapiens*, plus several other lineages of apes. The evolutionary path closest to our own includes the two species of chimpanzees, and it may include the Gorilla as well. Alternatively, the Gorilla might be first cousin to chimp-human sister lineages. Still further from humans is the Orang-Utan, and further still the gibbons. The mapping of ghost lineages suggests that Hominoidea arose in Africa between 20 and 30 million years ago. Although sparse, the fossil record of hominoids preserves a diversity of extinct species.[2] *Sivapithecus*, one of the oldest fossil hominoids, lived between 13 and 7 million years ago in Turkey, northern India, Pakistan, and China. Most anthropologists think that *Sivapithecus* was related to the modern Orang-Utan. More recent fossils mark the split between modern chimps and the lineage named Hominidae, which includes humans plus their closest extinct relatives.

Based on currently known fossils, the paths of hominids and other apes split in Africa between 5 and 10 million years ago. Nearly all early hominid fossils come from a narrow rift zone in eastern Africa that extends from Ethiopia south through Kenya, Tanzania, and into South Africa. Evidently, three distinct hominid species lived contemporaneously in east Africa. Among the oldest fossil hominids is *Australopithecus anamensis*, from the Awash district of Ethiopia, which is around 4.4 million years old.[3] A second species, *Australopithecus afarensis*, from the Fejej district of Ethiopia, is 3.9 to 4.1 million years old. A more recently discovered hominid named *Ardipithecus ramidus* was collected from Kenyan rocks that are between 3.9 million and 4.2 million years old. A short trackway of footprints made by someone walking bipedally across wet muddy ash along a lake shore in east Africa dates to about 3.7 million years old. These prints belong to a creature whose feet resemble human feet and whose hands were freed from their primitive role in locomotion. A modest adaptive radiation had begun long before the Pleistocene.[4]

The oldest hominids were bipedal and their brains were slightly enlarged, compared with the brains of chimps. The earliest are called australopithecines (a group of doubtful monophyly). They were just over 3 feet tall and weighed about 65 pounds—less than modern chimps. The volume of their brains measured just under a pint (400 to 450 cm^3),[5] which represents a real increase in brain size for the first hominids.

These hominids occur in rocks between about 4.4 and 1.0 million years old. By about 2.5 million years ago, fossils document a distinctive new species named *Homo habilis*. Its brain grew to more than a pint in volume (500 to 600 cm^3). Almost 4 feet tall, it weighed about 70 pounds. *Homo habilis* survived until 1.5 million years ago.

About 1.6 million years ago, fossils of still another species appear, named *Homo erectus*. Its brain

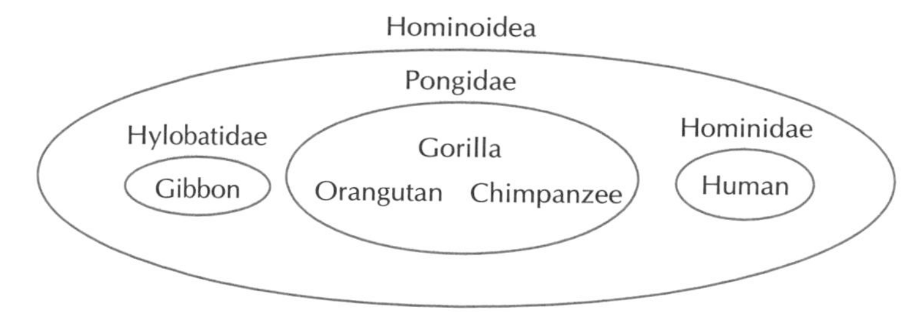

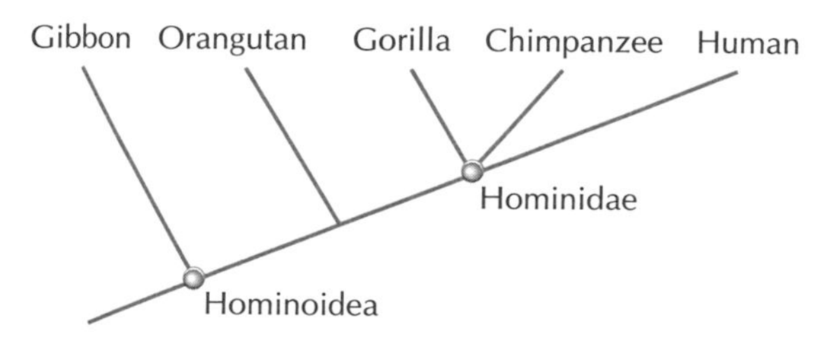

A traditional Linnaean classification of hominoids *(top)* compared with a phylogeny for hominoids *(bottom)*. Note that the term "Hominidae" denotes different assemblages of species in the two diagrams. Also note that the traditional classification reflects very little of the knowledge about relationships that is recorded in the phylogeny. Whether the Chimpanzee is closer to Gorilla or to *Homo* is controversial.

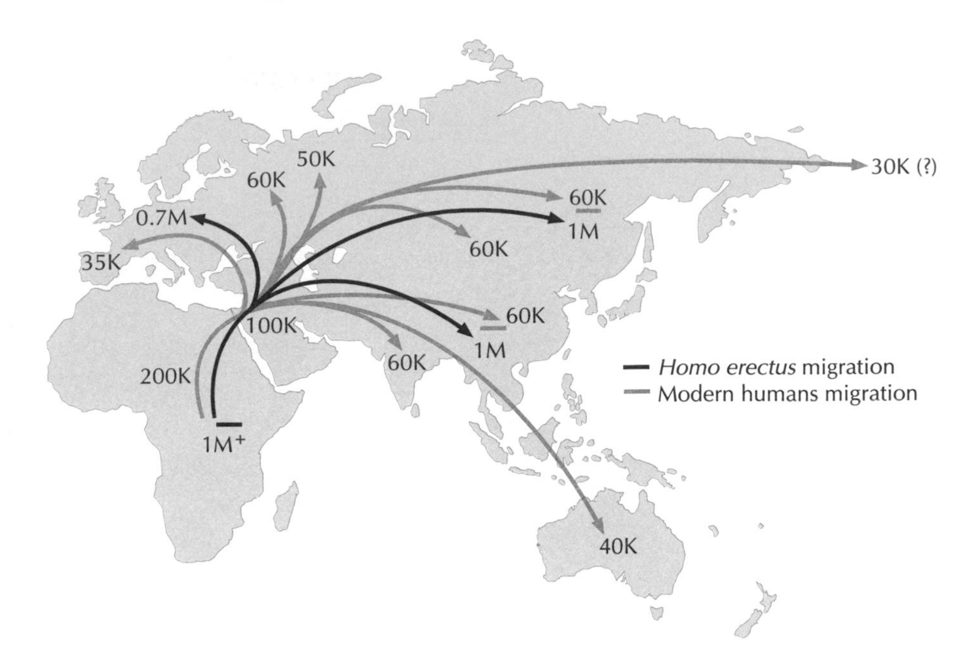

Alternative models to explain the distribution of humans over the globe. In the top figure, two migrations out of Africa are proposed, the first by *Homo erectus* and the second by *Homo sapiens,* following their separate origins in Africa. In the bottom map, *Homo erectus* moved out of Africa and locally evolved into *Homo sapiens.*

averaged nearly a quart (935 cm^3) in volume. This species is larger, averaging about 130 pounds. Brain size is partly a function of body size, but even correcting for the difference in size, a real increase in brain size and computing power evolved in *Homo erectus*. This was evidently the first hominid to disperse out of Africa, extending its range to Asia and Java by about 1 million years ago. By about 500,000 years ago it was living in Europe as well. Stone tools extend back about 2 million years, but so far there is no direct evidence about which of our relatives made and used them. Some of the Java sites were recently re-dated to between 27,000 and 53,000 years old, suggesting that *Homo erectus* survived much longer in southeast Asia than previously thought.[6] If so, it overlapped with our own species in time.

Homo sapiens, our own species, appears in the fossil record between 400,000 and 150,000 years ago. The exact date of the oldest *Homo sapiens* fossil is in dispute, as anthropologists argue over the identities of certain incomplete European fossils. But there is no mistaking the identity of more complete fossils, because *Homo sapiens* has a huge brain, with a volume approaching 1.5 quarts (between 1200 and 1600 cm^3)[7], and a body mass still around 130 pounds. With the emergence of *Homo sapiens*' huge brain came the accouterments of humans and human societies. The first evidence of "domestic" fire is about 500,000 years old. Art goes back about 35,000 years, and agriculture arose about 10,000 years ago.[8] The relationships among world populations are now being mapped out in detail, using information from both genetics and linguistic characteristics.

Elizabeth Vrba of Yale University and a group of colleagues document an episode of climatic deterioration, beginning in the Miocene and culminating in the Pleistocene, that corresponds with a major turnover in the fauna of the African Rift Valley during early hominoid history.[9] A rapidly

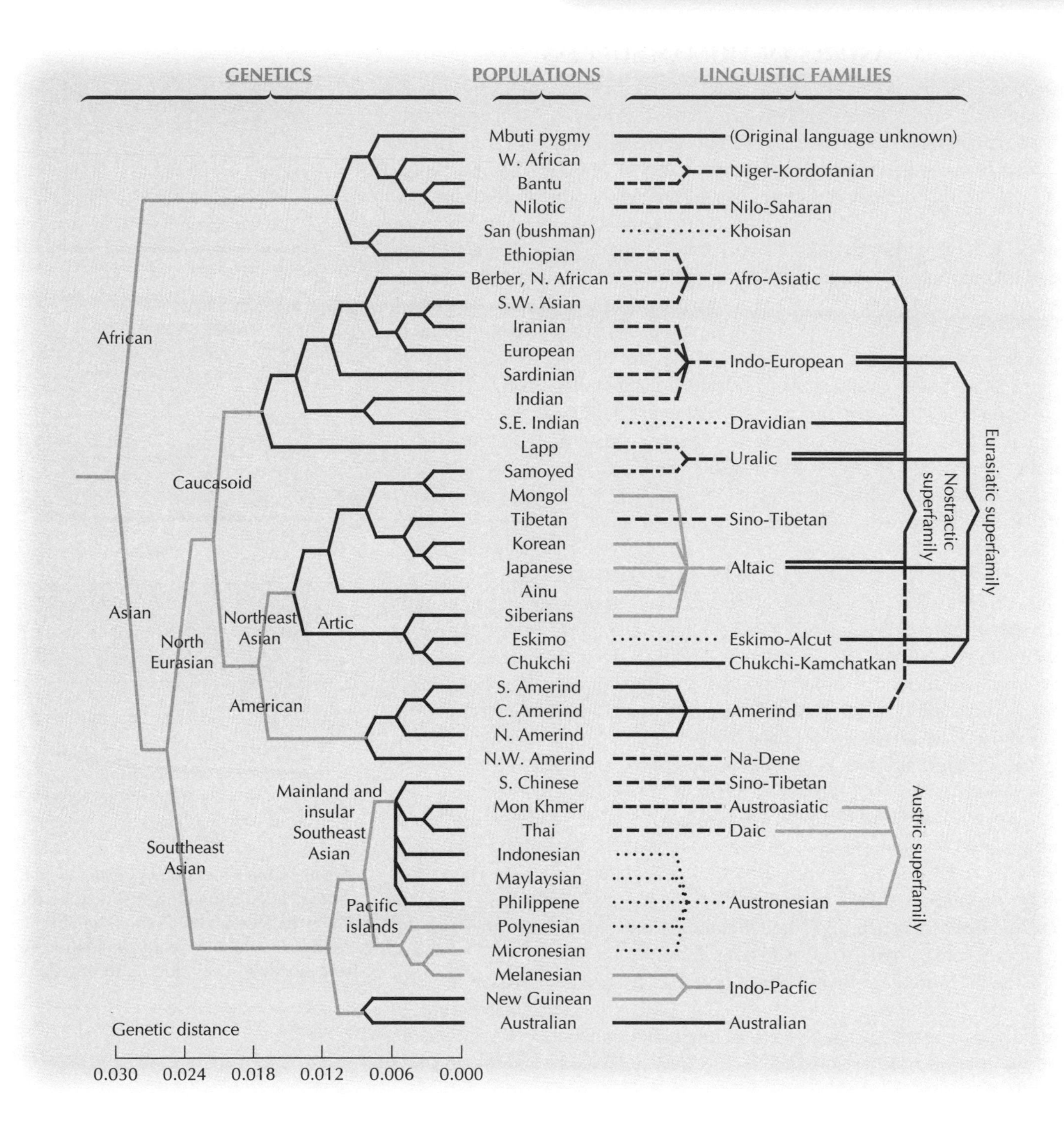

shifting environment may have been the substrate for human origins. Steven Stanley of Johns Hopkins University calls humans the children of the Ice Age,[10] arguing that a confluence of environmental and developmental factors sparked the emergence of *Homo*. However they did it, humans survived the Pleistocene extinctions to achieve unprecedented success by many measures of the term.

Information from genetics *(left)* and linguistic characteristics *(right)* lead to very similar conclusions on the relationships among world populations.

MEASURES OF HUMAN SUCCESS

One distinguishing feature of *Homo sapiens* is our repertoire and extensive use of tools. Tools and their effects leave an archeological record, and they document a rich history of human activity before there were written records. Intensive use of tools began sometime before 100,000 years ago. Demographer Joel Cohen of Rockefeller University has suggested that the discovery of how to make and use tools led to one of the first great spurts of human population growth, which peaked about 100,000 years ago.[11] Between 400,000 and 100,000 years ago, he estimates the combined population of the human species grew to 100,000.

Cohen describes that over the next few thousand years a drop in population levels may have occurred, based on analyses of the variability of DNA sequences in modern human populations. Human populations may have fallen to about 10,000 people, producing a bottleneck in DNA sequence variability that is still discernible in people today. The reduction probably occurred during the Pleistocene Ice Ages, about 80,000 years ago. The survivors lived in Africa and slowly expanded across the Old World. By 40,000 years ago, human populations had spread to Europe, Asia, and Australia. Before the end of the Pleistocene, by 12,000 years ago, people had discovered the New World and traveled through North America to the forests of South America.

A second surge in human population size began as people in different parts of the world discovered agricultural techniques and grew some of their own food. From about 10,000 to about 6,000 years ago, the global population rose to somewhere around 5 million people. Cohen estimates that before this surge, human populations grew at a rate that would lead them to double in approximately 40,000 years or more. Following the agricultural surge, global populations grew at a rate in which they doubled every 2,000 to 3,000 years.

A third surge in population growth began in the early eighteenth century. This episode coincided with the discovery of science and technology in Europe, and the renaissance of global navigation that distributed agricultural species cultivated by different peoples throughout the world. The global human population surged to 750 million people, and the population now doubled about every century. When Richard Owen named Dinosauria in 1842, the human census had just passed the 1 billion mark.

Cohen measured another surge in population growth shortly after World War II, and this one was the most significant surge thus far. It was prompted primarily by several developments in medicine and public health, which reduced mortality and prolonged life—the global human population reached 2.5 billion.

In the next 36 years, the global population doubled. A final surge occurred in the late 1960s, reaching an all-time peak in population growth rates. It was prompted by medical developments that led to increased human reproductive fertility. Today there are about 5.6 billion people on Earth. In the last 100 years, more people have been born than in all the rest of human history. Current human populations are growing at the rate of 2 percent per year, or 1 billion people every 12 years. Current projections estimate that the all-time high will be reached in 2135 A.D., with 11 to 12 billion people crammed on Earth. So successful are humans that we have already become 100 times more numerous than any other land animal in the history of Life in our 100-pound

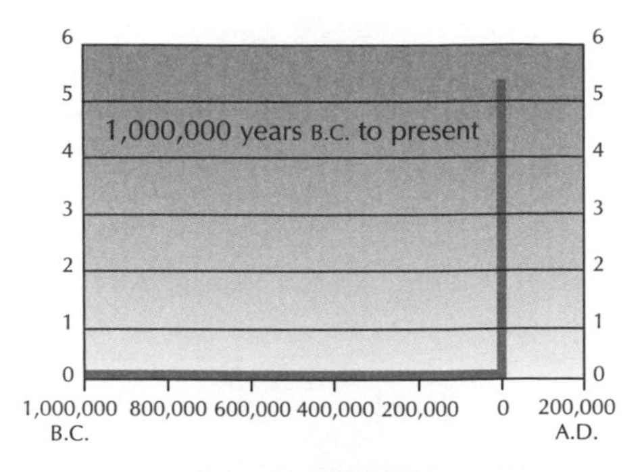

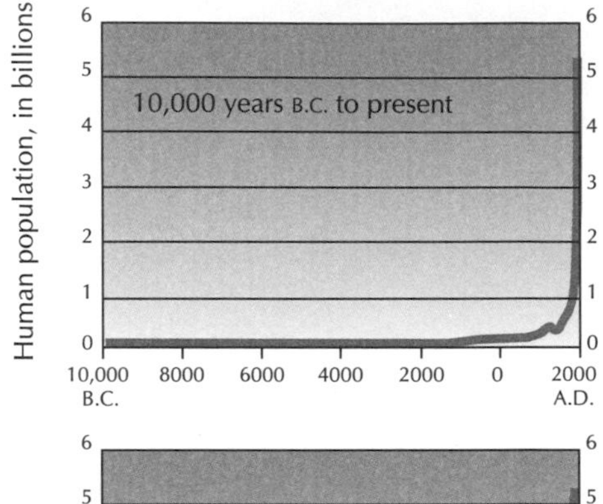

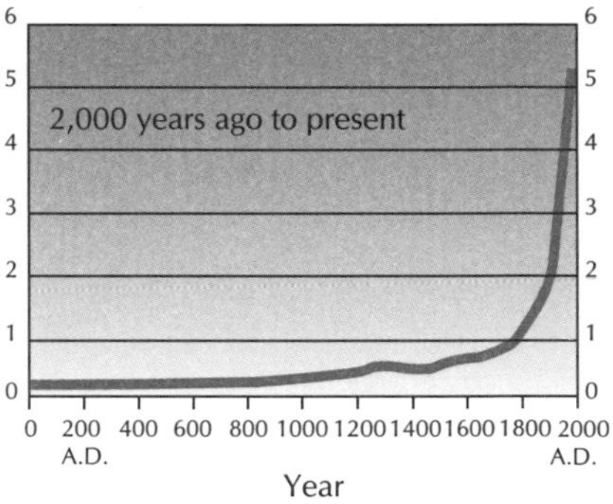

Demographer Joel Cohen's estimates of human population growth shown on three different time scales. Even when observed on a scale of 2,000 years, population size appears as an explosion *(bottom graph).*

weight class. Through our agricultural practices, it is estimated that humans capture 20 to 40 percent of the solar energy reaching the Earth's surface and consume it exclusively for our own needs.[12] *Homo sapiens* is one of Nature's greatest success stories, but at what cost?

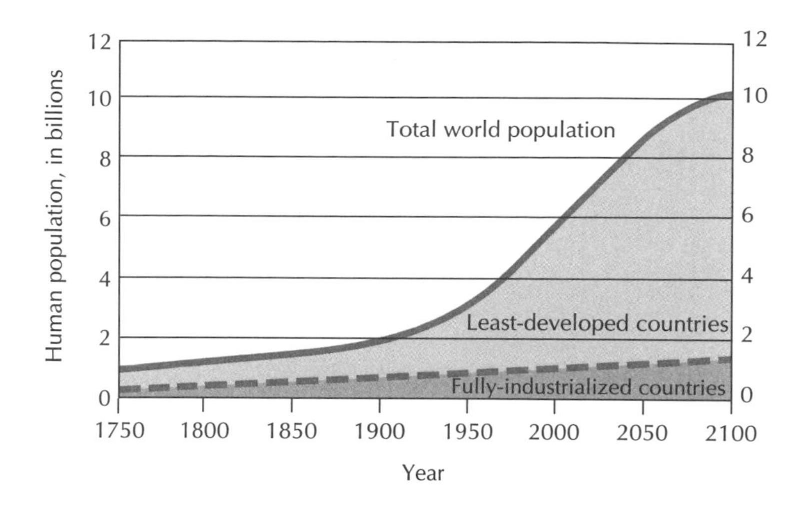

The projected peak human population will reach close to 10 billion people early in the twenty-second century. Most of these will be born in underdeveloped countries in the next century.

THE SPECTER OF 1662

About 300 miles west off the eastern coast of Madagascar in the Indian Ocean are three islands now called Mauritius, Réunion, and Rodriguez, which together form the Mascarene Islands. Mauritius was the first to be discovered, in the first half of the sixteenth century by a Portuguese navigator, but the entire chain remained uninhabited until 1598 when the Dutch took possession and established a colony on Mauritius. The colonists found birds that ". . . were of large size and grotesque proportions, the wings too short and feeble for flight, the plumage loose and decomposed, and the general aspect suggestive of gigantic immaturity."[13] Little did these colonists realize that their encounter with the Dodo bird would prove infamous.

Before Europeans landed on Mauritius, it was pristine, spectacularly beautiful, and uninhabited. Early Dutch, French, and English travelers described it as a Garden of Eden. Besides the Dodo, there were evidently flightless relatives on each of the other two Mascarene Islands. A second species known as the Rodriguez Solitaire lived on Rodriguez Island, but became extinct by the end of the eighteenth century. A third species, known as the White Dodo, may have lived on Réunion Island at the time of European settlement. This creature has been given several names, including *Raphus solitarius* and *Victoriornis imperialis*, but our entire record of it consists of a couple of oil paintings reportedly made from a live bird displayed in Amsterdam and contradictory tales by sixteenth- and seventeenth-century travelers.[14,15] If it ever existed, this bird had become extinct by the end of the seventeenth century. So far, no bones of it have been recovered.

The Dodo of Mauritius. This engraving was taken from a painting made before the Dodo went extinct.

The first great chronicler of the Dodo, H. E. Strickland, described in 1848 that "The history of these birds was as remarkable as their organization. About two centuries ago their native isles were first colonized by Man, by whom these strange creatures were speedily exterminated. So rapid and so complete was

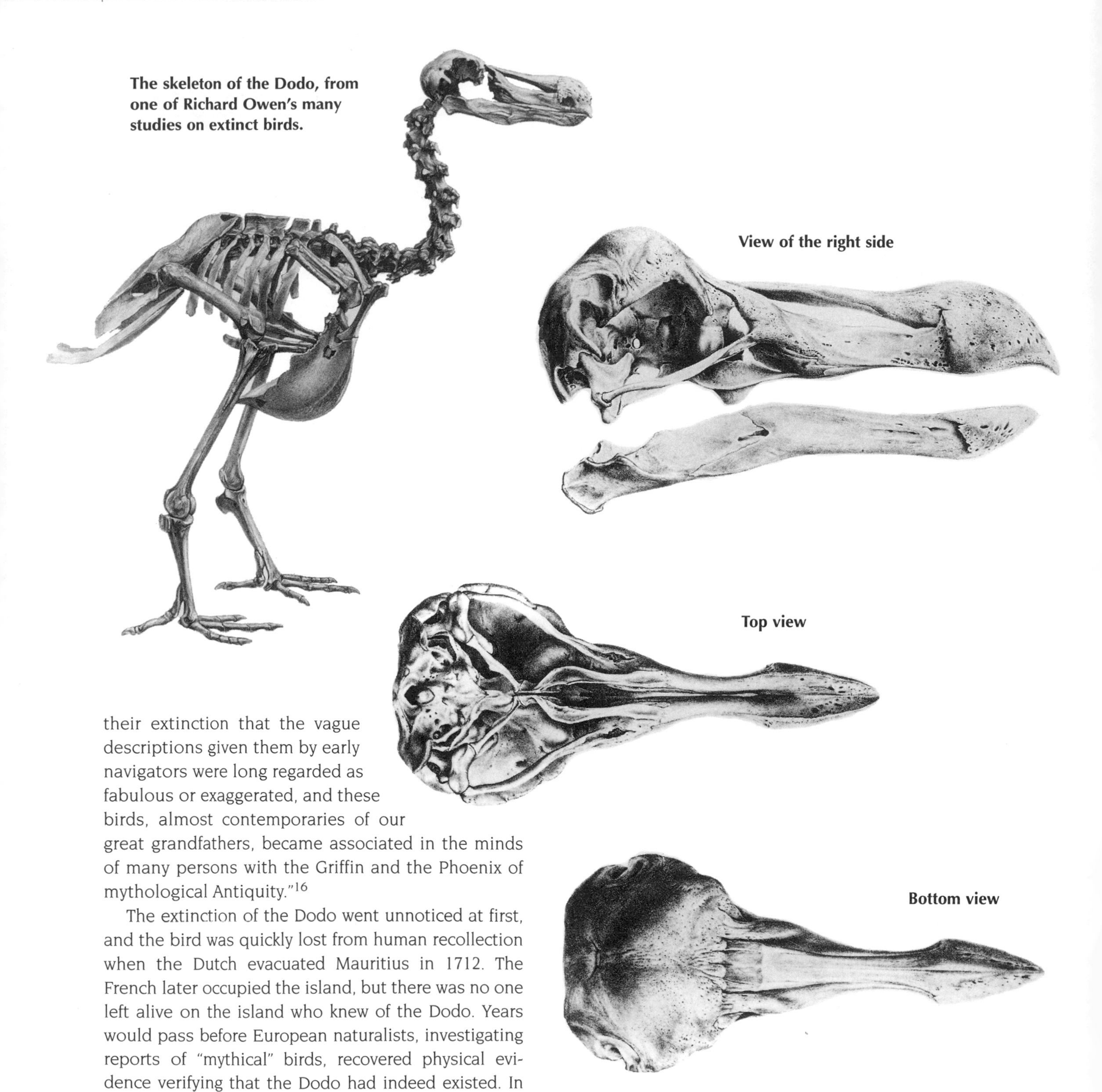

The skeleton of the Dodo, from one of Richard Owen's many studies on extinct birds.

Three views of the skull of the Dodo.

their extinction that the vague descriptions given them by early navigators were long regarded as fabulous or exaggerated, and these birds, almost contemporaries of our great grandfathers, became associated in the minds of many persons with the Griffin and the Phoenix of mythological Antiquity."[16]

The extinction of the Dodo went unnoticed at first, and the bird was quickly lost from human recollection when the Dutch evacuated Mauritius in 1712. The French later occupied the island, but there was no one left alive on the island who knew of the Dodo. Years would pass before European naturalists, investigating reports of "mythical" birds, recovered physical evidence verifying that the Dodo had indeed existed. In 1740, naturalists visiting the Mascarenes conducted an intense search for the Dodo, but no trace was found. Specimens eventually came into the possession of European naturalists. In 1758 Linnaeus named the

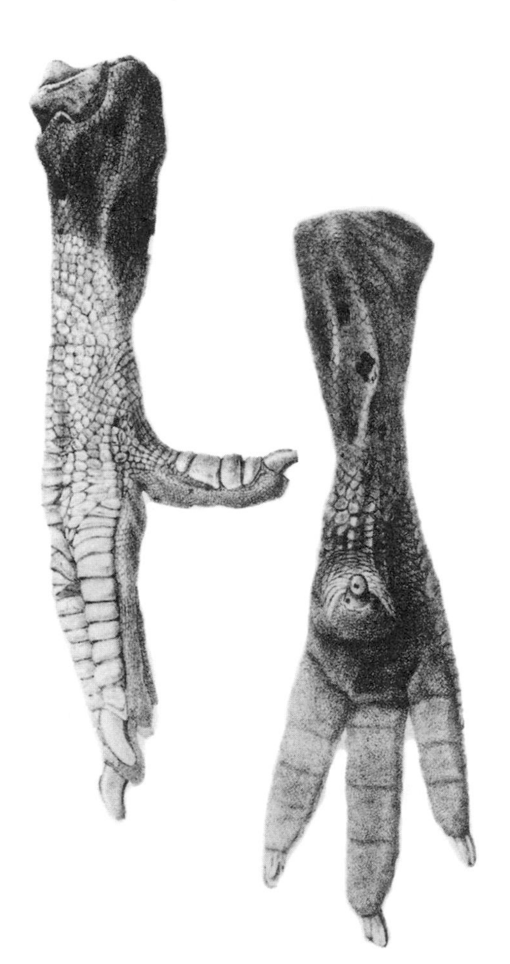

The head and foot of a Dodo specimen now in the British Natural History Museum.

species *Raphus cucullatus*, although he probably did not realize that the Dodo was extinct. Repeated attempts to locate a living Dodo on Mauritius failed, and by the mid-eighteenth century, it was all too evident that the bird was gone.

Once this was realized and the timing of events reconstructed, the culprits were obvious. In 1644, while the Dodo was still alive, a resident observed that "Here [in Mauritius] are Hogs of the China kind. These beasts do a great deal of damage to the inhabitants, by devouring all the young animals they can catch."[17] As Strickland was able to reconstruct, in the few short years following Dutch colonization, "This unfortunate and defenseless bird was slaughtered by the hundreds by the sailors who often for months lived on its flesh, but also often killed it for pure mischief, while finally the work of extermination was completed by the pigs, goats, and monkeys, introduced into the islands." The effects of domestic mammals may have exceeded the casualties inflicted directly by the colonists and sailors, who wrote that the birds were not especially tasty. The advent of domesticated animals proved more than the large flightless bird could withstand. The last reliable sighting[18] of a Dodo occurred in 1662. Since then, two-thirds of the other endemic birds of Mauritius have also disappeared.

Georges Cuvier had shown decades earlier that species once living in the distant geologic past had become extinct. By Richard Owen's day, thanks largely to Owen himself, it was established that species now alive could become extinct as well. And by that time, scientists were already pointing a finger at human culprits in the Holocene island extinction. In a textbook on paleontology published in 1860, Owen described the phenomenon of extinction this way:

> That species, or forms so recognized by their distinctive characters and the power of propagating them, have ceased to exist, and have successively passed away, is a fact no longer questioned. That they have been exterminated by exceptional cataclysmic changes of the Earth's surface

> has not been proved. That their limitations in time, in some instances or in some measure, may be due to constitutional changes accumulating by slow degrees in the long course of generations, is possible. But all hitherto observed causes of extirpation point either to continuous slowly operating geological changes, or to no greater sudden cause than the, so to speak, spectral appearance of mankind on a limited tract of land not before inhabited.[19]

From Owen's day to the present, both climate and human predation have been implicated as Pleistocene killing mechanisms. Numerous Pleistocene archeological sites leave no doubt that humans preyed on some now-extinct species. Although it is hard to overestimate the ingenuity and persistence with which humans have pursued other species into oblivion, it remains doubtful that human actions account for all the losses in the first pulse of extinction, owing to small human populations and primitive technology.

But during the second wave of extinction, in the Holocene, a huge and ominous record of evidence linking humans to catastrophic loss of bird species has accumulated in different parts of the world. One estimate suggests that, on average, one species became extinct every 83 years in the early days of the Holocene.[20] The rate of loss increased so that by the time the Dodo died out, another species of birds became extinct every 4 years, on average. At least 92 species of birds have become extinct since the Dodo, and the toll for this interval will probably rise as zooarcheologists study the remains of human habitation in Holocene sediments. The most recent studies of threatened and endangered birds suggest that by the end of this century, one species will become extinct every 6 months. Although human history did not begin this way, what E. O. Wilson calls a "tragic symmetry" arose in the Holocene, between the growth of human populations and the loss of biodiversity. Dinosaurs were especially hard-hit.

OWEN'S FIRST DISCOVERY OF EXTINCT DINOSAURS

In 1839, three years before publishing his first paper on Dinosauria, Owen reported that giant birds similar to the ostrich and emu once lived on New Zealand. Earlier tales had been told of gigantic birds that might still survive in remote areas never trodden by man. Traditions among the native Maori elders held that birds covered with "hair" had waylaid forest travelers, overpowering them before killing and devouring them. But no such birds had ever been seen there by Europeans.

Owen produced the first meager evidence that this wasn't just a myth. It consisted of only the broken shaft of a thigh bone. But with this fragment, Owen performed a feat of scientific deduction that astonished the world and catapulted him into the limelight. He wrote, "So far as my skill in interpreting an osseous fragment may be credited, I am willing to risk the reputation for it on the statement that there has existed, if there does not now exist, in New Zealand, a struthious bird nearly, if not quite, equal in size to the Ostrich, belonging to a heavier and more sluggish species."[21] From a mere scrap of bone, he reconstructed the whole animal, although he was roundly criticized by other scientists who found such a leap of faith outrageous.

The bone had been brought for sale at a cost of 39 guineas to the College of Royal Surgeons, where Owen's career began. The College's museum committee declined to buy the specimen despite Owen's pleas, but Owen persuaded a donor to purchase it for another natural history collection, so that it could be properly studied and published upon. In a paper that had met strong editorial resistance, he announced the discovery of a giant extinct New Zealand bird.[22] His critics demanded more substantial proof.

Recounting the story 40 years later, Owen describes that he had 100 extra copies of his article "distributed in every quarter of the islands of New Zealand where attention to such evidences was likely to be attracted."[23] In 1843, he received two shipments of bones in response. Pictured first in his imagination, Owen finally saw the bones whose existence he had predicted. They became the type specimen of *Dinornis struthoides*, and Owen launched a series of papers on the extinct birds of New Zealand. As additional specimens arrived, he described more than a dozen species of ratite extinct birds known as moas. Many years later, having risen to the Directorship of the British Museum (Natural History), Owen obtained the original scrap of moa bone for the national collections.

Even in 1843, New Zealand was largely unexplored by competent naturalists. It took several years to determine that all the moas were in fact extinct. As the European population there grew, it became impossible to deny that they were gone. But evidently, they hadn't been gone long. In 1878 a dried head with neck, legs, and skin with ligaments and feathers attached was found in a cave. These were purchased by Owen for the British Museum and became the type specimens of his new species *Megalapteryx didinus*. The feathers are described as very hairlike, some grayish-brown, some with a rust tinge, and others tipped in white. Several other pieces of mummified carcasses have been found since then.

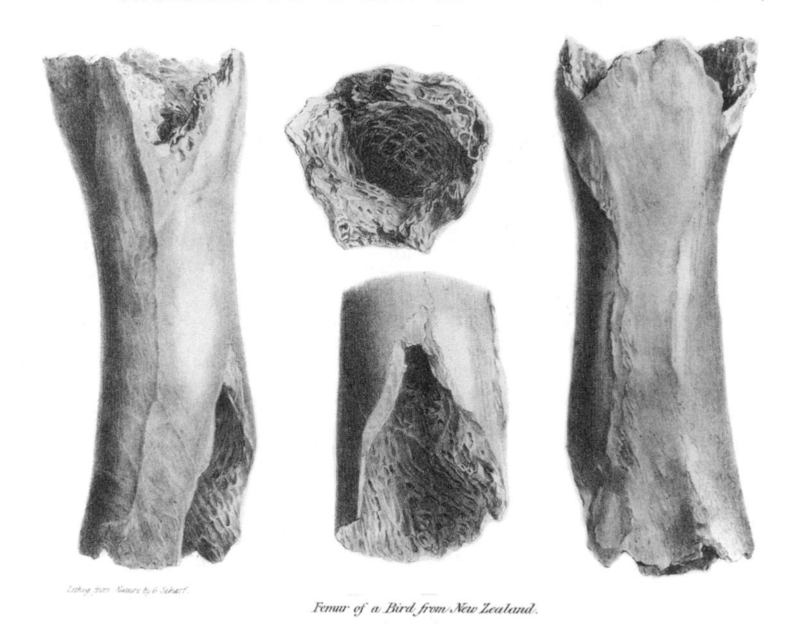

(Above) **The fragment of a thigh bone from which Richard Owen first deduced that large ostrich-like birds, now extinct, once lived in New Zealand.**

(At right) **Richard Owen, standing 6 ft 1 in tall, next to the skeleton of *Dinornis*, one of largest moas. In his right hand, Owen triumphantly holds the first fragmentary bone that revealed the presence of extinct moas in New Zealand.**

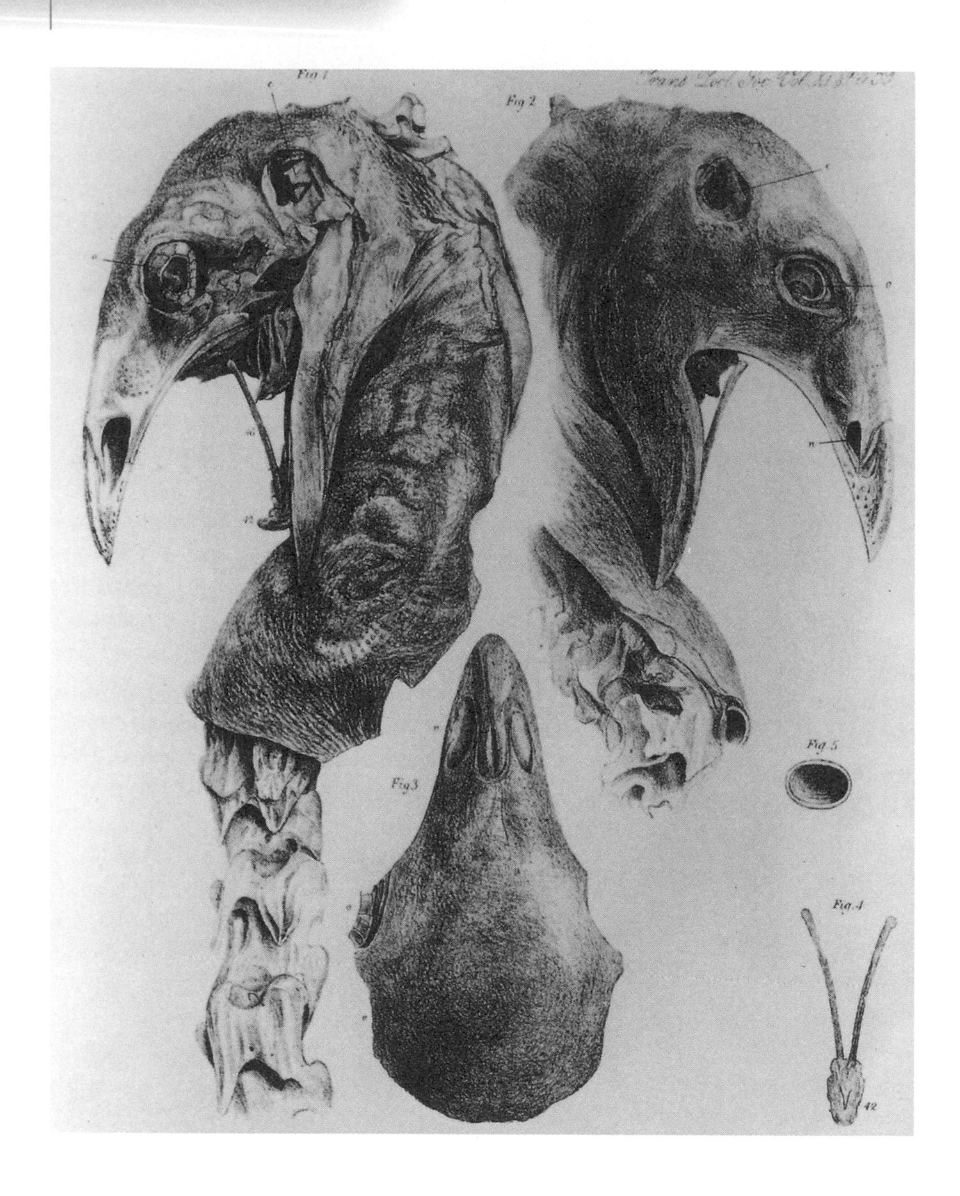

The mummified head of a moa. This specimen still exists in the Natural History Museum, London.

Fossil moas as old as the Miocene or Pliocene have now been recovered, but the tectonic history of New Zealand suggests that their flightless ancestors may have rafted away on the island during the Cretaceous. Long before the arrival of humans, moas had adaptively radiated on the islands of New Zealand. It has been argued that some species became extinct as the climate changed from being drier to much more rainy at the end of the Pleistocene. But there is much evidence that human hunters encountered the last of New Zealand's moas during the Holocene.

Before human habitation, there were approximately thirteen moa species, ranging from the size of a turkey to the tallest bird known—a towering 12 feet.[24] This is far more diversified than any living ratite lineage. The adaptive radiation of moas covered both islands of New Zealand, and the birds were abundant when the Maori people first landed on the North Island about 1,000 years ago.

The moa's flesh may be what convinced the Maoris to stay. Recent dating methods and finely calibrated chronological studies of the sites have enabled us to reconstruct a detailed crime scene.[25] At numerous archaeological sites, moa bones have been found with Maori kitchen refuse, along with bones of domestic dogs that were brought from their home to the north in Polynesia. Fires set by the Maoris in the grasses and brush near the beaches destroyed moa breeding colonies, and other evidence suggests heavy human predation on these birds. The Maoris hunted the moas for food and killed them in great numbers, leaving their butchered bones scattered all over New Zealand. The killing of large numbers of birds began around 1100 A.D., and over the next two hundred years, the diet of Maoris must have consisted predominantly of moas. The hunting sites slowly spread to the South Island, where the last were extirpated. Seeds, twigs, and other stomach contents preserved in one moa carcass provided a radiocarbon date of about 1330 A.D. for the moa's last meal. How much later they survived is unclear. The colonization of New Zealand began in the decades following the visit by Captain James Cook in 1769. There are reports by Europeans claiming to have seen live moas, but these are unverified. There are also some claims of butcher marks on moa bones that appear to have been made with iron blades. Although hard to substantiate, this could be evidence that Europeans were among the last humans to dine on moas.

Maori traditions recall the moa as resembling a cassowary, with a brightly colored neck and a comb on its head. The skulls of at least three different moa species preserve evidence of a crest, which was probably a sexual characteristic like the crests and wattles in their living relatives. Also recalled is that the female incubated the nest while the male supplied her with food.[26]

In 1844, Robert Fitzroy, former Captain of Darwin's famous voyage on the *Beagle* and then governor of New Zealand, interviewed an elder Maori named Kawane Paipai. He claimed to have taken part in a moa hunt. Fitzroy's account was later published by extinct bird chronicler Errol Fuller: "He remembered the birds being hounded, encircled and then speared to death, sometimes with weapons designed to snap easily once the body was stuck. Trapped moas defended themselves vigorously with terrible blows from their feet but while administering these, the monstrous bipeds were forced temporarily to support their weight on one leg. A party of hunters

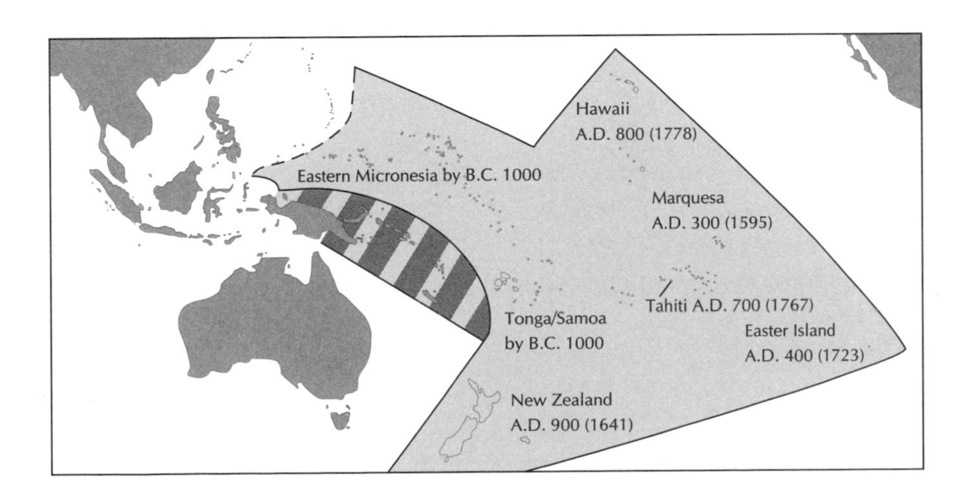

The dates of discovery of the western Pacific islands by Polynesians and Europeans. Dates of discovery for the latter are in parentheses.

would launch a frontal attack—a feint—while another crept behind waiting for the moment when the Moa raised a leg; then the party attacking from the rear would strike, knocking away the supporting leg. Once down, the victim was either dispatched immediately or such grievous wounds were inflicted that the final outcome was no more in doubt."[27]

OTHER ISLAND EXTINCTIONS

The combination of human overkill and environmental destruction by rats, cats, pigs, dogs, and goats devastated native birds on small islands. For ground nesting species, the effects of introduced mammals have been especially devastating. An additional source of pressure is deforestation. Declining diversity in island floras depleted certain food supplies and nesting territories, further reducing the diversity of island bird faunas. Nowhere has this damage been so severe as on the islands of the tropical Pacific, where high levels of avian diversity evolved during the Cenozoic.

The tragic history of Pacific island birds was recently investigated by David Steadman of the Florida Museum of Natural History, Storrs Olson of the U.S. National Museum of Natural History, and a group of collaborators. Steadman presented a chilling summary of their findings: "On tropical Pacific islands, a human-caused 'biodiversity crisis' began thousands of years ago and has nearly run its course."[28] The loss of bird life may have totaled between 2,000 and 8,000 species, representing at least a 20 percent decline in the number of bird species globally.

The islands of Melanesia were first occupied by humans 30,000 years ago. The islands of western Polynesia and Micronesia remained uninhabited until 3,500 years ago. Most islands of the Pacific and Indian Oceans were inhabited at least 1,000 years ago. Steadman describes the distinct signatures of human disturbance in the Holocene sediments deposited on and around the islands, including the accumulation of more clays as soils unbounded by vegetation washed off the island. Charcoal also appears in the sediments in abundance. Pollen from forest trees gave way to fern pollen, signaling the beginnings of deforestation. On the island of Mangaia, Steadman found this signature in sediments predating the earliest known cultural sites by 1,000 years.

Upon their arrival, the colonists began clearing forests, cultivating crops, and raising domesticated animals. The indigenous forests and grasslands with their many species were cleared and burned to make way for a few domestic crops, further disrupting the local ecology. Seabird colonies vanished from many islands, and numerous species became extinct. Some extinctions were due to human predation; others resulted from the loss of soil that removed all suitable breeding habitat for burrowing species, like the petrels. On the island of Ua Huka in the Marquesas, Steadman found evidence that breeding seabird species dropped from 22 to 4 during human occupation. On the whole, however, the landbirds suffered far more extensive losses.

Steadman reports that "Although the rate of extinction varied with ruggedness of terrain and size or performance of the prehistoric human population, we have no evidence that the process for prehistoric extinctions differed fundamentally from those that continue to deplete surviving species today. The differences are mainly technological (snares versus guns and stone adzes and fire versus chain saws and fire, for example)."[29]

On the Hawaiian Islands, 60 native bird species have become extinct since the arrival of Polynesians between 1,500 and 2,000 years ago. Roughly one-third of those remaining (20 to 25 species) disappeared in the two centuries following the arrival of Europeans, and two-thirds of the surviving species are now endangered. On these and other Pacific islands there is a consistent stratigraphic sequence in the lava tubes, sinkholes, stream and lake deposits that preserve extinct island animals. The oldest deposits contain fossils of only native species, including shorebirds and landbirds. The most common fossils are flightless rails, which had rapidly speciated as their ancestors flew from island to island, populating them with descendant species. Many islands hosted one or more endemic species. In higher and younger sediments, the bones of native species dwindle, and the bones of domestic animals predominate, in association with kitchen refuse. Finally, rats and iron appear in the record, indicating the arrival of Europeans, with their greater killing power.

In the first half of this century, biologists thought that the diversity of Pacific birds was well documented, but they had been unaware of the much higher levels of Holocene diversity. Current zooarcheological explorations of island prehistory are greatly extending our knowledge of organisms that had been prematurely considered to be well documented. Before discovering the great Holocene extinction of Pacific birds, biologists and biogeographers thought that today's distribution patterns for birds were natural. Now they are beginning to reinterpret the patterns and to recognize human effects. Throughout the Pacific, landbirds suffered much higher species-level extinction, but the seabirds were also affected. The modern patterns of global seabird distribution are unnatural—meaning that humans were now involved in the equation. Steadman points out that many ecologists continue to analyze these birds as if the modern ecosystems are natural. While some of the range losses for living species might be restored through conservation efforts, it is increasingly apparent that we are centuries too late to preserve any true reflection of the original Pacific avifaunas.

FUNK

To understand the full effects of humans on island birds, and the complex interplay between human and natural factors in recent extinctions, we consider one last example: the Great Auk, or Garefowl, that disappeared early in the nineteenth century. Auks are seabirds that resemble penguins, although they are allied with shorebirds such as gulls. Auks have a large head, short tail, and chunky body that is covered by a dense, waterproof, black and white plumage. They dive and swim for food, and all, save the Great Auk, can fly. They spend their entire lives in the sea and, unlike most island species, they seem to have wide geographic ranges. But once a year they congregate on islands to breed.

The Great Auk stood 3 feet tall with an ungainly upright stance on land. It occupied a niche in northern waters much like that of the penguins in the southern oceans. It hunted in coastal and open waters, mostly following schools of small fish. When swimming the Great Auks had penguinlike speed

The Great Auk stood upright on land, on the rare occasions that it came ashore.

The Great Auk spent most of its life at sea. It came ashore in great numbers, onto tiny islands in the north Atlantic, during the breeding season.

and dexterity. Their legs were short and powerful with a pistonlike motion, and while swimming on the surface their large, webbed feet provided strong thrust. Underwater, they used their short, pointed wings to "fly" and steered with their feet as they pursued fish and other prey. The Great Auk was the largest species of auk known at any time in the Quaternary. It is said to have issued a croaking and a gurgling noise.[30]

Although usually denizens of open water, Great Auks returned every spring in enormous flocks to court and breed, probably on the same island where they hatched. They congregated in great numbers for about two months on remote rocky islands in the north Atlantic to enact their courtship rituals. Once the chicks were seaworthy, they migrated north from their breeding grounds out to sea, where they spent the next nine or ten months.

Like many other seabirds, the Great Auk bred only on islands, where their enormous breeding success was probably related to the absence of predatory mammals, birds, snakes, and lizards of the mainland, which prey so successfully on ground-nesting birds. Breeding colonies in the north Atlantic were known on St. Kilda Island of the Outer Hebrides and possibly on the Orkney Islands, Lundy Island, and the Isle of Man. The Great Auks also bred near Iceland's southern coast, on the rocky tips of volcanic islands that rise up from the Mid-Atlantic Ridge. In the western Atlantic, Great Auks bred on Penguin and Wadham Islands off the south coast of Newfoundland. Their greatest known breeding colony was on Funk Island north of St. Johns, Newfoundland.[31,32]

In a textbook on paleontology published in 1860, Richard Owen alerted his audience to the dire status of this bird, based on the scanty reports to him at that time:

> The Great Auk (*Alca impennis*, L.) seems to be rapidly verging to extinction. It has not been specially hunted down, like the dodo and dinornis [a moa], but by degrees has become more scarce. Some of the geological changes affecting circumstances favorable to the well-being of the *Alca impennis*,

> have been matters of observations. The last great auks, known with anything like certainty to have been seen living, were two which were taken in 1844 during a visit made to the high rock, called "Eldey," or "Meelsoekten," lying off Cape Reykianes, the S.W. point of Iceland. This is one of three principal rocky islets formerly existing in that direction, of which the one specially named for this rare bird "Gierfugla Sker" sank to the level of the surface of the sea during a volcanic disturbance in or about the year 1830. Such disappearances of the fit and favourable breeding-places of the *Alca impennis* must form an important element in its decline toward extinction. The numbers of the bones of *Alca impennis* on the shores of Iceland, Greenland and Denmark attest to the abundance of the bird in former times.[33]

When Owen wrote those words the Great Auk was already extinct. The breeding colony lost as Gierfugla Sker sank beneath the waves was probably the last of its kind. Owen didn't realize that the minor volcanic belch that precipitated the end of the species was simply the last in a series of catastrophes to hit the breeding colonies. Owen was also unaware that in fact the Great Auk *was* "specially hunted down."

There is a long history of the Great Auk in the Quaternary record, and its bones are usually associated with human artifacts. Evidently, the Great Auk was hunted and eaten by humans for thousands of years before becoming extinct.[34] During the Pleistocene the Great Auk roamed the Mediterranean, where it has been found in fissure deposits at Gibraltar between 70,000 and 90,000 years old. Its bones are also found in 60,000-year- old sediments of southern Italy, in association with artifacts of the Aurignacian culture. The Great Auk reportedly appears in cave paintings of northern Spain, in addition to bones from sediments along the Spanish coast. But the Auk's southern limit retreated northwards as the Quaternary progressed.

The early Holocene record of the Great Auk is well represented in Scandinavia. A Swedish archaeological site contains Great Auk bones that are about 9,000 years old. Additional records are found in 6,500-year-old sites in Denmark, as well as sites of varying age in Norway and the northern British Isles. A British naturalist visiting the Orkney Islands in 1698 observed live birds and described the Great Auk as being the "stateliest as well as the largest, of all the Fowls here, and above the size of a Solan Goose, of a Black Colour, Red about the Eyes, a large White Spot under each Eye . . . stands stately, his whole body erected, his wings short; he Flyeth not at all."[35] He also described the egg, which was "twice as big as that of a Solan Goose, and is variously spotted black, green and dark . . . appears on the first of May and goes away about the middle of June." At most of the historic and prehistoric sites, the bones occur with kitchen refuse and tool marks sometimes scar the bones, providing ample evidence that Great Auks were eaten. But their bones occur in only small numbers compared to other prey species.

By the beginning of the nineteenth century, the Great Auk had become very rare. Two birds were captured alive in 1812 near the island of Papa Westera in the Orkneys and kept alive for several years in captivity. The skin of one of these is now in the British Museum (Natural History) collections, where it eventually came under the scrutiny of Richard Owen. The last specimen taken in Europe was on the Waterford coast of Ireland in 1834. The last report of a Great Auk in Europe was 1835 at Lundy—the Isle of Puffins. A resident reported seeing two large birds that stood upright—"up bold like"—these might have been Great Auks.[36]

Along eastern North America, bones of the Great Auk have been found in middens of American Indians from the Bay of Fundy to Cape Cod. The southernmost record is an American Indian pre-Columbian shell mound in Florida. In New England, seventeenth-century archaeological sites with Great Auk bones contain metal implements of Europeans. British navigators reported that it lived from Cape Cod northwards 300 years ago, but exactly when they disappeared from American waters is uncertain.[37]

Despite their wide range across the north Atlantic, their breeding was confined to a small number of sites. At sea the birds were excellent swimmers, reportedly staying submerged as long as a seal. They were extremely evasive, which may account for their rarity in coastal archeological sites. On land, however, the Great Auks were equally awkward, making easy prey for humans and dogs. They were especially vulnerable as they congregated to breed on tiny oceanic islands. Sailors and fishermen took advantage of this and caught large numbers to be salted for food or to bait hooks for the great predatory fish of the north Atlantic. In 1534 Jacques Cartier visited Bird Rocks in the Gulf of St. Lawrence, writing, "These islands were as full of birds as any field or meadow is of grasse, which there do make their nestes." They killed "above a thousand," and "we put into our boates so many of them as we pleased, for in lesse than one houre we might have filled thirtie such boats of them."[38]

The north Atlantic sailing routes established by European navigators brought humans to the breeding colonies of the Great Auk with lethal regularity and persistence. One by one, the great breeding colonies were decimated by sailors exploiting the rich waters of the north Atlantic. Hundreds of thousands of birds congregated in one of the largest breeding colonies, on Funk Island. It was among the last to go, and its survival may have resulted from its remote location and inhospitable terrain for landing. But by the late eighteenth century a market had grown in Europe and America for pillows made of Great Auk feathers, driving hunters to land on Funk Island and harvest a huge cache.

The hunters used a hut built by an earlier sealing crew as their shelter during the Auk's breeding season. They also built several holding pens. As the wary birds came to shore, they were surprised and herded into the pens. With their powerful bills the auks bit their captors with the tenacity of dogs, but in the end, the flightless birds had no real chance. In the pens, they were clubbed and tossed into cauldrons of boiling water. Parboiling loosened the feathers, rendering them easier to pluck. The fat of the birds was rich in flammable oils that were used to stoke the fires and keep the cauldrons boiling for more birds. The stripped carcasses were tossed aside in great heaps. Before the end of the eighteenth century the Great Auk had been extirpated on Funk Island.

One final incentive for human predation was the growing value of Great Auk eggs and skins, as they became scarce. The declining Dodo had sparked a collecting fad for skins and eggs among Europe's wealthiest amateur naturalists. Errol Fuller records that 80 skins and 75 eggs of the Great Auk sold for hundreds of pounds at Steven's Auction Rooms in London's Covent Garden during the nineteenth century.[39] Many late eighteenth-century seamen knew they could fetch a high price for a Great Auk, its skin or its egg. Thus, the remaining adults and their eggs were hunted down, and the few birds that had survived the cataclysmic loss of their breeding grounds finally disappeared.

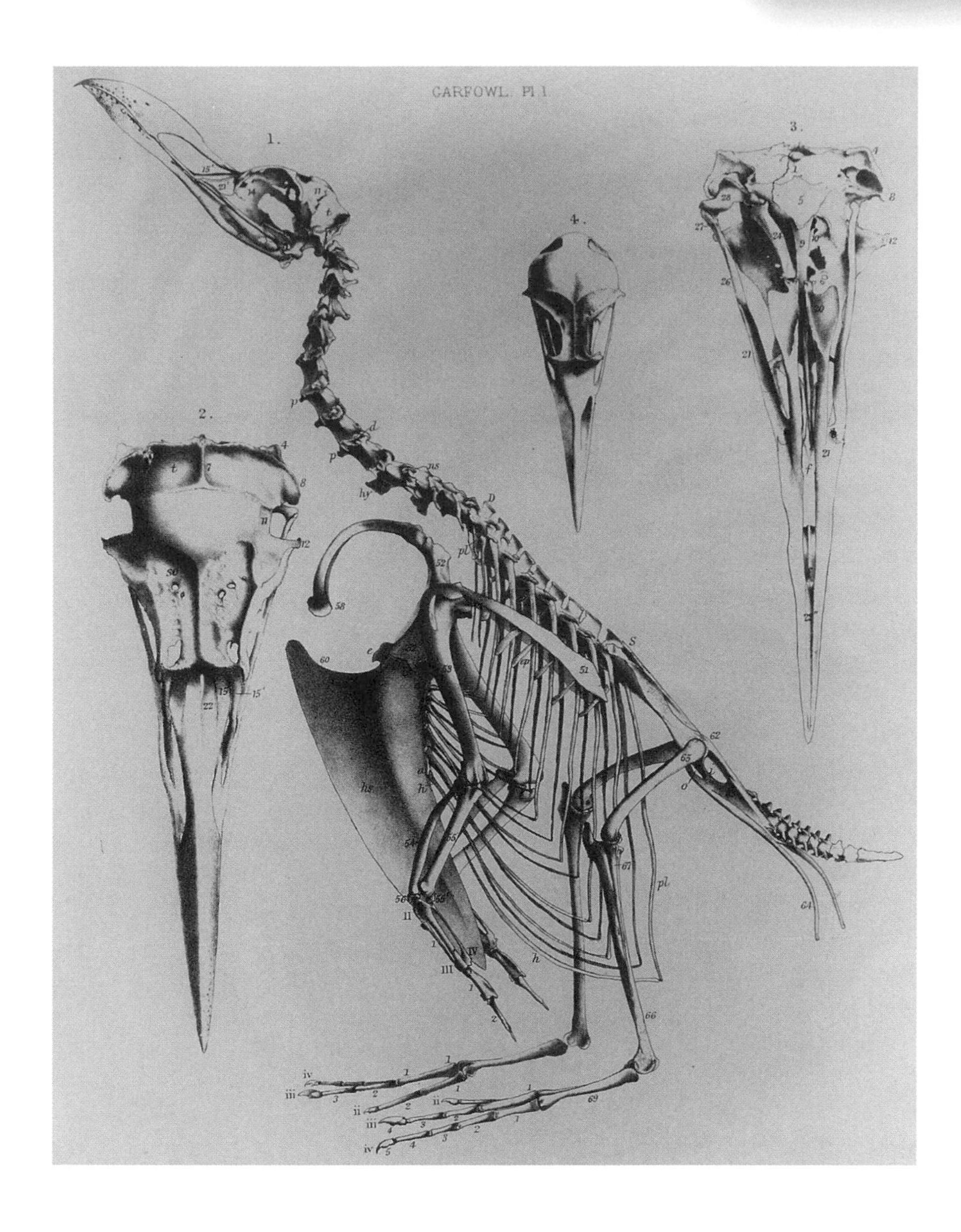

The skeleton of a Great Auk, from a study made by Richard Owen following its extinction.

The last two birds ever seen were spotted by sailors commissioned to find Great Auk specimens for private collectors. They scoured traditional breeding grounds on small islets off the southwest point of Iceland, where the submerged Gierfugla Sker had been specially named for the bird. Landing on Eldey Rock, on the morning of June 3, 1844, they saw among a mass of other seabirds a pair of Great Auks. The sailors immediately set upon the birds, which waddled away as fast as they could. But the birds were clubbed down, and it was rumored that an egg being incubated by the female was crushed underfoot in the excitement. Their carcasses were sold to a collector in Reykjavik. There was one, last, unconfirmed sighting of a Great Auk at the Grand Banks of Newfoundland in 1852, but that was it.[40] The Great Auk had passed from the realm of ornithologists to the realm of archeologists and paleontologists.

In 1864, more than twenty years after the last live specimen was collected, Richard Owen published a monograph on the skeleton of the Great Auk.[41]

Remains of the Great Auk were only very poorly represented in natural history museum collections at the time it became extinct. As the basis for much of his description, Owen had to settle for a specimen found by guano miners on some north Atlantic island, which came to him "dried, flattened featherless, and mummified." At the end of the nineteenth century, naturalists returned to Funk Island to collect a more extensive sample of bones. In a Report of the U.S. National Museum and Smithsonian Institutions, Frederick Lucas described his landing on Funk, July 22, 1887. "Here the Auk bred in peace for ages, undisturbed by man. . . . Here to day the bones of myriads of Garefowl lie buried in the shallow soil formed above their moldered bodies, and here, in this vast Alcine cemetery, are thickly scattered slabs of weathered granite, like so many crumbling tombstones marking the resting places of the departed Auks."[42]

Lucas described the stratigraphy of the island as having has two distinct layers. The lower and oldest lies directly on granite and weathered pebbles. Formed during the occupancy of the Great Auk, it contains many eggshell fragments, as well as charcoal that had leached down from the overlying layer. "The upper layer of the soil, also from 3 inches to one foot thick, has formed since the extermination of the Auk, principally by the growth and decay of vegetation nourished by their bodies. In fact it is possible, from the character of the plant growth above, to tell something of the probable abundance of Auk remains below; the thickness of the one indicating corresponding plenty of the other."[43]

Excavations in the region of the hut yielded the bones of "thousands of birds mixed together in inextricable confusion." The crania of some birds preserved the marks of cuts and blows, verifying local traditions about how the birds met their end.

The Great Auk was not the only bird to be extirpated on Funk Island. The Gannet, the Puffin, as well as the Common and Arctic Terns also bred there in great abundance when Funk Island was described as "a mountain of birds" by naturalists visiting the island in 1844. A few years later the Gannet and Common Tern were gone, and all species of birds breeding on Funk Island, with the exception of the Puffin, were drastically diminished by egg collectors. Puffins found security in their burrows and to them, at least, the extermination of the Great Auk presented a decided advantage by providing soil in which to dig their habitations.[44] Lucas and subsequent collectors prospected at the openings of burrows for the bones of the Great Auk that were unearthed by the digging Puffins.

A half-century after the Great Auk disappeared, Frederick Lucas wrote in a Smithsonian Report: "The circumstances that the bird, with suicidal persistence, resorted to the a few chosen breeding places, and that it was there found in great numbers, rendered its destruction not only possible but probable, and when the white man first set foot in America, the extinction of the Great Auk became merely a matter of time."[45] If these words seem overly dramatic, subsequent research on island birds only reinforces the message. A century later David Steadman summed up the second pulse of Quaternary extinction: "We expect extinction after people arrive on an island. Survival is the exception."[46]

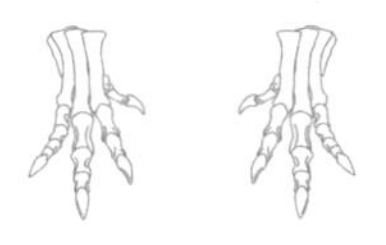

CHAPTER EIGHTEEN

The Third Wave

Island birds are especially vulnerable to extinction because they inhabit small areas, have small population sizes, or both. On continents, there are far greater areas to inhabit, and huge population sizes are more common. Continents generally have longer and more diverse geological histories than oceanic islands. With more time, area, climatic changes, topographic variation, and habitat diversity, the continental biota evolves with a different dynamic than do most island biotas. So continents are home to far greater numbers of species, and they are more strongly buffered against extinction.

Around 90 percent of the bird species that have become extinct in the Holocene were endemic to islands.[1] Many other island birds remain threatened or endangered today. A 1994 survey reported roughly 250 endangered bird species; over half are island endemics. But the proportion of endangered and threatened island birds is gradually diminishing, because most vulnerable island birds are already extinct and the number of endangered continental species is rising rapidly. Stanley Temple of the University of Wisconsin has observed that today the locus of bird extinction is shifting. "Although the majority of the world's endangered species still come from islands, it is clear that the process of extinction has been shifted geographically from relatively low-diversity island avifaunas to species-rich continental areas and habitats that support the world's greatest variety of bird life. As the numbers of endangered and extinct birds from these areas continue to expand, there will be an unprecedented reduction in the diversity of birds on Earth."[2] This third wave of Quaternary dinosaur extinction may ultimately prove the most severe of all.

NORTH AMERICA

In the last 200 years, North America has lost more bird species than any other comparable landmass.[3] There are three major sources of reduced survivorship, all familiar from our tour through of some of the world's islands: introduced species, human overkill, and habitat destruction. Each is exacerbated by human population growth.

James Greenway, of the American Museum of Natural History, studied the geography of extinction and its symmetry with human population growth in New England. Europeans arrived in about 1600, when the eastern half of the continent was heavily forested. Most was cleared by humans in the early years of westward expansion. Long before Europeans arrived, the forests had already been altered by American Indians, who frequently burned the undergrowth. But European tools and agricultural practices induced heavy deforestation. The original forest highlands were dominated by hickory, oak and chestnut trees. Lower hills and valleys supported maple and beech forests. Natural fires were also a regular part of these forests, creating many clearings among the trees. These forests have now been decimated. In their place stands secondary growth and agricultural land, vastly altering the original environment.

Greenway estimated that 431 million acres of original forest were cleared; only about 19 million acres remain. Another measure of this loss involves population density. In 1754, there were 24 acres of forest per person in Massachusetts. About 1790, as settlers began to move west toward the Appalachians, large tracts of forest were burned to augment agricultural areas. As populations grew, the forests shrank. In 1776 there were 17 acres per person, in 1800 ther were 11 acres, by 1830 only 8 acres, and by 1850 only 4. By 1850, virtually all the original forest was gone in Massachusetts. Within a few decades, virtually the entire United States east of the Mississippi had been cleared. Several bird species abundant in 1850 are now extinct, and a growing number are threatened or endangered.

American universities had established excellent traditions in natural history by the beginning of the nineteenth century. For more than 200 years the eastern United States has been studied by skilled natural histo-

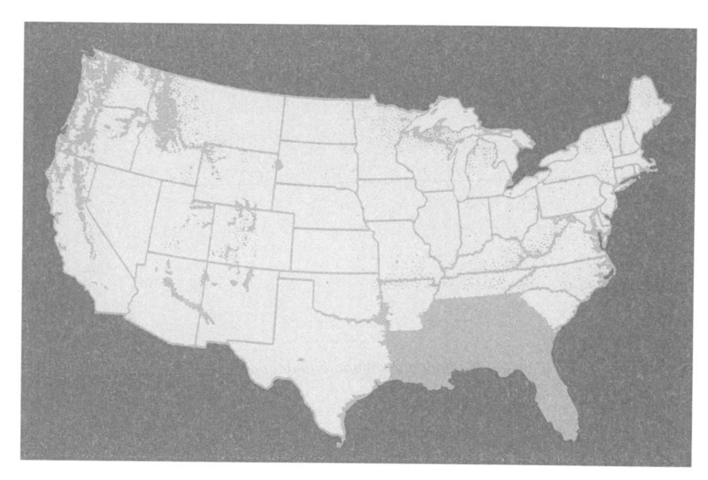

When European settlement of North America began in earnest in the early seventeenth century, forests stretched over half the continental United States. By the early twentieth century, only a sparse patchwork of original forests remained.

rians, observing and writing about birds. One of the greatest was a prolific writer, a meticulous illustrator, and the world's foremost authority on the living species of North American dinosaurs—John James Audubon. Several species that he observed and painted in the wild, at a time when they suffered no obvious threat, have since become extinct. In these cases, the trail of clues leading scientists to the cause of their extinction is preserved on paper rather than in stone. The testimony of Audubon and the subsequent generations of wildlife biologists he helped inspire vividly document the extinction of species on a continental scale.

MARTHA

Next to *Archaeopteryx* and the Dodo, the Passenger Pigeon (*Ectopistes migratorius*) is probably the world's most famous extinct bird. At the beginning of the nineteenth century, it was quite possibly the most abundant bird in North America, if not on Earth. It is hard to understand how it could be that the last wild specimen was shot in the early fall of 1899, and by then the species was all but extinct.[4,5]

Audubon prefaced his description of the Passenger Pigeon with the words:

> The multitudes of Wild Pigeons in our woods are astonishing. Indeed, after having viewed them so often, and under so many circumstances, I even now feel inclined to pause, and assure myself that what I am going to relate is fact. . . . In the autumn of 1813, I left my house at Henderson, on the banks of the Ohio, on my way to Louisville. In passing over the Barrens a few miles beyond Hardensburgh, I observed the Pigeons flying from north-east to south-west, in greater numbers than I thought I had ever seen them before, and feeling an inclination to count the flocks that might pass within the reach of my eye in one hour, I dismounted, seated myself on an eminence, and began to mark with my pencil, making a dot for every flock that passed. In a short time finding the task which I had undertaken impracticable, as the birds poured in countless multitudes, I rose, and counting the dots then put down, found that 163 had been made in twenty-one minutes.

In the early nineteenth century, the Passenger Pigeon congregated in flocks of millions of birds.

> I traveled on, and still met more the farther I proceeded. The air was literally filled with Pigeons; the light of noon-day was obscured as by an eclipse, the dung fell in spots, not unlike melting flakes of snow; and the continued buzz of wings had a tendency to lull my senses to repose.[6]

In 1810, Alexander Wilson, a preeminent ornithologist of the time, observed a huge flock in Kentucky, estimating that it contained 2.2 billion individuals.[7] A later ornithologist reported in 1871 on a breeding colony in Wisconsin, estimating at least 136 million birds were present in an area covering 850 square miles.[8] The roosts of these great flocks could also cover several square miles, and trees often collapsed under the sheer weight of roosting pigeons.

Given that the bird is now extinct, it is hard to believe that such accounts were true, regardless of their sources. But a close relative of the Passenger Pigeon also migrates in great flocks and breeds in immense colonies containing millions of birds. The Eared Dove (*Zenaida auriculata*) lives in northeastern Brazil and central Argentina, forming spectacular, wide-ranging flocks that confirm the estimates for the Passenger Pigeon.[9] It is a close analog to the Passenger Pigeon in many key evolutionary strategies that may have been responsible for the extinction of its cousin. American hunters and professional bird trappers are commonly indicted as primary suspects in the loss of the Passenger Pigeon. But from studying the Eared Dove and historic accounts of the Passenger Pigeon, it is evident that habitat alteration was a significant, if not predominant, factor in its extinction.

The Passenger Pigeon could fly at sustained speeds approaching 60 miles per hour, rapidly covering long distances. Its migrations represented great foraging expeditions. As Audubon described: "These are entirely owing to the necessity of procuring food, and are not performed with the view of escaping the severity of a northern latitude, or of seeking a southern one for the purpose of breeding. They consequently do not take place at any fixed period or season of the year. Indeed, it sometimes happens that a continuance of a sufficient supply of food in one district will keep these birds absent from another for years."[10]

As they entered new areas to feed, the birds were massacred and trapped with deadly efficiency as they roosted at night in their breeding grounds. Audubon described one site 40 miles long and 3 miles wide, where Pigeons roosted along the Green River in Kentucky, detailing the slaughter that was planned and executed.

> Many people were involved, one traveling over 200 miles and driving 300 hogs to be fattened on the pigeons that were slaughtered. "Here and there, the people employed in plucking and salting what had already been procured, were seen in the midst of large piles of these birds. The dung lay several inches deep, covering the whole extent of the roosting-place. Many trees two feet in diameter, I observed, were broken off at no great distance from the ground; and the branches of many of the largest and tallest had given way, as if the forest had been swept away by a tornado. Everything proved to me that the number of birds resorting to this part of the forest must be immense beyond conception. As the period of their arrival approached, their foes anxiously prepared to receive them. Some were furnished with iron-pots containing sulfur, others with torches of pine-knots, many with poles, and the rest with guns. . . . Suddenly there burst forth a general cry of "Here they come!" The noise which they made, though yet distant, reminded me of a hard gale at sea, passing through the rigging of a close-reefed vessel. As the birds arrived and passed over me, I felt a current of air that surprised me. Thousands were soon knocked down by the pole men. The birds continued to pour in. The fires were lighted, and a magnificent, as well wonderful and almost terrifying, sight presented itself. The Pigeons, arriving by thousands, alighted everywhere, one above another, until solid masses were formed on the branches all round. Here and there the perches gave way under the weight with a crash, and, falling to the ground, destroyed hundreds of the birds beneath, forcing down the dense groups with which every stick was loaded. It was a scene of uproar and confusion. I found it quite useless to speak, or even to shout to those who were nearest to me. Even the reports of the guns were seldom heard, and I was made aware of the firing only by seeing the shooters reloading.[11]

Professional netters trapped birds by the thousands. By the time the laws banning the practice took effect it was too late. From one nesting colony, two railroad cars per day filled with pigeons could be sent to eastern markets. Audubon visited New York in March of 1830 and found that pigeons "were so abundant in the markets, that piles of them met the eye in every direction." At the famous Wisconsin colony of 1871, 1.2 million birds were taken by 600 professional pigeon netters in one nesting. Reportedly, "hundreds" of sport hunters and professional trappers followed the birds across wide regions, taking a large toll as they went.[12] With such vivid testimony it might appear that the bird's extinction rests squarely on *Homo sapiens*.

But astonishingly, the number of birds killed by people may have amounted to only a few percent of these flocks.[13] The Eared Dove, whose populations range between 1 and 10 million birds, have suffered similar losses without significant long-term declines in population levels. In northern Argentina, an Eared Dove population estimated at 3 million is considered a pest, and attempts to extirpate it have included poisoned baits and year-round trapping. About 420,000 birds were exterminated in one application of poison, but no long-term effects were observed. So many birds breed several times each year that the species withstood the attacks by doubling or quadrupling their numbers every season.

Several key life-history characteristics of the Passenger Pigeon appear to have been responsible for their extinction.[14] The Passenger Pigeon may have descended from pigeons of the Great Plains, evolving into the hardwood forests and exploiting its profuse food supply. Whatever its prehistory, during historic times the Passenger Pigeon never spread beyond the eastern hardwood forests. The limiting factor was food. Its main staple was mast—the fruits of hardwood trees, particularly beech trees. Beech mast is higher in protein and carbohydrates than acorns and chestnuts, but the pigeons fed on these also. They foraged for accumulations of these fruits on the ground beneath the trees. The distribution of mast-producing trees naturally limited the pigeon's upper range to the southern part of Canada, and to the eastern half of North America.

Mast production is irregularly timed and distributed across eastern forests. Mast trees tend to have abundant crops every two to five years, but no regular cycle exists, except that the mast-producing trees of several species often synchronize their irregular production. Consequently, all the trees over widespread areas would produce in great abundance at the same time. Some biologists have argued that this satiates seed predators, like squirrels and jays, as well as pigeons, while leaving some seeds to produce new trees. Whatever the reason, forests present a patchwork of resources in any given season, with some regions producing abundant mast supplies and others not. During "migration," the Eared Dove wanders widely but loosely follows a linear course as it scours broad areas of the forest for food.

The Passenger Pigeon migrated in the same pattern. Its great flocks were well suited for discovering and exploiting the patches of high mast production. They would respond to sightings of other birds on the ground that might have discovered a mast cache. Hunters, accordingly, set out "stool pigeons", luring dozens or hundreds of birds into a trap. Covering vast areas, the pigeons scoured the ground for mast. The availability of food was critical for their breeding. Both the length of their breeding season and their spring nesting episodes were correlated with the supply of mast. Population levels were probably also determined by mast supply, as they are in the Eared Dove.

Their gregariousness and the size of their congregations greatly facilitated the discovery of food.

As settlers moved westward, old forests were completely cleared, transforming large areas into agricultural lands, especially where beech trees had lived in fertile bottom lands. Huge volumes of wood were also cut for fuel. The amount cut for fuelwood alone probably exceeded today's overall timber harvest. Deforestation accelerated in the 1870s with the invention of portable steam sawmills, and by the end of the nineteenth century most old tracts of original forest disappeared. Secondary growth that has been allowed to come back between the agricultural regions consists mostly of sprouts that rarely grow to maturity, because it is ready to cut for lumber in half the time it takes for the trees to grow into large-scale mast producers. The brushy understory that this type of forest fosters is difficult for the pigeons to move through as they look for mast. Domestic hogs competed with Passenger Pigeons for the mast that was produced by the trees that had not been cut. The decline of the forests coincided with the decline in the great breeding colonies. Evidently, their social organization was critical for locating food, because isolated pairs and small groups that survived into the 1890s eventually disappeared. Scattered sightings were not uncommon in the first few years of the twentieth century.[15]

In a way, Audubon predicted this fate. "Persons unacquainted with these birds might naturally conclude that [human predation] would soon put an end to the species. But I have satisfied myself, by long observations, that nothing but the gradual diminution of our forests can accomplish their decrease, as they not unfrequently quadruple their numbers yearly, and always at least double it."[16]

The last known individual of this species was named Martha by keepers in the Cincinnati Zoological Gardens, where Martha passed her last years. In late 1914, in the ornithological journal *The Auk*, an anonymous editor published this obituary: "*Ectopistes migratorius*, once one of the most notable species in the North American avifauna became extinct on September 1, by the death of the last surviving specimen, a female, which had lived for twenty-nine years in the aviary of the Cincinnati Zoölogical Garden. It is rarely possible to state the exact date of the extinction of a species as the process is usually a gradual one, but in view of the fruitless efforts extending over the past ten years to find evidence of the existence of wild Passenger Pigeons we may safely consider the passing of this last captive specimen as the extinction of the species."[17]

TOO ADAPTABLE?

The extinct Carolina Parakeet (*Conuropsis carolinensis*) suffered a similar fate, its last members dying in the same zoo only four years later. But its extinction was different. It was present in great abundance when Europeans arrived in North America, ranging from the Gulf of Mexico to the Great Lakes, and from the Mississippi River east to Florida. It had an orange face, a yellow head and neck, with green body plumage. It lived in nonmigratory flocks, nesting in holes of mature trees in old deciduous forests. They sometimes assembled sizable rookeries in huge hollows. Reportedly, they had a fondness for seeds of elm, maple, cypress, pine, and beech, as well as thistles and cockleburs.[18]

The Carolina Parakeet.

But their voracious appetites made them a pest. As the original forests were cut and burned, they turned to domestic seeds and fruits like a plague of locusts. Audubon recorded that whole flocks would descend on grain fields covering them "so entirely that they presented to the eye the same effect as if a brilliantly coloured carpet had been thrown over them." [19] They destroyed entire orchards in a "wanton and mischievous manner," a characterization in keeping with the large brains and intelligence of parrots generally. The demise of the Carolina Parakeet was hastened by its gregariousness. Instead of flying away, the flock would often hover over birds wounded by hunters, facilitating the flock's extermination in a single afternoon.

During 90 years of westward European expansion, the range of the Carolina Parakeet retreated rapidly toward the Mississippi. In their wake, frontier colonists left a broad swath of deforested lands barren of the parakeet's natural food. The last sightings of this bird generally coincide with the first wave of European population growth and the establishment of permanent settlements. The final sightings in Ohio were in 1832, in Indiana 1856, in Kentucky in 1878, in the central Mississippi drainage in 1857, from the upper Missouri River in 1881, from Florida in 1904, from Missouri in 1905, from Louisiana in 1910, and from Kansas in 1912.[20] Through human extermination and habitat destruction, the range of the Carolina Parakeet shrank to the size of one cage in the Cincinnati Zoo, before vanishing entirely.

THE CALIFORNIA CONDOR

The California Condor teeters close to extinction, due not so much to habitat destruction as direct human persecution. It is a giant member of the vulture family, which today consists of seven species ranging across the Western Hemisphere. The Andean Condor is the largest flying bird, but its northern cousin is only slightly smaller. Both have wingspans approaching nine feet. Their feet leave footprints in soft substrate that are seven inches across. Condors are the undisputed soaring champions of the land, as the albatross is over the water. Effortlessly riding thermals, they soar widely over the California scrub, searching for dead or dying animals, even along roadways and around dumps. Their hunting network is widespread, because the birds watch each other from great distances. If one goes down after food, others in the air know it right away. No record exists of a California Condor attacking a living animal. They are generally solitary but are also seen in pairs. It is not known whether they pair for life, but the same pair has been observed to return to the same nest season after season for many years. They survive in captivity for three decades. All members of the family are now protected, but the California bird remains critically endangered.

In the Pleistocene, the California Condor ranged over much of North America. Even in prehistoric times these birds extended east into Florida and have been found in caves and other sites in Texas, New Mexico, Arizona and Utah. Between 1,500 and 3,000 years ago the Condor nested in Texas and probably other southwestern states.[21] But the arrival of American settlers in the West led to a steady restriction of its range. By the 1800s, it disappeared from all but the western margin of its former range. During the nineteenth century, condors still soared majestically over the ground from British Columbia to Baja California. The last sighting of a condor in Mexico occurred about 1930. Since 1937, the condor has been confined to California, where its populations continue to diminish.

The decline of condor populations is related to the colonization of the west. Many condors were shot, possibly constituting the major source of mortality in these slowly reproducing birds. Individuals also died from eating poisoned carcasses of bears and coyotes that were targeted in massive "varmint" campaigns in California during the nineteenth century. Strychnine and lead poisoning, more than habitat decline, have both been cited as the source of condor mortality. By 1980, only about 23 birds remained in the wild, along with about 2 dozen captive birds in American zoos.[22] In 1987, the last of the wild birds were captured. They reproduced with some success and the numbers have rebounded to the point that by July 1994, the captive population had risen to 89. Birds are now being returned to the wild with some success in California. Six birds were also released in 1996 onto the Colorado Plateau of Arizona and southern Utah, to establish a second wild population. As of this writing, there are 17 condors in the wild and 104 in captivity.

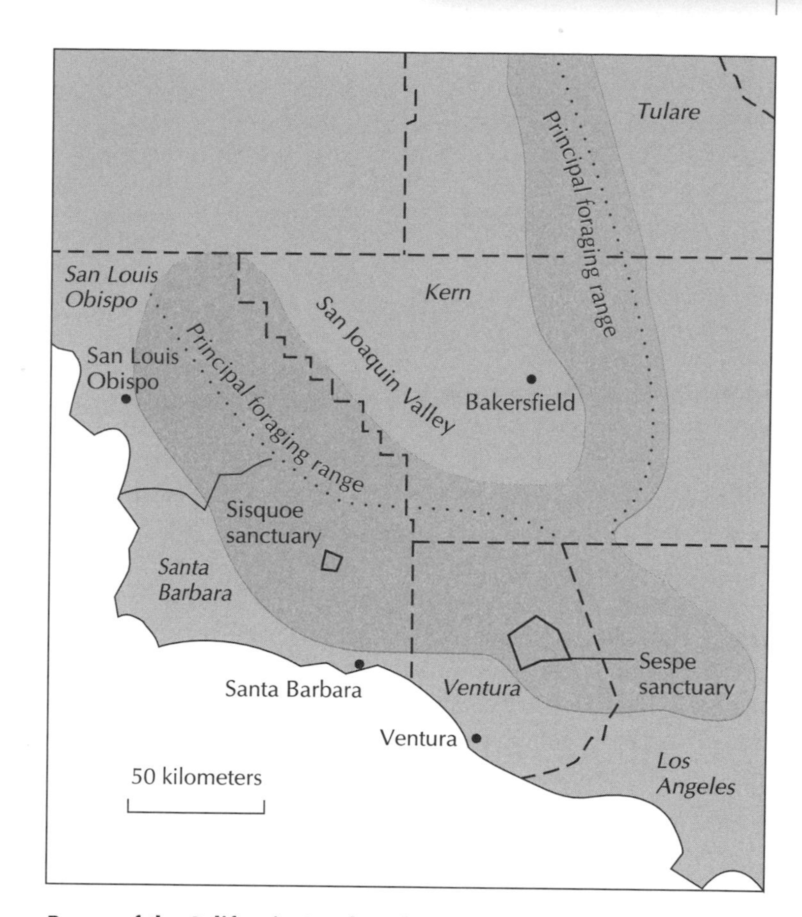

Range of the California Condor, since 1937.

GEOGRAPHY OF THE CONTINENTAL THREAT

Today across North America there are fifty threatened bird species.[23] Habitat disruption has catalyzed complex consequences that are beginning to catch up with North American birds. In the United States, numerous endangered bird species are threatened either directly or indirectly by disruption of woodlands and adjacent wetlands. Bachman's Warbler, Kirtland's Warbler, the Golden-cheeked Warbler, the Black-capped Vireo, the Red-cockaded Woodpecker, the Spotted Owl, and the Whooping Crane are among thirteen critical species, twelve endangered species, and twenty-one vulnerable species, many orinthologists believe that Bachman's Warbler is already extinct. In some cases the threatened habitat is in the tropics outside of North America, where the birds breed. In others, the threat is related to competition from introduced species or species that thrive in the disrupted habitat. Environmental toxicity also drives the decline, most strongly affecting birds at the top of the food chain.

Birds on other continents face the same problems, but far more species are at risk. The threat is most severe in the tropical forests of South America, central Africa, and southeastern Asia because this is where avian diversity is greatest. Until recently, these regions were spared the effects of European technology on

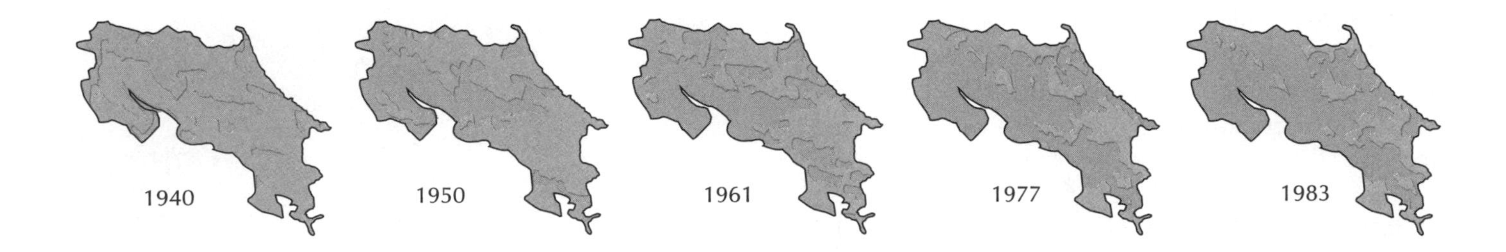

In the few short years between 1940 and 1983, Costa Rica was heavily deforested.

forested lands. But lately technology and foreign agricultural practices have struck with a vengeance.

Throughout tropical forests, human populations are growing very rapidly. Many inhabitants are impoverished, and providing fire for warmth and cooking represents a strong incentive to cut down trees. Wood is cheap compared to kerosene or bottled gas. By 1980, developing countries used wood and other "noncommercial" sources for 90 percent of all energy consumed. In many countries deforestation has yielded desertification—expansion of deserts into previously forested regions. Globally in 1987-1989, just over half the wood produced annually was used for fuel and charcoal. In Africa during this period, 89 percent of cut wood was burned for these purposes, and in Asia 74 percent. In Europe, however, only 15 percent of the wood cut went to fuel. European populations are now comparatively static, while in tropical countries populations are growing at rates exceeding the 2 percent per year global average.[24,25]

An insidious cycle commences with tropical deforestation. Forested uplands are cleared for new agricultural lands or fuel. Excess runoff from rain, which had been prevented by the forest's biota, triggers rapid soil erosion. Any remaining soil is soon leached of its nutrients by the farming practices. Having lost its agricultural value, it is abandoned, and another tract is burned. Down river, in developing countries without adequate flood control systems, flooding and crop destruction ensues from the increased upland runoff. And as the population grows, more of the land is bound in this chain of poverty. As forests are cleared, some of the carbon combined in the wood is released to the atmosphere as carbon dioxide. As E. O. Wilson writes "The net loss of tropical forest cover world-wide during 1850-1980 contributed between 90 and 120 billion metric tons of carbon dioxide to the Earth's atmosphere, not far below the 165 billion metric tons emanating from the burning of coal, oil, and gas."[26]

Joel Cohen estimates that developing countries converted 1.45 million kilometers (145 million hectares) of forest to farmland between 1973 and 1988. During this time, human populations increased by 1.2 billion people. This translates to deforestation of 0.12 hectare per person—a rectangle 50 meters by 24 meters or one-quarter of an American football field. If each additional person requires 0.12 hectare of land, each additional billion people require 1.2 million square kilometers of additional land, which would otherwise host wildlife.[27] Wilson observed the irony, that if nineteenth-century technology had been born amidst tropical rain forests instead of temperate-zone oaks and pines, there would be very little biodiversity left to save.

In 1964, a series of reports on biodiversity, known as the Red Data Books, were compiled to identify and document endangered birds throughout the globe to prevent their global extinction.[28] The thresholds for determining whether a species is endangered or merely threatened reflect a range of criteria. If there has been a rapid decline in their population or if their range is small and becoming fragmented, they may be critically endangered. The overall size of their population, range, and other factors are also involved. Critically endangered species are those judged to have a 50-50 chance of becoming extinct in the next five years. Across the map of avian phylogeny species are threatened with extinction.

In the rapidly shrinking forests of South America, many different lineages of woodland birds have species that are threatened or endangered. Today, Brazil ranks first among New World countries in the number of endangered birds, with 16 critically endangered species, 31 endangered, and 56 vulnerable—a total of 103 threatened species. Colombia is fourth with 62 threatened species, Peru, fifth, with 60 threatened species; Ecuador, seventh, with 50 species; Argentina, twelfth, with 40 threatened species; and most other countries host at least a few endangered species.

Through the loss of both wetlands and forest, many birds are threatened across southern Asia, Indochina, and over southeastern China. Human exploitation has also been a significant factor. China is second among continental countries in having 86 threatened species, India is third with 71 threatened species, and most Old World continental countries are inhabited by threatened birds.

EVOLUTION AND EXTINCTION

To Richard Owen, as to most pre-Darwinian scholars, species were static and incapable of change. Perhaps this is somewhat unfair, because there is evidence that Owen entertained the notion of limited transformation among species. Owen wrote of the *loss* of flight in the Dodo, for instance, admitting that its ancestors must have flown to Mauritius. But in the end, he rejected the idea that two major groups within the Linnean classification, like Reptilia and Aves, could be related to one another as ancestor and descendant. When push came to shove, he rejected evolution as a general principle. Owen refused to accept that there could be small dinosaurs or that they could evolve feathers and flight, so he could never realize his mistake in claiming that Dinosauria is extinct.

If a meteorite struck Yucatan 65 million years ago, its effects on the indigenous dinosaurs were only short-lived. Central America is one of the centers of avian

The colored areas of tropical South America have especially high numbers of endemic species.

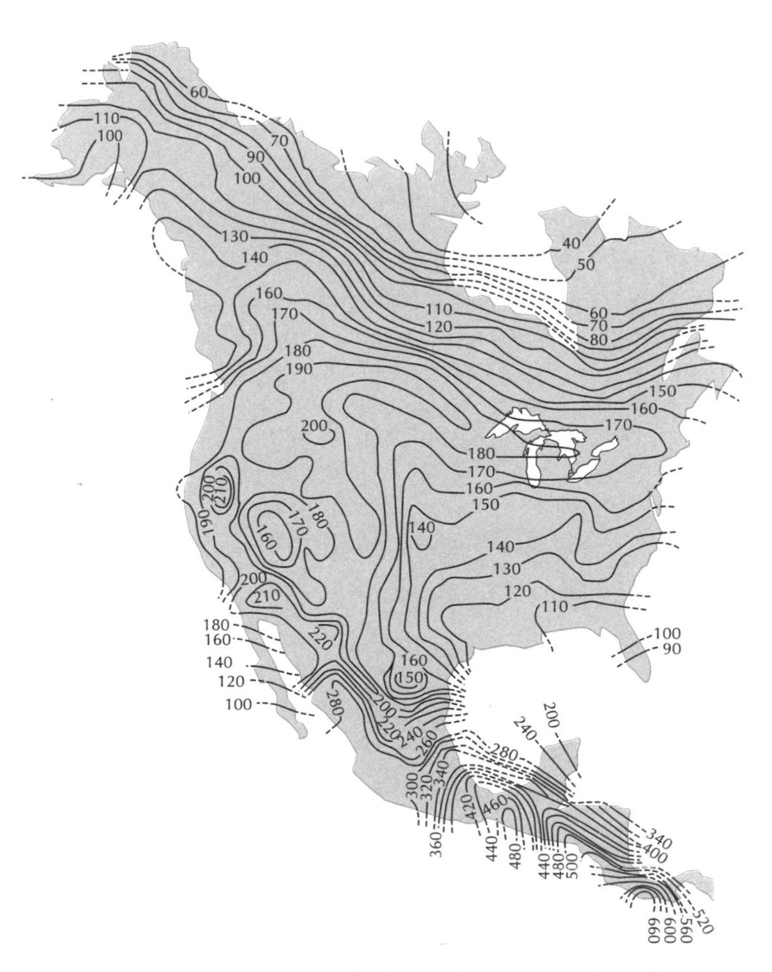

The contour lines on this map indicate the number of nesting bird species. Diversity rises by an order of magnitude moving from arctic regions toward the equator.

diversity today. Evidently, the evolution of today's great diversity of birds required great expanses of time, and we can now trace the history of Central American birds back across the K-T boundary and deep into time. The extinct Mesozoic dinosaurs, moas, and other flightless birds that Owen described are all parts of this same major highway system on the vertebrate phylogenetic map. Owen's first encounter with an extinct dinosaur was with the moa, and he unwittingly rediscovered the lineage when he named Dinosauria.

Richard Owen and his grandson, Richard, c. 1880.

Richard Owen was among the first to study extinct island birds, and he made them widely known within the scientific community. Although he recognized that some bird extinctions were caused by humans, he never suspected the full portent that scientists read into his discoveries today. Without understanding how species originate, that species evolve from other species and that time and diversity are linked, Owen never fully understood the meaning of the extinctions to which he was such an articulate witness. But in a way, he was also fortunate in his ignorance of the future for dinosaurs that we now contemplate.

With methods like cladistics that offer greater testability for scientific hypotheses, the major features of vertebrate phylogeny have now been mapped out in general form. If current phylogenetic maps are correct, it appears that dinosaurs and our own primate lineage co-existed for a time span exceeding 70 million years. Both lineages survived whatever happened at the end of the Cretaceous. They further withstood the Pleistocene wave of extinction that surged across the globe. But following in its wake have come second and third waves, the second engulfing tropical islands and the third now moving across the continents. Both of the Holocene waves have cascaded from the tragic symmetry between human population growth and the loss of biological diversity. The third wave of extinction has yet to reach its crest, and it probably won't until after human population growth has peaked and begun to decline. By current estimates this is still a century away. So in our museums and classrooms we are training the next generation of scientists to test our view of living dinosaurs with discoveries of their own, and to confront the idea that the emergence of human agricultural and technological societies presents the greatest threat that dinosaurs have ever faced.

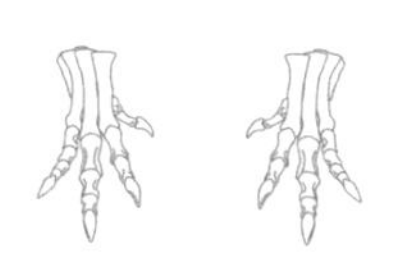

EPILOGUE

Destiny Hangs in the Balance

Our charge in this book has been to investigate all the issues surrounding the mystery of dinosaur extinction. In the process, we have explored not only what caused the extinction of the regal forms that dominated the Mesozoic world, such as *Tyrannosaurus* and *Velociraptor*, but also the destiny of their descendant relatives, the birds. We have laid out the evidence gleaned from the rock and fossil records so that you can judge for yourself what happened. From our perspective, however, this investigation has led to some startling conclusions.

It's hard for us to believe that, since we met at Berkeley, we may well have witnessed and even participated in the solution of long-standing mysteries involving dinosaur extinction. But it wasn't the solution that either of us envisioned when we started. Like many breakthroughs in science, solving the mystery turned on asking the right question. The final solution required us to ask, "Did *all* dinosaurs really go extinct at the K-T boundary?" This is a very different question than the one scientists had asked for more than 150 years: "What caused the extinction of dinosaurs at the K-T boundary?" To our surprise, the solution didn't involve catastrophic asteroid impacts or cataclysmic volcanic eruptions, and we realized that, even in the "modern scientific age," our world view could be jolted and turned upside down.

It was exhilarating to see the historic arguments of the nineteenth century replayed in our contemporary setting. Several aspects of the classical debates over catastrophism vs. uniformitarianism that accompanied the emergence of geology as a science were revisited by our colleagues studying the K-T boundary. Our fascination stemmed from the fact that issues of historic proportion were once again at stake. The modern discovery of evidence pointing to the likelihood of a great bolide impact at the end of the Cretaceous represents an historic breakthrough in terms of potential mechanisms for interpreting Earth and evolutionary history. This advancement in our knowledge will certainly provide us with an incentive to watch the sky much more closely in the future. Compelling evidence has also been recovered from the geologic record which establishes that an enormous episode of volcanic activity punctuated the end of the Cretaceous. However, the damaging consequences of these events could have had very similar effects on the Earth's environment and biota—atmospheric pollution, acid rain, initial climatic cooling, subsequent climatic heating, and so on. Due to the potential of these events to trigger similar "killing mechanisms," along with our inability to tell time precisely at catastrophic intervals in the Cretaceous, it is impossible to be certain what caused the extinction of non-avian dinosaurs.

Also replayed in two different centuries was the historic debate over the ancestry of birds. In the modern drama, *Deinonychus* replaced *Compsognathus*, and more evidence has been brought to bear on the question. But the plot is basically the same. Today, the great preponderance of evidence preserved in the skeletons, genes and growth patterns of fossil and living animals supports the conclusion that Mesozoic theropods were the ancestors of birds, just as most of the evidence did a century ago. More importantly, a classical battle over the theory of evolution was also replayed, framed in modern times as a battle between Linnaean taxonomists and phylogenetic systematists. Today's battlefield might seem different because nearly everyone claims to be an evolutionist. But conceptual tools forged by anti-evolutionists, like the theory of homoplasy, are still at the root of the conflict. Pre-Darwinian tools for ordering the diversity of Life, like classifications based on overall similarity instead of ancestry, catalyzed both generations of the debate. As before, these outdated tools and concepts perform poorly in explaining all the available evidence or making predictions about the future. The Darwinian revolution should have led to a fundamental change in the methods we use to classify the Earth's biota. However, it took more than a century for scientists to recognize the full implications of the theory of evolution and to forge a new set of tools.

Our investigation also provided unexpected insights into how our world views are shaped. Although we had learned about uniformitarianism and evolution as undergraduates, we didn't realize how many new discoveries would continue to cascade from these great ideas. We didn't appreciate the degree to which these ideas would continue to alter our world view as scientists struggled to understand humanity's place in Nature. That uniformitarianism and evolution continue to be hotly discussed and debated attests to their enormous explanatory power and the wealth of knowledge that has flowed from them. Detecting the impact at the end of the Cretaceous and tracking the ancestry of birds were as much an exercise in understanding the history of these two ideas as in sleuthing for new evidence in the geologic and biologic records. After all, Walter and Luis Alvarez recognized the great bolide impact from an almost immeasurably small amount of iridium, and Carl Gegenbaur connected the evolutionary paths of modern birds and Mesozoic dinosaurs using only one fossil—*Compsognathus*.

While students of Nature have always encountered surprises such as these, we can see that today's students face a very different future than we or our predecessors did. As naturalists circumnavigated the globe in the eighteenth and nineteenth centuries, the world was dazzled by the wealth of new species of living birds that were discovered. Students struggled to comprehend the seemingly unbounded diversity of Life that naturalists were documenting on the first large-scale scientific explorations of the world. Yet, despite the best efforts of Richard Owen and other great naturalists, the fossil record was only slowly revealing its secrets of ancient life forms.

Because the early naturalists did such a thorough job, by the time we were students, we were presented with a precise census of modern dinosaurian diversity and biogeography that has changed only slightly in the last two decades. Today, students can grasp the diversity of living birds from a single textbook of ornithology. With the new tools and techniques of cladistics, the new generation of naturalists is rebuilding many of the evolutionary paths and highways on our map of Life's history. Our knowledge concerning the fascinating evolutionary pathways of life continues to grow. At least two new species of living birds have been discovered in the year that it took to write this book, one in the Andean forests, the other in the Philippines. So, the centers of dinosaur evolution continue to yield new insights into dinosaurian diversity, as remote regions still poorly known to naturalists are finally explored. But with hundreds of thousands of amateur bird watchers and hundreds of trained naturalists studying the birds of the world, it is clear that nearly all of the species of birds alive today on the planet have been discovered and named. If we had wanted to discover and describe new species of living birds, we were born a century too late. Indeed the students of our generation often express their disappointment and sense of deep loss that Nature's frontiers, by the time we reached them, were so thoroughly explored. Compared to the prospects that faced a young Richard Owen, Charles Darwin, or John James Audubon, our own prospects often seem pitifully tame.

Although there is no shortage of challenges in documenting the modern biota, for us the most enthralling scientific frontier has been the past—the deep history of birds and their extinct dinosaurian relatives. Like the students in Richard Owen's time, we recognized that only by measuring what has gone before can we predict what might lie ahead. But during our careers, the pace of discovery of extinct birds has greatly exceeded nineteenth-century levels, so the ancient world is far more accessible to us than to students a century ago. With modern radioisotopic dating methods, detailed geological maps, and a host of other powerful technologies, we now have detailed chronologies and phylogenetic maps that trace the roots of our modern biota far back into time. The tumultuous pace of new discoveries has presented a challenging and rapidly shifting view of the history of birds and other dinosaurs. With that far richer picture of Life's history, we now look ahead with an entirely different perspective than students did in the last century. So great has been this shift in perspective that the students in today's classroom face a radically different situation from the one that we experienced just two decades ago at Berkeley. Having long gazed back over our shoulders into the Cretaceous in our efforts to understand the K-T mass extinction, we and other scientists have realized that Earth history has presented us with a profound irony in terms of the destiny of dinosaurs. Although a diverse array of magnificent animals was wiped out at the end of the Cretaceous, the worst for dinosaurs may, in fact, lie just ahead, staring us right in the face.

While a human role in the extermination of modern species has been acknowledged since Richard Owen's day, only recently have we come to appreciate the full scope of that role. Today's students will be the first generation of scientists to live with that realization from the earliest stages of their careers. They are witnessing one of the greatest global extinctions of all time. When we share the bleak irony of modern dinosaur extinction with this next generation of naturalists training in our classrooms, our students sometimes express their humiliation in belonging to the human species and the hopelessness of cleaning up our mess. These are new attitudes, and they reflect a huge shift in world view from when we were students.

But in spite of the ample cause for pessimism, it is important for today's students to recognize that they themselves did not create this situation, even if their human forebears did. More importantly, they must understand that they and their human descendants will shape and witness the ultimate destiny of dinosaurs. And since we now understand that so much of the modern extinction episode was triggered by human activities, it is evident that human solutions can also be found. It is also important for today's students to realize that their ability to preserve the modern diversity of dinosaurs, as well as biodiversity in general, will be predicated upon the skills and knowledge that they assemble today. Great universities and natural history museums have been established in many parts of the world, and so the organizational structure is largely in place to attack the next great frontiers of natural history. For the coming generation of students, these frontiers will include the heavens as well as the Earth itself, and the great issues for them will now include future biodiversity in addition to refining our own measures of the diversity of the present and past. So, the prospects for great challenges and heroic accomplishments by the next generation of scientists are rich indeed.

That is, if sufficient resources are provided for them. Across the United States, geology departments were shut down in many colleges and universities following a depression in oil prices and cutbacks in petroleum exploration over the last two decades. Richard Owen's Natural History Museum—once the world's greatest centers for the study of natural history—has slashed its curatorial and research staff owing to a budget crisis in England. For the same reasons, the great London Zoological Gardens were recently forced to cut back their public displays and to give away many of the rare and endangered animals in their collections. Research budgets are diminishing everywhere, and as biodiversity continues to decline at an alarming rate, the remaining opportunities for us to study Nature and plan for our planet's future are rapidly dwindling. Whereas a great, if aging, infrastructure of universities, museums, research laboratories, government agencies, and libraries is now in place to train a new generation of naturalists, what is lacking is a widespread political will to prioritize funding toward education in general and natural history in particular. The Texas state government, to cite one tragic example, now invests more public funding in building new prisons that in building new schools or maintaining schools already in operation.

Still there is time to educate our politicians and public, and to accomplish a great deal of good in the decades ahead. For endangered birds, projects like the on-going attempts to restore natural populations of California Condors and to rebuild natural populations of the Whooping Crane offer both hope and informative examples for future strategies. Experiments on the preservation of large diverse habitats are underway in Costa Rica and Irian Jaya, and these may lead to conservation of huge regions. The recent biotechnology revolution is sufficiently advanced that we can only say that it is impossible to predict where we will be a century from now or what role this exciting breakthrough will play in conservation biology. For students willing to take the risk of a long apprenticeship and uncertain job prospects, the next generation of natural historians will face the many of the greatest challenges and the greatest rewards in the history of human endeavors.

One of those rewards will be the romance of doing field work. Although we now work out of New York City and Austin, Texas, we still go into the field whenever we can to search for evidence of what happened at the end of the Cretaceous and to refine the evolutionary paths on the map of vertebrate phylogeny. The opportunities to explore natural history in the field and make new discoveries out in the badlands is the primary reason we put up with all those painful tests in school. It remains the most exhilarating work that we can imagine, especially as we come to appreciate the growing importance of natural history. The examples set by Owen, Darwin, Audubon, and the many naturalists since their time provide great models for future scientists by showing the immense impact that natural historians can have on the world view of all people. But in many ways, today's challenges offer far greater opportunities for more critical contributions than were available to our predecessors. Today we can read a far deeper significance into the future by exploring the geologic and evolutionary events of the past. Those lessons provide the greatest motivation that we have ever had to go into Nature, explore the history of our world, and take immediate action based on the knowledge gleaned from those efforts. One problem that needs to be addressed by us and our students has become painfully obvious in light of what we have tried to present in this book:

No matter what caused the K-T episode of extinction, dinosaurs have survived to the present-day. But the number of dinosaur species has dwindled alarmingly in the short time that humans have populated and exploited the world. Human proliferation across the Pacific islands, by everyone from Aboriginals to the British and American navies, wiped out more dinosaur species than whatever event or combination of events triggered the terminal Cretaceous extinctions. We are doing it again with the current wave of expansion that is decimating the world's tropical rainforests—the greatest haven for modern dinosaur diversity. So, in spite of the truly remarkable progress that the scientific community has made in unraveling the evolutionary history of dinosaurs, we peer out toward the future from a perplexing perspective. Wouldn't it be both ironic and tragic if, despite all our attempts to pin the atrocity on ancient catastrophes and cataclysms, dinosaurs were extinguished—not by the next extraterrestrial impact or volcanic eruption—but rather by the actions of our own human hands?

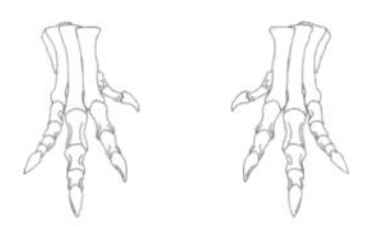

NOTES

CHAPTER ONE

The Seductive Allure of Dinosaurs

1. F. M. Gradstein et al., 1995, A Triassic, Jurassic, and Cretaceous time scale, in *Geochronology, timescales and global stratigraphic correlation*, eds. W. A. Berggren et al., Society of Economic Paleontologists and Mineralogists Special Publication 54, pp. 95–126.

CHAPTER TWO

Earlier Extinction Hypotheses

1. For a fairly comprehensive listing of published hypothetical causes, see R. Molnar and M. O'Reagan, 1989, Dinosaur extinctions, *Australian Natural History* 22:563; and M. J. Benton, 1990, Scientific methodologies in collision: The history of the study of the extinction of the dinosaurs, *Evolutionary Biology* 24:371–400.
2. J. D. Archibald, 1996, *Dinosaur extinction and the end of an era* (New York: Columbia University Press), p. 85.
3. K. R. Johnson and L. J. Hickey, 1990, *Megafloral change across the Cretaceous/Tertiary boundary in the northern Great Plains and Rocky Mountains*, Geological Society of America Special Paper 247, p. 441.
4. C. D. Ratcliff et al., 1994, Paleomagnetic record of a geomagnetic field reversal from Late Miocene mafic intrusions, southern Nevada, *Science* 266:412.
5. F. M. Gradstein et al., 1995, A Triassic, Jurassic and Cretaceous time scale, in *Geochronology, timescales and global stratigraphic correlation*, eds. W. A. Berggren et al., Society of Economic Paleontologists and Mineralogists Special Publication 54, pp. 95–126.
6. Archibald, 1996, p. 16.
7. Johnson and Hickey, 1990, p. 441.
8. J. H. Hutchison and J. D. Archibald, 1996, Diversity of turtles across the Cretaceous/Tertiary boundary in northeast Montana, *Paleogeography, Paleoclimatology, Paleoecology* 55:14–15.
9. J. A. Lillegraven and L. M. Ostresh, 1990, *Latest Cretaceous evolution of western shorelines of the Western Interior Seaway*, Geological Society of America Special Paper 243, pp. 10–11; and J. D. Obradovich and W. A. Cobban, 1975, A timescale for the Late Cretaceous of the western interior of North America, in *The Cretaceous system of the western interior of North America*, Geological Association of Canada Special Paper 13, p. 36.

CHAPTER FOUR

Enormous Eruptions and Disappearing Seaways

1. V. E. Courtillot, 1990, What caused the mass extinction? A volcanic eruption, *Scientific American* (October): 92.
2. P. R. Vogt, 1972, Evidence for global synchronism in mantle plume convection, *Nature* 240:338–342.
3. C. B. Officer and C. L. Drake, 1983, The Cretaceous-Tertiary transition, *Science* 219:1383–1390.
4. C. B. Officer and C. L. Drake, 1985, Terminal Cretaceous environmental events, *Science* 227:1161–1167.
5. C. B. Officer et al., 1987, Late Cretaceous and paroxysmal Cretaceous/Tertiary extinctions, *Nature* 326:143–149.
6. Courtillot, 1990, pp. 85–92. Another more extensive and detailed analysis of emissions and resulting environmental effects is provided in Officer et al., 1987, pp. 145–146. Estimated amounts and intensity of effects vary with those reported by Courtillot to some extent, but the interpreted general effects are similar.
7. A. K. Baksi, 1990, Timing and duration of Mesozoic-Tertiary flood-basalt volcanism, *Eos* 71:1835–1836.
8. A. R. Basu et al., 1993, Early and late alkali igneous pulses and a high ^{3}He plume origin for the Deccan flood basalts, *Science* 261:902–906.
9. R. S. White and D. P. McKenzie, 1989, Volcanism at rifts, *Scientific American* (July): 62–71.
10. Courtillot, 1990, p. 85.
11. W. J. Broad, 1995, Heavy volcanic eras were caused by plumes from the Earth's core, *New York Times* (August 22).
12. M. F. Coffin and O. Eldholm, 1993, Large igneous provinces, *Scientific American* (October): 42–49.
13. R. A. Kerr, 1994, Throttling back the great lava flows? *Science* 264:662–663.
14. Courtillot, 1990, pp. 91–92.
15. Basu et al., 1993, pp. 902–906.
16. White and McKenzie, 1989, pp. 62–71.
17. Broad, 1995.
18. ibid.
19. Coffin and Eldholm, 1993, p. 43.
20. Courtillot, 1990, pp. 85–92; see also note 6.
21. White and McKenzie, 1989, pp. 62–71.
22. Coffin and Eldholm, 1993, pp. 42–49.
23. Courtillot, 1990, p. 92; see also note 6.
24. Baksi, 1990, pp. 1835–1836.
25. Basu et al., 1993, pp. 902–906.
26. F. L. Sutherland, 1988, Demise of the dinosaurs and other denizens by cosmic clout, volcanic vapors, or other means, *Journal of the Proceedings of the Royal Society of New South Wales* 121:123–164.
27. ibid.
28. ibid.

29. ibid.
30. Basu et al., 1993, pp. 902–906.
31. Broad, 1995.
32. A. R. Basu et al., 1995, High ^{3}He plume origin and temporal-spatial evolution of the Siberian flood basalts, *Science* 269:822–825.
33. Broad, 1995.
34. White and McKenzie, 1989, pp. 62–71.
35. A. Hallam, 1987, End-Cretaceous mass extinction event: Argument for terrestrial causation, *Science* 237:1237–1242.
36. C. C. Swisher et al., 1993, ^{40}Ar/^{39}Ar dating and magnetostratigraphic correlation of the terrestrial Cretaceous-Paleogene boundary and Puercan Mammal Age, Hell Creek–Tullock Formations, eastern Montana, *Canadian Journal of Earth Sciences* 30:1981–1996.
37. Basu et al., 1993, pp. 902–906.
38. Officer et al., 1987, pp. 143–149.
39. See note 6.
40. See note 6.
41. Basu et al., 1993, pp. 902–906.
42. P. R. Renne and A. R. Basu, 1991, Rapid eruption of the Siberian traps flood basalts at the Permo-Triassic boundary, *Science* 253:176–179.
43. A. K. Baksi and E. Farrar, 1991, ^{40}Ar/^{39}Ar dating of the Siberian traps, USSR, *Geology* 19:461–46.
44. See note 6.
45. Coffin and Eldholm, 1993, pp. 42–49.
46. A. Hallam, 1989, Catastrophism in geology, in *Catastrophes and evolution: Astronomical foundations*, eds. W. Alvarez et al. (Cambridge: Cambridge University Press), pp. 25–55.
47. D. M. Raup, 1979, Size of the Permo-Triassic bottleneck and its evolutionary implications, *Science* 206:217–218.
48. Broad, 1995.
49. P. R. Renne et al., 1995, Synchrony and causal relations between Permian-Triassic boundary crises and Siberian flood volcanism, *Science* 269:1413–1416.
50. Basu et al., 1995, pp. 822–825.
51. Kerr, 1994, pp. 662–633.

CHAPTER FIVE

The Fatal Impact

1. W. Glen, 1994, *The mass extinction debates: How science works in a crisis* (Stanford, Calif.: Stanford University Press), p. 41.
2. D. A. Russell and W. Tucker, 1971, Supernovae and the extinction of dinosaurs, *Nature* 229:553–554.
3. W. Alvarez et al., 1989, Uniformitarianism and the response of earth scientists to the theory of impact crises, in *Catastrophes and evolution: Astronomical foundations*, eds. W. Alvarez et al. (Cambridge: Cambridge University Press), pp. 13–24.
4. W. B. N. Berry, 1968, *Growth of a prehistoric time scale* (San Francisco: W. H. Freeman).
5. A. Hallam, 1989, Catastrophism in geology, in *Catastrophes and evolution: Astronomical foundations*, eds. W. Alvarez et al. (Cambridge: Cambridge University Press), pp. 25–55.
6. Berry, 1968, pp. 22–45.
7. Hallam, 1989, pp. 25–55.
8. Berry, 1968, pp. 22–45.
9. Hallam, 1989, pp. 25–55.
10. Berry, 1968, pp. 22–45
11. C. Lyell, 1830–1833, *Principles of Geology* (London: John Murray), vols. 1–3.
12. S. J. Gould, 1965, Is uniformitarianism really necessary?, *American Journal of Science* 263:223–228.
13. L. Alvarez et al., 1980, Extraterrestrial cause for the Cretaceous-Tertiary extinction, *Science* 208:1095–1108; and L. Alvarez et al., 1979, Anomalous iridium levels in the Cretaceous/Tertiary boundary at Gubbio, Italy: Negative results of test for supernova origin, *Proceedings of the Cretaceous/Tertiary Boundary Events Symposium* II (Copenhagen: University of Copenhagen), p. 69.
14. W. Alvarez and F. Asaro, 1980, personal communication; W. Alvarez and F. Asaro, 1990, An extraterrestrial impact, *Scientific American* (October): 78; and W. Alvarez, 1997, *T. rex and the crater of doom* (Princeton, N.J.: Princeton University Press), pp. 59–81.
15. F. L. Sutherland, 1988, Demise of the dinosaurs and other denizens by cosmic clout, volcanic vapors or other means, *Journal of the Proceedings of the Royal Society of New South Wales* 121:130.
16. Alvarez and Asaro, 1990, p. 79.
17. A. R. Hildebrand and W. V. Boynton, 1991, Cretaceous ground zero, *Natural History* (June): 47–49.
18. Sutherland, 1988, p. 130.
19. Alvarez et al., 1979, p. 69; Alvarez, 1997, pp. 73–74.
20. R. A. F. Grieve, 1989, Hypervelocity impact cratering: A catastrophic terrestrial geologic process, in *Catastrophes and evolution: Astronomical foundations*, eds. W. Alvarez et al. (Cambridge: Cambridge University Press), p. 72.
21. Hildebrand and Boynton, 1991, pp. 47–49; Alvarez, 1997, p. 8.
22. P. Rich and T. H. Rich, 1993, Australia's polar dinosaurs, *Scientific American* (July): 50–55.
23. E. M. Brouwers et al., 1987, Dinosaurs on the North Slope, Alaska: High latitude, latest Cretaceous environments, *Science* 237:1608–1610; and W. A. Clemens and L. G. Nelms, 1993, Paleoecological implications of Alaskan terrestrial vertebrate fauna in the latest Cretaceous time at high paleolatitudes, *Geology* 21:503–506.
24. R. G. Prinn and B. F. Fegley, 1987, Bolide impacts, acid rain and biospheric traumas at the Cretaceous-Tertiary boundary, *Earth and Planetary Science Letter* 83:1–15; W. B. McKinnin, 1992, Killer acid at the K/T boundary, *Nature* 376:386–387; M. Palmer, 1991, Acid rain at the K/T boundary, *Nature* 352:758; Glen, 1994, p. 20; and Hildebrand and Boynton, 1991, p. 49.

25. W. Wolbach et al., 1985, Cretaceous extinctions: Evidence for wildfires and search for meteoric material, *Science* 230:167–170; W. Wolbach et al., 1990, *Major wildfires at the Cretaceous/Tertiary boundary*, Geological Society of America Special Paper 247, pp. 391–400.
26. O. B. Toon et al., 1982, *Evolution of an impact-generated dust cloud and its effects on the atmosphere*, Geological Society of America Special Paper 190, pp. 187–200.
27. T. J. Aherns et al., 1983, Impact of an asteroid or comet in the ocean and the extinction of terrestrial life, Proceedings of the 13th Lunar and Planetary Science Conference, Part 2, *Journal of Geophysical Research* 88:A799–A806.
28. M. R. Rampino and T. Volk, 1988, Mass extinctions, atmospheric sulfur and climatic warming at the K/T boundary, *Nature* 322:63–65.
29. Alvarez and Asaro, 1990, p. 79.
30. Glen, 1994, p. 9.
31. L. Alvarez, 1983, Experimental evidence that an asteroid impact led to the extinction of many species 65 million years ago, *Proceedings of the National Academy of Sciences*, p. 639.
32. J. D. Archibald, 1996, *Dinosaur extinction and the end of an era* (New York: Columbia University Press), p. 47.
33. Alvarez, 1983, p. 639.
34. Archibald, 1996, p. 47.
35. Archibald, 1996, p. 47; C. L. Pillmore et al., 1994, Footprints in the rocks: New evidence that dinosaurs flourished on land until the terminal Cretaceous impact event, *Lunar and Planetary Institute Contributions* 825:90.
36. Glen, 1994, p. 10.
37. W. H. Zoeller et al., 1983, Iridium enrichment in airborne particles from Kilauea volcano: January 1983, *Science* 222:1118–1121.
38. A. Hallam, 1987, End-Cretaceous mass extinction event: Argument for terrestrial causation, *Science* 238:1240.
39. V. Courtillot, 1990, A volcanic eruption, *Scientific American* (October): 87–88.
40. C. B. Officer and C. L. Drake, 1983, The Cretaceous-Tertiary transition, *Science* 219:1389–1390.
41. J.-P. Toutain, 1989, Iridium-bearing sublimates at a hot-spot volcano (Piton de la Fournaise, Indian Ocean), *Geophysical Research Letters* 16:1391–1394.
42. Alvarez and Asaro, 1990, p. 80.
43. Hildebrand and Boynton, 1991, p. 48.
44. John McHone et al., 1989, Stishovite at the Cretaceous-Tertiary boundary, Raton, New Mexico, *Science* 243:1182–1184; and Grieve, 1989, pp. 71–73.
45. Hallam, 1987, p. 1240.
46. Courtillot, 1990, pp. 87–88.
47. Officer and Drake, 1983, pp. 1389–1390.
48. Zoeller et al., 1983, p. 1120.
49. Alvarez and Asaro, 1990, p. 80.
50. Grieve, 1989, pp. 71–73.
51. F. T. Kyte et al., 1980, Siderophile-enriched sediments from the Cretaceous-Tertiary boundary, *Nature* 288:651.
52. Officer and Drake, 1983, pp. 1389–1390.
53. F. F. Bohor et al., 1984, Mineralogical evidence for an impact event at the Cretaceous-Tertiary boundary, *Science* 224:867–869; and B. F. Bohor, 1990, *Shocked quartz and more: Impact signatures in Cretaceous/Tertiary boundary clays*, Geological Society of America Special Paper 247, pp. 335–342.
54. Grieve, 1989, pp. 57–79.
55. R. Molnar and M. O'Reagan, 1989, Dinosaur extinctions, *Australian Natural History* 22:562.
56. Hildebrand and Boynton, 1991, p. 48.
57. C. B. Officer et al., 1987, Later Cretaceous and paroxysmal Cretaceous/Tertiary extinctions, *Science* 326:145; and Courtillot, 1990, p. 88.
58. M. R. Owen and M. H. Anders, 1988, Evidence from cathodoluminescence for non-volcanic origin of shocked quartz at the Cretaceous-Tertiary boundary, *Nature* 334:145–147.
59. McHone, 1989, pp. 1182–1184.
60. B. F. Bohor, 1990, pp. 335–342.
61. J. Smit and J. Klaver, 1981, Sanidine spherules at the Cretaceous-Tertiary boundary indicate a large impact event, *Nature* 292:47; and Alvarez and Asaro, 1990, p. 80.
62. Alvarez and Asaro, 1990, p. 80.
63. Officer et al., 1987, p. 144; and Courtillot, 1990, p. 88.
64. H. Sigurdsson et al., 1991a, Glass from the Cretaceous/Tertiary boundary in Haiti, *Nature* 349:482–487; H. Sigurdsson et al., 1991b, Geochemical constraints on source region of Cretaceous/Tertiary impact glasses, *Nature* 353:839–842; and V. Sharpton et al., 1992, New links between the Chicxulub impact structure and the Cretaceous/Tertiary boundary, *Nature* 359:819–821.
65. Wolbach et al., 1985, pp. 167–170; and Wolbach et al., 1990, pp. 391–400.
66. D. Heymann et al., 1994, Fullerines in the Cretaceous-Tertiary boundary layer, *Science* 265:645–647.
67. Archibald, 1996, pp. 133, 140–141.
68. D. B. Carlisle and D. R. Braman, 1991, Nanometre-size diamonds in the Cretaceous/Tertiary boundary of Alberta, *Nature* 352:708–709.
69. Glen, 1994, p. 11.
70. K. Zahnle and D. Grinspoon, 1990, Comet dust as a source of amino acids at the Cretaceous/Tertiary boundary, *Nature* 348:157–160.
71. Glen, 1994, p. 12.
72. ibid.
73. J. Bourgeois et al., 1988, A tsunami deposit at the Cretaceous-Tertiary border in Texas, *Science* 241: 567–570. For the history of the discovery of tsunami deposts at the K-T boundary, see W. Alvarez, 1997, pp. 106–129.
74. F. Maurrasse et al., 1991, Impacts, tsunamis, and the Haitian Cretaceous-Tertiary boundary layer, *Science* 252: 1690–1693.

75. J. B. Lyons and C. B. Officer, 1992, Mineralogy and petrology of the Haiti Cretaceous/Tertiary section, *Earth and Planetary Science Letters* 109:205–224. Several other papers dealing with these potential tsunami deposits appear in G. Ryder et al., eds., 1996, *The Cretaceous-Tertiary event and other catastrophes in Earth history*, Geological Society of America Special Paper 307, pp. 151–242.
76. Archibald, 1996, p. 133.
77. D. M. Raup and J. J. Sepkoski, 1982, Mass extinction in the marine fossil record, *Science* 215:1501–1503; D. M. Raup and J. J. Sepkoski, 1984, Periodicity of extinctions in the geologic past, *Proceedings of the National Academy of Sciences* 81: 801–805; Hallam, 1989, pp. 38–40.
78. W. Alvarez and R. A. Muller, 1984, Evidence from crater ages for periodic impacts on the Earth, *Nature* 308:718–720.
79. M. Davis et al., 1984, Extinction of species by periodic comet showers, *Nature* 308:715–717.
80. D. P. Whitmire and A. A. Jackson, 1984, Are periodic mass extinctions driven by a distant solar companion?, *Nature* 308:713–715; and D. P. Whitmire and J. J. Matese, 1985, Periodic comet showers and Planet X, *Nature* 313:36–38.
81. M. R. Rampino and R. B. Strothers, 1984, Terrestrial mass extinctions, cometary impacts, and the Sun's motion perpendicular to the galactic plane, *Nature* 308:709–712; and R. D. Schwartz and P. B. James, 1984, Periodic mass extinctions and the Sun's oscillation about the galactic plane, *Nature* 308:712–713.
82. Hallam, 1987, p. 1240.
83. A. Hoffman, 1985, Patterns of family extinction depend on definition and geologic timescale, *Nature* 315:659–662.
84. J. J. Sepkoski and D. M. Raup, 1986, Was there 26-Myr periodicity of extinctions?, *Nature* 321:533.
85. S. M. Stigler and M. J. Wagner, 1987, A substantial bias in non-parametric tests for periodicity in geophysical data, *Science* 238:940–942.
86. J. A. Kitchell and D. Pena, 1984, Periodicity of extinctions in the geologic past: Deterministic versus stochastic explanations, *Science* 266:689–689.
87. C. Patterson and A. B. Smith, 1987, Is the periodicity of extinctions a taxonomic artifact?, *Nature* 330:248–251.
88. Glen, 1994, chapter 11, Mass extinctions: Fact or fallacy? (discussion with John C. Briggs), pp. 230–236

CHAPTER SIX

Direct Evidence of Catastrophe

1. R. A. F. Grieve, 1989, Hypervelocity impact cratering: A catastrophic geologic process, in *Catastrophes and evolution: Astronomical Foundations*, eds. W. Alvarez et al. (Cambridge: Cambridge University Press), p. 65; and B. M. French, 1990, Twenty-five years of the impact-volcanic controversy, *Eos* (April 24): 411.
2. Grieve, 1989, p. 62.
3. W. Glen, 1994, *The mass-extinction debates: How science works in a crisis* (Stanford, Calif.: Stanford University Press), p. 165.
4. J. E. Lyne and M. Tauber, 1995, Origin of the Tunguska event, *Nature* 375:638.
5. Glen, 1994, p. 41.
6. F. L. Sutherland, 1988, Demise of the dinosaurs and other denizens by cosmic clout, volcanic vapors or other means, *Journal of the Proceedings of the Royal Society of New South Wales* 121:134.
7. W. Alvarez and F. Asaro, 1990, An extraterrestrial impact, *Scientific American* (October): 82; and W. Alvarez, 1997, *T. rex and the crater of doom* (Princeton, N.J.: Princeton University Press), pp. 95–96.
8. Glen, 1994, p. 13.
9. R. Kerr, 1993, New crater age undercuts killer comets, *Science* 262:659; and G. A. Izett et al., 1993, The Manson impact structure: ^{40}Ar/^{39}Ar age, *Science* 262:729.
10. G. T. Penfield and Z. A. Camargo, 1981, Definition of a major igneous zone in the central Yucatan platform with aeromagnetics and gravity, *Abstract of the Annual Meeting of the Society for Exploration Geophysicists* 51:37; see also Alvarez, 1997, pp. 109–137.
11. A. R. Hildebrand and G. T. Penfield, 1990, A buried 180 km-diameter probable impact crater on the Yucatan Peninsula, Mexico, *Eos* 71:1425; A. R. Hildebrand et al., 1991, *Geology* 19:867–871; and K. O. Pope et al., 1991, Mexican site for K/T impact crater?, *Nature* 351:105.
12. A. R. Hildebrand et al., 1995, Size and structure of the Chicxulub Crater revealed by horizontal gravity gradients and cenotes, *Nature* 376:415–417.
13. V. L. Sharpton et al., 1993, Chicxulub multiring impact basin: Size and other characteristics derived from gravity analysis, *Science* 261:1564–1567.
14. Hildebrand et al., 1995, pp. 415–417.
15. Grieve, 1989, pp. 64–70.
16. H. J. Melosh, 1995, Around and around we go, *Nature* 376:386–387.
17. Sharpton et al., 1993, pp. 1564–1567.
18. Hildebrand et al., 1995, pp. 415–417.
19. C. Swisher et al., 1992, Coeval ^{40}Ar/^{39}Ar ages of 65.0 million years ago from Chicxulub crater melt rock and Cretaceous-Tertiary boundary tektites, *Science* 257:954–958.
20. D. DePaolo et al., 1983, Rb-Sr, Sm-Nd, K-Ca, O and H isotopic study of Cretaceous-Tertiary boundary sentiments, Caravaca, Spain: Evidence for an oceanic impact, *Earth and Planetary Science Letters* 64:356–373.
21. M. R. Owens and M. H. Anders, 1988, Evidence for cathodoluminescence for non-volcanic origin of shocked quartz at the Cretaceous-Tertiary boundary, *Nature* 334:145–147; R. Kerr, 1987, Searching land and sea for the dinosaur killer, *Science* 237:856–857; and Alvarez, 1997, pp. 92–95, 96–98.
22. H. Sigurdsson et al., 1991a, Glass from the Cretaceous-Tertiary boundary in Haiti, *Nature* 349:482–487.
23. H. Sigurdsson et al., 1991b, Geochemical constraints on source region of Cretaceous/Tertiary glasses, *Nature* 353:839–842.

24. T. E. Krogh et al., 1993, U-Pb ages of single shocked zircons linking distal K/T ejecta to the Chicxulub crater, *Nature* 366:731–734; and J. D. Blum, 1993, Zircon can take the heat, *Nature* 366:718.

25. Glen, 1994, pp. 15–16; and Grieve, 1989, pp. 73–74.

26. V. L. Masaytis and M. S. Mashchak, 1984, Impact events at the Cretaceous-Tertiary boundary, *Doubleday Earth Science Selections* 265:195–197.

CHAPTER SEVEN

Patterns of Extinction and Survival

1. A. Hallam, 1989, Catastrophism in geology, in *Catastrophes and evolution: Astronomical foundations*, eds. W. Alvarez et al. (Cambridge: Cambridge University Press), pp. 44–45.

2. J. Smit, 1977, Discovery of a planktonic foraminiferal association between the *Abathomphalus mayaroensis* Zone and the "*Globigerina*" *eubina* Zone at the Cretaceous/Tertiary boundary in the Barranco del Gredero (Caravaca, SE Spain), *Proceedings of the Koniglijke Nederlandsch Akademie van Wetenschappen* B80:280–301; J. Smit, 1979, The Cretaceous/Tertiary transition in the Barranco del Gredero, Spain, in *Cretaceous-Tertiary Boundary Events*, eds. W. K. Christensen and T. Birkelund, University of Copenhagen Proceedings 2:156–163; J. Smit, 1982, *Extinction and evolution of planktonic foraminifera after a major impact at the Cretaceous/Tertiary boundary*, Geological Society of America Special Paper 190, pp. 329–352; and J. Smith and J. Hertogen, 1980, An extraterrestrial event at the Cretaceous-Tertiary boundary, *Nature* 285:198–200/

3. W. Alvarez and F. Asaro, 1990, An extraterrestrial impact, *Scientific American* (October): 81.

4. J. A. Kitchell et al., 1986, Biological selectivity of extinction: A link between background and mass extinction, *Palaios* 1:501–511.

5. N. MacLeod and G. Keller, 1991a, Hiatus distributions and mass extinctions at the Cretaceous/Tertiary boundary, *Geology* 19:497–501; and N. MacLeod and G. Keller, 1991b, How complete are Cretaceous/Tertiary boundary sections?..., *Geological Society of America Bulletin* 103:66–77.

6. R. Kerr, 1994, Testing an ancient impact's punch, *Science* 263:1371–1372; and J. D. Archibald, 1996, *Dinosaur extinction and the end of an era* (New York: Columbia University Press), pp. 189–190.

7. G. Ryder et al., 1996, *The Cretaceous/Tertiary boundary event*, Geological Society of America Special Paper 307, pp. 151–242.

8. J. J. Sepkoski, 1990, *The taxonomic structure of periodic extinction*, Geological Society of America Special Paper 247, pp. 33–44.

9. A. Hallam, 1987, End-Cretaceous mass extinction event: Argument for terrestrial causation, *Science* 238:1237–1242.

10. S. D'Hondt et al., 1994, Surface-water acidification and extinction at the Cretaceous-Tertiary boundary, *Geology* 22, pp. 983–986.

11. S. D'Hondt, 1997, personal communication.

12. W. Glen, 1994, *Mass extinction debates: How science works in a crisis* (Stanford, Calif.: Stanford University Press), p. 23.

13. K. Johnson and L. Hickey, 1990, *Megafloral change across the Cretaceous-Tertiary boundary in the northern Great Plains and Rocky Mountains*, Geological Society of America Special Paper 247, pp. 433–444; K. Johnson et al., 1989, High resolution leaf-fossil record spanning the Cretaceous-Tertiary boundary, *Nature* 340:708–711; K. Johnson, 1992, Leaf-fossil evidence for extensive floral extinction at the Cretaceous-Tertiary boundary, North Dakota, *Cretaceous Research* 13:91–117; D. J. Nichols, 1992, Plants at the Cretaceous-Tertiary boundary, *Nature* 356:295; and C. Hotton, 1984, Palynofloral changes across the Cretaceous-Tertiary boundary in east central Montana, *Abstracts of the International Palynofloral Conference* 6:66.

14. J. Wolfe, 1991, Paleobotanical evidence for a June "impact winter" at the Cretaceous/Tertiary boundary, *Nature* 352:420–423.

15. L. J. Hickey and L. J. McSweeney, 1992, Plants at the K/T boundary, *Nature* 356:295–296; and Nichols, 1992, p. 295.

16. R. Askin et al., 1994, Seymour Island: A southern high-latitude record across the KT boundary ..., *Lunar and Planetary Institute Contributions*, no. 825, pp. 7–8.

17. W. J. Zinsmeister et al., 1989, Latest Cretaceous/earliest Tertiary transition on Seymour Island, Antarctica, *Journal of Paleontology* 63:731–738.

18. G. Keller, 1993, The Cretaceous-Tertiary boundary transition in the Antarctic Ocean and its global implications, *Marine Micropaleontology* 21:1–45.

19. D'Hondt, 1997, personal communication.

20. K. Johnson, 1993, Extinction at the antipodes, *Nature* 366:511–512.

21. C. J. Hollis, 1993, Radiolarian faunal change through the K-T transition in eastern Marlborough, New Zealand, *Geological Society of America Abstracts with Programs* 25:295.

22. P. H. Schultz and S. D'Hondt, 1996, Cretaceous-Tertiary (Chicxulub) impact angle and its consequences, *Geology* 24: 963–967; and W. J. Broad, 1996, Asteroid's shallow angle may have sent inferno over northern kill zone, *New York Times* (Nov. 26).

23. D. Jablonski and D. M. Raup, 1995, Selectivity of end-Cretaceous marine bivalve extinctions, *Science* 268:389–391.

24. P. M. Dodson, 1991, Maastrichtian dinosaurs, *Geological Society of America Abstracts with Programs* 23:184–185; and Archibald, 1996, p. 16.

25. D. A. Russell, 1979, The enigma of the extinction of the dinosaurs, *Annual Review of Earth and Planetary Sciences* 7:163–182.

26. P. M. Sheehan and T. A. Hansen, 1986, Detritus feeding as a buffer to extinction at the end of the Cretaceous, *Geology* 14:868–870; and P. M. Sheehan et al., 1996, *Biotic selectivity during the K/T and late Ordovician extinction events*, Geological Society of America Special Paper 307, pp. 477–490.

27. J. D. Archibald and L. Bryant, 1990, *Differential Cretaceous-Tertiary extinctions of nonmarine vertebrates: Evidence from northeastern Montana*, Geological Society of America Special Paper 247, pp. 549–562; Archibald, 1996; J. H. Hutchison and J. D. Archibald, 1986, Diversity of turtles across the

Cretaceous/Tertiary boundary in northeastern Montana, *Paleogeography, Paleoclimatology, Paleoecology* 55:1–22; D. L. Lofgren, 1995, The Bug Creek problem and the Cretaceous-Tertiary transition at McGuire Creek, Montana, *University of California Publications in Geological Sciences* 140:1–200; W. A. Clemens, J. D. Archibald, and L. J. Hickey, 1981, Out with a whimper not a bang, *Paleobiology* 7:293–298; and J. D. Archibald and W. A. Clemens, 1982, Late Cretaceous extinctions, *American Scientist* 70:377–385.

28. P. Currie, 1997, personal communication.
29. Archibald, 1996, pp. 116–204.
30. D'Hondt, 1997, personal communication.
31. A. G. Smith et al., 1981, Phanerozoic paleocontinental world maps (Cambridge: Cambridge University Press), p. 63.
32. D'Hondt, 1997, personal communication.

CHAPTER EIGHT

Our Hazy View of Time at the K-T Boundary

1. J. F. Mount et al., 1986, Carbon and oxygen isotope stratigraphy of the upper Maastrichtian, Zumaya, Spain: A record of oceanographic and biologic changes at the end of the Cretaceous Period, *Palaios* 1:87–92.
2. P. W. Signor and J. H. Lipps, 1990, *Sampling bias, gradual extinction patterns and catastrophes in the fossil record*, Geological Society of America Special Paper 247, pp. 291–296.
3. K. W. Flessa and D. Jablonski, 1983, Extinctions are here to stay, *Paleobiology* 9:315–321.
4. J. D. Archibald, 1996, *Dinosaur extinction and the end of an era* (New York: Columbia University Press), pp. 68–71.
5. Archibald, 1996, p. 33.
6. D. A. Russell, 1984, The gradual decline of dinosaurs—fact or fallacy?, *Nature* 307:360–361.
7. P. M. Sheehan et al., 1991, Sudden extinction of the dinosaurs: Latest Cretaceous, upper Great Plains, *Science* 254:835–839.
8. Archibald, 1996, pp. 165–169.
9. L. Dingus, 1984, Effects of stratigraphic completeness on interpretations of extinction rates across the Cretaceous-Tertiary boundary, *Paleobiology* 10:420–438.
10. S. D'Hondt et al., 1996, *Planktic foraminifera, asteroids, and marine production: Death and recovery at the Cretaceous-Tertiary boundary*, Geological Society of America Special Paper 307, pp. 303–317.
11. P. M. Sadler, 1981, Sediment accumulation rates and the completeness of stratigraphic sections, *Journal of Geology* 89:569–584; and L. Dingus and P. M. Sadler, 1982, The effects of stratigraphic completeness on estimates of evolutionary rates, *Systematic Zoology* 31:400–412.
12. Dingus, 1984, pp. 420–438.
13. ibid.
14. M. A. Anders et al., 1987, A new look at sedimentation rates and the completeness of the stratigraphic record, *Journal of Geology* 95:1–14.
15. Dingus, 1984, pp. 424–427, 433.
16. F. L. Sutherland, 1988, Demise of the dinosaurs and other denizens by cosmic clout, volcanic vapors or other means, *Journal of the Proceedings of the Royal Society of New South Wales* 121:150.
17. W. Alvarez and F. Asaro, 1990, An extraterrestrial impact, *Scientific American* (October): 84.
18. W. Alvarez, 1997, *T. rex and the crater of doom* (Princeton, N.J.: Princeton University Press), pp. 144–146.
19. V. Courtillot et al., 1996, *The influence of continental flood basalts on mass extinctions: Where do we stand?*, Geological Society of America Special Paper 307, pp. 513–526.
20. Archibald, 1996, pp. 198–206.
21. W. J. Broad, 1994, New theory would reconcile rival views on dinosaur's demise, *New York Times* (December 24).
22. C. C. Swisher et al., 1992, Coeval 40Ar/39Ar ages of 65.0 million years ago from Chicxulub crater melt rock and Cretaceous-Tertiary boundary tektites, *Science* 257:954–958.
23. A. R. Basu et al., 1993, Early and late alkali igneous pulses and a high ^{3}He plume origin for the Deccan flood basalts, *Science* 261:902–906.
24. Courtillot et al., 1996, p. 513.
25. At least they will be better than our current estimates based on geologic evidence independent of the extinction scenarios themselves or current estimates based on interpolated rates of sedimentation that completely ignore the possibility of gaps or nonconformities in the sequence of rocks spanning the K-T boundary. Given that many K-T sections are known to contain such gaps, assigning ages to layers based on interpolated rates of sedimentation is especially foolhardy, as noted in Sadler, 1981, pp. 569–584; Dingus and Sadler, 1982, pp. 400–412; and Dingus, 1984, pp. 420–438.

CHAPTER NINE

Living Dinosaurs?

1. J. H. Ostrom, 1969a, Osteology of *Deinonychus antirrhopus*, an unusual theropod from the Lower Cretaceous of Montana, *Bulletin of the Peabody Museum of Natural History* 30:1–165.
2. ibid.
3. ibid.
4. A. Feduccia, 1996, *The origin and evolution of birds* (New Haven, Conn.: Yale University Press).
5. R. F. Ewer, 1965. The anatomy of the thecodont reptile *Euparkeria capensis* Broom, *Philosophical Transactions of the Royal Society of London*, series B, 248:379–435.
6. ibid.
7. R. Broom, 1913a, Note on *Mesosuchus browni* Watson and on a new South African Triassic pseudosuchian, *Records of the Albany Museum* 2 (5):395–396; and R. Broom, 1913b, On the South African pseudosuchian *Euparkeria* and allied genera, *Proceedings of the Zoological Society of London* 1913:619–633.
8. G. Heilmann, 1927, *The origin of birds* (New York: D. Appleton).
9. Feduccia, 1996.
10. J. H. Ostrom, 1969b, A new theropod dinosaur from the Lower Cretaceous of Montana, *Postilla* 128:1–17.

11. J. H. Ostrom, 1973, The ancestry of birds, *Nature* 242:136; J. H. Ostrom, 1974, *Archaeopteryx* and the origin of flight, *Quarterly Review of Biology* 49:27–47; J. H. Ostrom, 1975, The origin of birds, *Annual Reviews of Earth and Planetary Sciences* 3:55–77; J. H. Ostrom, 1976, *Archaeopteryx* and the origin of birds, *Biological Journal of the Linnean Society of London* 8:91–182; J. H. Ostrom, 1979, Bird flight: How did it begin?, *American Scientist* 67:46–56; and J. H. Ostrom, 1994, On the origin of birds and of avian flight, in *Major features of vertebrate evolution*, eds. D. R. Prothero and R. M. Schoch, Short Courses in Paleontology No. 7 (Knoxville: University of Tennessee), pp. 160–177.

12. H. von Meyer, 1861a, Vogel-Ferden und Palpipes priscus von Solenhofen, *Neues jharbuch für mineralogie, geologie, und paläontologie* (Stuttgart), p. 561; H. von Meyer, 1861b, *Archaeopteryx lithographica* aus dem lithographischen Schiefer von Solenhofen, *Palaeontographica* 10:53–56, 1 plate; H. von Meyer, 1861c, *Archaeopteryx lithographica* (Vogel-feder) und *Pterodactylus* von Solenhofen, *Neues jharbuch für mineralogie, geologie, und paläontologie* (Stuttgart), pp.678–679; H. von Meyer, 1862a, On the *Archaeopteryx lithographica* from the lithographic slate of Solenhofen, *Annals and Magazine of Natural History, London* 9 (third series): 366–370.

13. K. W. Barthel, N. H. M. Swinburne, and S. Conway Morris, 1990, *Solenhofen: A study in Mesozoic palaeontology* (Cambridge: Cambridge University Press); and J. H. Ostrom, 1991, Solenhofen: A study in Mesozoic palaeontology, *Journal of Vertebrate Paleontology* 11:528–529.

14. ibid.

15. von Meyer, 1861a.

16. G. R. deBeer, 1954, Archaeopteryx lithographica: A *study based on the British Museum specimen* (London: British Museum publication 224); and R. Owen, 1863, On the *Archaeopteryx* of von Meyer, with a description of the fossil remains of a long-tailed species, from the lithographic stone of Solenhofen, *Philosophical Transactions of the Royal Society of London* 153:33–47.

17. P. Wellnhofer, 1990, *Archaeopteryx, Scientific American* (May): 70–77.

18. ibid.

19. Ostrom, 1996b.

20. F. Heller, 1959, Ein dritter *Archaeopteryx*-Fund aus den Solnhofener Plattenkalken von Langenaltheim/Mfr., *Erlanger Geologische Abhandlungen* 31:3–25.

21. A. Charig et al., 1986, *Archaeopteryx* is not a forgery, *Science* 232:622–626.

22. J. H. Ostrom, 1970, *Archaeopteryx*: Notice of a new specimen, *Science* 170:537–538.

23. F. X. Mayr, 1973, Ein neuer *Archaeopteryx*-Fund, *Paleontologisches Zeitschrift* 47:17–24.

24. P. Wellnhofer, 1988, A new specimen of *Archaeopteryx*, *Science* 240:1790–1792; and P. Wellnhofer, 1993, Das siebte exemplar von *Archaeopteryx* aus den Solnhofener Schichten, *Archaeopteryx* 11:1–48.

25. P. Wellnhofer, 1992, Missing *Archaeopteryx*, *News Bulletin of the Society of Vertebrate Paleontology* 155:53–54.

26. F. Hoyle and C. Wickramasinghe, 1986, Archaeopteryx, *the primordial bird: A case of fossil forgery* (London: Christopher Davies).

27. A. Charig, 1986, Unnatural selection, *Science* 233:1213–1214.

28. ibid.

29. Broom, 1913.

CHAPTER TEN

Dinosaurs Challenge Evolution

1. R. Owen, 1842, *Report on British fossil reptiles*, 1841, Report of the Eleventh Meeting of the British Association for the Advancement of Science held at Plymouth (July 1841), p. 103.

2. L. B. Halstead and W. A. S. Sarjeant, 1993, *Scrotum humanum* Brookes—the earliest name for a dinosaur?, *Modern Geology* 18:221–224.

3. Halstead and Sargeant, 1993, 221–224.

4. ibid.

5. W. Buckland, 1824, Notice on the *Megalosaurus* or great fossil lizard of Stonesfield, *Transactions of the Geological Society, London* (series 2) 1:390–396.

6. G. Mantell, 1825, Notice on the *Iguanodon*, a newly discovered fossil reptile, from the sandstone of Tilgate forest, in Sussex, *Philosophical Transactions of the Royal Society* 115:179–186.

7. Owen, 1842.

8. ibid., p. 103.

9. ibid., p. 204.

10. ibid., p. 196.

11. ibid., p. 201.

12. ibid., p. 202.

13. C. Darwin, 1859, *On the origin of the species, by means of natural selection, or the preservation of favoured races in the struggle for life* (London: John Murray), pp. 116–117.

14. Darwin, 1859.

15. T. H. Huxley, 1868, On the animals which are most nearly intermediate between birds and reptiles, *Annals and Magazine of Natural History* 2:66–75.

16. Darwin, 1859.

17. Owen, 1842.

18. R. Owen, 1863, On the *Archaeopteryx* of von Meyer, with a description of the fossil remains of a long-tailed species, from the lithographic stone of Solenhofen, *Philosophical Transactions of the Royal Society of London* 153:33–47.

19. A. Wagner, 1862, On a new fossil reptile supposed to be furnished with feathers, *Annals and Magazine of Natural History* 9 (third series):261–267. This is a translation of the original article from 1861.

20. Owen, 1863.

21. K. Schmidt-Nielsen, 1971, How birds breathe, *Scientific American* 225 (December): 72–79,

22. Owen, 1863.

23. J. Cracraft, 1986, The origin and early diversification of birds, *Paleobiology* 12:383–399; and A. Desmond, 1982, *Archetypes and ancestors* (London: Blond and Briggs).

24. Huxley, 1868.

25. Huxley, 1868.

26. Huxley, 1868.

27. A. Wagner, 1861, Neue beiträge zur Kenntnis der urweltlichen fauna des lithographisches Schiefers. II. Schildkröten und Saurier, *Abhandlungen der (Könighichen preussichen) Akademie der Wissenschaft*, IX Band.

28. C. Gegenbaur, 1863, Vergleichend—anatomische bemerkungen über das Fusskelet der Vögel, *Archiv für anatomie, physiologieund wissenschaftliche medicin* 1863:450–472.

29. T. H. Huxley, 1870, Further evidence of the affinity between the dinosaurian reptiles and birds, *Quarterly Journal of the Geological Society of London* 26:12–31.

30. Huxley, 1868.

31. Huxley, 1870.

32. T. H. Huxley, 1867, On the classification of birds; and on the taxonomic value of the modifications of certain of the cranial bones observable in that class, *Proceedings of the Zoological Society of London* 1867:415–472.

33. Huxley, 1870.

34. H. F. Osborn, 1900, Reconsideration of the evidence for a common dinosaur-avian stem in the Permian, *American Naturalist* 34:777–799.

35. Huxley, 1870.

CHAPTER ELEVEN

Dinosaurs and the Hierarchy of Life

1. C. Patterson, 1981, Significance of fossils in determining evolutionary relationships, *Annual Review of Ecology and Systematics* 12:195–223.

2. J. A. Gauthier, A. G. Kluge, and T. Rowe, 1988, Amniote phylogeny and the importance of fossils, *Cladistics* 4:105–209; and M. J. Donoghue et al., 1989, Importance of fossils in phylogeny reconstruction, *Annual Review of Ecology and Systematics* 20:431–460.

3. A. G. Kluge, 1989, A concern for evidence and a phylogenetic hypothesis of relationships among Epicrates (Boidae, Serpentes), *Systematic Zoology* 38:7–25.

4. G. J. Nelson, 1969a, Gill arches and the phylogeny of fishes, with notes on the classification of vertebrates, *Bulletin of the American Museum of Natural History* 141:475–552; G. J. Nelson, 1969b, Infraorbital bones and their bearing on the phylogeny and geography of osteoglossomorph fishes, *American Museum Novitates* 2394:1–37; and G. J. Nelson, 1973, Relationships of clupeomorphs, with remarks on the structure of the lower jaw in fishes, in *Interrelationships of fishes*, eds. P. H. Greenwood, R. S. Miles, and C. Patterson (London: Academic Press), pp. 333–349.

5. D. E. Rosen and C. Patterson, 1969, The structure and relationships of the paracanthopterygian fishes, *Bulletin of the American Museum of Natural History* 11:357–474; D. E. Rosen, 1973, Interrelationships of higher eutelostean fishes, in *Interrelationships of fishes*, eds. P. H. Greenwood, R. S. Miles, and C. Patterson (London: Academic Press), pp. 397–513; and D. E. Rosen, 1982, Teleostean interrelationships, morphological function and evolutionary inference, *American Zoologist* 22:261–273.

6. C. Patterson, 1973, Interrelationships of holosteans, in *Interrelationships of fishes*, eds. P. H. Greenwood, R. S. Miles, and C. Patterson (London: Academic Press), pp. 233–305; C. Patterson and D. E. Rosen, 1977, Review of ichthyodectiform and other Mesozoic teleost fishes and the theory and practice of classifying fossils, *Bulletin of the American Museum of Natural History* 158:81–172; C. Patterson, 1982, Morphology and interrelationships of primitive actinopterygian fishes, *American Zoologist* 22:241–259; and C. Patterson, 1994, Bony fishes, in *Major features of vertebrate evolution*, eds. D. R. Prothero and R. M. Schoch, Short Courses in Paleontology no. 7 (The Paleontological Society), pp. 57–84.

7. B. Schaeffer, 1987, Deuterostome monophyly and phylogeny, in *Evolutionary biology*, eds. M. K. Hecht, B. Wallace, and G. T. Prance (New York: Plenum), pp. 179–235.

8. J. G. Maisey, 1984, Chondrichthyan phylogeny: A look at the evidence, *Journal of Vertebrate Paleontology* 4:359–371; J. G. Maisey, 1986, Heads and tails: A chordate phylogeny, *Cladistics* 2:201–256; J. G. Maisey, 1994, Gnathostomes (jawed vertebrates), in *Major features of vertebrate evolution*, eds. D. R. Prothero and R. M. Schoch, Short Courses in Paleontology no. 7 (The Paleontological Society), pp. 38–56.

9. J. A. Gauthier et al.,1989, Tetrapod phylogeny, in *The hierarchy of life*, eds. B. Fernholm, H. Bremer, and H. Jornvall, Nobel Symposium 70 (Amsterdam: Excerpta Medica), pp. 337–353.

10. Gauthier, Kluge, and Rowe, 1988, pp. 105–209.

11. T. Rowe, 1988, Definition, diagnosis, and origin of Mammalia, *Journal of Vertebrate Paleontology* 8:241–264; Gauthier, Kluge, and Rowe, 1988, pp. 105–209; and Gauthier et al, 1989, pp. 337–353.

12. T. Rowe, 1996a, Coevolution of the mammalian middle ear and neocortex, *Science* 273:651–654; T. Rowe, 1996b, Brain heterochrony and evolution of the mammalian middle ear, in *New perspectives on the history of life*, eds. M. Ghiselin and G. Pinna, Memoirs of the California Academy of Sciences no. 20, pp. 71–96.

13. Gauthier et al., 1989, pp. 337–353; and J. A. Gauthier, 1994, The diversification of amniotes, in *Major features of vertebrate evolution*, eds. D. R. Prothero and R. M. Schoch, Short Courses in Paleontology no. 7 (The Paleontological Society), pp. 129–159.

14. R. Estes and G. K. Pregill, 1988, The phylogenetic relationships of the lizard families: Essays commemorating Charles L. Camp (Palo Alto, Calif.: Stanford University Press).

15. A. D. Walker, 1990, A revision of *Sphenosuchus acutus* Haughton, a crocodylomorph reptile from the Elliot Formation (Late Triassic or Early Jurassic) of South Africa, *Philosophical Transactions of the Royal Society of London* B330:1–120; and A. D. Walker, 1985, The braincase of *Archaeopteryx*, in *The beginnings of birds*, eds. M. K. Hecht, J. H. Ostrom, G. Viohl, and P. Wellnhofer (Eichstätt: Freunde des Jura-Museums), pp. 123–134.

16. L. D. Martin, 1991, Mesozoic birds and the origin of birds, in *Origins of the higher groups of tetrapods: Controversy and consensus*, eds. H.-P. Schultze and L. Treub (Ithaca, N.Y.: Cornell University Press), pp. 485–540.

17. M. J. Benton and J. Clark, 1988, Archosaur phylogeny and the relationships of the Crocodylia, in *The phylogeny and classifica-*

tion of the tetrapods, vol. 1, *Amphibians, Reptiles and Birds*, ed. M. Benton (Oxford: Clarendon Press), pp. 295–338.

18. M. A. Norell, 1989, The higher level relationships of the extant Crocodylia, *Journal of Herpetology* 23: 325–335; and M. A. Norell, J. M. Clark, and J. H. Hutchison, 1994, The Late Cretaceous alligatoroid *Brachychampsa montana* (Crocodylia): New material and putative relationships, *American Museum Novitates* 3116:1–26.

19. C. A. Brochu, 1997, Phylogenetic systematics and taxonomy of Crocodylia, Ph.D. dissertation, University of Texas at Austin.

20. J. A. Gauthier, 1986, Saurischian monophyly and the origin of birds, in *The origin of birds and the evolution of flight*, ed. K. Padian, Memoirs of the California Academy of Sciences no. 8, pp. 1–55; F. E. Novas, 1989, The tibia and tarsus in Herrerasauridae (Dinosauria, incertae sedis) and the origin and evolution of the dinosaurian tarsus, *Journal of Paleontology* 63:677–690; F. E. Novas, 1992, Phylogenetic relationships of basal dinosaurs, the Herrerasauridae, *Paleontology* 35:51–62; and F. E. Novas, 1996, Dinosaur monophyly, *Journal of Vertebrate Paleontology* 16:723–741.

21. K. Padian, 1984, The origins of pterosaurs, in *Third symposium on Mesozoic ecosystems: Short papers* (Tübingen: Attempto Verlag), pp. 163–169.

22. P. Sereno and A. B. Arcucci, 1990, The monophyly of crurotarsal archosaurs and the origin of bird and crocodylian ankle joints, *Neues jahrbuch für geologie und palaontologie, abhandlungen* 180:21–52; and P. Sereno and F. E. Novas, 1994, Dinosaurian precursors from the Middle Triassic of Argentina: *Lagerpeton chanarensis*, *Journal of Vertebrate Paleontology* 16:723–741.

23. Novas, 1989, pp. 677–690; P. Sereno and F. E. Novas, 1992, The skull and neck of the basal theropod *Herrerasaurus ischigualastensis*, *Journal of Vertebrate Paleontology* 13:451–476; P. Sereno, 1994, The pectoral girdle and forelimb of the basal theropod Herrerasaurus ischigualastensis, *Journal of Vertebrate Paleontology* 13: 425–450; and H.-D. Sues, 1990, Staurikosauridae and Herrerasauridae, in *The Dinosauria*, eds. D. B. Weishampel, P. Dodson, and H. Osmólska (Berkeley: University of California Press)., pp. 143–147.

24. P. A. Murray and R. A. Long, 1989, Geology and paleontology of the Chinle Formation, Petrified Forest National Park and vicinity, Arizona, and a discussion of vertebrate fossils of the southwestern Upper Triassic, in *Dawn of the age of dinosaurs in the American Southwest*, eds. S. G. Lucas and A. P. Hunt (Albuquerque, N.Mex.: New Mexico Museum of Natural History).

CHAPTER TWELVE

The Evolutionary Map for Dinosaurs

1. J. A. Gauthier, 1984, A cladistic analysis of the higher systematic categories of the Diapsida, Ph.D. dissertation, Department of Paleontology, University of California, Berkeley.

2. K. Padian and D. J. Chure, 1989, *The age of dinosaurs*, Short Courses in Paleontology no. 2 (The Paleontological Society); D. B. Weishampel, P. Dodson, and H. Osmólska, eds., *The Dinosauria* (Berkeley: University of California Press); and D. E. Fastovsky and D. B. Weishampel, 1996, *The evolution and extinction of the dinosaurs* (New York and Cambridge: Cambridge University Press).

3. P. Sereno, 1984, The phylogeny of Ornithischia: A reappraisal, in *Third symposium on Mesozoic terrestrial ecosystems: Short papers*, eds. W. Reif and F. Westphal (Tübingen: Attempto Verlag), pp. 219–226; D. B. Norman, 1984, A systematic appraisal of the reptile order Ornithischia, in *Third symposium on Mesozoic terrestrial ecosystems: Short papers*, eds. W. Reif and F. Westphal (Tübingen: Attempto Verlag), pp. 157–162; M. R. Cooper, 1985, A revision of the ornithischian dinosaur *Kangnasaurus coetseei* Haughton, with a classification of the Ornithischia, *Annals of the South African Museum* 95:281–317; J. A. Gauthier, 1986, Saurischian monophyly and the origin of birds, in *The origin of birds and the evolution of flight*, ed. K. Padian, Memoirs of the California Academy of Sciences no. 8, pp. 1–55; P. Sereno, 1986, Phylogeny of the bird-hipped dinosaurs (Order Ornithischia), *National Geographic Research* 2:234–256; and D. B. Weishampel and L. M. Witmer, 1990a, Basal Ornithiuschia, in *The Dinosauria*, eds. D. B. Weishampel, P. Dodson, and H. Osmólska (Berkeley: University of California Press), pp. 416–425.

4. D. B. Weishampel and L. M. Witmer, 1990b, Heterodontosauridae, in *The Dinosauria*, eds. D. B. Weishampel, P. Dodson, and H. Osmólska (Berkeley: University of California Press), pp. 486–497; H.-D. Sues and D. B. Norman, 1990, Hypsilophodontidae, Tenontosaurus, and Dryosauridae, in *The Dinosauria*, eds. D. B. Weishampel, P. Dodson, and H. Osmólska (Berkeley: University of California Press), pp. 498–509; D. B. Norman and D. B. Weishampel, 1990, Iguanodontidae and related ornithopoda, in *The Dinosauria*, eds. D. B. Weishampel, P. Dodson, and H. Osmólska (Berkeley: University of California Press), pp. 510–533; and D. B. Weishampel and J. R. Horner, 1990, Hadrosauridae, in *The Dinosauria*, eds. D. B. Weishampel, P. Dodson, and H. Osmólska (Berkeley: University of California Press), pp. 534–561.

5. Sereno, 1984, pp. 219–226; Sereno, 1986, pp. 234–256; T. Maryanska and H. Osmólska, 1985, On ornithischian phylogeny, *Acta Palaeontologia Polonica* 46:119–141; and H.-D. Sues and P. Galton, 1987, Anatomy and classification of the North American Pachycephalosauria (Dinosauria, Ornithischia), *Paleontographica* (A) 178:1–40.

6. T. Maryanska and H. Osmólska, 1975, Protoceratopsidae (Dinosauria) of Asia, *Acta Palaeontologia Polonica* 33:133–182; Sereno, 1986, 234–236; P. Sereno, 1990, Psittacosauridae, in *The Dinosauria*, eds. D. B. Weishampel, P. Dodson, and H. Osmólska (Berkeley: University of California Press), pp. 579–592; P. Dodson and P. J. Currie, 1990, Neoceratopsia, in *The Dinosauria*, eds. D. B. Weishampel, P. Dodson, and H. Osmólska (Berkeley: University of California Press), pp. 593–618; C. A. Forster et al., 1993, A complete skull of *Chasmosaurus mariscalensis* (Dinosauria, Ceratopsidae) from the Aguja Formation (Late Campanian) of west Texas, *Journal of Vertebrate Paleontology* 13 (2):161–170; and T. M. Lehman, 1996, A horned dinosaur from the El Picacho Formation of west Texas, and a review of ceratopsian dinosaurs from the American Southwest, *Journal of Paleontology* 70:494–508.

7. W. P. Coombs, D. B. Weishampel, and L. M. Witmer, 1990, Basal Thyreophora, in *The Dinosauria*, eds. D. B. Weishampel, P. Dodson, and H. Osmólska (Berkeley: University of California Press), pp. 427–434; P. M. Galton, 1990, Stegosauria, in *The Dinosauria*, eds. D. B. Weishampel, P. Dodson, and H. Osmólska (Berkeley: University of California Press), pp.

435–455; W. P. Coombs and T. Maryanska, 1990, Aankylosauria, in *The Dinosauria*, eds. D. B. Weishampel, P. Dodson, and H. Osmólska (Berkeley: University of California Press), pp. 456–483; Sereno, 1986, pp. 234–256; Gauthier, 1986, pp. 1–55; and P. Sereno and Z. Dong, 1992, The skull of the basal stegosaur Huayangosaurus taibaii and a cladistic diagnosis of Stegosauria, *Journal of Vertebrate Paleontology* 12:318–343.

8. Gauthier, 1986, pp. 1–55.
9. Gauthier, 1986, pp. 1–55; P. M. Galton, 1990, Basal Sauropodomorpha—Prosauropoda, in *The Dinosauria*, eds. D. B. Weishampel, P. Dodson, and H. Osmólska (Berkeley: University of California Press), pp. 320–344; and J. S. Macintosh, 1990, Sauropoda, in *The Dinosauria*, eds. D. B. Weishampel, P. Dodson, and H. Osmólska (Berkeley: University of California Press), pp. 345–401.
10. J. A. Jensen, 1985, Three new sauropod dinosaurs from the Upper Jurassic of Colorado, *Great Basin Naturalist* 45:697–709.
11. T. Rowe, 1989a, Origin and early evolution of theropods, in *Dinosaurs*, ed. K. Padian, Short Courses in Paleontology no. 2 (The Paleontological Society).
12. P. C. Sereno et al., 1993, Primitive dinosaur skeleton from Argentina and the early evolution of Dinosauria, *Nature* 361:64–66.
13. R. Owen, 1863, On the *Archaeopteryx* of von Meyer, with a description of the fossil remains of a long-tailed species, from the lithographic stone of Solenhofen, *Philosophical Transactions of the Royal Society of London* 153:33–47.
14. A. Feduccia, 1980, *The age of birds* (Cambridge, Mass.: Harvard University Press).
15. Gauthier, 1986, pp. 1–55.
16. T. Rowe and J. Gauthier, 1990, Ceratosauria, in *The Dinosauria*, eds. D. B. Weishampel, P. Dodson, and H. Osmólska (Berkeley: University of California Press), pp. 151–168; and T. Rowe, 1989b, A new species of the theropod dinosaur *Syntarsus* from the Early Jurassic Kayenta Formation of Arizona, *Journal of Vertebrate Paleontology* 9:125–136.
17. E. H. Colbert, 1995, *The little dinosaurs of Ghost Ranch* (New York: Columbia University Press).
18. Gauthier, 1986, pp. 1–55.
19. X.-J. Zhao and P. J. Currie, 1993, A large crested theropod from the Jurassic of Xinjiang, People's Republic of China, *Canadian Journal of Earth Sciences* 30:2027–2036.
20. P. J. Currie and X.-J. Zhao, 1993, A new carnosaur (Dinosauria, Theropoda) from the Jurassic of Xinjiang, People's Republic of China, *Canadian Journal of Earth Sciences* 30:2037–2081.
21. R. A. Coria and L. Salgado, 1995, A new giant carnivorous dinosaur from the Cretaceous of Patagonia, *Nature* 377:224–226.
22. P. C. Sereno et al., 1994, Early Cretaceous dinosaurs from the Sahara, *Science*, 266:267–271.
23. T. R. Holtz, Jr., 1996, Phylogenetic taxonomy of the Coelurosauria (Dinosauria, Theropoda), *Journal of Paleontology* 70:536–538.
24. R. Barsbold and H. Osmólska, 1990, Ornithomimosauridae, in *The Dinosauria*, eds. D. B. Weishampel, P. Dodson, and H. Osmólska (Berkeley: University of California Press), pp. 225–243.
25. R. Barsbold, T. Maryanska, and H. Osmólska, 1990, Oviraptorosauria, in *The Dinosauria*, eds. D. B. Weishampel, P. Dodson, and H. Osmólska (Berkeley: University of California Press), pp. 249–258.
26. M. A. Norell et al., 1994, A theropod dinosaur embryo and the affinities of the Flaming Cliffs dinosaur eggs, *Science* 266:779–782.
27. M. A. Norell et al., 1995, A nesting dinosaur, *Nature* 378:774–776.
28. Gauthier, 1986, pp. 1–55.
29. J. A. Gauthier and K. Padian, 1985, Phylogenetic, functional, and aerodynamic analysis of the origins of birds and their flight, in *The beginnings of birds*, eds. M. K. Hecht, J. H. Ostrom, G. Viohl, and P. Wellnhofer, Proceedings of the First International *Archaeopteryx* Conference (Eichstätt: Fruende des Jura Museums), pp. 185–197.
30. Gauthier, 1986, pp. 1–55.
31. S. Chatterjee, 1991, Cranial anatomy and relationships of a new Triassic bird from Texas, *Philosophical Transactions of the Royal Society of London* (B) 332:277–342; and S. Chatterjee, 1995, The Triassic bird *Protoavis*, *Archaeopteryx* 13:15–31.
32. J. H. Ostrom, 1979, Bird flight: How did it begin?, *American Scientist* 67:46-56; Ostrom, 1994, pp. 160–177.
33. G. R. Caple, R. T. Balda, and W. R. Willis, 1983, The physics of leaping animals and the evolution of pre-flight, *American Naturalist* 121:455–467.
34. Gauthier and Padian, 1985, pp. 185–197; J. A. Gauthier and K. Padian, 1989, The origin of birds and the evolution of flight, in *The age of dinosaurs*, eds. K. Padian and D. J. Chure, Short Courses in Paleontology no. 2 (The Paleontological Society).

CHAPTER THIRTEEN
Death by Decree

1. R. Owen, 1842, *Report on British fossil reptiles*, 1841, Report of the Eleventh Meeting of the British Association for the Advancement of Science held at Plymouth (July 1841), pp. 60–204.
2. Owen's views on evolution changed somewhat over his lifetime, and they have been explored in depth by A. Desmond, 1975, *The hot-blooded dinosaurs* (London: Blond & Briggs); A. Desmond, 1982, *Archetypes and ancestors* (London: Blond & Briggs); A. Desmond, 1989, *The politics of evolution* (Chicago: University of Chicago Press); and N. A. Rupke, 1994, *Richard Owen—Victorian naturalist* (New Haven, Conn.: Yale University Press).
3. E. Mayr, 1982, *The growth of biological thought* (Cambridge, Mass., and London: Belknap Press).
4. T. M. Fries, 1903, *Linné* (Stockholm: Lefnadsteckning and Company).
5. C. Linnaeus, 1753, *Species plantarum* (Halmiae).

6. C. Linnaeus, 1758, *Systema naturae*, 10th ed. (Stockholm).

7. D. L. Hull, 1965, The effect of essentialism on taxonomy: Two thousand years of stasis, *British Journal for the Philosophy of Science*, 60:314–326.

8. W. K. Gregory, 1919, The orders of mammals, *Bulletin of the American Museum of Natural History* 56:28.

9. T. Rowe, 1988, Definition, diagnosis and origin of Mammalia, *Journal of Vertebrate Paleontology* 8:241–246; T. Rowe and J. Gauthier, 1992, Ancestry, paleontology, and definition of the name Mammalia, *Systematic Biology* 41:372–378; and K. de Queiroz, 1994, Replacement of an essentialistic perspective on taxonomic definitions as exemplified by the definition of "Mammalia," *Systematic Biology* 43:497–510.

10. Linnaeus, 1758.

11. C. Darwin, 1859, *On the origin of species, by means of natural selection, or the preservation of favoured races in the struggle for life* (London: John Murray).

12. K. de Queiroz, 1988, Systematics and the Darwinian revolution, *Philosophy of Science* 55:238–259; and K. de Queiroz, 1985, The ontogenetic method for determining character polarity and its relevance to phylogenetic systematics, *Systematic Zoology* 34:280–299.

13. ibid.

14. W. Hennig, 1966, *Phylogenetic systematics* (Urbana: University of Illinois Press).

15. J. A. Gauthier, 1986, Saurischian monophyly and the origin of birds, in *The origin of birds and the evolution of flight*, ed. K. Padian, Memoirs of the California Academy of Sciences no. 8, pp. 1–47.

16. K. de Queiroz and J. A. Gauthier, 1990, Phylogeny as a central principle in taxonomy: Phylogenetic definitions of taxon names, *Systematic Zoology* 39:307–322; K. de Queiroz and J. A. Gauthier, 1992, Phylogeny taxonomy, *Annual Review of Ecology and Systematics* 23:449–480; and K. de Queiroz and J. A. Gauthier, 1994, Toward a phylogenetic system of biological nomenclature, *Trends in Ecology and Evolution* 9:27–31.

CHAPTER FOURTEEN
The Road to Jurassic Park

1. J. M. Starck, 1993, Evolution of avian ontogenies, *Current Ornithology* 10:275–366.

2. B. K. Hall, 1992, *Evolutionary developmental biology* (London, New York, and Tokyo: Chapman and Hall).

3. C. Darwin, 1859, *On the origin of species, by means of natural selection, or the preservation of favoured races in the struggle for life* (London: John Murray).

4. C. Gegenbaur, 1863, Vergleichendend-anatomische Bemerkungen über das Fusskelet der Vögel, *Archive für anatomie, physiologie und wissenschaftliche medicin* 1863:450–472.

5. K. Padian, 1996, Early bird in slow motion, *Nature* 382:400–401; J. Ruben, Reptilian physiology and the flight capacity of *Archaeopteryx*, *Evolution* 45:1–17.

6. L. Hou et al., 1995, The oldest beaked bird is from the "Jurassic" of China, *Nature* 377:616–618; A. Feduccia, 1996, *The origin and evolution of birds* (New Haven, Conn.: Yale University Press); and L. Hou et al., 1996, Early adaptive radiation of birds: Evidence from fossils from northeastern China, *Science* 274:1164–1167.

7. L. Chiappe, 1995, The first 85 million years of avian evolution, *Nature* 378:349–355.

8. J. Jensen and K. Padian, 1989, Small pterosaurs and dinosaurs from the Uncompahgre Fauna (Brush Basin Member, Morrison Formation: Tithonian), Last Jurassic, Western Colorado, *Journal of Paleontology* 63:364–373.

9. J. A. Gauthier, 1986, Saurischian monophyly and the origin of birds, in *The origin of birds and the evolution of flight*, ed. K. Padian, Memoirs of the California Academy of Sciences no. 8, pp. 1–47.

10. Chiappe, 1995, pp. 349–355.

11. H. Gruneberg, 1963, *The pathology of development: A study of inherited skeletal disorders in animals* (New York: Wiley).

12. A. Hampé, 1959, Contribution a l'étude du development et de la regulation des deficiences et des excedents dans la patte de l'embron de poulet, *Archives des anatomie microscopique et morphologique experiments* 48:345–478.

13. G. B. Müller, 1989, Ancestral patterns in bird limb development: A new look at Hampé's experiment, *Journal of Evolutionary Biology* 1:31–47.

14. P. Altangerel et al., 1993, Flightless bird from the Cretaceous of Mongolia, *Nature* 362:623–626; and P. Altangerel et al., 1994, Skeletal morphology of *Mononykus olecranus* (Theropoda: Avialae) from the Late Cretaceous of Mongolia, *American Museum Novitates* 3105:1–29.

15. C. Patterson, 1993, Bird or dinosaur?, *Nature* 365:21–22; and P. Wellnhofer, 1995, New data on the origin and early evolution of birds, Comptes rendu des séances de l'Académie des sciences, series II, 319:299–308.

16. Z. Zhou, 1995, Is *Mononykus* a bird?, *Auk* 112:958–963.

17. Gruneberg, 1963.

18. F. Novas, 1996, Alvarezsauridae, Cretaceous basal birds from Patagonia and Mongolia, in *Proceeding of the Gondwanan Dinosaur Symposium* (Memoirs of the Queensland Museum 39), eds. F. E. Nova and R. E. Molnar, pp. 675–702.

19. F. Novas, 1997, Anatomy of *Patagonykus puerti* (Theropoda, Avialae, Alvarezsauridae), from the Late Cretaceous of Patagonia, *Journal of Vertebrate Paleontology* 17:137–166.

20. Padian, 1996, pp. 1–17.

21. J. Sanz et al., 1996, An early Cretaceous bird from Spain and its implications for the evolution of flight, *Nature* 382:442–445.

22. Chiappe, 1995, pp. 349–355.

23. J. L. Sanz, J. F. Bonaparte, and A. Lacasa, 1988, Unusual Early Cretaceous birds in Spain, *Nature* 331:433–435; and Chiappe, 1995, pp. 349–355.

24. E. V. Kurochkin, 1995, The assemblage of Cretaceous birds in Asia, in *Sixth symposium on Mesozoic terrestrial ecosystems and biota*, ed. A. L. Sun (Beijing: China Ocean Press), pp. 203–208.

25. Chiappe, 1995, pp. 349–355.

26. Feduccia, 1996; and Hou et al., 1996, pp. 1164–1167.
27. Chiappe, 1995, pp. 349–355.
28. ibid.
29. ibid.
30. Kurochkin, 1995, pp. 203–208; and Chiappe, 1995, pp. 349–355.
31. We follow the nomenclature of Gauthier (1986), who was the first to define the meaning of the name "Aves" to refer to the last common ancestor of modern avian species and all its descendants. Other authors commonly apply this name as a substitute for Ornithurae.
32. E. J. Kollar and C. Fisher, 1980, Tooth induction in chick epithelium: Expression of quiescent genes for enamel synthesis, *Science* 207:993–995.
33. R. T. Bakker, 1968, The superiority of dinosaurs, *Discovery* 3:11–22; R. T. Bakker, 1975, The dinosaur renaissance, *Scientific American* 232:58–78; and R. T. Bakker, 1986, *The dinosaur heresies* (New York: William Morrow).
34. A. F. Bennett and B. Dalzell, 1973, Dinosaur physiology: A critique, *Evolution* 27:170–174; A. Bennett and J. Ruben, 1979, Endothermy and activity in vertebrates, *Science* 206:649–654; and J. O. Farlow , P. Dodson, and A. Chinsamy, 1995, Dinosaur biology, *Annual Review of Ecology and Systematics* 26:445–471.
35. Farlow, Dodson, and Chinsamy, 1995, pp. 445–471.
36. R. E. Barrick and W. J. Showers, 1994, Thermophysiology of *Tyrannosaurus rex*: Evidence from oxygen isotopes, *Science* 265:222–224.
37. J. Ruben, 1995, The evolution of endothermy in mammals, birds, and their ancestors, *Journal of Experimental Biology*; and J. Fishman, 1995, Were dinos cold-blooded after all? The nose knows, *Science* 270:735–736.
38. A. Chinsamy and P. Dodson, 1995, Inside a dinosaur bone, *American Scientist* 83:174–180; and Farlow, Dodson, and Chinsamy, 1995, pp. 445–471.
39. A. Chinsamy, L. M. Chiappe, and P. Dodson, 1995, Mesozoic avian bone microstructure: Physiological implications, *Paleobiology* 21:561–574.
40. S. R. Woodward et al., 1994, DNA sequence from Cretaceous period bone fragments, *Science* 266: 1229–1232.
41. S. B. Hedges and M. H. Schweitzer, 1995, Detecting dinosaur DNA, *Science* 268:1191.
42. H. N. Poinar et al., 1996, Amino acid racemization and the preservation of ancient DNA, *Science* 272:864–866.

CHAPTER FIFTEEN
Crossing the Boundary

1. D. Weishampel, P. Dodson, and H. Osmólska, 1990, *The Dinosauria* (Berkeley: University of California Press).
2. A. Feduccia, 1995, Explosive evolution in Tertiary birds and mammals, *Science* 267:637–638; A. Feduccia, *The origin and evolution of birds* (New Haven, Conn.: Yale University Press); and L. Hou et al., 1996, Early adaptive radiation of birds: Evidence from fossils from northeastern China, *Science* 274:1164–1167.
3. J. Cracraft, 1981, Toward a phylogenetic classification of the Recent birds of the world (Class Aves), *Auk* 98:681–714; J. Cracraft, 1986, The origin and early diversification of birds, *Paleobiology* 12:383–399; and J. Cracraft, 1988, The major clades of birds, in *The phylogeny and classification of Tetrapods*, Systematics Association special volume 35A, ed., M. J. Benton (Oxford: Clarendon Press).
4. L. Chiappe, 1995, The first 85 million years of avian evolution, *Nature* 378:349–355.
5. A. Cooper and D. Penny, 1997, Mass survival of birds across the Cretaceous-Tertiary boundary: Molecular evidence, *Science* 275:1109–1113.
6. M. A. Norell, 1992, Taxic origin and temporal diversity: The effect of phylogeny, in *Extinction and phylogeny*, eds. M. J. Novacek and Q. D. Wheeler (New York: Columbia University Press); and M. A. Norell, 1993, Tree-based approaches to understanding history: Comments on ranks, rules, and the quality of the fossil record, *American Journal of Science* 293A:407–417.
7. R. Owen, 1839. Observations on the fossils representing the Thylacotherium prevostii, Valenciennes, with reference to the doubts of its mammalian and marsupial nature recently promulgated; and on the Phascolotherium bucklandi, *Geological Transactions*, series 2, 6:186–265; and R. Owen, 1871, *Monograph of the fossil Mammalia of the Mesozoic formations* (London: Paleontological Society), monograph 24.
8. J. A. Lillegraven, Z. Kielan-Jaworowska, and W. A. Clemens, 1979, *Mesozoic mammals: The first two-thirds of mammalian history* (Berkeley: University of California Press); and F. S. Szalay, M. J. Novacek, and M. C. McKenna, 1993, *Mammal phylogeny* (New York: Springer Verlag).
9. F. A. Jenkins, Jr., et al., 1997, Haramiyids and Triassic mammalian evolution, *Nature* 385:715–718.
10. L. A. Nessov, 1987, Results of search and study of Cretaceous and early Paleocene mammals on the territory of the USSR [in Russian], *Ezhegodnik Vsesoyouznovo Paleontologichesko-vo Obschestva* 30:199–219; see also a review of this work in J. D. Archibald, 1996a, *Dinosaur extinction and the end of an era* (New York: Columbia University Press).
11. J. D. Archibald, 1996b, Fossil evidence for Late Cretaceous origin of "hoofed" mammals, *Science* 272:1150–1153.
12. Archibald, 1996a.
13. J. I. Noriega and P. Tambussi, 1995, A Late Cretaceous Presbyornithidae (Aves: Anseriformes) from Vega Island, Antarctic Peninsula: Paleobiogeographic implications, *Ameghiniana* 32:57–61.
14. S. L. Olson, 1985, The fossil record of birds, in *Avian biology*, vol. 8, eds. D. S. Farner, J. R. King, and K. C. Parkes, pp. 79–252; S. L. Olson, 1994, A giant Presbyornis (Aves: Anseriformes) and other birds from the Paleocene Aquia formation of Maryland and Virginia, *Proceedings of the Biological Society of Washington* 107:429–435; and P. Ericson, 1992, Evolution and systematic elements of the Paleogene family Presbyornithidae, *Frankfurt III Symposium of the Society of Avian Paleontology and Evolution* 1:16.
15. E. V. Kurochkin, 1995, The assemblage of Cretaceous birds in Asia, in *Sixth symposium on Mesozoic terrestrial ecosystems and*

biota, ed. A. L. Sun (Beijing: China Ocean Press), pp. 203–208.

16. S. L. Olson, 1992, *Neogaeornis wetzeli* Lambrecht: A Cretaceous loon from Chile, *Journal of Vertebrate Paleontology* 12:122–124.
17. Olson, 1985, pp. 79–252.
18. S. L. Olson and D. Parris, 1987, The Cretaceous birds of New Jersey, Smithsonian contributions to *Paleobiology* 63:1–22.
19. Olson, 1985, pp. 79–252; and Kurochkin, 1995, pp. 203–208.
20. Feduccia, 1996, p. 165.
21. Cooper and Penny, 1997, pp. 1109–1113.
22. Archibald, 1996a.
23. ibid.
24. D. E. Fastovsky and P. M. Sheehan, 1994, Habitat versus asteroid fragmentation in vertebrate extinctions at the KT boundary: The good, the bad, and the untested, in New developments regarding the KT event and other catastrophes in Earth history, *Lunar and Planetary Institute Contributions* 825:36–37.
25. D. E. Archibald and L. Bryant, 1990, Differential Cretaceous-Tertiary extinctions of nonmarine vertebrates: Evidence from northeastern Montana, in *Global catastrophes in Earth history: An interdisciplinary conference on impacts, volcanism, and mass mortality*, eds. V. L. Sharpton and P. Ward, Geological Society of America Special Paper 247.
26. Weishampel, Dodson, and Osmólska, 1990.

CHAPTER SIXTEEN

Diversification and Decline

1. C. G. Sibley and B. L. Monroe, Jr., 1990, *Distribution and taxonomy of birds of the world* (New Haven, Conn.: Yale University Press); and B. L. Monroe, Jr., and C. G. Sibley, 1993, *A world checklist of birds* (New Haven, Conn.: Yale University Press).
2. M. F. Coffin and O. Eldholm, 1993, Large igneous provinces, *Scientific American* (October): 42–49; and R. L. Larson, 1995, The mid-Cretaceous superplume episode, *Scientific American* (February): 82–86.
3. E. O. Wilson, 1992, *The diversity of life* (Cambridge, Mass.: Belknap Press); D. Quammen, 1996, *The song of the dodo* (New York: Scribner); and H. W. Menard, 1986, *Islands* (New York: W. H. Freeman).
4. J. Cracraft, 1981, Toward a phylogenetic classification of the Recent birds of the world (Class Aves), *Auk* 98:681–714; J. Cracraft, 1988, The major clades of birds, in *The phylogeny and classification of the Tetrapods*, Systematics Association special volume 35A, ed. M. J. Benton (Oxford: Clarendon Press); and L. Chiappe, 1995, The first 85 million years of avian evolution, *Nature* 378:349–355. See also C. G. Sibley and J. E. Ahlquist, 1990, *Phylogeny and classification of birds* (New Haven, Conn.: Yale University Press).
5. Sibley and Monroe, 1990.
6. ibid.
7. S. L. Olson, 1985, The fossil record of birds, in *Avian biology*, vol. 8, eds. D. S. Farner, J. R. King, and K. C. Parkes, pp. 79–256.
8. Sibley and Monroe, 1990.
9. Olson, 1985.
10. E. V. Kurochkin, 1995, The assemblage of Cretaceous birds in Asia, in *Sixth symposium on Mesozoic terrestrial ecosystem and biota*, ed. A. L. Sun (Beijing: China Ocean Press), pp. 203–208.
11. Sibley and Monroe, 1990.
12. L. M. Witmer and K. D. Rose, 1991, Biomechanics of the jaw apparatus of the gigantic Eocene bird *Diatryma*: Implications for diet and mode of life, *Paleobiology* 17:95–120.
13. D. W. Stedman, 1995, Prehistoric extinctions of Pacific island birds: Biodiversity meets zooarchaeology, *Science* 267:1123–1131.
14. Sibley and Monroe, 1990.
15. ibid.
16. Olson, 1985, pp. 79–256.
17. Sibley and Monroe, 1990.
18. ibid.
19. J. C. Greenway, 1967, *Extinct and vanishing birds of the world*, 2nd rev. ed. (New York: Dover).
20. Sibley and Monroe, 1990.
21. ibid.
22. ibid.
23. Olson, 1985, pp. 79–256.
24. A. Cooper and D. Penny, 1997, Mass survival of birds across the Cretaceous-Tertiary boundary: Molecular evidence, *Science* 275:1109–1113.
25. J. Imbrie and K. P. Imbrie, 1979, *Ice ages—Solving the mystery* (Cambridge, Mass.: Harvard University Press).
26. R. W. Graham et al. (FAUNMAP Working Group), 1996, Spatial response of mammals to late Quaternary environmental fluctuations, *Science* 272:1601–1606.
27. E. L. Lundelius, Jr., et al., 1983, Terrestrial faunas, in *Late quaternary environments of the United States*, vol. 1, *The Late Pleistocene*, ed. S. C. Porter (Minneapolis: University of Minnesota Press), pp. 311–333; and R. W. Graham and E. Lundelius, Jr., 1984, Coevolutionary disequilibrium and Pleistocene extinctions, in *Quaternary extinctions: A Prehistoric revolution*, eds. P. S. Martin and R. L. Klein (Tucson: University of Arizona Press), pp. 223–249.
28. Lundelius et al. 1983, pp. 311–333.
29. P. S. Martin, 1967, Prehistoric overkill, in *Pleistocene extinctions: The search for a cause*, eds. P. S. Martin and H. E. Wright (New Haven, Conn.: Yale University Press), pp. 75–120; P. S. Martin, 1973, The discovery of America, *Science* 179:969–974; and P. S. Martin and R. L. Klein, 1984, *Quarternary extinctions: A prehistoric revolution* (Tucson: University of Arizona Press).
30. Graham et al., 1996, pp. 1601–1606.
31. M. W. Beck, 1996, On discerning the cause of Late Pleistocene megafaunal extinctions, *Paleobiology* 22:91–103.
32. T. Flannery, 1994, *The future eaters* (Chatswood, New South Wales: Reed Books); E. L. Lundelius, Jr., 1989, The implications of disharmonious assemblages for Pleistocene extinctions, *Journal of Archaeological Science* 1989:407–417.

33. E. L. Lundelius, Jr., 1981, Climatic implications of Late Pleistocene and Holocene faunal associations in Australia, *Alcheringa* 7:125–149.
34. Greenway, 1967.

CHAPTER SEVENTEEN

The Real Great Dinosaur Extinction

1. J. D. Archibald, 1996, Fossil evidence for a Late Cretaceous origin of "hoofed" mammals, *Science* 272:1150–1153.
2. R. Lewin, 1993, *The origin of modern humans* (New York: W. H. Freeman).
3. J. Kappelman et al., 1996, Age of *Australopithecus afarensis* from Fejej, Ethiopia, *Journal of Human Evolution* 30:139–146.
4. E. O. Wilson, 1992, *The diversity of life* (Cambridge, Mass.: Belknap Press).
5. J. Kappelman, 1996, The evolution of body mass and relative brain size in fossil hominids, *Journal of Human Evolution* 30:243–276.
6. C. Swisher et al., 1996, Latest *Homo erectus* of Java: Potential contemporaneity with *Homo sapiens* in southeast Asia, *Science* 274:1870–1874.
7. Kappelman, 1996, 243–276.
8. Lewin, 1993.
9. E. S. Vrba et al., 1995, *Paleoclimate and evolution, with emphasis on human origins* (New Haven, Conn.: Yale University Press).
10. S. M. Stanley, 1996, *Children of the ice ages* (New York: Harmony).
11. J. Cohen, 1995a, *How many people can the Earth support?* (New York: W. W. Norton); and J. Cohen, 1995b, Population growth and Earth's human carrying capacity, *Science* 269:341–346.
12. Wilson, 1992.
13. H. E. Strickland and A. G. Melville, 1848, *The dodo and its kindred* (London: Reeve, Benham, and Reeve), p. 10.
14. E. Fuller, 1987, *Extinct birds* (London: Viking/Rainbird).
15. J. C. Greenway, 1967, *Extinct and vanishing birds of the world*, 2nd rev. ed. (New York: Dover).
16. Strickland and Melville, 1848, p. 10.
17. Strickland and Melville, 1848, p. 28.
18. D. Quammen, 1996, *The song of the dodo* (New York: Scribner); and H. W. Menard, 1986, *Islands* (New York: W. H. Freeman).
19. R. Owen, 1860, *Palaeontology, or a systematic summary of extinct animals and their geological relations* (Edinburgh: Adam and Charles Black), p. 399.
20. Wilson, 1992.
21. R. Owen, 1839a, Exhibited bone of an unknown struthious bird from New Zealand, *Proceedings of the Zoological Society of London* 7:169–171; and R. Owen, 1839b, Notice of a fragment of the femur of a gigantic bird of New Zealand, *Transactions of the Zoological Society of London* 3:29–32.
22. R. Owen, 1879, *Memoirs on the extinct wingless birds of New Zealand, with an appendix on those of England, Australia, Newfoundland, Mauritius, and Rodriguez*, 2 vols. (London: John van Voorst).
23. ibid., vol. 1, p. v.
24. J. Cracraft, 1976, The species of moas (Aves: Dinornithidae), in Collected papers in avian paleontology honoring the 90th birthday of Alexander Wetmore, ed. S. L. Olson, Smithsonian Contributions to *Paleobiology* 27:189–205.
25. M. M. Trotter et al., 1984, Moas, men and middens, in *Quaternary extinctions: A prehistoric revolution*, eds. P. S. Martin and R. L. Klein (Tucson: University of Arizona Press), pp. 708–727; and A. Anderson, 1989, *Prodigious birds: Moas and moa-hunting in prehistoric New Zealand* (New York: Cambridge University Press).
26. Fuller, 1987.
27. ibid.
28. D. W. Stedman, 1995, Prehistoric extinctions of Pacific island birds: Biodiversity meets zooarchaeology, *Science* 267:1123–1131.
29. ibid.
30. Fuller, 1987.
31. Greenway, 1967.
32. F. A. Lucas, 1888, The expedition to Funk Island, with observations upon the history and anatomy of the Great Auk, *Report of the U.S. National Museum for 1888*, pp. 493–529.
33. Owen, 1860, p. 400.
34. Greenway, 1967.
35. ibid, p. 281.
36. Fuller, 1987.
37. Greenway, 1967.
38. Lucas, 1888, p. 498.
39. Fuller, 1987.
40. Lucas, 1888, p. 498.
41. R. Owen, 1864, Description of the skeleton of the Great Auk or Garefowl, *Transactions of the Zoological Society of London* 5:317–335.
42. Lucas, 1888, p. 507.
43. ibid, p. 508.
44. ibid, p. 510.
45. ibid, p. 515.
46. Stedman, 1995, p. 1130.

CHAPTER EIGHTEEN

The Third Wave

1. J. C. Greenway, 1967, *Extinct and vanishing birds of the world*, 2nd rev. ed. (New York: Dover).
2. S. A. Temple, 1986, The problem of avian extinctions, *Current Ornithology* 3:460.
3. Greenway, 1967, p. 35.
4. Greenway, 1967.

5. E. Fuller, 1987, *Extinct birds* (London: Viking/Rainbird).

6. J. J. Audubon, 1827–1838, *The Birds of America*, 4 vols. (London); and J. J. Audubon, 1831–1839, *Ornithological biography*, 5 vols. (Edinburgh).

7. A. Wilson, 1808–1814, *American ornithology; or the natural history of the birds of the United States*, 9 vols. (Philadelphia).

8. Greenway, 1967.

9. E. H. Bucher, 1992. The causes of extinction of the passenger pigeon, *Current Ornithology* 9:1–36.

10. Audubon, 1827–1838; and Audubon, 1831–1839.

11. ibid.

12. Fuller, 1987.

13. Bucher, 1992, pp. 1–36.

14. ibid.; and Greenway, 1967.

15. Greenway, 1967.

16. Audubon, 1827–1838; and Audubon, 1831–1839.

17. Anonymous, 1914, News and notes: Ectopiosts migratorius ... etc., *Auk* 31:566.

18. Greenway, 1967; and Fuller, 1987.

19. Greenway, 1967.

20. ibid.

21. N. F. R. Snyder and H. A. Snyder, 1989, Biology and conservation of the California Condor, *Current Ornithology* 6:175–267.

22. N. J. Collar, M. J. Crosby, and A. J. Stattersfield, 1994, *Birds to watch 2: The world list of threatened birds*, Birdlife Conservation Series no. 4 (Bird Life International).

23. J. Cohen, 1995a, *How many people can the Earth support?* (New York: W. W. Norton); and J. Cohen, 1995b, Population growth and Earth's human carrying capacity, *Science* 269:341–346.

24. E. O. Wilson, 1992, *The diversity of life* (Cambridge, Mass.: Belknap Press).

25. ibid., p. 308.

26. Cohen, 1995a; and Cohen, 1995b.

27. Collar, Crosby, and Stattersfield, 1994.

CREDITS

The skeletal drawings in this book were enhanced by Fine Line Illustrations, based on computerized illustrations supplied by Timothy Rowe.

Layout done by Paula Jo Smith.

CHAPTER ONE

Page 4 — Mary Evans Picture Library.

Page 5 — American Museum of Natural History.

CHAPTER TWO

Page 13 — © 1985 Mark Hallett. All rights reserved.

Page 15 — © 1985 Mark Hallett. All rights reserved.

Page 16 (*top*) — Neg. no. 2A 21980. Photo by Denis Finnin. Courtesy Department of Library Services, American Museum of Natural History.

Page 16 (*bottom*) — Neg. no. 2A 21316. Photo by Ben Blackwell/Denis Finnin. Courtesy Department of Library Services, American Museum of Natural History.

Page 17 (*top*) — Neg. no. 2A 5378. Courtesy Department of Library Services, American Museum of Natural History.

Page 17 (*bottom*) — Neg. no. 602317. Photo by Mick Ellison. Courtesy Department of Library Services, American Museum of Natural History.

Page 18 — From *Dinosaur Extinction and the End of an Era* by David Archibald. Copyright © 1996 by Columbia University Press. Reprinted with permission of the publisher.

Page 19 — Adapted from Lillegraven, J. A. and L. M. Ostresh,"Late Cretaceous evolution of western shorelines, Western Interior Seaway," *Geol. Soc. Am. Sp. Pap.* 243, pp. 10–11.

CHAPTER THREE

Page 22 — Katharina Perch-Nielsen.

Page 22 (*lower right*) — Allan A. Ekdale, University of Utah.

Page 23 — From *The Elements of Paleontology* by Rhona M. Black. Copyright © 1970 and 1988 by Cambridge University Press. Reprinted with the permission of Cambridge University Press.

Page 23 (*left*) — The British Museum (Natural History).

Page 23 — Chip Clark.

Page 24 (*top*) — Neg. no. 2A 23737. Photo by Jackie Beckett. Courtesy Department of Library Services, American Museum of Natural History.

Page 24 (*upper middle*) — Neg. no. 2A 23741. Photo by Jackie Beckett. Courtesy Department of Library Services, American Museum of Natural History.

Page 24 (*lower middle*) — Neg. no. 35041. Courtesy Department of Library Services, American Museum of Natural History

Page 24 (*bottom*) — Neg. no. 315714. Courtesy Department of Library Services, American Museum of Natural History.

Page 25 (*top*) — Kirk R. Johnson.

Page 25 (*bottom*) — Douglas J. Nichols.

Page 26 (*upper left*) — Douglas Henderson.

Page 26 (*upper right*) — Douglas Henderson.

Page 26 (*bottom*) — By Ely Kish. Reproduced with permission of the Canadian Museum of Nature, Ottawa, Canada.

Page 27 (*top*) — Neg. no. 338590. Photo by Ben Blackwell/ Denis Finnin. Courtesy Department of Library Services, American Museum of Natural History.

Page 27 (*bottom*) — Neg. no. 35425. Courtesy Department of Library Services, American Museum of Natural History.

Page 28 — From *What Color is that Dinosaur* by Lowell Dingus. Illustrations by Stephen Quinn. Copyright © 1994 and reprinted by permission of The Millbrook Press, Inc.

Page 29 — Douglas Henderson.

CHAPTER FOUR

Page 31 — Adapted from Press, F. and R. Siever, 1978, *Earth*, fig. 1-10.

Page 33 (*top*) — Adapted from Moores, E. M. and R. J. Twiss, 1995, *Tectonics*, p. 7.

Page 33 (*bottom*) — Adapted from Press, F. and R. Siever, 1978, *Earth*, p. 461.

Page 34 — Adapted from Moores, E. M. and R. J. Twiss, 1995, *Tectonics*, p. 6.

Page 35 (*top*) — Dr. Keith Cox photo from White and McKenzie, July 1989, *Scientific American*, p. 63.

Page 35 (*bottom*) — Adapted from White and McKenzie, July 1989, *Scientific American*, p. 68.

Page 40 — Lowell Dingus.

Page 42 — Neg. no. 39137, Courtesy Department of Library Services, American Museum of Natural History.

CHAPTER FIVE

Page 44 — Neg. no. 28000. Courtesy Department of Library Services, American Museum of Natural History.

Page 45 — Scottish National Portrait Gallery.

Page 46 — Reprinted from Bonney, T. G., 1895, *Charles Lyell and Modern Geology*.

Page 47 — Alessandro Montanari.

Page 48 (*top*) — Gerta Keller.

Page 48 (*bottom*) — Adapted from Courtillot, V., October 1990, "A volcanic eruption," *Scientific American*, p. 91.

Page 49 — William K. Hartmann.

Page 50 — Saxon Donnelly photo from Booth, G., 1982, "The case of the missing dinosaurs," *California Monthly*.

Page 51 — Lowell Dingus.

Page 52 — Adapted from Rich, P. and T. H. Rich, July 1993, "Australia's polar dinosaurs," *Scientific American*, p. 52.

Page 53 — Adapted from Clemens, W. A. and L. G. Nelms, 1993,"Paleoecological implications of Alaskan terrestrial vertebrate fauna in the latest Cretaceous time at high paleolatitudes," *Geology*, v. 21, pp. 503–506.

Page 54 — Adapted from Alvarez, W. and F. Asaro, October 1990, "An extraterrestrial impact," *Scientific American*, p. 79.

Page 56 — Adapted from Alvarez, L., 1983, "Experimental evidence that an asteroid impact led to the extinction of many species 65 million years ago," *Proc. Nat. Acad. Sci., USA*, p. 630.

Page 57 — Photo by J. D. Griggs, U.S. Geological Survey, January 31, 1984, reprinted from Decker, R. and B. Decker, 1989, *Volcanoes*, plate 1.

Page 59 (*top*) — Bruce F. Bohor.

Page 59 (*bottom*) — U.S. Geological Survey, reprinted from Press, F. and R. Siever, 1978, *Earth*, p. 531.

Page 61 (*left*) — Bruce F. Bohor.

Page 61 (*right*) — Bruce F. Bohor.

Page 62 — Wendy S. Wolbach.

Page 63 — Adapted with permission from Bourgeois, J. et al., "A tsunami deposit at the Cretaceous-Tertiary boundary in Texas," *Science*, v. 241. Copyright ©1988 American Association for the Advancement of Science.

Page 65 — Adapted from Hallam, A., 1989, Catastrophism in geology, In Alvarez, W., et al., *Catastrophes and Evolution Astronomical Foundations*, p. 39.

CHAPTER SIX

Page 68 — William K. Hartmann.

Page 69 — Adapted from *Nature*, Pope K. O. et al., "Mexican site for K/T impact crater?," *Nature* v. 351, p. 105, Copyright © 1991 Macmillan Magazines Limited.

Page 70 — Adapted from Swisher, C. et al., 1992, "Coeval $^{40}Ar/^{39}Ar$ ages of 65.0 million years ago from Chicxulub crater melt rock and Cretaceous-Tertiary boundary tektites," *Science*, v. 257, p. 955.

Page 71 — Adapted from Grieve, R. A. F., 1989, "Hypervelocity impact cratering: A Catastrophic geologic process," In Alvarez, W. et al., *Catastrophes and Evolution: Astronomical Foundations*, pp. 64–65.

Page 72 — Reprinted with permission from *Nature*, Hildebrand et al., "Size and structure of the Chicxulub Crater revealed by horizontal gravity gradients and cenotes," *Nature* v. 376, p. 415, Copyright ©1995 Macmillan Magazines Limited.

CHAPTER SEVEN

Page 81 — From "Cretaceous-Tertiary (Chicxulub) impact angle and its consequences," *Geology*, v. 24, Schultz, P. H. and S. D'Hondt. Reproduced with permission of the publisher, the Geological Society of America, Boulder, Colorado USA. Copyright © 1996, The Geological Society of America, Inc. (GSA). All rights reserved.

Page 81 — From *Dinosaur Extinction and the End of an Era* by David Archibald. Copyright © 1996 by Columbia University Press. Reprinted with permission of the publisher.

Page 86 — From *Dinosaur Extinction and the End of an Era* by David Archibald. Copyright © 1996 by Columbia University Press. Reprinted with permission of the publisher.

CHAPTER EIGHT

Page 93 — Adapted from Hallam, A., 1989, "Catastrophism in geology," In Alvarez, W., et al, *Catastrophes and Evolution: Astronomical Foundations*, p. 42.

Page 97 — Reprinted from Dingus, L., 1984, "Effects of stratigraphic completeness on interpretations of extinction rates across the Cretaceous-Tertiary boundary," *Paleobiology*, v. 10, p. 423.

CHAPTER NINE

Page 108 (*top*) — Texas Memorial Museum.

Page 108 (*bottom*) — Courtesy of John Ostrom.

Page 109 (*top*) — Texas Memorial Museum.

Page 109 (*bottom*) — Neg. no. 2A 21966. Photo by Jackie Beckett. Courtesy Department of Library Services, American Museum of Natural History.

Page 110 — Texas Memorial Museum.

Page 111	Neg. no. 2A 23730. Photo by Jackie Beckett/Denis Finnin. Courtesy Department of Library Services, American Museum of Natural History.
Page 112 (*top*)	Texas Memorial Museum.
Page 113 (*top*)	Courtesy of John A. Wilson.
Page 113 (*bottom*)	Texas Memorial Museum.
Page 114	Texas Memorial Museum.
Page 117	H. von Meyer, 1862. *Archaeopteryx* lithographica from the Lithographic Slate of Solenhofen. *Annals and Magazine of Natural History*, London, 9: 366–370.
Pages 118–119	Texas Memorial Museum.
Page 120	Texas Memorial Museum.
Pages 122–123	Texas Memorial Museum.

CHAPTER TEN

Page 126	Courtesy of A. Charig.
Page 127 (*top*)	R. Owen, 1854. *Geology and Inhabitants of the Ancient World*. London.
Page 127 (*bottom*)	Texas Memorial Museum.
Page 128	Texas Memorial Museum.
Page 130	Courtesy Department of Library Services, American Museum of Natural History.
Page 131	Courtesy Department of Library Services, American Museum of Natural History.
Page 132	Courtesy of A. Charig.
Pages 135–136	Texas Memorial Museum.
Page 137	F. B. Gill, 1995, *Ornithology*. W. H. Freeman & Company, New York.
Page 138	Courtesy of John Ostrom.

CHAPTER ELEVEN

Page 144	Texas Memorial Museum.
Pages 145–146	Texas Memorial Museum.
Page 147 (*left*)	Texas Memorial Museum.
Page 147 (*right*)	Modified from H. Rahn, A. Ar, and C. V Paganelli,"How bird eggs breath," *Scientific American*, February, 1979.
Page 148	Texas Memorial Museum.
Pages 150-153 (*top*)	Texas Memorial Museum.
Page 151 (*bottom*)	Neg. no. 2A 23740. Courtesy Department of Library Services, American Museum of Natural History.
Page 154	Neg. no. 2A 23736. Photo by Denis Finino. Courtesy Department of Library Services, American Museum of Natural History.
Page 155	From T. Rowe, 1996. "Brain heterochrony and evolution of the mammalian middle ear," pp. 71–96 In *New Perspectives on the History of Life*. M. Ghiselin and G. Pinna (eds.), California Academy of Sciences.
Pages 156–160	Texas Memorial Museum.
Pages 162–165	Texas Memorial Museum.

CHAPTER TWELVE

Pages 171–179	Texas Memorial Museum.
Page 181	Texas Memorial Museum.
Page 183	Photo by T. Rowe.
Pages 184–186	Texas Memorial Museum.
Pages 189-194	Texas Memorial Museum.

CHAPTER THIRTEEN

Page 197	American Museum of Natural History.
Page 198	Texas Memorial Museum.
Pages 200–202	Texas Memorial Museum.
Page 204	R. Lewin 1997. *Patterns in Evolution: the New Molecular View*. Scientific American Library.

CHAPTER FOURTEEN

Page 208	Texas Memorial Museum.
Page 209	H. Rahn, A. Ar, and C. V Paganelli, "How bird eggs breath," *Scientific American*, February, 1979.
Pages 210–211	Texas Memorial Museum.
Pages 213–215	Texas Memorial Museum.
Page 216 (*top*)	Neg. no. 2A 21967.Courtesy Department of Library Services, American Museum of Natural History.
Page 216 (*bottom*)	Texas Memorial Museum.
Page 217	Texas Memorial Museum.
Page 218	T. A. McMahon and J. T. Bonner, 1983. *On Size and Life*. Scientific American Library.
Pages 219–221	Texas Memorial Museum.
Page 222 (*bottom*)	Neg. no. 31511. Courtesy Department of Library Services, American Museum of Natural History.
Pages 222–224 (*top*)	Texas Memorial Museum.
Page 226	Texas Memorial Museum.

CHAPTER FIFTEEN

Page 230	Redrawn from on A. Feduccia, 1996. *The Origin and Evolution of Birds*. Yale University Press, New Haven.
Pages 232,–233	Texas Memorial Museum.
Page 235	Texas Memorial Museum.
Pages 236–237	*Birds, A Picture Sourcebook*, 1980. Edited and arranged by Don Rice. Van Nostrand Reinhold Company, New York.

Pages 238–239 Texas Memorial Museum.

Page 243 Texas Memorial Museum.

CHAPTER SIXTEEN

Page 246 Texas Memorial Museum.

Page 249 Based on Cracraft, J. 1988. The major classes of birds. pp. 339–361, In M. J. Benton (ed.), *The Phylogeny and Classification of the Tetrapods*, Systematics Association Special volume 35A, Oxford, Clarendon Press.

Page 253 (*bottom*) Neg. no. 310104. Courtesy Department of Library Services, American Museum of Natural History

Pages 250–257 *Birds, A Picture Sourcebook*, 1980. Edited and arranged by Don Rice. Van Nostrand Reinhold Company, New York.

Page 257 (*bottom*) Texas Memorial Museum.

CHAPTER SEVENTEEN

Page 267 Texas Memorial Museum.

Pages 268–269 R. Lewin, 1993. *The Origin of Modern Humans*. Scientific American Library, New York.

Page 270 Redrawn from J. E. Cohen, 1995. *How Many People Can the Earth Support*? W. W. Norton & Company, New York.

Page 271 (*top*) Based on A. P. Dobson, 1996. *Conservation and Biodiversity*. Scientific American Library, New York.

Page 271 (*bottom*) H. E. Strickland and A. G. Melville, 1848. *The Dodo and its Kindred*. London.

Page 272 (*left*) R. Owen, 1971. Notes on the articulated skeleton of the Dodo in the British Museum. *Transactions of the Zoological Society of London*, 7: 513–525.

Page 272 (*right*) H. E. Strickland and A. G. Melville, 1848. *The Dodo and its Kindred*. London.

Page 273 H. E. Strickland and A. G. Melville, 1848. *The Dodo and its Kindred. London.*

Pages 275–276 R. Owen, 1879. *Memoirs on the extinct wingless birds of New Zealand, with an appendix on those of England, Australia, Newfoundland, Mauritius and Rodriguez*, 2 volumes, London.

Page 277 W. H. Menard, 1986. *Islands*. Scientific American Library, New York.

Page 279 *Birds, A Picture Sourcebook*, 1980. Edited and arranged by Don Rice. Van Nostrand Reinhold Company, New York.

Page 280 J. Gould, 1832–1837. *Birds of Europe*. 5 volumes, London.

Page 283 R. Owen, 1865. "Description of the skeleton of a Great Auk or Garefowl." *Transactions of the Zoological Society of London*, 7: 317–335.

CHAPTER EIGHTEEN

Page 287 A. P. Dobson, 1996. *Conservation and Biodiversity*. Scientific American Library, New York.

Page 288 *The Illustrated Sporting and Dramatic News*, July 3, 1875.

Page 291 *Birds, A Picture Sourcebook*, 1980. Edited and arranged by Don Rice. Van Nostrand Reinhold Company, New York.

Page 293 A. P. Dobson, 1996. *Conservation and Biodiversity*. Scientific American Library, New York.

Page 294 A. P. Dobson, 1996. *Conservation and Biodiversity*. Scientific American Library, New York.

Page 295 (*top and bottom*) A. P. Dobson, 1996. *Conservation and Biodiversity*. Scientific American Library, New York.

Page 296 Rev. R. Owen, 1894. *The Life of Richard Owen by his Grandson*. 2 vols., London.

INDEX

*Page numbers in **boldface** indicate illustrations.*

Acetabulum, *see* hip socket
Acid rain, 27, 88–89
  and extinctions, 53, 76, 78, 79
  and impact hypothesis, 86–87
  tree tolerance of, 78
Acidity (of water)
  and extinctions, 78
Actinistia, **151,** 152
Actinopterygian lineage, 151
Adaptive radiations
  birds, 229–231, 243, **246,** 247, 263
  moas, 276
Africa
  human evolution, 267
  magma plumes, 37
  threatened bird species, 293
African Rift Valley, 268
Afrovenator, 186
Age of Dinosaurs, 5, 57, 231
  mammalian diversity, 235
  see also Cretaceous Period; Mesozoic Era
Age of Fishes, 231
Age of Mammals, 47, 231, **243**
Age of Reptiles, 231
  mammals in, 231, 233
Agriculture, 268, 270
Alamosaurus, 86
Alaska
  dinosaur fossils, 52–53
Alba Patera (Mars), 101
Albatross, 238
Alberta (Canada)
  boundary clay, 62, 63
Albertosaurus, 86
Alca impennis, *see* Great Auk
Aldabra Islands, 247
Aliwal North (South Africa), 112, 113
Allergic drowning
  and dinosaur extinction hypothesis, 13
Alligators, 14, 83, 86, 87, 90
Allosauroidea, 186
Allosaurus, 5, 185, 186
Altangerel, Perle, 215
Altmühl River, 116
Alula, 218
Alvarez, Luis, 47–50, **51,** 54, 57, 58, 60
Alvarez, Walter, 47–50, **51,** 57, 58, 60, 65, 69, 99
Alvarezsauridae, alvarezsaurids, **216,** 221, 242
Alvarezsaurus, 217
Amazon Basin
  birds, 247
Amber, 227
Ambiortus, 213, 222
Amino acids
  in K-T boundary clay, 63
Ammonites
  extinction of, 23, 77, 80
Amniota, **147,** 148, 154–155
Amniotes, 147, 148, 159, 167
  phylogeny, 143, 156
Amniotic egg, 147
Amphibia, 154, 198
Amphibians, 154
Anatotitan, 15, 27, 82, 86
Ancestry
  and phylogenetic system of classification, 204–205
Andean Condor, 292
Anders, Mark, 60, 73, 98
Anderson, Don, 38
Andes, 247, 248, 251
Andesites, 73
Angiosperms
  and dinosaur extinction, 12
Animalia kingdom, 198
Ankylosaurs, 177
Ankylosaurus, 5, 86
Anseriformes, anseriforms, **236,** 237, 239, **249,** 250–251
Apatosaurus, 5, 6, 15, **16**
Ape lineages, 266, 267
Apodiformes, **249, 254,** 255
Arboreality, 220
Archaeopteryx lithographica, 9, 115–117, 124, 134–138, 140, 152, 170, 182, 185, 188, 190, 193–194, 207, 218–220, 231
  classification of, 201, **202**
  compared to *Deinonychus*, 123
  and evolution, 133–134
  and origin of birds, 212–213
Archibald, David, 54, 62, 80, 81, 83, 86–88, 94, 100, 234, 235, 241, 242, 266
Archosauria, archosaurs, 111, 113, 140, 158–162, 167
  see also Euparkeria capensis
Archosauriformes, 159, **160**
Arctic Ocean, 52
Arctic Tern, 284
Ardeidae, 253
Argentina
  endangered birds, 295
Argentinosaurus, 16
Aripithecus ramidus, 267
Arkansas
  tsunami deposits, 64
Arthropods, 149
Asaro, Frank, **51,** 57, 58, 69, 99
Ascending process, 185
Asia
  Homo erectus, 268
  threatened birds, 293, 295
Askin, Rosemary, 80
Asteroids
  impacts, 68
  see also impact extinction hypothesis
Asthenosphere, **32**
Atavisms, 210, 212
Atlantic Ocean, 282
  Great Auks, 280
Aublysodon, 86
Audubon, John James, 287–289, 291
Auks, *see* Great Auk
Australia
  hot-spot volcanism, 38
  mammal extinctions, 261
  megafaunal extinction, 262
  see also Dinosaur Cove
Australopithecines, 267
Australopithecus afarensis, 267
Australopithecus anamensis, 267
Aves, 111, 133, 198, 223, 240
Avialae, avialians, 188, 190–191, 194
Avian diversity hypothesis, 230–231, 233
Avocet, **237**

Bachman's Warbler, 293
Bakker, Bob, 225
Bare-legged owl, 255
Basalt, 3
  see also flood basalts
Bass Basin, 38
Bastard wing (alula), 218
Basu, Asish, 37, 39, 42
Bavaria, 116, 138
Beak, 157, 223
Bekov, George, 58
Belemnites, 78
Beloc (Haiti), 73
Benton, Michael, 161
Bering Land Bridge, 90
Bipedalism, 113, 115, 161
  hominids, 267
Bird-dinosaur hypothesis, **114,** 186, 212–213
Birds
  adaptive radiations, 229–231, 243, 263
  aerobic metabolism, 226
  and ancestral archosaur, 160
  and Avialae, 190–191
  clavicles, 135, 139–140
  Cretaceous diversity mapping, 235–240
  and crocodylians, 160–161
  and dinosaurs, 9, 108-109, 111, 112, 138–141, 149, 166–170, 174, 177, 179, 187, 194–196, 205–206

Birds *continued*
- diversification, 230–231, 233, 243, 245–259, 266
- diversion of humans from, 155–156
- endangered, 293–295
- and endothermy, 226
- evolution, 149–150, 207–219
- evolution from reptiles, 9
- extinctions, 274, 295
- feet, 182, 191, 210–211, 220–221
- fish-eating, 240
- flapping, 217
- flight, **114,** 189
- flightless, 216, 252, 279
- foot, 182, 210–211, 220-221
- furcula, 135–136, 139–140
- genetic memory, 209
- ghost lineages, 239, 240
- ground-up hypothesis, **114,** 115
- Hell Creek formation fossil record, 86
- hindlimb, 166
- hip socket, 136
- history of, 132-133, 229–231
- island extinctions, 278–279, 285-286, 296
- K-T boundary, **239**
- and mammals, 156
- and maniraptorans, 188–190
- Mesozoic, 207–208, 230, 231, 235-236
- opposite, 220
- origins of, 108, 111–124, 192
- and ornithomimosaurs, 187
- Pacific Basin diversification, **246,** 247
- Palaeognathae, 248–250
- phylogeny of modern, **248**
- protobirds, 114, 133
- Quaternary extinctions, 261
- running, 193
- saurischian dinosaurs, 179
- seabirds, 248, 278
- shorebirds, 248, 252, 253
- skeletal structure, 135–136, 139–140
- and tetanurines, 183–185
- and theropods, **209**
- threatened, 293–295
- tree-down hypothesis, 114–115
- turbinates, 226
- walk, 218–219
- *see also Archaeopteryx*; carinate birds; Ornithurae, ornithurines

Black-capped Vireo, 293
Blending inheritance, 131
Body size
- and extinction rates, 82–83, 86

Bohor, Bruce, 59, 60
Bone growth
- and endothermy, 227

Bones
- avian, 182
- theropods, 182

Boslough, Mark, 100
Boundary clays, *see* K-T transition boundary clays
Bourgeois, Joanne, 63
Bowfins, 83
Brachiosaurus, 15
Brain
- Aves, 223, **224**
- hominids, 267–268

Braman, Dennis, 62
Brazil
- endangered birds, 295

Brazos River (Texas), 63
Briggs, John, 66
Brochu, Chris, 161
Brooke, Robert, 126
Broom, Robert, 113, 114, 140
Brown, Alfred, 112, 113
Brown, Barnum, 109–110
Bryant, Laurie, 83, 242
Buckland, William, 127
Bug Creek sequence (Montana), 98
Bustards, **249, 252**
Buttonquail, **249,** 255

Caecilians, 154
California Condor, 292–293
Cambrian period, **5**
Camels, 261
Cameroon line, 38
Caprimulgiformes, **249**
Caravaca, Spain, 61, 97, 98
Carbon enrichment, 61–63
Carbonaceous chondrite meteorites, 57, 58, 63
Carboniferous period, **5**
Carinate birds, **217,** 221–222, 225–227
Carlisle, David, 62
Carnivorous dinosaurs, 13, 16
- *see also* theropod dinosaurs

Carnosaurs, 52
Carnotaurus, 215
Carolina Parakeet (*Conuropsis carolinensis*), 291–292
Carpals, 123
Cartier, Jacques, 281
Caruncle, 223
Cataracts
- and dinosaur extinction hypothesis, 14

Catastrophic extinction scenario, 47
Catastrophic theory, 103
Catastrophism, school of, 44–46
Cathodo-luminescence, 60
Cenozoic Era, **5**
Central America
- avian diversity, 295
- deforestation, **294**

Ceratopsi, ceratopsians, 175–176
Ceratosauria, ceratosaurs, 182–183
Ceratosaurus nasicornis, 183
Chameleon, 158
Chaoyangia, 213
Charadriiformes, charadriiforms, **237**–239, 252, 253
Charcharodontosaurus, 86
Charcoal
- in K-T boundary clay, 61–62

Charig, Alan, 122
Chatterjee, Sankar, 192
Chiappe, Luis, 213, 215, 231
Chicken, domestic, **238**
Chicxulub Crater (Mexico), **68,** 69–74, 80, 99
- asymmetry, 80–81
- impact shock waves, 100, 101
- time of impact, 102

Chile, 238
Chimpanzees, 266, 267
China, 185, 186
- threatened birds, 295

Chindesaurus, 166
Chinsamy, Anusuya, 227
Chirostenotes, 86
Choana, **152,** 153
Choanata, 152–153
Chondrichthyes, chondrichthians, 151
Chondritic meteorites, 57, 58, 63
Chron 29r, 48
Ciconiformes, **249,** 251–252
Cladistics, 8–9, 66, 145, 148, 170
Cladogram, *see* phylogenetic map
Clams
- extinctions, 23, 76–78, 80

Clark, James, 161, 215
Classifications
- and evolutionary history, 202–203

Clavicles
- avian, 135, 139–140

Clemens, William, 50, 52–54, 83, 241
Climate
- and dinosaur extinction hypothesis, 14, 18–19
- and extinction, 53
- increased warming, 260
- as killing mechanism, 274

Coccolithophora, coccoliths, **22,** 53, 78
- extinctions, 76, 77

Code of Taxonomy, 203
Coelophysis, 183
Coelurosauria, 186–187
Cohen, Joel, 270, 294
Colbert, Edwin, **183**
Coleville River, 52
Coliidae, **249, 254,** 255
Colombia
- endangered birds, 295

Colombian grebe, 251
Columbia River Basalts, 42
Columbiformes, 253
Comets
- and extinctions, 65
- impact hypothesis, 58, 68

Common Tern, 284
Comoros Islands, 152
Compsognathus longipes, 120, 121, 138, 141, 170, 185, 195, 212
Condors, 261, 292–293
Confuciusornis, 212, 220
Conifers, 53
Continental breakup, *see* rifting
Continental plates, **33**
Continents
- distribution of, **18**

and endangered species extinctions, 285–286
plate tectonic activity, 32
Convergent evolution, *see* homoplasy, theory of
Coombs, Walter, 177
Cooper, Alan, 240
Coprolites, 13
Coquette Hummingbird, **254**
Coraciiformes, **249,** 256–257
Coracoid, **217,** 218
Coral Sea Hot Spot, **33,** 38
Corals
extinctions, 78
Coria, Rudolfo, 186
Cormorants, **249,** 251
Costa Rica, **294**
Courtillot, Vincent, 34, 36, 40, 58, 60, 61, 100
Cracraft, Joel, 231
Cranes, **249,** 252, 253, 261
Crater formation process, 70–71
Crater impact
frequency of, 68
Crested Grebe, **252**
Cretaceous bird diversity
mapping, 235–240
Cretaceous Long Normal period, 36
Cretaceous Period, 6
North America, 19
see also K-T transition boundary
Crichton, Michael, 209, 227
Crocodile, 83, 86, 87, 89, 90
Crocodylians
and birds, 111, 160–161
evolution, 113, 161
gait, 160
Crowned Pigeon, **253**
Crystals, fractured, 59–60
Cuban Bee hummingbird, 255
Cuckoo lineage, *see* Cuculiformes
Cuculiformes, **249,** 256
Curry, Philip, 185
Cuvier, Georges, 44–46, 128, 129, 233, 273

D-layer, 36
Darkness
and dinosaur extinction, 51–53
Darwin, Charles, 129–133, 182, 195, 196, 212, 230
and natural classification, 199–201
Darwin, Erasmus, 129
Darwinian evolution, 130–133
Davis, Marc, 65
De Laubenfels, M. W., 68
De Queiroz, Kevin, 203–204
Deccan Traps, **26, 33**–35, 37, 39, 55, 57, 93, 99–101
eruption of, 40, 41
time of eruptions, 102
Deforestation
Costa Rica, **294**
and North American extinctions, 286, **287,** 291
tropical forests, 294
Deinocheirus mirificus, 187
Deinonychus antirrhopus, **108**–111, 115, 123, 124, 140, 141, 152, 170, 188–190, 195, 212, 213, 221
Denmark, 61, 281
DePaolo, Don, 73
Developmental history
and evolutionary history, 209–212
Devonian period, **5**
Diatoms, 77
Diatryma, 252
Diatryma gigantea, **253**
Diatrymidae, 252
Dimethylsulfide, 53
Dinoflagellates, 77, 78
Dinornis struthoides, 275
Dinosaur
coinage of word, 125
Dinosaur-bird connection, 212–213
Dinosaur Cove (Victoria, Australia), 51–52
Dinosaur dung, *see* coprolites
Dinosaur evolution, 6, 113, 149–150, 162
map of, 169–194
sequence of, 5–6
time scale, **5**
Dinosaur extinction, 7, 86, 90, 99, 101–104
at boundary events, 86
debate over causes, 8
patterns, 87, 89
supernovas, 44
theories, 41, 103–104
time scale, **5**
see also extinction hypotheses
Dinosauria, **4,** 137–138, 159, 163–167, 170, 171, 174, 195, 204–206, 296
homoplasy theory, 140
K-T boundary, 241–242
Dinosauromorpha, dinosauromorphs, 162, 171, 173
Dinosaurs
adaptive radiation, 230
ancestral, 166–167
armored, 5, 177
and bird flight, 189
and birds, 9, 108–112, 138–141, 149, 166–170, 174, 177, 179, 187, 194–196, 205–206
divergence of humans from, 155–156
diversity of, 94
duckbill, 5, 6, 17, 52
ectothermy, 225
eggs and extinction, 16
endothermy, 224–226
evolutionary maps, 169–194
feet, **128,** 210–211
femur, 127–128
fossil evidence, 4–5, 6, 81–82
gait, 128
ground-up hypothesis, 115
hand, 163–165, 189
herbivorous, 13
hip socket, 165–166
horned, 5, 6, **26,** 27, 175
limb structure, 128
and lizards, 158–159
K-T boundary, 241–242
lineages, 94
North America fossil record, 94
posture, 4–5
and primate lineage, 296
running, 115
and saurians, 158
skeleton, **163**
and theory of evolution, 128–130
see also bird-dinosaur hypothesis
Diplodocus, 5
Dipnoi, 153
Disease
and dinosaur extinction hypothesis, 13–14
Diversification, 131
birds, 230–231, 233, 243, 245–259, 266
mammals, 243
see also adaptive radiations
Diversity
biological, 263
birds, 245–259
Cretaceous birds, 235–240
decline in, 260–263
and isolation, 246–248
mapping, 231–233
DNA
ancient, 227–228
Dodo, 253, 271–273
Dodson, Peter, 81, 175
Dollo's law, 140
Drake, Charles, 34, 40, 56, 58–61
Dromaeosauridae, dromaeosaurs, 188, 193
Dromaeosaurus, 16, 83, 86
Duckbills, 5, 6, 17, 27, 28, 52
D'Hondt, Steven, 79, 80, 86, 89, 95

Eagles, **249,** 253–255
Eared Dove (*Zenaida auriculata*), 288–290
Earth
formation of, 6, 32–33
origin of, **5**
time scale of and evolutionary history, **5**
Earthquakes
and plate tectonics, 33
Echinoderms
extinctions, 66, 78
Ectothermy, 224, 225
Ecuador
endangered birds, 295
Edmontia, 86
Edmontosaurus, 82, 86
Egg tooth, 223
Eggs, dinosaur
and extinction hypothesis, 16
Egrets, **249, 253**
El Kef (Tunisia), 77
Elliot, W. Crawford, 58
Emu, 187
Enantiornithes, enantiornithines, 220–221, 236, 242–243

Endotherms, 224–227
Endothermy, 224–226
Environmental disruption
  and extinction, 40–42
Environmental effects
  of impact hypothesis, 51–53, 99
Eocene period, **5**
Eoraptor, 181
Epidemics
  and dinosaur extinction hypothesis, 13, 14
Ethiopia
  human evolution, 267
Eudimorphodon, 162
Euparkeria capensis, 112–114, 140,159
Europe
  Homo erectus, 268
  mammal extinctions, 261
Evolution
  and *Archaeopteryx*, 133–134
  Darwinian, 130–133
  and extinction, 295–296
  human, 266–269
  progressive, 129
  see also cladistics
Evolution, convergent, *see* homoplasy, theory of
Evolution, independent, *see* homoplasy, theory of
Evolution theory, 202
  and dinosaurs, 128–130
Evolutionary history
  and classification, 202–203
  and developmental history, 209–212
  time scale of, **5**
Evolutionary maps
  dinosaurs, 169–194
Evolutionary relationships, *see* phylogeny reconstruction
Extinction
  birds, 274, 295
  and evolution, 295–296
  gradual patterns of, 80
  island birds, 285–286, 196
  at K-T boundary, 99
  killing mechanisms, 76–79
  marine environments, **23,** 76–78, 80
  mass, 41, 65–66, 260–263
  patterns of on continents, 82–90
  periodic, 65–66
  seabirds, 278
  stepwise, 92–94
Extinction hypotheses, 7, 12–20, 21–25
  testing of, 21–25
  see also impact extinction hypothesis; seaways, continental; volcanic extinction hypothesis
Extinction phenomenon, 273–274
Extinction scenarios, 26–30
Extraterrestrial extinction hypothesis, *see* impact extinction hypothesis

Facultative bipedal running, 113
Falconiformes, **249,** 253–255
Falcons, **249,** 253, 255
Fastovsky, David, 94, 242
FAUNMAP database, 261–262
Feathers
  evolution of, 114, 115
Feduccia, Alan, 220, 230, 240, 242, 244
Feet
  birds, 182, 191, 210–211, 220–221
  development from fins, 153–154
  dinosaurs, **128,** 210–211
  evolutionary and developmental history, 210–211
  ornithodirans, 161
  ornithothoracine birds, 219
  theropods, 181–182, 186
Femur, 123
  dinosaur, 127–128, 165–166
Fenestrae, 156, 158
  archosaurs, 159
Ferns, 53
  and dinosaur extinction hypothesis, 12–13
Fibula, 123
  ornithurine birds, 214
Finches, 230
Fins
  into hands and feet, 153–154
  paired, 151
Fire, domestic, 268
Fish species
  bony, 86, 87
  extinctions, 23, 66, 78
  ray-finned, 83, 90, 151
Fitzroy, Robert, 277
Flapping birds, 217
Flessa, Karl, 93
Flight
  maniraptorans, 191
  origin and evolution of, 114, 115, 192–194, 212, 217–220
  Ornithurae, 213–215
Flight stroke, 193
Flightless birds
  Diatrymidae, 252
  rails, 279
  see also Dodo; Ratitae, ratites
Flood basalts, 38
  origin of, 36
  Siberian Traps, 41
Flowering plants, *see* angiosperms
Fluorine, 55
Food chain
  and survival and extinction, 53, 77, 83
Footprints, fossil, *see* trackways
Foraminifera, foraminifera (forams), 53
  extinctions, 76–78
  gradual extinction, 80
Forelimbs
  dwarfed, 215–217
  evolution of, 164
  maniraptorans, 189–**191**
  tetanurines, 183
  tetropods, **122,** 123
Forest fires
  and impact hypothesis, 61–62
Forests, *see* deforestation; tropical forests
Forster, Catherine, 175
Fossil record
  Cretaceous dinosaurs, 81–82
  and Darwinian evolution, 131–132
  and dinosaur history, 4–6
  and geologic ages, 231
  and ghost lineages, 232
  Mesozoic, 235–236
  North American dinosaurs, 94
  and phylogeny reconstruction, 142–143
  theropods, 182
Fossils
  and history of dinosaurs, 6
  and stepwise extinctions, 93
  and volcanic rock, 5
Fracture patterns of rocks, 60
Fractured crystals
  impact hypothesis, 59–60, 71
France
  impact craters, 60
Freezing temperatures
  and extinctions, 86
Frenchman Formation, **26**
Freshwater
  acidity of and extinctions, 78
Frills (dinosaurs), 176
Frogmouths, 255
Frogs
  evolutionary pathway, 154
  survival rates, 83, 86
Fuller, Errol, 277, 282
Fullerines, 63
Fulmar, **237**
Funk Island, 280–282, 284
Furcula, 135–136, 139–140, 183–184

Gait
  dinosauromorphs, 162
  parasagittal, 159–160
Galagos, 266
Galápagos Islands, 230, 247
Galápagos Islands cormorant, 251
Gallant, René, 44
Galliformes, **249**–251
Gallinuloididae, 261
Galton, Peter, 177
Gannet, 284
Gansus, 213
Garefowl, *see* Great Auk
Gars, 83
Gauthier, Jacques, 9, 143, 155, 170, 171, 179, 182, 188, 190, 192–194, 203–205, 213
Gaviiformes, gaviiforms, 237–239, **249,** 251
Gegenbaur, Carl, 138, 212
Gene flow, 246
Genealogy
  and classification, 199–200, 202
Genetic memory, 209
Genetics
  and linguistic characteristics, **269**
Geochemical fingerprinting, 73

Geologic time, 5–6
ages, 231
determining at K-T boundary, 92–104
Quaternary, 260
Geological fingerprints, 55
Germany
impact craters, 60
Ghost lineages, 232
birds, 239, 240
mammalian groups, 235
Passeriformes, 257
primate, 266
Ghost Ranch (New Mexico), 183
Giant deer, 261
Giant Panda, 164
Gibbons, 266
Gibraltar, 281
Gierfugla Sker, 281, 283
Gila monster, 158
Glassy spheres
K-T boundary clay, 61
Gliding, 114, 192
Global warming, 87
Glyptodonts, 261
Gnathostomata, gnasthostome, 151, 159
Golden-cheeked Warbler, 293
Gondwana, 250
Gorilla, 266
Gould, Stephen Jay, 46
Gradualism, school of, *see* uniformitarianism
Graham, Russell, 261
Great Auk, 279–284
Grebes, **249,** 251, **252**
Greeley, Ronald, 101
Greenhouse effect, 53, 62, 87
Greenland, 38, 281
Greenland Hot Spot, **33**
Greenwald, Mike, **51**
Greenway, James, 286
Grieve, Richard, 58, 59
Grinspoon, David, 63
Griphosaurus, see Archaeopteryx
Ground sloths, 261
Ground-up hypothesis, **114,** 115, 192
Gruiformes, **249,** 252, **253**
Gubbio, Italy, 47, 48
geologic intervals, 98
iridium concentration in K-T boundary, **56**
Gulf of Mexico, *see* Chicxulub crater
Gustavus III of Sweden, 196–197

Häberlein, Ernst, 119
Häberlein, Karl, 118, 121, 134
Habitat destruction, 286
Habitat fragmentation, 89
Habitual quadruped, 113
Hagstrum, John, 100, 101
Haiti
K-T boundary layer, 61, 73, 74
tsunami deposits, 64
Hallam, Anthony, 39, 40, 45, 56, 58, 60, 61, 66, 76–79
Hallux, 123, 191
Hand
avialians, 191
development from fins, 153–154
dinosaurs, 163–165, 189
and flight, 193
human, **163**
ornithurine birds, 213–214
saurischian dinosaurs, 179
tetanurines, 184
theropods, 181
Haramiyidae, harimiyids, 234
Harpymimus, 187
Hawaiian-Emperor chain, 38
Hawaiian-Emperor Hot Spot, **33**
Hawaiian Islands, 38, 39
bird extinctions, 279
Hawkins, Waterhouse, **4**
Hawks, 255
Hay fever
and dinosaur extinction hypothesis, 13
Head
theropods, **181**
Hedges, Blair, 227
Heilmann, Gerhard, 113
Helium-3, 38–39
Hell Creek Formation (Montana), **26,** 40, 50, **51,** 94, 241, 242
Hell Creek Volcanic Sediment, **33**
Hellas Plenitia (Mars), 101
Heller, Florian, 119
Hennig, Willi, 203, **204**
Herbivorous dinosaurs
extermination of, 13
Heritable characteristics, 145–146
Herons, **249,** 253
Herrerasaurus, 166, 171–172
Hesperornis, 140, 216, 226–227
Hesperornithiformes, 222, 236, 242
Heymann, Dieter, 62
Hickey, Leo, 79
Hierarchy of relationships, 148
Hildebrand, Alan, 69, 70, 72
Hindlimbs
birds, 166
evolution of, 164, 165
Hip socket (acetabulum)
birds, 136
dinosauromorphs, 162
dinosaurs, 165–166
Hoffman, Antoni, 66
Hollis, Chris, 80
Holocene, 260–263
Hominidae, hominids, 267
Hominoidea, hominoids, 266–267
Linnean classification, **267**
Homo erectus, 267–268
Homo habilis, 267
Homo sapiens, 266, 268
tools, 270
Homoplasy, theory of, 111, 139–141, 148, 161, 170, 230–231, 243
and phylogeny reconstruction, 145
Hornbills, **249,** 256–257
Horned dinosaurs, 5, 6, **26,** 27, 175
Horner, John, 174
Hot-spot volcanism, 36–39, 55–58, 99
Hotton, Carol, 79
Hou, Lian-Hai, 212
Hoyle, Sir Fred, 121–122
Human evolution, 266–269
Human overkill hypothesis, 261, 262, 286
Human population growth, 270, **271,** 286
Humans
evolutionary history of, **5,** 6, 155–156
Humerus, 123
Hummingbirds, **249, 254,** 255
Hunter, John, 136, 182
Hut, Piet, 65
Hutchison, Howard, 83
Hutton, James, 45, 128, 129
Huxley, Thomas, 131–132, 137–139, 141, 143, 162, 179, 182, 194, 195, 205, 212, 215
Hylaeosaurus armatus, 127–128
Hypselosaurus, 16, **17**
Hypsilophodontids, 52
Hypsilophodonts, 17

Iberomesornis, 220
Ibis, 249, 252
Ice Ages, 260
extinctions, 103
Ice cap formation, 42
Iceland, 280, 281
Ichthyodectids, **151**
Ichthyornis, 140, **221,** 222
Ichthyornithiformes, **221,** 222, 236, 242
Ichthyosaurs
extinction of, 23, 78
Iguanodon, **4,** 127–128, 205
Impact craters, 74
Impact extinction hypothesis, 7–8, 43–44, 47–66, 75, 92
asteroid or comet, 58
crater evidence, 68–74
and dinosaur extinction, 101
environmental effects, 51–53
evidence, 67–68
and extinctions of continental vertebrates, 86
fractured crystals, 59–60
global wildfires, 87
gradual extinctions, 80
length of extinction mechanisms, 95
multiple impacts, 93, 95
and sediment layers and fossils, 96
and stepwise extinctions, 93
testing of, 21–25
time of impact, 102
tsunamis, 64
and volcanic and marine regression hypotheses, 99–103
Impact extinction scenario, 29–30, 53, 75–76
Independent evolution, *see* homoplasy, theory of
India
threatened birds, 295
volcanic activity, 26, 34
see also Deccan Traps

Indian Ocean, 37
  birds, 247
  volcanic activity, 34
Indochina
  threatened birds, 295
Inoceramids
  extinction of, 23, 78
Insects, 148–149, 247
Introduced species, 286
Ireland, 281
Iridium, 47, 48
  in comets, 58
  in K-T boundary clay, **48–51,** 54–58, 92
  and periodic extinctions, 65–66
Iridium anomalies, 55, 63, 79, 80, 92–94
Island extinctions, 278–279
  birds, 285–286, 296
  see also Great Auk; moa
Islands
  continental, 248
  oceanic, 247
Isle of Puffins, 281
Isle of Man, 280
Isolation
  diversity, 246–248
Italy
  boundary clay, 61
  see also Gubbio, Italy
Ivory-billed woodpecker, **257**
Izett, Glen, 69

Jablonski, David, 81, 93
Jamaican Streamertail, 255
James, Philip, 65
Java
  Homo erectus, 268
Jaw
  flexible, 180
  ornithischians, 172
Jenkins, Farish, 234
Jensen, Jim, 213
Johnson, Kirk, 79, 80
Judith River Formation, 94, 241, 242
Jurassic Park (film), 209, 227
Jurassic period, **5,** 6

K-T transition boundary, **5, 22**
  bird and mammal diversity, 243–244
  bird lineages, **239**
  bird survivorship, **239,** 240
  birds, 236–237
  Dinosauria, 241–242
  dinosaurs, 243
  evolutionary history during, 102–103
  extinction and survival patterns, 82–90
  extinction causes, 88 (table)
  extinction episode, 41
  fossil pollen and leaves, 79
  geologic intervals, 97
  geologic time frame, 96
  impact craters, 69–74
  mammal survival, 241
  marine organism extinctions, 76–77
  mass extinctions at, 99
  Milankovitch cycles, 95
  ocean temperature, 88
  plant extinctions, 80
  species extinction, **22**
  steplike extinction events, 92
  telling time at, 94–95
  time frame of, 91–104
  time of impact, 102
  tsunami deposits, 63–64
  vertebrate survival and extinction rates, 84–85 (table)
  victims and survivors, 76–90
  volcanism, 37–41, 79
  zircon, 73–74
K-T transition boundary clays, 47–49, **51,** 54–55, 73, 74
  amino acids in, 63
  chemical composition of, 57–59
  diamonds in, 62–63
  fractured (shocked) quartz crystals, 59–60
  glassy spheres, 61
  iridium concentrations, **48–51,** 54–58, 92
  mudstone, 63–64
  sandstone, 63–64
  sediment layers, 96–99
  soot and charcoal, 61–62
  spinel crystals, 60, **61**
  tektites, 73
  zircon, 73–74
Kastner, Miriam, 57
Keel (in birds), 213, 221
Keller, Gerta, 77, 80
Kentucky
  Passenger Pigeon, 288, 289
Kilauea volcano, 55, **57,** 58
Killing mechanisms, 90, 102
  K-T transition extinction hypotheses, 76–79
  Pleistocene, 274
  Quaternary extinctions, 260
  and timescales, 96
Kingfishers, **249,** 256
Kirtland's Warbler, 293
Kitchell, Jennifer, 66, 77
Kiwi, 137
Kluge, Arnold, 143
Knudsen's farm (Alberta, Canada), 62
Kreide, *see* Cretaceous Period
Krogh, Thomas, 73
Kunk, Michael, 69
Kyte, Frank, 58, 73
Kyzylkum Desert, 234

Lagosuchus, 162
Lamarck, Jean-Baptiste, 129, 131
Lampreys, 151
Lance Formation, **26,** 40
Lava flows, 34–35
Leaellynasaura, 52
Leaves, fossil
  and extinction rates, 79–80
Lehman, Tom, 175
Lemurs, 266
Lepidosauria, lepidosaurs, 158
Leptoceratops, 86
Lesothosaurus, **173**
Liaonang province, China, 186
Life
  origin of, **5**
Lift (flight), 193
Limb structure
  and classification, 147
Linguistics
  and genetics, **269**
Linnaeus, Carolus, 126, 133, 196–199, 203, 273
Linnean Commission, 203
Linnean system of classification, 196–201, 203, 205, 206
  hominoids, **267**
Lipps, Jere, 93
Lithography, 116–117
Lithosphere, **32**
Little White Egret, **253**
Lizards, 158
  and dinosaurs, 158–159
  extinctions, 23, 83, 86, 87, 89, 90
Lobsters
  gradual extinction, 80
Locomotion
  crocodylian history, 161
  synapsids, 156
  tetrapods, 154
Lofgren, Donald, 83
Loons, **236**–238, 240, **249,** 251
Lorises, 266
Lucas, Frederick, 284
Luck, Jean-Marc, 58
Lundelius, Ernest, 261
Lundy Island, 280
Lungfish, **152,** 153
Lyell, Charles, 45–46, 128

Macaws, **249, 255**
McHone, John, 60
McKenzie, Dan, 39
MacLeod, Norman, 77, 80
Madagascar, 248
Magma plumes, 36–39, 41
Magnetic field (Earth's)
  reversal of, 17–18, 36, 48
Maisey, John, 150
Malay Archipelago, 247
Mammalia, 197, 198
Mammals
  Age of Dinosaurs, 235
  archaic hoofed, 90
  and birds, 156
  and dinosaur extinction hypothesis, 14–15
  diversification, 230, 243
  extinction rates, 83
  extinctions, 261
  ghost lineages, 235
  K-T transition survival, 241
  lineages, 266
  Mesozoic, 233–235

placental, 86, 87, 89, 241
turbinates, 226–227
zhelestid, 266
Mammoths, 261
Mangaia, 278
Maniraptorans, 187–192, 213
and birds, 194
hands, 193
Manson Crater (Iowa), 69
Mantell, Gideon, 127
Mantle (Earth), 32
Maoris, 276–277
Mapping, *see* phylogenetic maps
Marginocephalia, marginocephalians, 174–177
Mariana Trench, 48
Marine organisms
extinctions, **23,** 76–78, 80
Mars
craters, 101
Marsh, O. C., 140, 173, 187
Marsupial species
extinction of, 23, 77, 83, 86, 87, 89, 90
Martin, Larry, 160, 192, 212
Martin, Paul, 261
Maryanska, Teresa, 177
Mascarene Islands, 271–273
Mass extinctions
periodicity, 65–66
Quaternary, 260–263
Mast, 290, 291
Mastodons, 261
Maupertuis, Pierre de, 43
Mauritius, 271–272
Maurrasse, Florentin, 64
Megalapteryx didinus, 275
Megalosaurus bucklandii, 126–128, 185, 205
Melanesia, 247, 278
Mesozoic Era, **5,** 6
fossil record: birds, 235–236
mammals, 233–235
see also Age of Dinosaurs; Cretaceous Period; Triassic Period
Metabolism, aerobic, 226
Metacarpals, 123
Metasequoia, 53
Metatarsals, 123
Meteor Crater (Arizona), 59, 68
Meteorites
carbonaceous chondrite, 57, 58, 63
impacts, 44, 47–48, 68–69
see also tektites
Methodological uniformitarianism, 46
Mexico
tsunami deposits, 64
see also Chicxulub Crater
Meyer, Georges, 57
Michel, Helen, **51**
Micronesia, 247, 278
Mid-Atlantic Ridge, 33
Migration
of dinosaurs, 53
Milankovitch, Milutin, 95
Milankovitch cycles, 95
Miocene period, **5**
Missouri Breaks (Montana), 50
Moa, 275–278, 296
Moa-nalos, 216
Mongolia, 237, 251
Monkeys, 266
Monolophosaurus, 185
Mononykus olecranus, 215–217
Monroe, Burt, 245
Montanari, Alessandro, **51,** 61
Morgan, W. Jason, 36
Mosasaurs, **24**
extinctions, 78
Moschops capensis, **42**
Mount, Jeffrey, 92
Mount Everest, 35
Mount Whitney, 64
Mousebirds, **249, 254,** 255
Mudminnows, 83
Mudstone
K-T boundary, 63–64
Müller, Friedrich, 120
Muller, Richard, 65
Multituberculates
extinctions, 23, 77, 83, 86, 87, 89
K-T boundary, 241
Musophagidae, **249, 254,** 255

Nanotyrannus, 86
Natural selection, 130–131
Nautilus, **23**
Neck
saurischian dinosaurs, 179
theropods, **181**
Nelson, Gareth, 9, 150, 203
Neogaeornis wetzeli, 238
Neognathae, neognath lineage, 189, **238,** 239, 248–259
Nessov, Lev, 234, 266
New England
extinctions, 286
New Guinea Eagle, 255
New Mexico
K-T boundary clay, 60
New Zealand
K-T boundary clay, 61, 80
moa, 274–277
Nichols, Doug, 79
Nighthawks, **249,** 255
Nightjars, **249,** 255
Noguerornis, 219
Norell, Mark, 161, 215, 232
Norman, David, 174
North America, 248
bird species loss, 286–293
Cretaceous dinosaurs, 82
dinosaur decline, 242
dinosaur fossil record, 82
endangered bird species, 293
extinction patterns of vertebrates, 83
fossil localities, 261–262
Great Auk, 282
ice age extinctions, 103
impact extinctions, 81
K-T extinctions, 241
loss of vertebrate diversity, 260–261
marsupials, 241
multituberculates, 241
plant extinction, **25**
shocked quartz, 81
threatened bird species, 293
see also Western Interior Seaway
North Tasman Sea
hot-spot volcanism, 38
Norway, 281

Obligate bipeds, 115
Occam's Razor, 7, 145
Oceanic plates, **33**
Oceans
distribution of, **18**
extinctions and survival in, 76–77
Officer, Charles, 34, 40, 56, 58–61, 64
Oilbirds, **249,** 255
Oligocene period, **5**
Olson, Storrs, 238, 278
Oort Cloud, 65
Opitsch, Eduard, 119. 121
Opposite birds, 220
Orangutan, 266, 267
Ordovician period, **5**
Orientale Basin, 71
Orkney Islands, 280, 281
Ornithischia, ornithischians, 5, 170–177, 181
decline, 242
extinctions, 86, 87
Ornithodira, ornithodirans, 161
Ornithomimids, 52
Ornithomimosaur, 187
Ornithomimus, 86
Ornithopoda, ornithopods, 173–174
Ornithothoraces, ornithothoracans, 217–221
Ornithurae, ornithurines, 213–215
Osmium, 58
Osteichthyes, osteichthyans, 151, 198
Ostrich, 187, 239, 248, **250**
Ostrom, John, **108**–111, 115, 120, 123, 124, 140, 141, 170, 187, 188, 190, 192, 195, 213
Outer Hebrides, 280
Oviraptor, oviraptorids, 17, 187–188
Owen, Michael, 60, 73
Owen, Richard, **4,** 121, 125–138, 140, 156, 159, 162, 165, 166, 170, 174, 182, 194–196, 205, 218, 225, 231, 233, 272–275, 280, 283–284, 295, 296
Owlet-frogmouths, 255
Owls, **249,** 253–255
Ozone layer
and K-T volcanism, 79

Pachycephalosauria, pachycephalosaurs, 5, 175
Pachycephalosaurs, 15, 82, 86
Pacific Basin, **246,** 248, 249
extinctions, 262
Pacific islands, **277**

Pacific Ocean
  birds, 247
  hot-spot volcanism, 38
  island extinctions, 278–279
Paddlefish, 83
Padian, Kevin, 170, 192, 193, 205, 213
Paleocene period, **5**
Paleognathae, paleognath lineages, 189, **238,** 239, 248–250
Paleozoic Era, **5**
Pangaea, 167, 248
Parasagittal gait, 159–160
Parasagittal plane, 166
Paris Basin, 44
Paronychodon, 86
Parrots, 240, **249,** 250, 255–256
Passenger Pigeon (*Ectopistes migratorius*), 287–291
Passeriformes, passeriforms, 240, **249,** 257
  ghost lineage, 257
  major groups (table), 258–259
Patagopteryx, 221
Patterson, Colin, 66, 142–143, 150, 203
Pelicaniformes, **249,** 251
Pelicans, **249,** 251
Pelvis
  dinosaurs, 165
  maniraptorans, 190
  ornithischians, 172
  theropods, 181
Pena, Daniel, 66
Penfield, Glen, 69
Penguin Island, 280
Penguins, **249,** 251
Penny, David, 240
Perch, 83
Perforate acetabulum, 165
Permian Triassic boundary, 42, 99
Peru
  endangered birds, 295
Petrified Forest National Park 166
Petromyzontida, 150–151
pH scale, 78
Phalanges, 123
Phylogenetic maps, 142–149, 208–209, 232
  how to read, **144**
  vertebrates, **200**
Phylogenetic system of classification, **202**–206
Phylogeny reconstruction, 142–149, 205
Piciformes, **249,** 257
Pike, 83
Piton de la Fournaise, 37, 56–57
Placental mammals
  K-T boundary, 241
  extinction pattern, 86, 87, 89
Planet X, 65
Plankton, 23
  and extinctions, 76–77, 79
Plantae kingdom, 198
Plantain-eaters, **249, 254,** 255
Plants
  acid rain and extinctions, 78
  extinctions, 23, **25**
  impact survival and extinction, 79–80
Plate tectonics, 31–32
  and volcanic activity, 38
Playfair, John, 45
Pleistocene period, **5,** 260–263
Plesiosaurs, **24**
  extinctions, 23, 78
Pliocene period, **5**
Plott, Robert, 126, 127
Podicipediformes, **249,** 251, **252**
Poinar, Hendrik, 227
Pollen
  and dinosaur extinction hypothesis, 13
  and extinction rates, 79–80
  fossil, **25**
Pollutants
  from volcanic activity, 27, 40–41
Polymerase chain reaction (PCR), 227
Polynesia, 247, 278
Pope, Kevin, 69
Popigay crater (Siberia), 74
Potoos, **249,** 255
Prairie Chicken, **250**
Precambrian era, **5**
Predation, human
  as killing mechanism, 274
Predentary, 172, **173**
Prefrontal bone, 188, 189
Presbyoris, **236,** 237, 250
Primate lineage, 235, 266
  and dinosaurs, 296
  ghost lineage, 266
Principle of Parsimony, *see* Occam's Razor
Procellariiformes, procellariforms, **237**–239, **249**
Progressive evolution, 129
Protoavis texensis, 170, 192
Protobirds, 114, 133
Protoceratops, 188
Protofeathers, 186
Pseudoextinction, 241
Psittaciformes, **249,** 255–256
Pterosauria, pterosaurs, **24, 25,** 112, 120, 161–162
  and birds, 111, 112
  evolution of, 113
  extinctions, 77
  wings and flight, 136, 162
Pubis
  dinosaurs, 172, 173
  maniraptorans, 190, **191**
Puerto Rican Emerald hummingbird, 255
Puffin, 284
Pygostyle, 218, **219**
Pythons, 158

Quartz crystals
  K-T boundary clay, 59–60
Quaternary time period, 260–263
Quetzalcoatlus,162

Racemization, 227
Radioisotopic dating, 5, 92, 95
Radiolarians
  extinctions, 78, 80
Radius, 123
Rails, **249,** 252–253, 279
Rampino, Michael, 65
Rancho La Brea tar pits, 261
Raphus cucullatus, 273
Raphus solitarius, 271
Raptors, 263
Ratites (Ratitae), 248–250, 262–263
Raton Basin (Colorado), 73
Raup, David, 65, 66, 81
Ray-finned fishes, 151
  extinction and survival rates of boundary event, 83, 90
Rays, 151
  extinctions, 83, 86, 87, 89, 90
Recapitulations, 211–212, 214
Recent period, **5**
Red Data Books, 294
Red Deer River Valley (Canada), 98
Red-cockaded Woodpecker, 293
Renne, Paul, 42, 99
Reptiles
  evolution of, 129–130
  evolution of birds from, 9
  extinctions, 23, 82–83
  lineage pathways, 156–159
Reptilia, 155, 198
Resemblances, evolutionary, 148–149
Réunion Island, 37, 39, 271
Rhea, **238**
Rhenium, 58
Rhodium, 58
Rich, Tom, 51
Ricqlès, Armand de, 227
Rift Valley System (East Africa), 89
Rifting, 39
Rifts (fractures), 32
Roadrunners, **256**
Rocky Mountains, 18
  ancestral, 26, 39, **40**
Rodriguez Island, 271
Rodriguez Solitaire, 271
Rosen, Donn, 150, 203
Rostral bone, 175, **176**
Ruben, John, 226
Rudistid clams
  extinction, 23, 78
Running
  bipedal, 113, 162
  birds, 193
  dinosaurs, 115
Russell, Dale, 44, **51,** 94
Russia
  impact craters, 60
Ruthenium, 58

Saber-toothed cats, 261
Sacrum
  dinosaurs, 165
Sadler, Peter, 96
Saigado, Leonardo, 186
St. Helens, 60
Saint-Hilaire, Etienne Geoffroy, 129

St. Kilda Island, 280
Salamanders
evolutionary pathway, 154
survivals, 83, 86, 90
Salt gland, 188–189, 238
San Andreas Rift, 33
San Juan Basin (New Mexico), 98
Sandstone
K-T boundary, 63–64
Sanz, José, 218
Sarcopterygian lineage, 152, 159
Sarcopterygii, 151–152
Sauria, saurians, 158, 159
Saurischia, saurischians, 5, 170–172, **178**–180
and birds, 194
decline, 242
extinctions, 86, 87
Sauriurae, 220, 221
Sauropodomorpha, sauropodomorphs, 180–192
Sauropods, 6, 15–17
Saurornithoides, 16
Scandinavia, 281
Scelidosaurs, **177**
Schaeffer, Bobb, 150
Schroeder, Eric, 261
Schultz, Peter, 80
Schwartz, Richard, 65
Schweitzer, Mary, 227
Science and technology, 270
Scleromochlus, 161
Scollard Formation, **26**
Screen washing, 234
Scutes, 177
Sea lions, 247
Seabirds
diversification, 248
extinctions, 278
Seals, 247
Seawater
acidity of and extinctions, 78
Seaways, continental, 31
distribution of, **18**
retreat of, hypothesis, 21, 26, 27, 39, 78, 89, 95, 99–103
Sediment layers
and geologic time, 96–99
and K-T boundary, 96–99
Seeley, Harry, 139, 145, 215
Seismosaurus, 15
Self, Stephen, 35
Semilunate carpal, 123, 190
Senefelder, Alois, 116
Seposki, J. J., 65, 66
Sereno, Paul, 174, 186
Sexual selection, 131
Seychelles Islands, 34, 39, 57
Seymour Island, 80, 238, 251
Sharks, 151
extinctions, 83, 86, 87, 89, 90
Sharpton, Virgil, 61, 70, 72
Sheehan, Peter, 83, 94
Shock waves
Chicxulub impact, 100
Shocked (fractured) quartz crystals, 59–60
impact hypothesis, 81
Shorebirds (Chadriiformes), 252, 253
diversification, 248
Siberia
Popigay crater, 74
Tunguska event, 68
Siberian Traps, 41–42, 99
Sibley, Charles, 245
Sierra Nevada, 248
Signor, Phil, 93
Sigurdsson, Haraldur, 61, 73
Silurian period, **5**
Single-celled life
extinction of, **22**
formation of , 6
Sink holes
Chicxulub crater, **69,** 70
Sinraptor, 185, 186
Sister lineages, 232
Sivapithecus, 267
Skeleton
tubular, 182
Skull
fenestrae, 156
Smeared anomalies, 92–94
Smectite, 58
Smit, Jan, 61, 76
Smith, Alan, 89
Smith, Andrew, 66
Snails
extinctions, 76–78, 83, 86
gradual extinction, 80
Snakes, 158
Solnhofen Limestone, 116–117
Soot
in boundary clay, 61–62
South pole
rates of extinction for fossil pollen, 80
South America, 238
bird species, **295**
fossils, 166
magma plumes, 37
mammal extinctions, 261
threatened bird species, 293, 295
South Atlantic Ocean, 37
K-T boundary, 95
Spain, 281
see also Caravaca, Spain
Speciation, 130, 246–247
Sphenisciformes, **249,** 251
Spiders, 247
Spinel crystals
K-T boundary clay, 60, **61**
Spotted Owl, 293
Stalling (flight), 218
Stanley, Steven, 269
Steadman, David, 278–279, 284
Stegoceras, 86
Stegosaurs, 177
Stegosaurus, 5, 6
Steller's Sea Eagle, **254**
Sternum
birds, 213, 221
Stevns Klint (Denmark), **22, 51,** 58, 98
Stigler, Stephen, 66
Stishovite, 60
Stonesfield Slate (Britain), 233
Storks, **249,** 253
Strickland, H. E., 271, 273
Strigiformes, **249,** 253–255
Strothers, Richard, 65
Sturgeons, 83
Stygimoloch, 86
Subduction zones, 33
Mariana Trench, 48
Substantive uniformitarianism, 46
Sullivan, Robert, 94
Supernovas, 44
Superposition, law of, 44, 92
Supersaurus, 15, 180
Supracoracoideus muscle, 218
Survival
K-T boundary, 241
patterns of on continents, 82–90
vertebrates at K-T boundary, 84–85 (table)
Sutherland, F. L., 37, 99
Swanson, Donald, 35
Swifts, **249,** 255
Swisher, Carl, 40, 73
Sydenham Park, London, **4**
Synapsida, synapsids, 155–156
Systema Naturae (Natural System), *see* Linnean system of classification

Tail
maniraptorans, 190
ornithothoracine birds, 218–219
Tapirs, 261
Tarsals, 123
Tarsirs, 266
Tarsometatarsus, 211
Tasman rift, 38
Teeth
Mesozoic mammals, 234
ornithischians, 172
ornithomimosaur, 187
ornithopods, 174
sauropodormorphs, 180
tetanurines, 183, **184**
Tektites, **48,** 71, 73
Teleosts
extinction and survival rates, 83
Temperature
and extinction, 14, 27–28, 53
Temple, Stanley, 286
Tendons, ossified, 173
Teratornis, 261
Terns, **249,** 284
Tertiary Period, **5**
see also K-T transition boundary
Testudines, 157–158
Tetanurae, tetanurines, 182–192
and birds, 194
Tetrapoda, tetrapods, **147,** 148, 153, 159, 167
amniotic egg, 148
and birds, 133

Tetrapoda, tetrapods *continued*
bones, 123
evolutionary time scale, **5**
fingers and toes, 154–155
forelimb, **122,** 123
limbs, 164–165
lineage, **146,** 147
locomotion, 154
Thecodont hypothesis of avian origins, 112, **114,** 160
Thecodontia, thecodonts, 111, 140
Therizinosaurids, 17
Theropoda, theropods, **17,** 180–192, 263
and avian foot development, 210–211
and birds, 192, 194, **209,** 263
evolution of flight, 193
see also *Mononykus olecranus*
Thescelosaurus, 86
Thumb, opposable, **164**
Thyreophora, **176,** 177
Tibia, 123, 214
Timescales, 92
Tinamiformes, tinamous, **249,** 250
Titanis, 252, 261
Toba volcano, 60
Toomey, Rickard, 261
Torosaurus, 86
Tortoises, 230, 247
Toucans, **249,** 257
Tournicidae, **249,** 255
Toutain, Jean-Paul, 57
Trackways, 5, 54
dinosaurs, **128,** 225
three-toed, 162
Trees
acid rain tolerance, 78
mast-producing, 290, 291
Trees-down hypothesis, 114–115, 193–194
Triassic period, **5,** 6
Triceratops, 5, 6, 15, **26**–28, 40, 82, 86
extinction scenario, 30
Tristan da Cunha, **33,** 37
Trochanters, 166
Trogons, **249,** 257
Troodon, 16, 52, 83, 86
Troodontids, 17
Tropical forests
deforestation, 294
threatened bird species, 293–295
Tsunamis
deposits, 63–64
and extinction hypotheses, 30
Tube noses, 238
Tucker, Wallace, 44
Tunguska event, 68
Turacos, **249,** 255
Turbidity currents, 64
Turbinates, 221, 226–227
Turtles
extinctions, 89, 90
survival rates, 83, 86, 87
see also Testudines

Tyrannosauridae, tyrannosaurids, 187
dwarfed forelimbs, 215
Tyrannosaurs, 52
Tyrannosaurus rex, 5, 6, 15, 27, 28, 82, 86, 186, 187, 221
extinction scenario, 30

Ua Huka, 278
Ulna, 123
Ultrasaurus, 15
Ultraviolet radiation
and extinctions, 79
Ungulates, 234–235
Uniformitarianism, 44–47, 103
Ussher, Bishop, 45
Uzbekistan, **235,** 238

Vega Island, 237
Velociraptor, **16,** 86, 188, 189, 221
Vertebrae
dinosaurs, 179
Vertebrata, **145, 147,** 148, 154
Vertebrates
ancestral, 159, 167
classes, 198
classification, **145–147**
evolution of, 6, 9
evolutionary time scale, **5**
freshwater survival, 90
loss of diversity, 260–261
phylogenetic map, 150–156
survival and extinction across K-T boundary, 84–85 (table)
Vickers-Rich, Pat, 51
Victoriornis imperialis, 271
Viohl, Gunter, 120
Vogt, Peter, 34
Volcanic activity, 31
antipodal, 100–101
and avian diversity, 247
and dinosaur extinction, 101
environmental effects, 40–41
extinction scenario, 26–28
and extinctions, 34–42
and iridium, 55–57
K-T transition, 37
and plate tectonics, **33**
Siberian Traps, 41–42
Volcanic ash, 39–40
Volcanic eruptions
and extinction hypotheses, 21–23
Volcanic extinction hypothesis, 7, 8, 75
determining time frames, 95
and impact and marine regression hypotheses, 99–103
iridium, 56
iridium anomalies, 93
rock fracture patterns, 60
and stepwise extinctions, 93
testing of, 21–25
Volcanic extinction hypothesis scenarios, 53
extinction and survival patterns and rates, 87–89
sediment layers, 96

Volcanic rock
and fossils, 5
radioisotopic dating, 5
Von Meyer, Hermann, 117–118, 120
Vrba, Elizabeth, 268

Wadham Island, 280
Wagner, Andreas, 121, 134, 136, 138
Wagner, Melissa, 66
Walker, Alick, 160
Walking
birds, 218–219
Walvis Ridge, 37, 39
Warm-bloodedness, 224–227
Weishampel, David, 174
Wellnhofer, Peter, 121
Western Interior Seaway, 18–19, 25–27
and impact extinction scenario, 30
White, Robert, 39
White Dodo, 271
Whitmire, Daniel, 65
Whooping crane, 293
Wickramasinghe, Chandra, 121–122
Wildfires, 53
impact hypothesis and extinctions, 62, 87
Williams, David, 101
Wilson, Alexander, 288
Wilson, E. O., 263, 274, 294
Wingless syndrome, 216–217
Winglessness, 216
Wings
insects, 148–149
pterosaurs and birds, 162
Wishbone, *see* furcula
Wolbach, Wendy, 61–62
Wood Ibis, **252**
Woodpeckers, **249,** 257
Woodward, Scott, 227
Woolly rhino, 261
World populations
genetics and linguistic characteristics, **269**
Wrist, 164
ornithurine birds, 213–214
Wrynecks, **249,** 257

Xiphactinus audax, **151**

Yale *Deinonychus* Quarry (Montana), 110
Yucatan Peninsula, *see* Chicxulub crater

Zahnle, Kevin, 63
Zhelestid mammals, 266
Zhou, Zhang, 215
Zhou, Zhongue, 212
Zinsmeister, William, 80
Zircon
K-T boundary layer, 73–74
Zoeller, William, 55, 58